Advances in
Heterocyclic
Chemistry

Volume 14

Editorial Advisory Board

A. Albert
A. T. Balaban
J. Gut
J. M. Lagowski
J. H. Ridd
Yu. N. Sheinker
H. A. Staab
M. Tišler

Advances in

HETEROCYCLIC CHEMISTRY

Edited by

A. R. KATRITZKY

A. J. BOULTON

School of Chemical Sciences
University of East Anglia
Norwich, England

Volume 14

Academic Press · New York and London · 1972

COPYRIGHT © 1972, BY ACADEMIC PRESS, INC.
ALL RIGHTS RESERVED
NO PART OF THIS BOOK MAY BE REPRODUCED IN ANY FORM,
BY PHOTOSTAT, MICROFILM, RETRIEVAL SYSTEM, OR ANY
OTHER MEANS, WITHOUT WRITTEN PERMISSION FROM
THE PUBLISHERS.

ACADEMIC PRESS, INC.
111 Fifth Avenue, New York, New York 10003

United Kingdom Edition published by
ACADEMIC PRESS, INC. (LONDON) LTD.
24/28 Oval Road, London NW1

LIBRARY OF CONGRESS CATALOG CARD NUMBER: 62-13037

PRINTED IN THE UNITED STATES OF AMERICA

Contents

CONTRIBUTORS vii

PREFACE ix

Recent Advances in the Chemistry of Mononuclear Isothiazoles

K. R. H. WOOLDRIDGE

I. Synthesis of Mononuclear Isothiazoles 2
II. The Physiochemical Properties of Isothiazoles 14
III. The Chemical Properties of Isothiazoles 16
IV. Reactions Involving Cleavage of the Isothiazole Ring . . 34
V. Photochemical Reactions of Isothiazoles 37
VI. Biologically Active Isothiazoles 37

Benzisothiazoles

MICHAEL DAVIS

I. Introduction 43
II. 1,2-Benzisothiazoles 44
III. 2,1-Benzisothiazoles 63
IV. Selenium and Tellurium Analogs 76
V. Tables 76

Recent Advances in Pyrazine Chemistry

G. W. H. CHEESEMAN AND E. S. G. WERSTIUK

I. Introduction 99
II. Physical and Spectroscopic Properties 105
III. General Synthetic Methods 112
IV. General Chemical Properties 122
V. Substituted Pyrazines 127
VI. Reduced Pyrazines 182
VII. Pyrazine N-Oxides 192
VIII. Biological Activity 208

Heterocycles by Ring Closure of Ortho-Substituted t-Anilines (the t-Amino Effect)

O. METH-COHN AND H. SUSCHITZKY

I. Introduction 211
II. Interactions of the Ortho Substituent with the Nitrogen in t-Anilines 212
III. Interactions of the Ortho Substituent with the α-Methylene Group in t-Anilines 225

1,2-Dihydroisoquinolines

S. F. Dyke

I. Introduction 279
II. Formation 280
III. Stability 289
IV. Detection and Estimation 294
V. Reactions 294

Benzo[c]thiophenes

B. Iddon

I. Introduction 331
II. Theoretical Aspects 333
III. Hydrobenzo[c]thiophenes 335
IV. Benzo[c[thiophene 350
V. Alkyl and Aryl Derivatives of Benzo[c]thiophene . . . 355
VI. Hydrobenzo[c]thiophene 2-Oxides and 2,2-Dioxides . . . 360
VII. 2-Thiophthalide, Phthalic Thioanhydride, and Related Compounds 368
Note Added in Proof 381

Author Index 383

Cumulative Index of Titles 405

Contributors

Numbers in parentheses indicate the pages on which the authors' contributions begin.

G. W. H. CHEESEMAN, *Chemistry Department, Queen Elizabeth College, University of London, London, England* (99)

MICHAEL DAVIS, *Department of Chemistry, La Trobe University, Bundoora, Victoria, Australia* (43)

S. F. DYKE, *School of Chemistry and Chemical Engineering, Bath University of Technology, Bath, England* (279)

B. IDDON, *Department of Chemistry and Applied Chemistry, University of Salford, Salford, Lancashire, England* (331)

O. METH-COHN, *Department of Chemistry and Applied Chemistry, University of Salford, Salford, England* (211)

H. SUSCHITZKY, *Department of Chemistry and Applied Chemistry, University of Salford, Salford, England* (211)

E. S. G. WERSTIUK, *Biochemistry Department, McMaster University, Hamilton, Ontario, Canada* (99)

K. R. H. WOOLDRIDGE, *The Research Laboratories, May & Baker Ltd., Dagenham, Essex, England* (1)

Preface

Volume 14 of this serial publication comprises six chapters of which four deal with general accounts of ring systems: benzisothiazoles (M. Davis), 1,2-dihydroisoquinolines (S. F. Dyke), benzo[c]thiophenes (B. Iddon), and pyrazines (G. W. H. Cheeseman and E. S. G. Werstiuk). One chapter updates a previous review in this series and is concerned with the rapidly expanding chemistry of mononuclear isothiazoles (K. R. H. Wooldridge). The remaining chapter (O. Meth-Cohn and H. Suschitzky) deals with the cyclizations of ortho-substituted t-anilines.

A. R. KATRITZKY
A. J. BOULTON

Recent Advances in the Chemistry of Mononuclear Isothiazoles

K. R. H. WOOLDRIDGE

The Research Laboratories, May & Baker Ltd., Dagenham, Essex, England

I. Synthesis of Mononuclear Isothiazoles	2
A. Preparation of Isothiazoles from Bicyclic Systems	2
B. Preparation of Isothiazoles Involving Oxidative Formation of a Nitrogen–Sulfur Bond	3
C. Preparation of Isothiazoles from Olefins	7
D. Preparation of Isothiazoles from Acetylenes	8
E. Preparation of Isothiazoles from β-Mercaptoacrylonitriles and Related Compounds	10
F. Preparation of Isothiazoles from 1,2-Dithiolium Salts	12
G. Preparation of Isothiazoles from β-Aminocrotonate and Thiophosgene	14
II. The Physiochemical Properties of Isothiazoles	14
A. Ultraviolet Spectra	14
B. Infrared Spectra	14
C. Nuclear Magnetic Resonance Spectra	15
D. Mass Spectra	15
III. The Chemical Properties of Isothiazoles	16
A. Electrophilic Substitution Reactions	16
B. Metallation Reactions	18
C. Alkyl- and Arylisothiazoles	20
D. Aminoisothiazoles	21
E. Halogenoisothiazoles	24
F. Hydroxy- and Alkoxyisothiazoles	26
G. Isothiazolecarboxylic Acids, Amides, and Nitriles	28
H. Isothiazole Aldehydes and Ketones	29
I. Isothiazole Thiols, Alkylthiols, and Sulfides	31
J. Quaternary Isothiazoles	32
IV. Reactions Involving Cleavage of the Isothiazole Ring	34
V. Photochemical Reactions of Isothiazoles	37
VI. Biologically Active Isothiazoles	37

Following publication of the first synthesis of a mononuclear isothiazole in 1956, this ring system has attracted considerable interest. The aim of the present review is to present a coherent survey of the chemistry of isothiazoles in the light of the numerous developments which have taken place since the subject was last reviewed in 1965.[1]

I. Synthesis of Mononuclear Isothiazoles

In this section an attempt has been made to rationalize the numerous routes now available to the isothiazole ring system.

A. Preparation of Isothiazoles from Bicyclic Systems

The parent ring system was first prepared by Adams and Slack[2, 3] by the oxidation of 5-aminobenz[*d*]isothiazole to isothiazole-4,5-dicarboxylic acid, which was degraded stepwise to isothiazole (**1**).

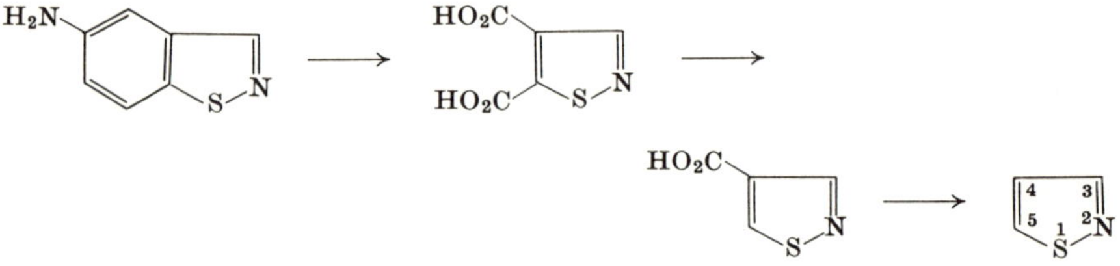

(**1**)

This preparation is of no practical value, but it is of interest that a route (Scheme 1) involving the chlorination of cyanodithioformate probably passes through a bicyclic intermediate.[4, 5]

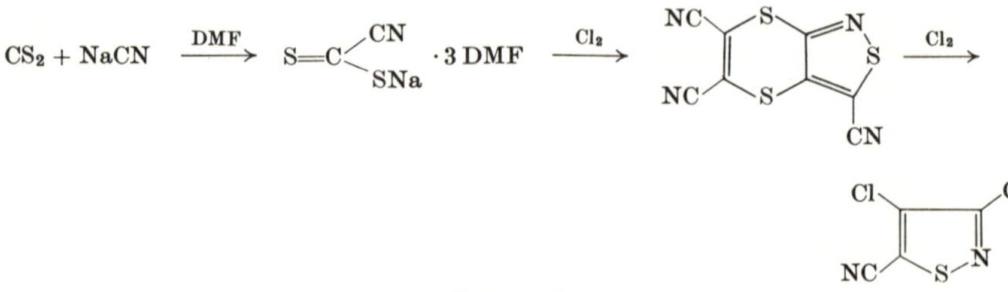

Scheme 1

---

[1] R. Slack and K. R. H. Wooldridge, *Advan. Heterocycl. Chem.* **4**, 107 (1965).
[2] A. Adams and R. Slack, *Chem. Ind.* (*London*) 1232 (1956).
[3] A. Adams and R. Slack, *J. Chem. Soc.* 3061 (1959).
[4] Pennsalt Chem. Corp., U.S. Patent 3,341,547 (1967) [*Chem. Abstr.* **68**, 114,596 (1968)].
[5] H. E. Simmons, R. D. Vest, D. C. Blomstrom, J. R. Roland, and T. L. Cairns, *J. Amer. Chem. Soc.* **84**, 4746 (1962).

B. Preparation of Isothiazoles Involving Oxidative Formation of a Nitrogen–Sulfur Bond

There are several syntheses which depend on the oxidation of compounds that may be represented by "imino-enethiols" (**2**) as one of their tautomeric forms. Oxidation may be carried out with peracids,

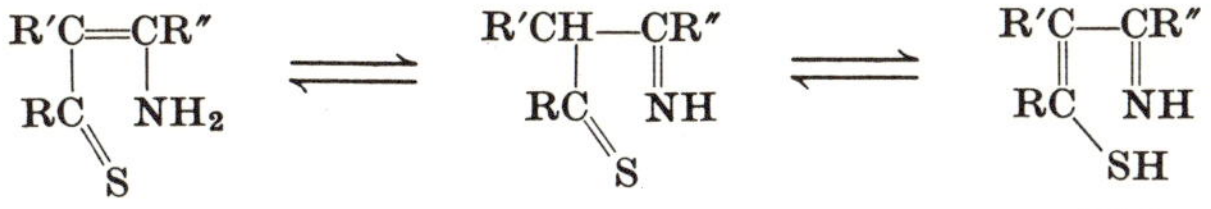

high potential quinones, sulfur, or more commonly, halogens (Scheme 2).

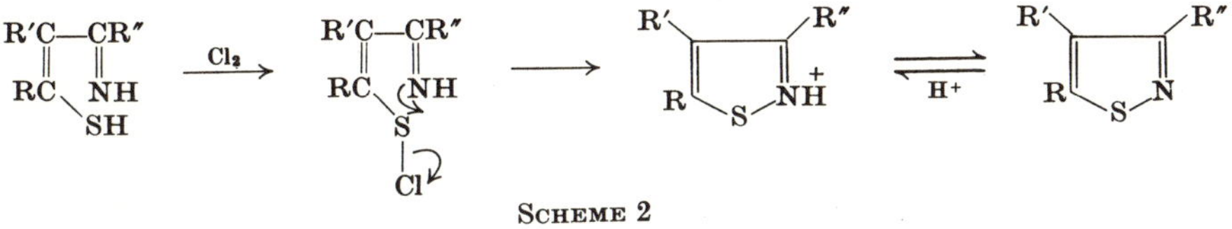

Scheme 2

This route has many ramifications depending on the nature of R and R″ in Scheme 2; thus R may be alkyl, aryl, amino (thioamide), alkylthio, or thiol and R″ may be alkyl, aryl, amino (amidine), hydroxy (amide), or alkoxy (imino ether), although as yet not all of the possible combinations of these groups has been fully examined. In practice, the utility of the route depends on the availability of the requisite intermediates, and the situation is further complicated by the numerous routes which have been devised to these precursors.

Simple 5-aminoisothiazoles are readily prepared by oxidation of β-iminothioamides,[3, 6–9] and more complex 5-aminoisothiazoles with electronegative substituents in the 4-position are available following

---

[6] J. Goerdeler and H. W. Pohland, *Chem. Ber.* **94**, 2950 (1961); **98**, 4040 (1965).

[7] Bristol-Banyu Research Inst., U.S. Patent 3,341,518 (1967) [*Chem. Abstr.* **68**, 95810 (1968)].

[8] R. E. Smith, Ph.D. Thesis, University of N. Carolina (1966).

[9] T. Taito, S. Nakagawa, and K. Takahashi, *Chem. Pharm. Bull.* **16**, 148 (1968).

the development of elegant syntheses of the appropriate iminothio-amides by Goerdeler and his colleagues[6, 10-15] (Scheme 3).

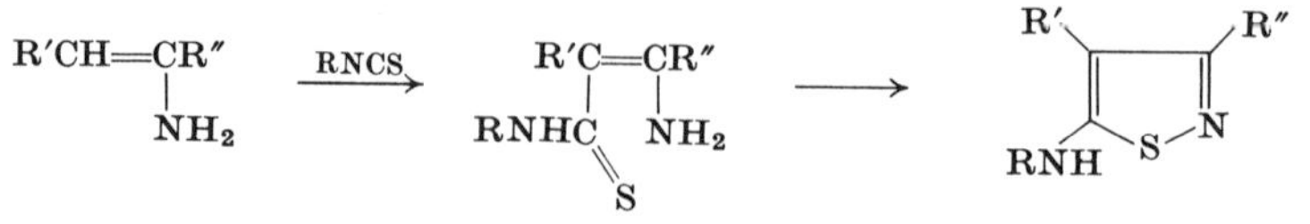

$$R'CH=CR''\ |\ NH_2 \xrightarrow{RNCS} \begin{array}{c} R'C=CR'' \\ | \quad | \\ RNHC \quad NH_2 \\ \| \\ S \end{array} \longrightarrow \begin{array}{c} R' \quad R'' \\ \diagup \quad \diagdown \\ RNH \quad S \quad N \end{array}$$

R = Me, Ph, CO$_2$Ph, COPh, CONHR''', SO$_2$Ph
R' = CN, CO$_2$Et, CONH$_2$, COMe
R'' = Me, Ph, OH, NH$_2$, OAlk

SCHEME 3

The synthesis (Scheme 4) of 4-cyano-3,5-dimethylmercaptoiso-thiazole by Gewald[16] may be regarded as a variant of this method.

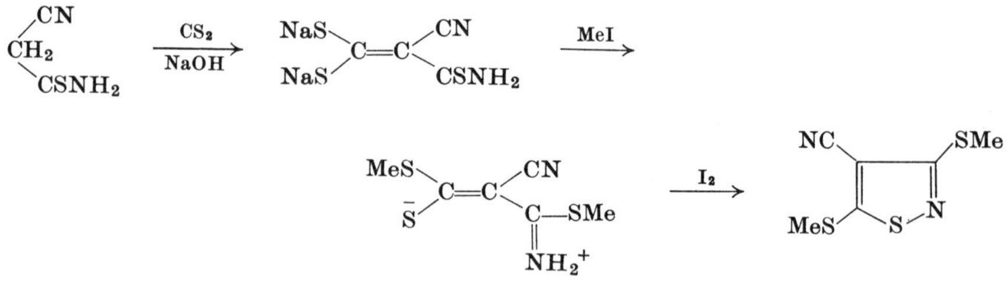

$$\begin{array}{c} CN \\ CH_2 \\ CSNH_2 \end{array} \xrightarrow[NaOH]{CS_2} \begin{array}{c} NaS \quad CN \\ \diagdown C=C \diagup \\ NaS \quad CSNH_2 \end{array} \xrightarrow{MeI}$$

$$\begin{array}{c} MeS \quad CN \\ \diagdown C=C \diagup \\ S^- \quad C-SMe \\ \| \\ NH_2^+ \end{array} \xrightarrow{I_2} \begin{array}{c} NC \quad SMe \\ \diagup \quad \diagdown \\ MeS \quad S \quad N \end{array}$$

SCHEME 4

Faust[17] prepared 5-methylmercaptoisothiazole hydrobromide (3) from a similar intermediate.

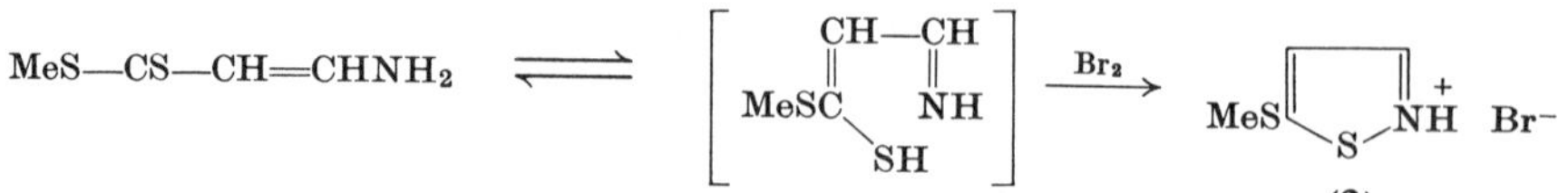

$$MeS-CS-CH=CHNH_2 \rightleftharpoons \left[\begin{array}{c} CH-CH \\ \| \quad \| \\ MeSC \quad NH \\ \diagdown \\ SH \end{array} \right] \xrightarrow{Br_2} \begin{array}{c} \\ MeS \quad S \quad NH \end{array}^+ Br^-$$

(3)

[10] J. Goerdeler and H. Horn, *Chem. Ber.* **96**, 1551 (1963); see also Ciba, Belgian Patent 740,363 (1969) [*Chem. Abstr.* **73**, 56118 (1970)].
[11] J. Goerdeler and H. W. Pohland, *Chem. Ber.* **96**, 526 (1963).
[12] J. Goerdeler and U. Keuser, *Chem. Ber.* **97**, 3106 (1964).
[13] J. Goerdeler, *Angew. Chem.* **76**, 654 (1964).
[14] J. Goerdeler and U. Krone, *Chem. Ber.* **102**, 2273 (1969).
[15] J. Goerdeler, *Angew. Chem. Int. Ed. Eng.* **2**, 693 (1963).
[16] K. Gewald, *Chem. Ber.* **101**, 383 (1968).
[17] J. Faust, *Z. Chem.* **7**, 306 (1967).

Goerdeler and Mittler[18] employed the appropriate thiones to prepare 3-amino-, 3-hydroxy-, and 3-methoxy-5-phenylisothiazoles [Eq. (1)].

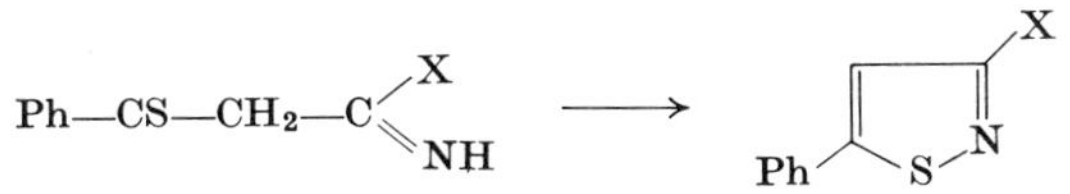

$$Ph-CS-CH_2-C{\overset{X}{\underset{NH}{\diagdown}}} \longrightarrow \underset{Ph}{\text{isothiazole}} \qquad (1)$$

X = NH$_2$, OH, OMe

A method involving treatment of a β-iminoketone with phosphorus pentasulfide and sulfur[19, 20] may well involve the initial formation of an iminothione which undergoes oxidation with sulfur (Scheme 5).

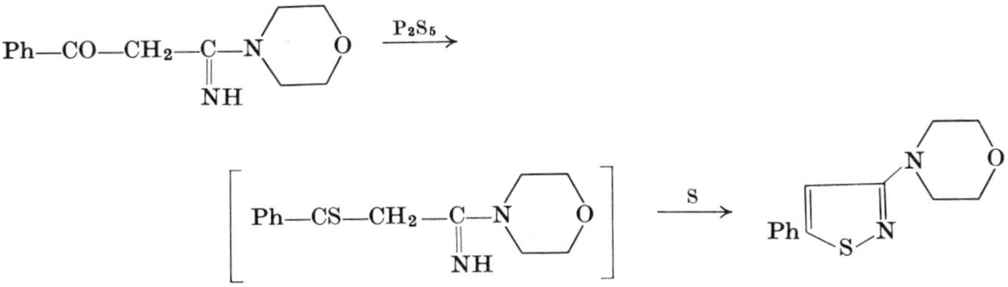

SCHEME 5

An analogous reaction utilizes the ready availability of certain isoxazoles to give isothiazoles with the same substitution pattern.[21] The intermediate enaminoketones are treated with phosphorus pentasulfide and chloranil or sulfur to give moderate yields of isothiazoles (Scheme 6).

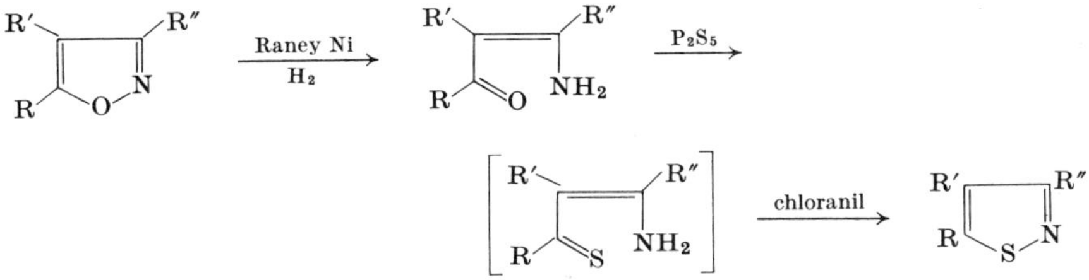

R and R″ = Me or Ph, R′ = H or electronegative substituent

SCHEME 6

[18] J. Goerdeler and W. Mittler, *Chem. Ber.* **96**, 944 (1963).
[19] A. Bruno and G. Purrello, *Gazz. Chim. Ital.* **96**, 1009 (1966).
[20] A. Bruno and G. Purrello, *Gazz. Chim. Ital.* **96**, 986 (1966).
[21] D. N. McGregor, U. Corbin, J. E. Swigor, and L. C. Cheney, *Tetrahedron* **25**, 389 (1969).

The synthesis (Scheme 7) employed by Crenshaw and co-workers[22,23] may be rationalized as proceeding via a thione intermediate. In a later variation,[24] a Vilsmeier–Haack reaction was

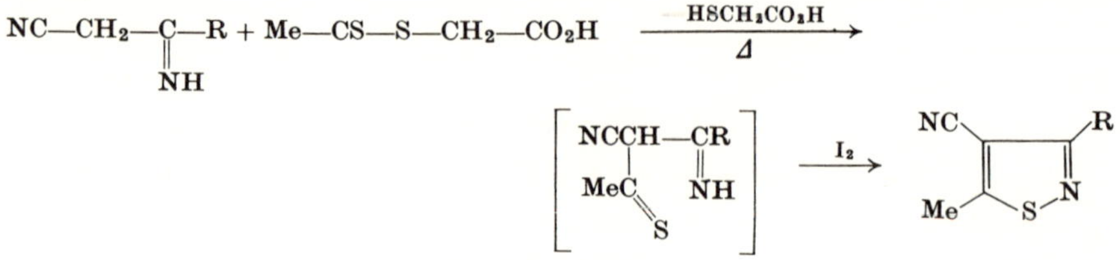

$$\text{NC—CH}_2\text{—C(=NH)—R} + \text{Me—CS—S—CH}_2\text{—CO}_2\text{H} \xrightarrow[\Delta]{\text{HSCH}_2\text{CO}_2\text{H}}$$

$$\left[\begin{array}{c}\text{NCCH—CR}\\ \text{MeC(=S) NH}\end{array}\right] \xrightarrow{I_2} \underset{\text{Me}}{\overset{\text{NC}\quad\quad\text{R}}{\underset{S-N}{\bigcirc}}}$$

R = Me or Ph
SCHEME 7

employed to acylate aminoacrylonitriles to give intermediates which, on treatment successively with sodium hydrosulfide and iodine, gave 4-cyanoisothiazoles (Scheme 8).

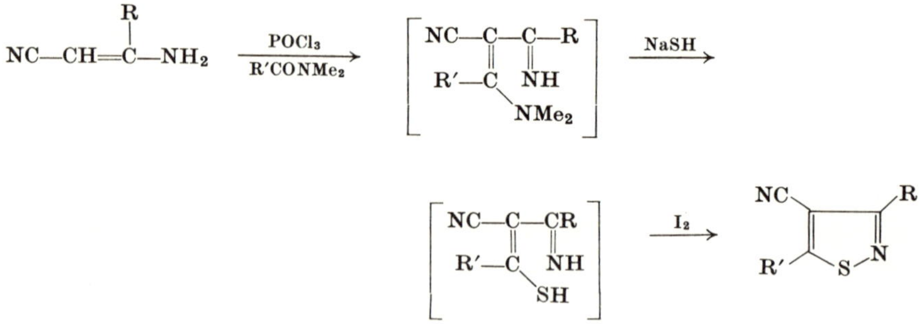

$$\text{NC—CH=C(R)—NH}_2 \xrightarrow[\text{R'CONMe}_2]{\text{POCl}_3} \left[\begin{array}{c}\text{NC—C—C—R}\\ \text{R'—C NH}\\ \text{NMe}_2\end{array}\right] \xrightarrow{\text{NaSH}}$$

$$\left[\begin{array}{c}\text{NC—C—CR}\\ \text{R'—C NH}\\ \text{SH}\end{array}\right] \xrightarrow{I_2} \underset{\text{R'}}{\overset{\text{NC}\quad\quad\text{R}}{\underset{S-N}{\bigcirc}}}$$

R and R' = methyl or phenyl
SCHEME 8

5-Iminoisothiazoles,[15,25] isothiazolin-5-thiones,[26] and isothiazolin-3-ones[27,28] have been prepared by oxidation of the appropriate N-substituted intermediates. Isothiazolin-3-ones have been chlorin-

---

[22] R. R. Crenshaw, J. M. Essery, and A. T. Jeffries, *J. Org. Chem.* **32**, 3132 (1967).
[23] Bristol Myers, Netherlands Patents 6,710,248, 671,0,249, 6,710,250 (1968) [*Chem. Abstr.* **71**, 101,850 (1969)].
[24] R. R. Crenshaw and R. A. Partyka, *J. Heterocycl. Chem.* **7**, 871 (1970).
[25] J. Goerdeler and J. Gnad, *Chem. Ber.* **98**, 1531 (1965).
[26] R. Mayer, H. J. Hartmann and J. Jentzsch, *J. Prakt. Chem.* **31**, 312 (1966).
[27] J. Faust, *Z. Chem.* **8**, 170 (1968).
[28] Ciba, French Patent 2,012,336 (1970) [*Chem. Abstr.* **72**, 79026 (1970)].

ated with oxalyl chloride or phosgene to 3-chloroisothiazolium salts which may be thermally degraded to 3-chloroisothiazoles,[27] although the yields are poor (Scheme 9).

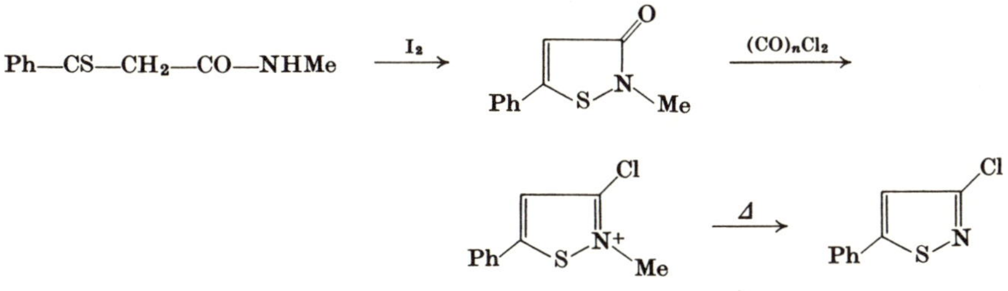

Ph—CS—CH$_2$—CO—NHMe $\xrightarrow{I_2}$ [Ph, S, N-Me, =O ring] $\xrightarrow{(CO)_nCl_2}$

[Ph, S, N$^+$-Me, Cl ring] $\xrightarrow{\Delta}$ [Ph, S, N, Cl ring]

SCHEME 9

The syntheses by Naito and co-workers of 4-hydroxy, 4-cyano, and related isothiazoles from α-iminoketones or α-iminonitriles and sulfuryl or thionyl chloride[29, 30] may be envisaged as a variant of the general method if it is assumed that the reagents introduce a sulfur atom and also provide chlorine for oxidative cyclization (Scheme 10).

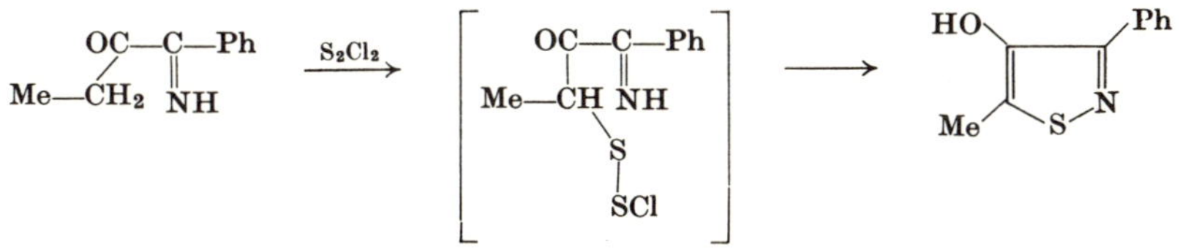

Me—CH$_2$—NH / OC—C—Ph $\xrightarrow{S_2Cl_2}$ [Me—CH—NH—S—SCl / OC—C—Ph] $\longrightarrow$ [Me, S, N, HO, Ph ring]

SCHEME 10

C. PREPARATION OF ISOTHIAZOLES FROM OLEFINS

Hübenett and his colleagues observed that a mixture of propylene, sulfur dioxide, and ammonia in the presence of a catalyst at 200° gave isothiazole in 65% yield possibly via an intermediate thionylimide.[31, 32]

---

[29] T. Naito, S. Nakagawa, J. OKumuria, K. Takahashi, and K. Kasai, *Bull. Chem. Soc. Jap.* **41**, 959 (1968).
[30] T. Naito, S. Nakagawa, J. OKumuria, K. Takahashi, K. Masuko, and Y. Narita, *Bull. Chem. Soc. Jap.* **41**, 965 (1968).
[31] F. Hübenett, F. H. Flock, and H. Hofmann, *Angew. Chem. Int. Ed. Engl.* **1**, 508 (1962); Hans J. Zimmer Verfahrenstechnik, U.S. Patent 2,257,409 (1966).
[32] F. Hübenett, F. H. Flock, W. Hansel, H. Heinze, and H. Hofmann, *Angew. Chem. Int. Ed. Engl.* **2**, 714 (1963).

Isobutylene gives 4-methylisothiazole[31] [Eq. (2)], which is not available by any other route. 1- or 2-Butene gave a mixture of 3- and

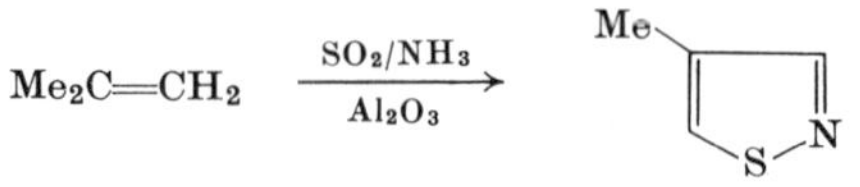

$$Me_2C=CH_2 \xrightarrow[Al_2O_3]{SO_2/NH_3} \underset{S}{\underset{\|}{\overset{Me}{\diagdown}}}\!\!\!\underset{N}{\diagup} \qquad (2)$$

5-methylisothiazoles, but side reactions occur with higher olefins leading to increasing yields of thiophenes.[18] 4-Cyanoisothiazole and the three monophenylisothazoles have also been prepared by this method.[33]

D. Preparation of Isothiazoles from Acetylenes

Probably the most convenient laboratory preparation of isothiazole was devised by Wille and his collaborators[34] (Scheme 11) and depends on the cyclization with liquid ammonia of the cis-addition product from propargyl aldehyde and sodium thiosulfate or thiocyanate.[34, 35]

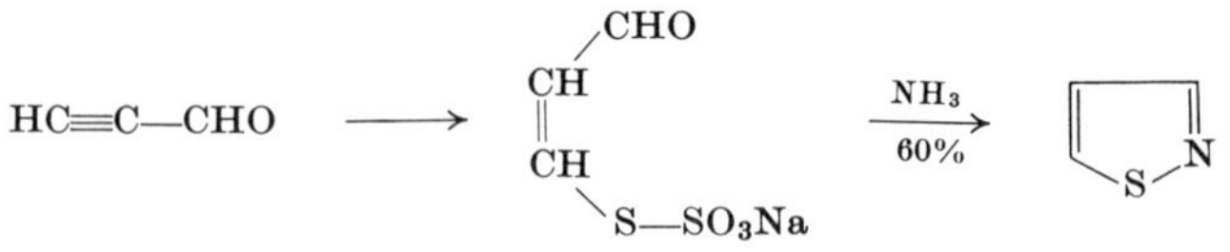

$$HC\equiv C-CHO \longrightarrow \underset{S-SO_3Na}{\overset{CHO}{\underset{\|}{CH}}\atop{CH}} \xrightarrow[60\%]{NH_3} \text{(isothiazole ring)}$$

SCHEME 11

A higher yield (90%) may be obtained by ring closure with potassium thiohydroxylamino-S-sulfonate (Scheme 12). The same reagent may be condensed with propargyl aldehyde to give an intermediate which forms isothiazole on treatment with alkali, possibly via a transient thiooxime derivative.[36]

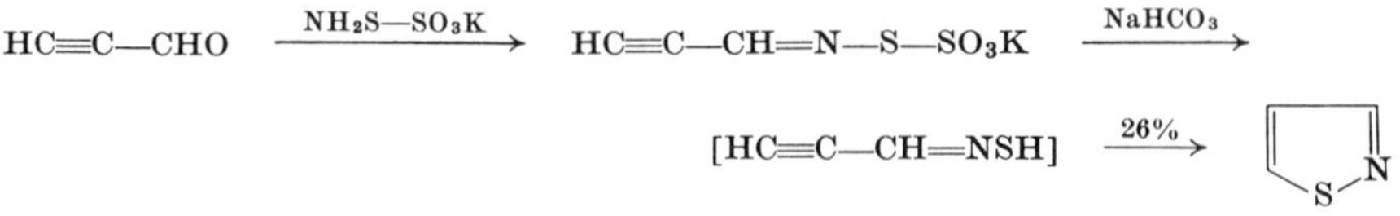

$$HC\equiv C-CHO \xrightarrow{NH_2S-SO_3K} HC\equiv C-CH=N-S-SO_3K \xrightarrow{NaHCO_3}$$

$$[HC\equiv C-CH=NSH] \xrightarrow{26\%} \text{(isothiazole ring)}$$

SCHEME 12

---

[33] F. Hübenett and H. Hofmann, *Angew. Chem. Int. Ed. Engl.* **2**, 325 (1963).
[34] F. Wille, *Angew. Chem. Int. Ed. Engl.* **1**, 335 (1962).
[35] R. Raap, *Can. J. Chem.* **44**, 1324 (1966).
[36] D. Buttimore and R. Slack, unpublished data.

3-Phenylisothiazole was prepared by Beringer, Prijs, and Erlenmeyer[37] using a variation of Wille's synthesis (Scheme 13).

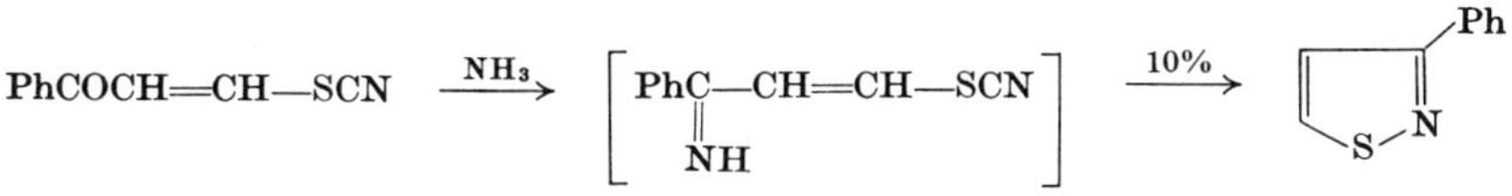

PhCOCH=CH—SCN $\xrightarrow{NH_3}$ [PhC(=NH)—CH=CH—SCN] $\xrightarrow{10\%}$ [3-phenylisothiazole structure]

SCHEME 13

Crow and Leonard extended this method to prepare 3-hydroxyisothiazoles and 3-isothiazolones (Scheme 14).[38,39] More detailed

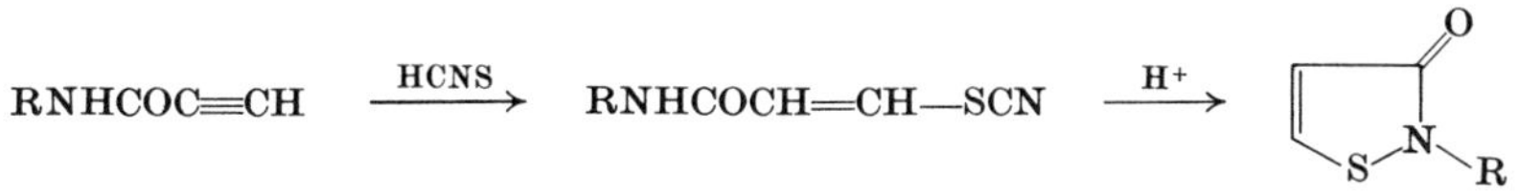

RNHCOC≡CH $\xrightarrow{HCNS}$ RNHCOCH=CH—SCN $\xrightarrow{H^+}$ [3-isothiazolone structure]

R = H, Me, Et
SCHEME 14

studies demonstrated that the reaction could be reversed by cyanide ion and was dependent on two kinetically distinct mechanisms (Scheme 15).[40,41]

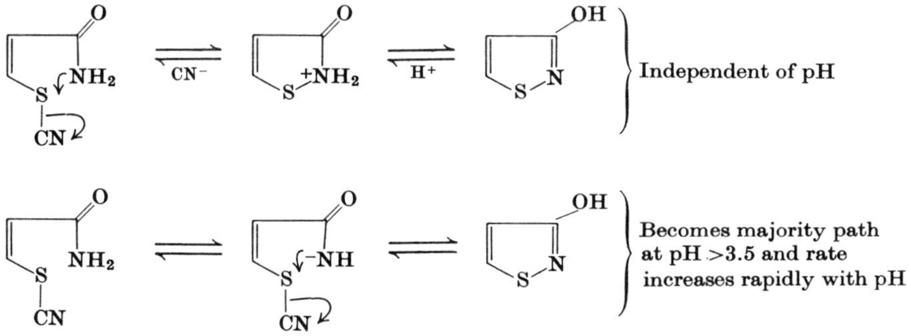

} Independent of pH

} Becomes majority path at pH >3.5 and rate increases rapidly with pH

SCHEME 15

Franz and Black[42] obtained good yields of 3-phenylisothiazole esters by thermal decomposition of 5-phenyl-1,3,4-oxathiazol-2-one

[37] M. Beringer, B. Prijs, and H. Erlenmeyer, *Helv. Chim. Acta* **49**, 2466 (1966).
[38] W. D. Crow and N. J. Leonard, *Tetrahedron Lett.* 1477 (1964).
[39] W. D. Crow and N. J. Leonard, *J. Org. Chem.* **30**, 2660 (1965).
[40] W. D. Crow and I. Gosney, *Aust. J. Chem.* **19**, 1693 (1966).
[41] W. D. Crow and I. Gosney, *Aust. J. Chem.* **20**, 2729 (1967).
[42] J. E. Franz and L. L. Black, *Tetrahedron Lett.* 1381 (1970).

(4) in the presence of acetylene mono- or dicarboxylic esters. Presumably this involved dipolar addition to an intermediate benzonitrile sulfide.

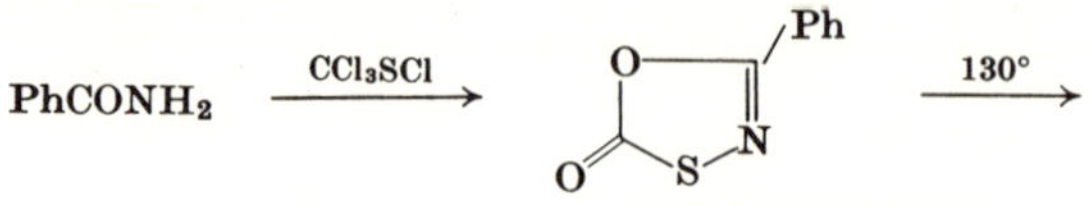

$$PhCONH_2 \xrightarrow{CCl_3SCl} (4) \xrightarrow{130°} [Ph\overset{+}{C}\equiv N-\bar{S}] \xrightarrow[>90\%]{MeOCO-C\equiv C-COOMe} \text{(isothiazole product)}$$

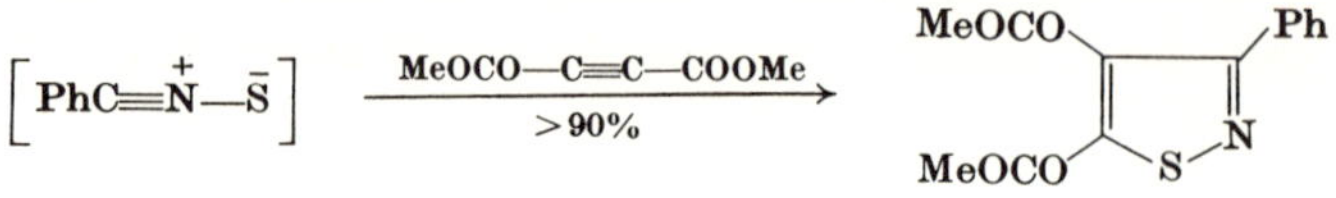

E. Preparation of Isothiazoles from β-Mercaptoacrylonitriles and Related Compounds

A number of important isothiazole syntheses depend on oxidative cyclization of substituted β-mercaptoacrylonitriles possibly by the following mechanism (Scheme 16), e.g., with chlorine as the oxidant.

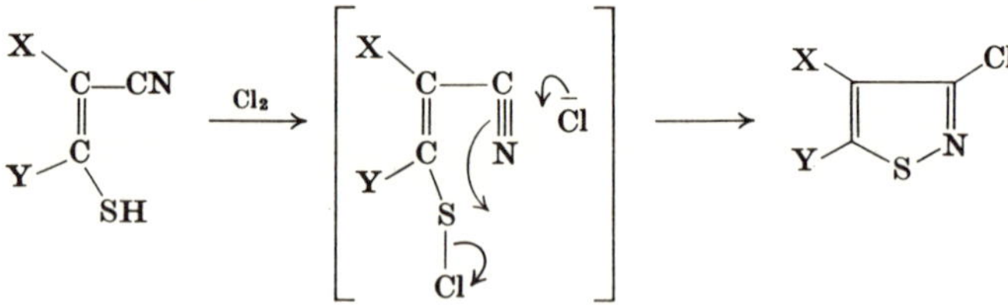

SCHEME 16

Thus, Söderbäck[43] and Hatchard[44,45] independently obtained the disodium salt of 4-cyano-3,5-dimercaptoisothiazole from the readily available dicyanoethylenedithiolate and sulfur (Scheme 17). Pianka[46]

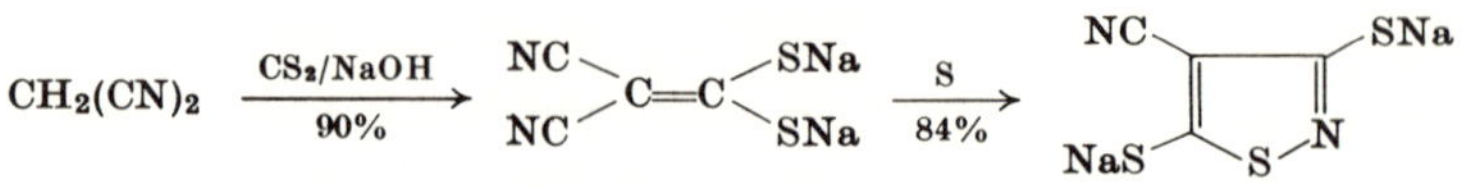

$$CH_2(CN)_2 \xrightarrow[90\%]{CS_2/NaOH} \underset{NC}{\overset{NC}{>}}C=C\underset{SNa}{\overset{SNa}{<}} \xrightarrow[84\%]{S} \text{(isothiazole)}$$

SCHEME 17

[43] E. Söderbäck, *Acta Chem. Scand.* **17**, 362 (1963).
[44] W. R. Hatchard, *J. Org. Chem.* **28**, 2163 (1963); du Pont, U.S. Patent 3,118,901 (1964) [*Chem. Abstr.* **60**, 15878 (1964)].
[45] W. R. Hatchard, *J. Org. Chem.* **29**, 665 (1964).
[46] M. Pianka, *J. Sci. Food Agr.* **19**, 507 (1968).

reported higher yields under modified reaction conditions, and Gewald[47] obtained the isothiazole directly from malononitrile, carbon disulfide, and sulfur. The analogous 4-phenyl-,[48] 4-ethoxycarbonyl-,[47,49] and carboxamidoisothiazole[45,50] have been prepared from the appropriate ethylenedithiolates.

The use of chlorine as the oxidant has given 4-cyano-3,5-dichloroisothiazole,[51] and hydrogen peroxide forms a 3-hydroxy derivative in good yield (Scheme 18)[44,52]

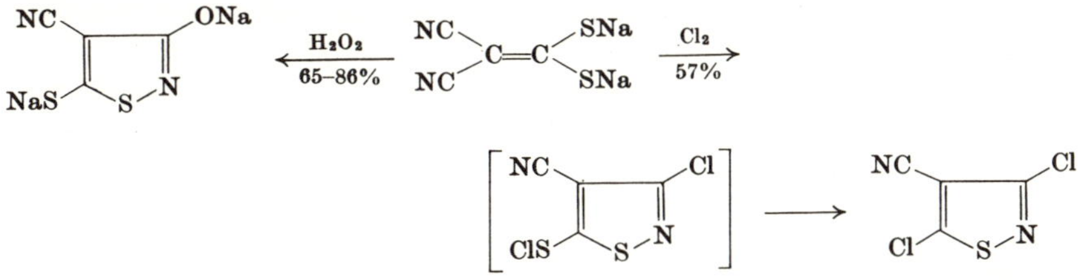

SCHEME 18

Hartke and Peshkar[53,54] employed chloramine as the oxidant to give a series of 3-amino-4-cyanoisothiazoles [Eq. (3)].

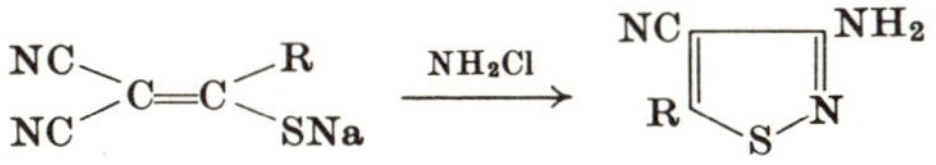

$$R = Me, OEt, SMe, CH_2Ph \tag{3}$$

5-Amino-3-chloro-4-cyanoisothiazole has been prepared by a further variation [Eq. (4)].[55]

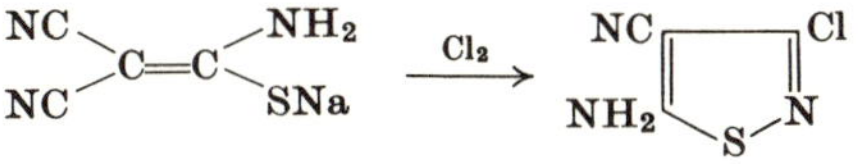

$$\tag{4}$$

[47] K. Gewald, *J. Prakt. Chem.* **31**, 214 (1966).
[48] M. Davis, G. Snowling, and R. W. Winch, *J. Chem. Soc. C* 124 (1967).
[49] E. Söderbäck, *Acta Chem. Scand.* **19**, 549 (1965).
[50] E. Söderbäck, *Acta Chem. Scand.* **24**, 228 (1970).
[51] W. R. Hatchard, *J. Org. Chem.* **29**, 660 (1964).
[52] Farbenfabriken Bayer, A.G. German Patent 1,814,249 (1970) [*Chem. Abstr.* **73**, 120753 (1970)].
[53] K. Hartke and L. Peshkar, *Angew. Chem. Int. Ed. Engl.* **6**, 83 (1967).
[54] K. Hartke and L. Peshkar, *Arch. Pharm.* **301**, 611 (1968).
[55] Merck Aktiengesellschaft, French Patent 2,014,527 (1970).

The synthesis of 3-chloroisothiazole from β-mercaptopropionitrile and chlorine[56] (Scheme 19) may be envisaged as occurring by a similar mechanism with an additional chlorination–dehydrochlorination stage either before or after cyclization.

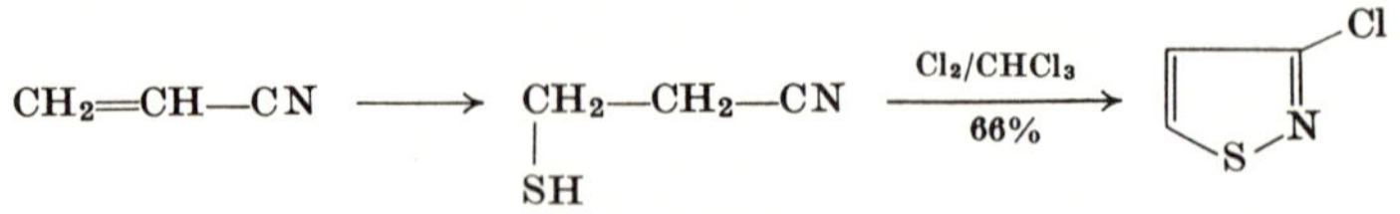

$$CH_2{=}CH{-}CN \longrightarrow \underset{SH}{CH_2{-}CH_2{-}CN} \xrightarrow[66\%]{Cl_2/CHCl_3} \text{[3-chloroisothiazole]}$$

SCHEME 19

The synthesis by Naito and co-workers[57] of 3-chloroisothiazoles from acrylonitriles and sulfur chlorides or thionyl chlorides (Scheme 20) may be rationalized by postulating initial substitution by the reagent, the intermediate then cyclizing under the influence of chlorine donated by excess of reagent.

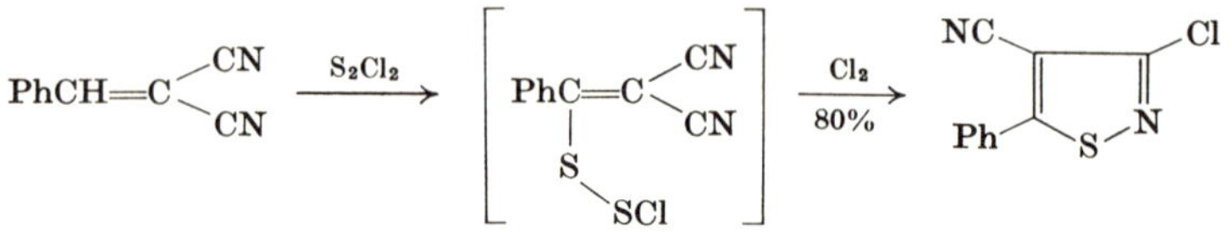

$$PhCH{=}C(CN)_2 \xrightarrow{S_2Cl_2} \left[PhC({=}C(CN)_2){-}S{-}SCl \right] \xrightarrow[80\%]{Cl_2} \text{[4-CN-3-Cl-5-Ph-isothiazole]}$$

SCHEME 20

F. PREPARATION OF ISOTHIAZOLES FROM 1,2-DITHIOLIUM SALTS

1,2-Dithiolium salts, particularly with aromatic substituents, readily form isothiazoles on treatment with ammonia[58–62] and a mechanism (Scheme 21) has been proposed by Olofson et al.[60]

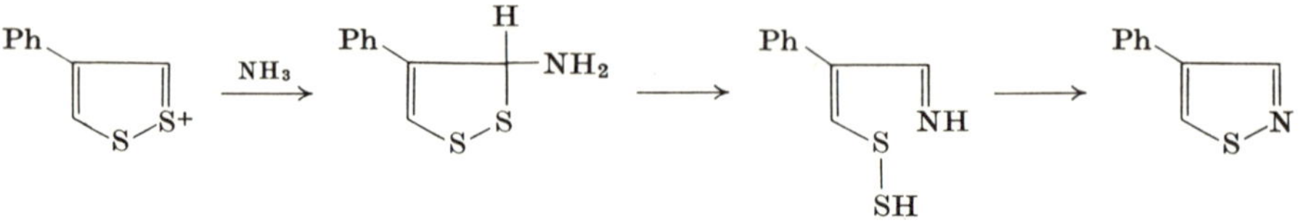

SCHEME 21

[56] Phillips Petroleum Company, U.S. Patent 3,285,930 (1966) [*Chem. Abstr.* **66**, 28766 (1967)].
[57] S. Nakagawa, J. Okumura, F. Sakai, H. Hoshi, and T. Naito, *Tetrahedron Lett.* 3719 (1970).
[58] D. Leaver and W. A. H. Robertson, *Proc. Chem. Soc.* 252 (1960).
[59] D. Leaver, D. M. McKinnon, and W. A. H. Robertson, *J. Chem. Soc.* 32 (1965).
[60] R. A. Olofson, J. M. Landesberg, R. O. Berry, D. Leaver, W. A. H. Robertson, and D. M. McKinnon, *Tetrahedron* **22**, 2119 (1966).
[61] H. Newman and R. B. Angier, *Chem. Commun.* 353 (1967).
[62] G. Purrello, *Gazz. Chim. Ital.* **96**, 1000 (1966).

The use of methylamine instead of ammonia gave an intermediate which did not cyclize spontaneously, but required oxidation with iodine (Scheme 22).[63]

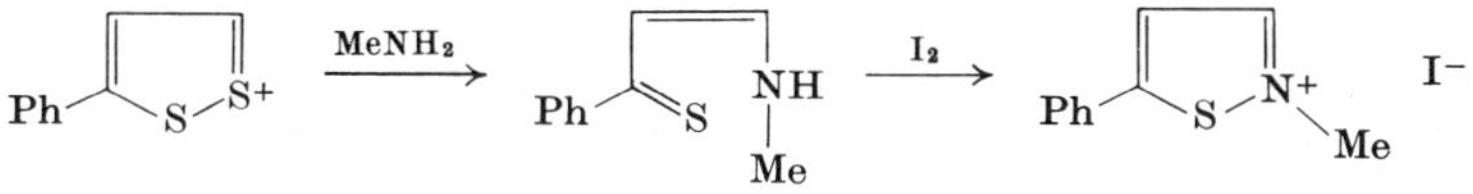

SCHEME 22

The 5-methylmercapto analog gave an isothiazole-3-thione which formed a quaternary isothiazole on treatment with methyl iodide (Scheme 23).[64]

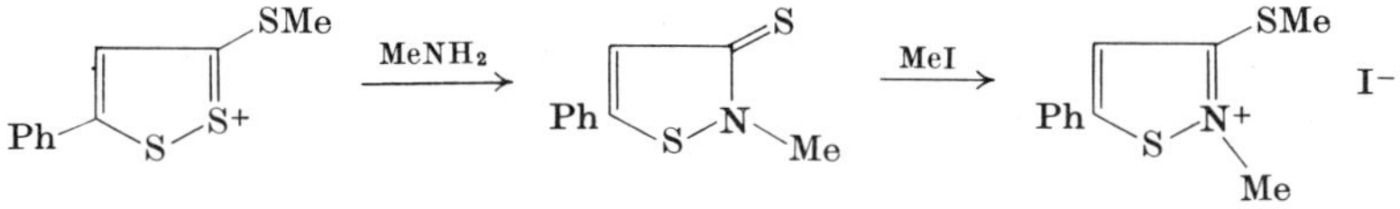

SCHEME 23

Aminodithiolium salts and aniline or morpholine gave the corresponding 3-phenylamino- or morpholinoisothiazoles or, in the presence of base, a 3-mercaptoisothiazole (Scheme 24).[65]

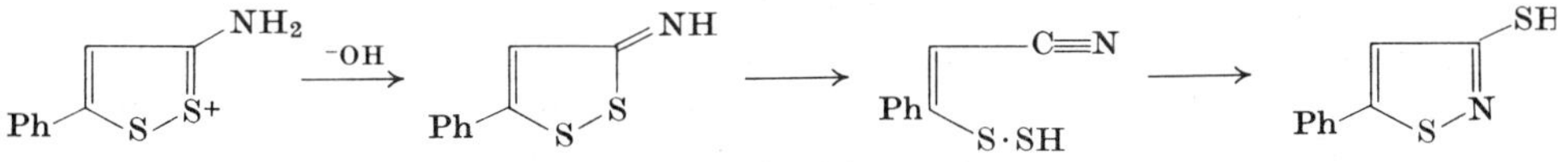

SCHEME 24

3-Aminoisothiazoles have been prepared from acyliminodisulfides (Scheme 25).[66]

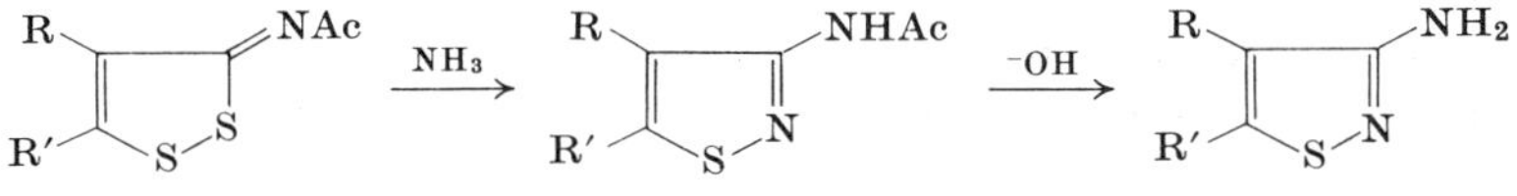

SCHEME 25

[63] D. M. McKinnon and E. A. Robak, *Can. J. Chem.* **46**, 1855 (1968).
[64] G. Le Coustumer and Y. Mollier, *C. R. Acad. Sci. C* **270**, 433 (1970).
[65] P. Condorelli, G. Pappalardo, and B. Tornetta, *Ann. Chim. (Rome)* **57**, 471 (1967).
[66] Farbenfrabriken Bayer, A.G., Belgian Patents 738,961 and 747,203 (1970) [*Chem. Abstr.* **73**, 120610 (1970)].

G. Preparation of Isothiazoles from β-Aminocrotonate and Thiophosgene

During the course of his synthesis of colchicine, Woodward and his collaborators devised a preparation [Eq. (5)] of 4-methoxycarbonyl-3-methylisothiazole from β-aminocrotonate and thiophosgene.[67, 68]

$$\text{Me---C}=\text{CH} \cdot \text{CO}_2\text{Me} \quad \xrightarrow{\text{CSCl}_2} \quad \begin{array}{c} \text{MeOCO} \quad \text{Me} \\ \diagdown \diagup \\ \text{S---N} \end{array} \quad (5)$$
$$\underset{\text{NH}_2}{|}$$

This route has been extended to the preparation of 4-acetyl- and 4-carboxyisothiazoles.[69]

II. The Physiochemical Properties of Isothiazoles

A. Ultraviolet Spectra

Isothiazole itself has an absorption maxima at 242 mμ and the bathochromic effects of substituents have been correlated.[70–72] There is some evidence that adjacent bulky substituents decrease the intensity of absorption as a result of steric hindrance of conjugation.[69]

B. Infrared Spectra

Isothiazoles normally give bands at or near 1510, 1400, 1340, and 810 cm^{-1},[1] although the 1400 and 1340 cm^{-1} bands may be weak and difficult to assign because of the complexity in this region.[65] A detailed

---

[67] R. B. Woodward, *Harvey Lect. Ser.* **59**, 31 (1965).
[68] J. Z. Gougoutas, Ph.D. Thesis, Harvard University (1964).
[69] S. Rajappa, A. S. Akerkar, and V. S. Iyer, *Indian J. Chem.* **7**, 103 (1969).
[70] D. Buttimore, D. H. Jones, R. Slack, and K. R. H. Wooldridge, *J. Chem. Soc.* 2032 (1963).
[71] D. H. Jones, R. Slack, and K. R. H. Wooldridge, *J. Chem. Soc.* 3114 (1964).
[72] M. P. L. Caton, D. H. Jones, R. Slack, and K. R. H. Wooldridge, *J. Chem. Soc.* 446 (1964).

assignment of fundamental vibration frequencies has been made for isothiazole itself on the basis of high resolution measurements of the vapor,[73] and thermodynamic functions have been calculated assuming a rigid harmonic oscillator model.[74] Microwave studies have shown that, not unexpectedly, the isothiazole molecule is planar.[75]

C. Nuclear Magnetic Resonance Spectra

The proton magnetic resonance spectrum of isothiazole shows a broad one-proton singlet at $\delta = 8.54$ ppm (assigned to the 3-H), a one-proton multiplet at 7.26 ppm (4-H) and a one-proton doublet at 8.72 ppm (5-H). The coupling constants are $J_{3H-4H} = 1.7$ Hz and $J_{4H-5H} = 4.7$ Hz.[76,77] The broadening of the signal due to the 3-H has been attributed to nuclear quadruple interaction involving ^{14}N.[78,79]

NMR studies have shown that the 5-proton exchanges rapidly with deuterium in the presence of added base,[60,80] whereas the 3-proton exchanges very much less readily (the rate constants are in the ratio of 400:1) and the 4-proton virtually not at all.[80]

D. Mass Spectra

The fragmentation pattern of isothiazole appears to parallel that of the thiazole molecule.[81] The molecular ion is prominent and most isothiazoles give fragments corresponding to 5 and 6.[82-86] There is

---

[73] S. Califano, F. Piacenti, and G. Sbrana, *Spectrochim. Acta* **20**, 339 (1964).
[74] T. R. Manley and D. A. Williams, *Spectrochim. Acta* **24**, 361 (1968).
[75] J. H. Griffiths, A. Wardley, V. E. Williams, N. L. Owen, and J. Sheridan, *Nature (London)* **216**, 1301 (1967).
[76] R. C. Anderson, *J. Heterocycl. Chem.* **1**, 279 (1964).
[77] H. A. Staab and A. Mannschreck, *Chem. Ber.* **98**, 1111 (1965).
[78] J. M. Lehn, *Z. Anal. Chem.* **235**, 10 (1968).
[79] J. P. Kintzinger and J. M. Lehn, *Mol. Phys.* **14**, 133 (1968).
[80] J. A. White and R. C. Anderson, *J. Heterocycl. Chem.* **6**, 199 (1969).
[81] B. J. Millard, *J. Chem. Soc. C* 1231 (1969).
[82] T. Naito, *Tetrahedron* **24**, 6237 (1968).
[83] J. C. Poite, R. Vivaldi, A. Bonzom, and J. Roggero, *C. R. Acad. Sci. C* **268**, 12 (1969).
[84] M. Kojima and M. Maeda, *Chem. Commun.* 386 (1970).
[85] F. T. Lee and G. P. Volpp, *J. Heterocycl. Chem.* **7**, 415 (1970).
[86] F. T. Lee, B. W. Li, and G. P. Volpp, *J. Heterocycl. Chem.* **7**, 941 (1970).

$$\underset{R_1}{\overset{R_2}{\diagup}}\underset{S}{\diagdown}\overset{R_3}{\underset{N}{\diagup}} \xrightarrow{-R_3C\equiv N} \underset{\underset{+}{S}}{R_1-C\overset{\diagdown\diagup}{=}C-R_2} \quad (5)$$

$$\searrow$$

$$R_3CNS^+$$
(6)

some evidence to suggest that methylisothiazoles under ring expansion [Eq. (6)].[81]

$$\underset{\overset{+}{S}-N}{\boxed{}}\!\!-Me \longrightarrow \left[\underset{S-NH}{\boxed{}}\right]^{\overset{+}{\bullet}} \quad (6)$$

III. The Chemical Properties of Isothiazoles

The isothiazoles possess the typical properties of a heterocyclic aromatic system without the ring lability so characteristic of the analogous isoxazoles.[87] The chemistry reflects the relative inertness of the 3-position, the susceptibility of the 4-position to electrophilic attack, and the susceptibility of the 5-position to nucleophilic attack. The ring nitrogen is only weakly basic, but can be induced to form quaternary derivatives, and the N–S bond may be cleaved under certain circumstances.

A. ELECTROPHILIC SUBSTITUTION REACTIONS

According to electron density calculations,[3, 88] isothiazole in the ground state possesses negative charges on the nitrogen atom and the 4-carbon. Isothiazole is weakly basic ($pK_a = -0.51 \pm 0.04$ at 25°[89]) and would be almost completely protonated in strong acid media, but nevertheless it readily undergoes nitration with a mixture of nitric and sulfuric acids to give 4-nitroisothiazole in over 90% yield,[32]

[87] N. K. Kochetkov and S. D. Sokolov, *Advan. Heterocycl. Chem.* **2**, 365 (1963).
[88] R. Phan-Tan-Luu, L. Bouscasse, E. J. Vincent, and J. Metzger, *Bull. Soc. Chim. Fr.* 3283 (1967).
[89] D. F. Muggleton and B. J. Ward, personal communication.

and high yields are also obtained with alkyl substituents in the 3- or 5-positions, or bromo or acetamido groups in the 5-position.[3, 32] 5-Amino-3-methylisothiazole, however, forms an N-nitro derivative.[3] 3-Phenylisothiazole nitrates predominantly in the meta position of the substituent[90] in contrast to 4-phenylisothiazole where ortho and para substitution occurs[91] (Scheme 26).

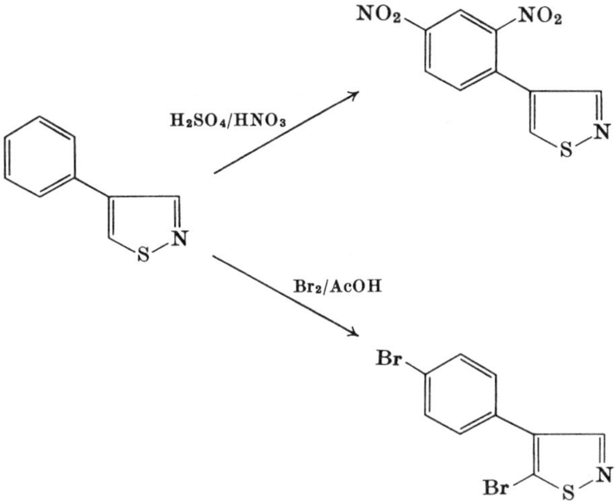

SCHEME 26

Isothiazole and alkylisothiazoles may be brominated under various conditions,[18, 70, 92] but yields are often poor possibly because of the formation of perbromo compounds. Isothiazoles with electron-releasing substituents, however, undergo facile bromination and high yields have been reported for isothiazoles with amino,[3, 7, 9, 93–95] hydroxy,[18] or alkoxy[93, 96] substituents in the 3- or 5-position. An isothiazole with an electron-withdrawing substituent, 3-methylisothiazole-5-carboxylic acid, has been brominated in 76% yield.[70]

[90] T. Naito, S. Nakagawa, and K. Takahashi, *Chem. Pharm. Bull.* **16**, 160 (1968).
[91] J. H. Finley and G. P. Volpp, *J. Heterocycl. Chem.* **6**, 841 (1969).
[92] F. Piacenti, P. Bucci, and P. Pino, *Chimi. Ind. (Milan)* **46**, 207 (1964).
[93] R & L Molecular Research, U.S. Patent 3,311,611 (1967) [*Chem. Abstr.* **68**, 59570 (1968)].
[94] R. G. Micetich and R. Raap, *J. Med. Chem.* **11**, 159 (1968).
[95] R. Raap and R. G. Micetich, *J. Med. Chem.* **11**, 70 (1968).
[96] I. D. H. Stocks, J. A. Waite, and K. R. H. Wooldridge, *J. Chem. Soc. C* 1314 (1971).

Isothiazoles are sulfonated readily with oleum[97] or sulfur trioxide,[98] but formylation with dimethylformamide and phosphorus oxychloride, and acylation under Friedel–Crafts conditions, failed.[70]

On the basis of spectroscopic studies Anderson[76] suggested that a 4-methyl substituent would supply electrons more efficiently to the 5-position that the 3-position and, in fact, under forcing conditions 4-methylisothiazole will undergo sulfonation to 4-methylisothiazole-5-sulfonic acid in poor yield.[32, 77] 4-Hydroxy-3-phenylisothiazole is chlorinated in the 5-position[29] and 4-phenylisothiazole is brominated in the heterocyclic ring as well as in the phenyl substituent[91] (Scheme 26). Naito et al.[99] reported that 3-phenylisothiazole-4-carboxylic acid is brominated in the 5-position, but no yields or conditions were given.

B. Metallation Reactions

A study of hydrogen–deuterium exchange rates showed that the 5-proton exchanges very rapidly under basic conditions indicating the formation of a relatively stable anion.[67, 80] Under the appropriate conditions, therefore, isothiazoles form lithium derivatives which are of considerable preparative value, leading to a wide variety of substituents in the 5-position usually in good yield.[7, 9, 37, 67, 69, 72, 93, 94, 99–104] The lithium derivatives have usually been prepared from butyllithium in tetrahydrofuran and ether or hexane at $-70°$,[72] but diethylaminoethyllithium[105] or 2-biphenyllithium[67] have also been employed in order to minimize ring cleavage. Under certain conditions 4-butylmercapto-2-oxobut-3-ene (7) is the main product from 3-methyl-

---

[97] D. L. Pain and E. W. Parnell, *J. Chem. Soc.* 7283 (1965).

[98] Hans J. Zimmer Verfahrenstechnik, German Patent 1,208,303 (1966) [*Chem. Abstr.* **64**, 8190 (1966)].

[99] T. Naito, S. Nakagawa, K. Takahashi, K. Fujisawa, and H. Kawaguchi, *Antimicrob. Ag. Chemother.* **94**, 162 (1968).

[100] R & L Molecular Research, U.S. Patent 3,271,407 (1966) [*Chem. Abstr.* **66**, 28767 (1967)].

[101] A. Layton and E. Lunt, *J. Chem. Soc.* 611 (1968).

[102] R. Raap and R. G. Micetich, *Can. J. Chem.* **46**, 1057 (1968). R & L Molecular Research, U.S. Patent 3,464,999 (1969) [*Chem. Abstr.* **72**, 21725 (1970)].

[103] R. G. Micetich, *Can. J. Chem.* **48**, 2006 (1970).

[104] FMC Corp., S. African Patent 662,869 (1966) [*Chem. Abstr.* **67**, 100136 (1968)].

[105] G. P. Volpp, quoted by J. Z. Gougoutas, Ph.D. Thesis, Harvard University (1964).

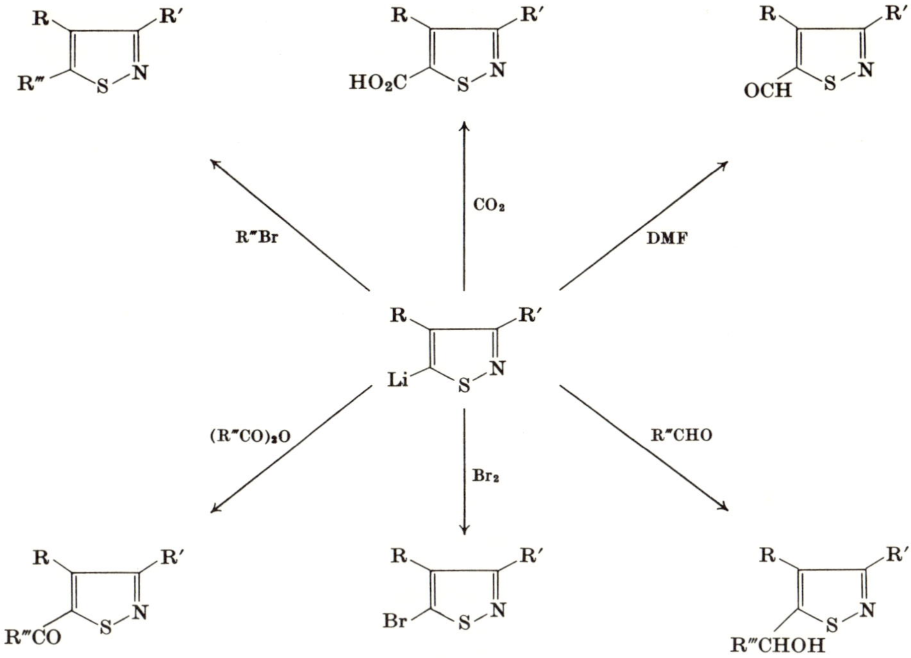

SCHEME 27

isothiazole and butyl- or phenyllithium,[105] and it has also been isolated as a by-product during the preparation of 5-formyl-3-methylisothiazole.[72]

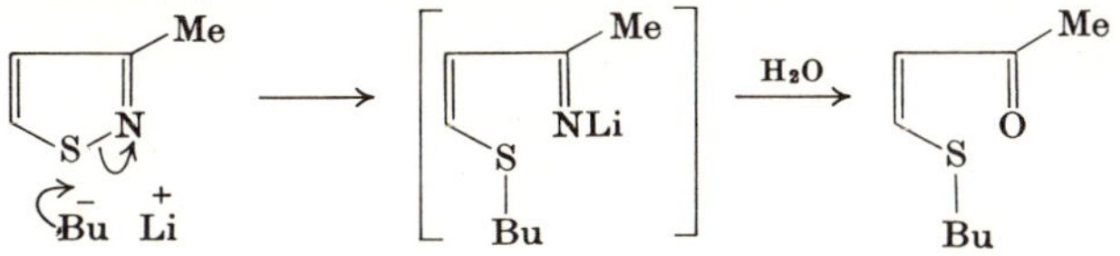

(7)

3,5-Dimethylisothiazole formed a similar ring cleavage product, and 4-methylisothiazole, although lithiated predominantly in the 5-position, gave a small proportion of a nitrile, possibly via an isothiazolyl-3-lithium intermediate (Scheme 28).[103] A 4-bromo

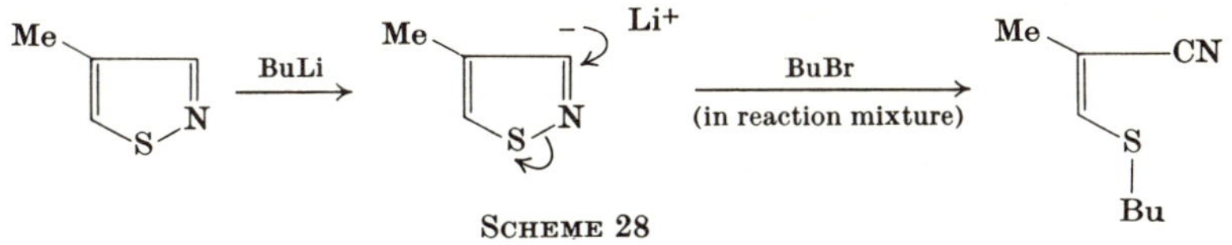

SCHEME 28

substituent is usually inert toward butyllithium, but one case has been reported of the formation of a 4-lithio derivative when the 3- and 5-positions were blocked (Scheme 29).[9]

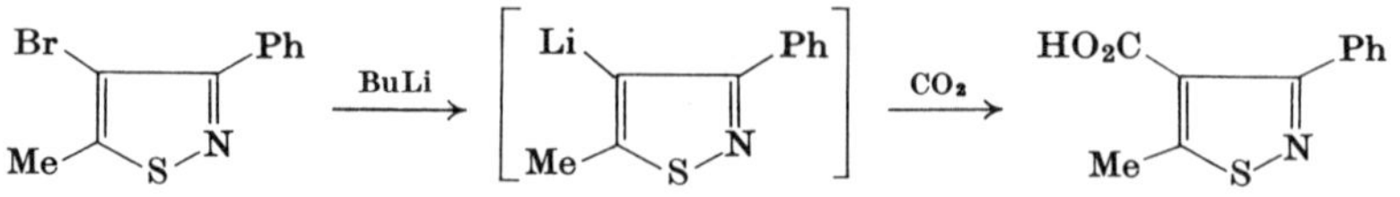

SCHEME 29

On treatment with mercuric acetate in aqueous acetic acid at 100°, isothiazole gives a 48% yield of a dimercuriacetate, which with bromine affords 4,5-dibromoisothiazole. Unlike lithiation, a nitro group does not interfere, and mercuriation provides a convenient route to 5-bromo-4-nitroisothiazole (Scheme 30).[106]

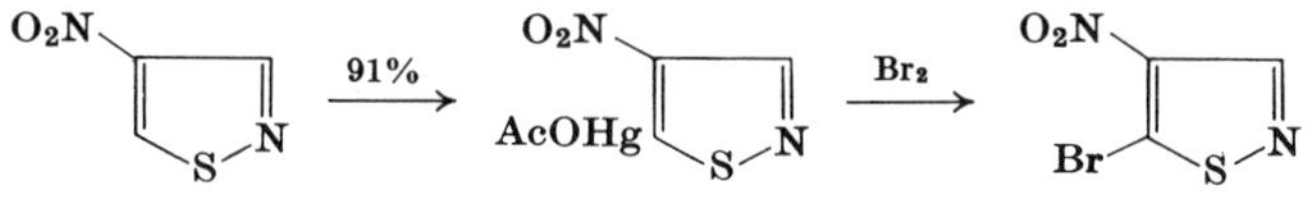

SCHEME 30

C. Alkyl- and Arylisothiazoles

Most of the simple monoalkyl- and arylisothiazoles may be prepared by direct ring synthesis either by the olefin route (Section I,C) or Wille's acetylenic route (Section I,D). 5-Alkylisothiazoles are accessible from isothiazolyllithium derivatives, and 3-alkyl- and arylisothiazoles are conveniently prepared by deamination of the readily available 5-aminoisothiazoles.[37, 70, 95] Arylisothiazoles have also been prepared from aminoisothiazoles by the Gomberg–Hey route.[107]

Phenylation of isothiazole by decomposition of benzoyl peroxide at 110° gave a mixture of the 3-, 4-, and 5-isomers in the ratio 47:9:44. 3- and 4-Methylisothiazole were phenylated predominantly in the 5-position and 5-methylisothiazole in the 3-position.[108, 109]

[106] D. Buttimore and R. Slack, personal communication.
[107] M. S. Grant, D. L. Pain, and R. Slack, *J. Chem. Soc.* 3842 (1965).
[108] H. J. Dou, J. C. Poite, G. Vernin, and J. Metzger, *Tetrahedron Lett.* 779 (1969).
[109] J. C. Poite, G. Vernin, G. Loridan, H. J. Dou, and J. Metzger, *Bull. Soc. Chim. Fr.* 3912 (1969).

Methyl groups in the 3- and 4-positions are unreactive toward aldehydes,[68, 110] but 5-methylisothiazoles condense with 3-nitrobenzaldehyde. This property has been used to separate pure 3,4-dimethylisothiazole from a mixture with 3,5-dimethylisothiazole.[110]

4-Alkylisothiazoles have been oxidized in the vapor phase to 4-formylisothiazole,[32, 111] and 3- and 4-methylisothiazoles may be oxidized to isothiazolecarboxylic acids.[100, 112]

Side-chain halogenation of methylisothiazoles takes place in good yield and preparations by this method of 3-bromomethyl,[67, 68, 95, 113] 3-dibromomethyl,[71] and 4-chloromethylisothiazoles[32] have been reported.

D. AMINOISOTHIAZOLES

3-Aminoisothiazoles have been prepared directly from acyliminodisulfides (see Section I, F) and by oxidative cyclization of thiobenzoylacetamidine[18] or a dicyanoethylene derivative[53, 54] (see Section I, E). 3-Aminoisothiazole itself was prepared in 62% yield by Hofmann degradation of the 3-carboxyamide.[112]

4-Aminoisothiazoles have usually been prepared by reduction of 4-nitroisothiazoles which are available by direct nitration. Catalytic hydrogenation has been accomplished successfully, particularly with more heavily substituted derivatives,[3, 112] whereas chemical methods appear preferable for simple nitroisothiazoles.[3, 72, 112] The poor yield originally reported for the reduction of 4-nitro- to 4-aminoisothiazole[72] was probably due to ring cleavage and has been improved from 35 to 65% by carrying out the reaction at 10° rather than at 40°.[114] 4-Amino-3-methylisothiazole has been prepared from the readily available 4-acetyl-3-methylisothiazole by formation of the oxime, followed by Beckmann rearrangement, and also from 3-methylisothiazole-4-carboxylic acid by Curtius degradation.[69] 4-Amino-3,5-dichloroisothiazole was prepared from 4-cyano-3,5-dichloroisothiazole by hydrolysis to the amide, followed by Hofmann degradation.[8]

[110] H. Hofmann, *Ann.* **690**, 147 (1965).
[111] Merck & Co. Inc., British Patent 1,113,705 (1968 [*Chem. Abstr.* **69**, 59226 (1968)].
[112] A. Holland, R. Slack, T. F. Warren, and D. Buttimore, *J. Chem. Soc.* 7277 (1965).
[113] D. Buttimore, M. P. L. Caton, J. D. Renwick, and R. Slack, *J. Chem. Soc.* 7274 (1965).
[114] K. R. H. Wooldridge, unpublished data.

5-Aminoisothiazoles are readily available by direct ring synthesis (see Section I,B), and most of those described in the literature have been prepared in this way. However, those methods do not lead to 3-chloroisothiazoles, and 5-amino-3-chloro-4-cyanoisothiazole was prepared in excellent yield from 3,5-dichloro-4-cyanoisothiazole and ammonia.[8, 51] 5-Aminoisothiazole itself has been prepared by Curtius degradation of isothiazole-5-carboxylic acid (Scheme 31),[115] although a triketotriazine has also been obtained under Curtius conditions.[116]

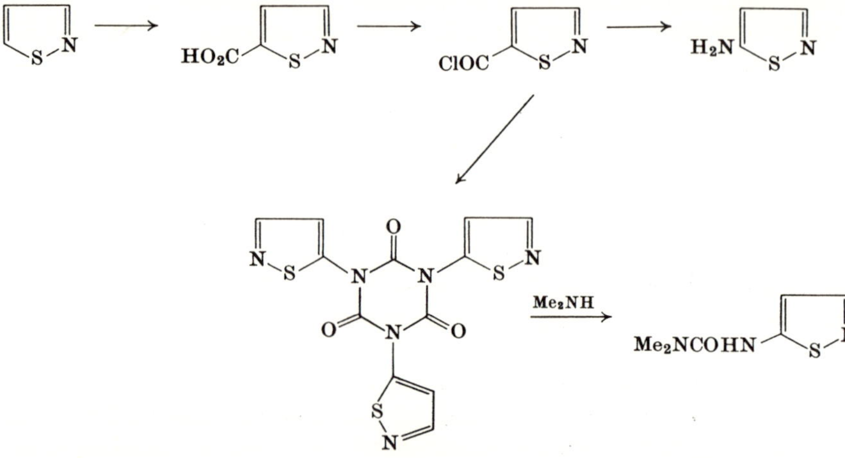

SCHEME 31

4,5-Diamino-3-methylisothiazole has been prepared by reduction of 5-acetamido-3-methyl-4-nitroisothiazole and subsequent hydrolysis (Scheme 32).[112] 4,5-Diamino-3-chloroisothiazole was prepared from 4-amino-3,5-dichloroisothiazole and ammonia.[8]

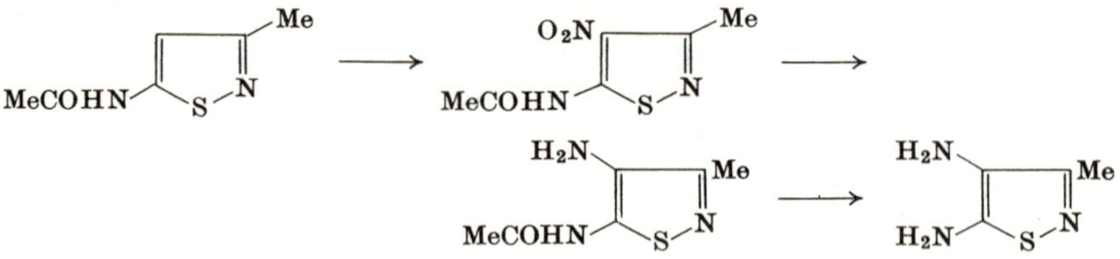

SCHEME 32

[115] D. H. Jones and K. R. H. Wooldridge, unpublished data.
[116] F.M.C. Corp., Belgian Patent 718,708 (1969) [*Chem. Abstr.* **72**, 100,679 (1970)].

A 3,5-diaminoisothiazole derivative was prepared by Goerdeler and Keuser[12] by ring synthesis (Scheme 33).

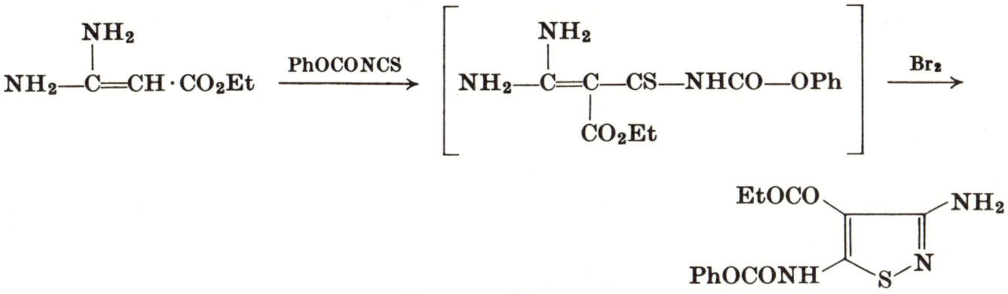

$$NH_2-\underset{\underset{NH_2}{|}}{C}=CH \cdot CO_2Et \xrightarrow{PhOCONCS} \left[NH_2-\underset{\underset{CO_2Et}{|}}{\overset{\overset{NH_2}{|}}{C}}=C-CS-NHCO-OPh \right] \xrightarrow{Br_2}$$

$$\underset{PhOCONH}{\overset{EtOCO}{\diagdown}} \underset{S}{\overset{}{\diagup}} \overset{NH_2}{\underset{N}{\diagdown}}$$

SCHEME 33

3-, 4-, and 5-Aminoisothiazoles have pK_a's of 2.49, 3.58, and 2.70, respectively,[89] and the aminoisothiazoles, in general, behave as would be expected for weak amines. Thus, they form acetyl derivatives, sulfonamides, ureas, and thioureas. They may be diazotized, and the diazonium salts undergo the Sandmeyer reaction to give halogenoisothiazoles,[3, 6-8, 37, 70, 72, 93, 94, 117, 118] reductive deamination,[3, 7, 8, 37, 70, 93, 118] the Meerwein reaction,[119] and the Gomberg–Hey reaction.[107] Solid diazonium fluoroborates have been isolated in some cases, but they were only stable for a few minutes in the solid state.[118] 5-Aminoisothiazoles with electron-withdrawing substituents in the 4-position have given N-nitroso compounds on diazotization.[94, 118] 5-Amino-3-methylisothiazole undergoes the Skraup reaction [Eq. (7)] although in poor yield.[120]

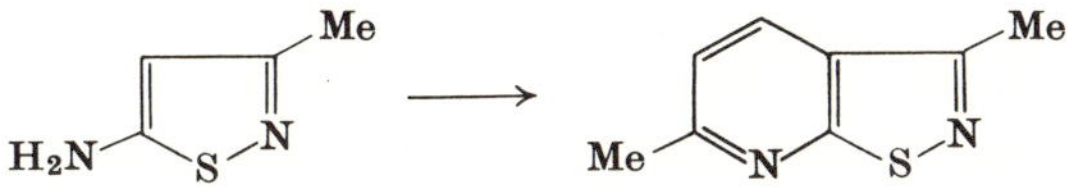

$$H_2N \diagup S \diagdown N \diagup Me \longrightarrow Me \diagup N \diagdown S \diagup N \diagup Me \quad (7)$$

3- and 5-Aminoisothiazoles can theoretically exist in imino tautomeric forms, but their chemical reactions suggest that the amino forms predominate.

---

[117] Z. Machón, *Diss. Pharm. Pharmacol.* **21**, 135 (1969).
[118] J. Goerdeler and M. Roegler, *Chem. Ber.* **103**, 112 (1970).
[119] B. A. Bennett, D. H. Jones, R. Slack, and K. R. H. Wooldridge, *J. Chem. Soc.* 3834 (1965).
[120] M. R. Willson and V. Rogers, personal communication.

E. HALOGENOISOTHIAZOLES

3-Chloroisothiazole is available by chlorinative cyclization of β-mercaptopropionitrile (see Section I, E). 3-Chloro-5-phenylisothiazole has been prepared by the Sandmeyer reaction from 3-amino-5-phenylisothiazole[118] and also by thermal decomposition of the 3-chloroisothiazolium compound formed by chlorination of N-methyl-5-phenyl-3-isothiazolone.[27] 3-Chloro-4-methylisothiazole was obtained, together with 4-chloromethylisothiazole, by free radical halogenation of 4-methylisothiazole.[32] 3-Fluoro-5-phenylisothiazole has been prepared by decomposition of the appropriate diazonium fluoroborate.[118]

4-Chloro- and 4-bromoisothiazoles are formed by direct electrophilic substitution (see Section III, A) and by the Sandmeyer reaction from 4-aminoisothiazoles (see Section III, D).

5-Halogenoisothiazoles are readily prepared by the action of halogens on the 5-lithium derivatives (see Section III, B) or by the Sandmeyer reaction from 5-aminoisothiazoles.

3,4-Dichloro-5-methylisothiazole,[57] 3,4,5-trichloroisothiazole,[57] 3,4-dichloro-5-cyanoisothiazole,[4,5] and 3,5-dichloro-4-cyanoisothiazole[44, 51] have been prepared by direct ring synthesis. 4,5-Dibromoisothiazole has been prepared from the dimercuriacetate[106] and 4,5-dibromo-3-methylisothiazole by bromination of 5-bromo-3-methylisothiazole or by the Sandmeyer reaction from 5-amino-4-bromo-3-methylisothiazole.[70]

The halogen atoms in the various positions of isothiazole show interesting differences in reactivity, which is of considerable synthetic value. Thus, the 5-halogen, particularly when activated by an electron-withdrawing group in the 4-position, readily undergoes nucleophilic displacement to give isothiazoles with hydroxy,[96] alkoxy,[51, 93, 94, 96] alkylthio,[9, 51, 99] amino,[8, 51] cyano,[3, 112] and hydrazino[117, 121, 122] substituents. Disulfides,[44, 121] sulfides,[86, 121, 123] and thiols[51] have also been prepared this way. A 3-halogen, however, even when activated is less reactive than a 5-halogen, and replacement may be accompanied by ring cleavage [Eq. (8)].[51] A preliminary communication has

[121] M. P. L. Caton, G. C. J. Martin, and D. L. Pain, *J. Chem. Soc. C* 776 (1971).
[122] Merck-Anlagen-Gesellschaft, Belgian Patent 745,155 (1970 [*Chem. Abstr.* **73**, 77238 (1970)].
[123] M. P. L. Caton and R. Slack, *J. Chem. Soc. C* 1402 (1968).

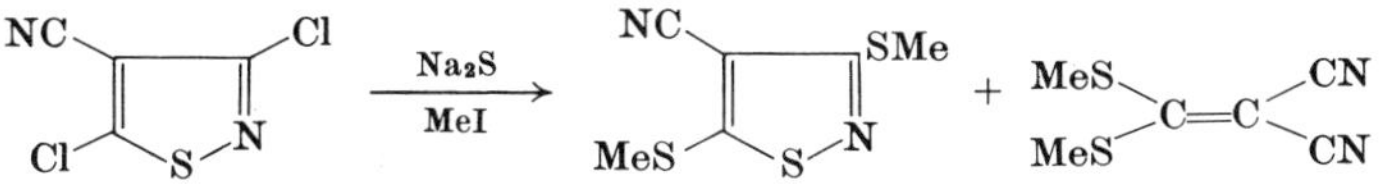

described the displacement of chlorine by methoxy in 3-chloro-4-cyano-5-phenylisothiazole, but the yield is not quoted.[57]

The 4-position is the most benzenelike and a 4-halogen resists nucleophilic attack with the exception of the formation of nitriles with cuprous cyanide[9, 70, 93, 94, 99] and one case of the formation of a lithium derivative (see Section III,B). Halogen exchange has been observed during the diazotization of a 5-amino-4-bromoisothiazole in the presence of concentrated hydrochloric acid.[119]

De Bie and van der Plas[124] treated 5-bromo-3-methylisothiazole with four equivalents of potassium amide in liquid ammonia at $-33°$ and obtained an 85% yield of 4-bromo-3-methylisothiazole, together with a small amount of 3-methylisothiazole. No aminoisothazoles were detected suggesting that an isothiazolyne was not an intermediate, and the following mechanism was proposed (Scheme 34).

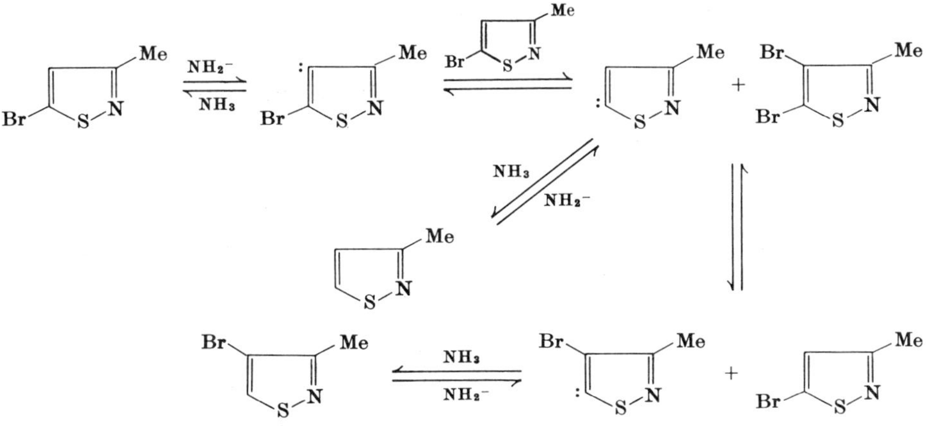

SCHEME 34

This mechanism is supported by the observation that an equimolar mixture of 3-methylisothiazole and 4,5-dibromoisothiazole gave a quantitative yield of 4-bromo-3-methylisothiazole under the same conditions.

[124] D. A. de Bie and H. C. van der Plas, *Tetrahedron Lett.* 3905 (1968).

F. Hydroxy- and Alkoxyisothiazloes

3-Hydroxy- and alkoxyisothiazoles are readily available by direct ring synthesis[12, 18, 28, 38, 44, 52, 125] (see Section I) and, therefore, other methods have been little explored. However, 3-hydroxy-5-phenylisothiazole has been prepared from the corresponding diazonium compound,[18,118] and 4-cyano-3-methoxy-5-phenylisothiazole from 3-chloro-4-cyano-5-phenylisothiazole and sodium methoxide.[57] 3-Isothiazolones have been obtained from penicillin sulfoxide[126] and 1,4-thiazepines which are related to penicillins (Scheme 35).[127]

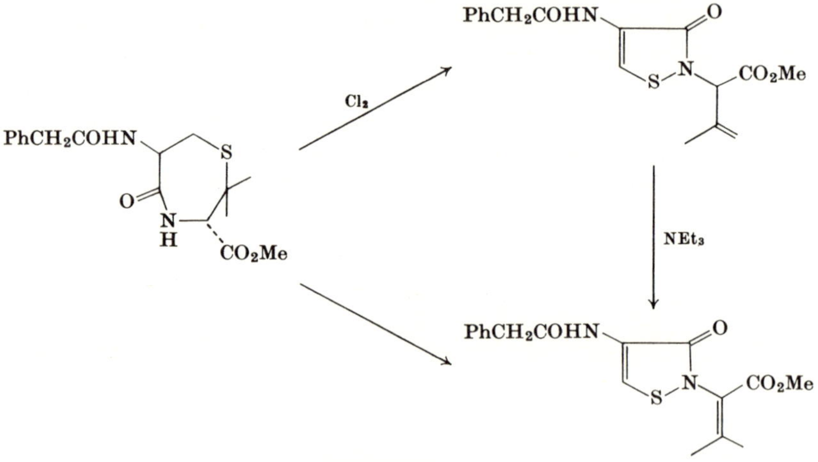

SCHEME 35

4,5-Dihydroisothiazol-3-ones have been described [Eq. (9)].[128]

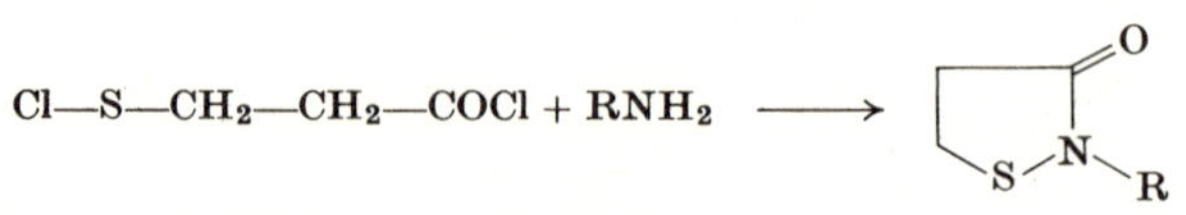

$$Cl—S—CH_2—CH_2—COCl + RNH_2 \longrightarrow \qquad (9)$$

4-Hydroxyisothiazoles have been prepared from aminoketones and thionyl or sulfuryl chlorides,[29] but have not been isolated from decomposition of 4-diazonium compounds.[29, 114]

[125] J. Goerdeler, *Angew. Chem.* **74**, 498 (1962).
[126] G. E. Gutowski, *Tetrahedron Lett.* 1779 (1970).
[127] N. J. Leonard and G. E. Wilson, *Tetrahedron Lett.* 1471 (1964); N. J. Leonard and G. E. Wilson, *J. Amer. Chem. Soc.* **86**, 5307 (1964).
[128] A. Lüttringhaus and R. Schneider, *Angew. Chem. Int. Ed. Engl.* **3**, 67 (1964); A. Lüttringhaus and R. Schneider, *Ann.* **679**, 123 (1964).

A 5-ethoxyisothiazole has been prepared by ring synthesis[53] and 5-hydroxy- and alkoxyisothiazoles by nucleophilic displacement of bromine.[94, 96] 5-Hydroxyisothiazoles have also been obtained from 5-diazonium compounds, but conditions are critical.[96, 118]

UV studies suggested that 3-hydroxyisothiazole exists as such in nonpolar solutions, whereas the NH form predominates in more polar solvents.[129] 3-Hydroxy-5-phenylisothiazole, however, appears to exist in the OH form even in methanolic solution.[18]

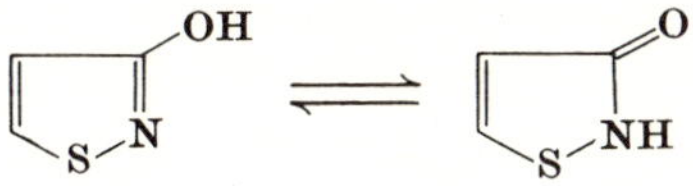

Chan and Crow[130] studied the acylation of 3-hydroxyisothiazole in some detail. In aprotic solvents, acylation is subject to kinetic control leading almost exclusively to 3-acyloxyisothiazoles. On heating or standing, a reversible O → N migration occurs and the final equilibrium position depends on the nature of the acyl group, larger groups tending to favor the oxygen atom. However, acylation with isatoic anhydride gave the O-anthaniloyl derivative, together with isothiazolo[2,3b]quinazolin-9-one formed by cyclization of the N-anthaniloyl derivative (Scheme 36). Under the experimental condi-

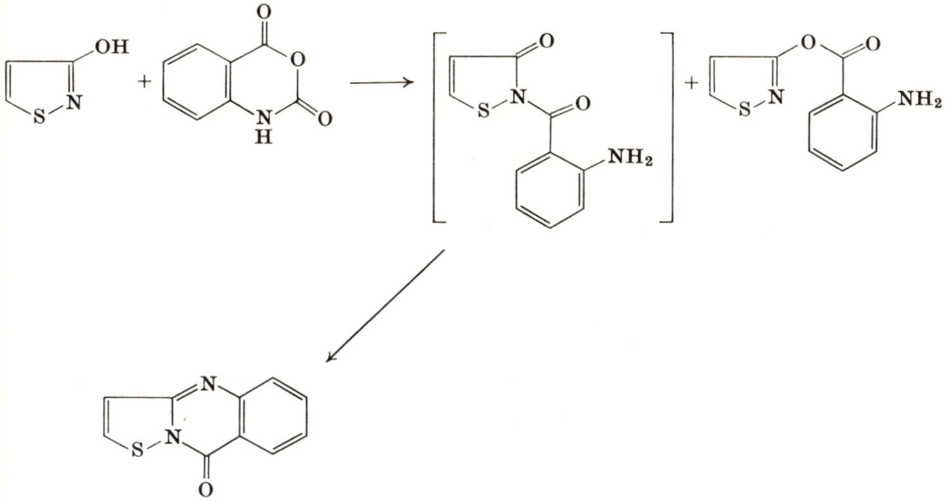

SCHEME 36

[129] A. W. K. Chan, W. D. Crow, and I. Gosney. *Tetrahedron* **26**, 2497 (1970).
[130] A. W. K. Chan and W. D. Crow, *Aust. J. Chem.* **21**, 2967 (1968).

tions employed, the favored N → O migration competed with dehydration to the quinazolinone, and it was shown by control experiments that O → N migration did not occur.[131]

The action of diazomethane on 3-hydroxyisothiazole gave approximately equal proportions of O- and N-methyl products,[129] whereas 3-hydroxy-5-phenylisothiazole gave exclusively an N-methyl derivative.[18] Nucleophilic reagents lead to ring cleavage,[132–134] which will be discussed in a later section (see Section IV).

Little information is available on 4-hydroxyisothiazoles except that they appear to be methylated and acylated normally.[29]

5-Hydroxyisothiazoles can theoretically exist in a number of tautomeric forms and preliminary studies[96] suggest that substituted 5-hydroxyisothiazoles exist predominantly in the NH form with substantial contributions from zwitterionic species [Eq. (10).

Diazomethane and 4-bromo-5-hydroxy-3-methylisothiazole gave the O-methyl derivative exclusively, but the 4-nitro analog did not react.[96]

G. ISOTHIAZOLECARBOXYLIC ACIDS, AMIDES, AND NITRILES

Isothiazole-3-carboxylic acids are conveniently prepared by oxidation of 3-methylisothiazoles with chromium trioxide in sulfuric acid.[112] Permanganate oxidation is less satisfactory, although 3-bromomethylisothiazole has been oxidized to the acid in 47% yield.[113] The amides, prepared via the acid chlorides and ammonia, may be dehydrated to the 3-nitriles by phosphorus oxychloride.[71]

Many isothiazole-4-carboxylic acids, esters, amides, and nitriles have been prepared by direct ring synthesis (see Section I), but another widely employed route is the sequence bromo compound, nitrile, amide, or acid (Scheme 37)[7, 70, 93, 95, 107]

[131] A. W. K. Chan and W. D. Crow, *Aust. J. Chem.* **22**, 2497 (1969).
[132] W. D. Crow and I. Gosney, *Aust. J. Chem.* **22**, 765 (1969).
[133] A. W. K. Chan, W. D. Crow, and I. Gosney, *Tetrahedron* **26**, 1493 (1970).
[134] W. D. Crow and I. Gosney, *Tetrahedron* **26**, 1463 (1970).

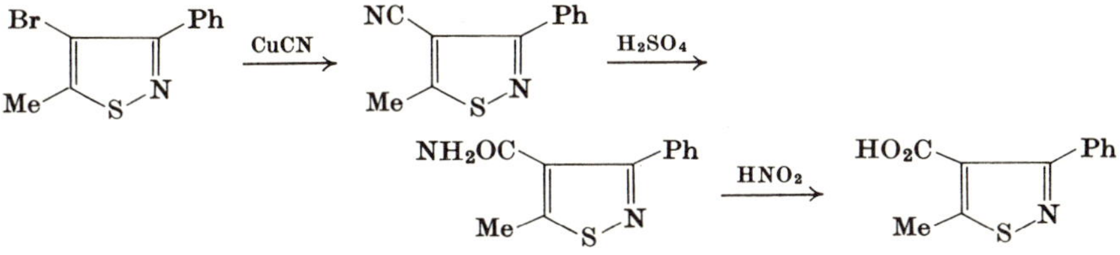

Sec. III.H.] CHEMISTRY OF MONONUCLEAR ISOTHIAZOLES 29

SCHEME 37

Isothiazole-5-carboxylic acids have usually been synthesized from the lithium derivatives and carbon dioxide, or by alkaline hydrolysis of nitriles obtained from 5-bromo compounds and cuprous cyanide.[3, 70]

There are marked differences in the thermal stability of the carboxyl groups in the various positions in the ring. Isothiazole-5-carboxylic acids decarboxylate readily at or near the melting points[3, 42, 70] and isothiazole-3-carboxylic acid behaves similarly.[113] The 4-carboxylic acids however are considerably more stable and require heating with copper salts[105, 107] or prolonged acid treatment[25] to effect decarboxylation.

Chemically, the isothiazolecarboxylic acids all behave normally and form acid chlorides, amides, esters, hydrazides, etc. 4-Esters have been reduced to hydroxymethylisothiazoles with lithium aluminum hydride[67, 69] and the Arndt–Eistert reaction has been employed to prepare 4- and 5-acetic acids.[95, 100, 135] Isothiazole-3-acetic acid was prepared from 3-bromomethylisothiazole via the nitrile.[95, 100, 113] Isothiazolyl-3- and -5-alanines have been prepared from the bromomethylisothiazoles by the acetamidomalonate route[119] and 4-isothiazolylglyoxylic acid by alkaline treatment of α-azido-4-isothiazolylacetic acid [Eq. (11)].[136]

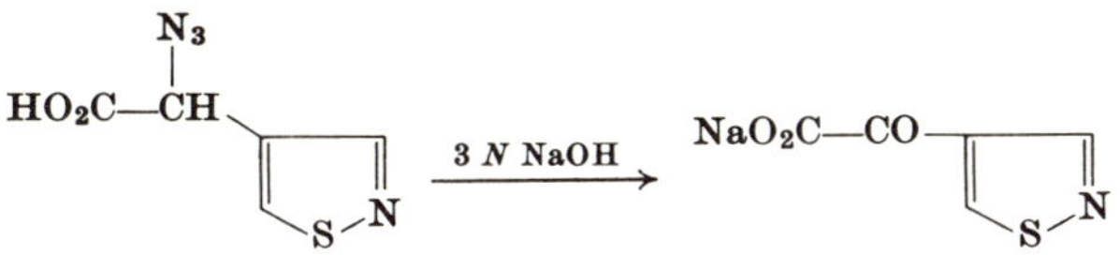

$$HO_2C-CH(N_3)-\text{[isothiazole]} \xrightarrow{3\,N\,\text{NaOH}} NaO_2C-CO-\text{[isothiazole]} \qquad (11)$$

H. ISOTHIAZOLE ALDEHYDES AND KETONES

3-Formyl- and 4-bromo-3-formylisothiazole have been prepared from the dibromomethylisothiazoles obtained by side-chain halo-

[135] Bristol Myers, U.S. Patent 3,294,783 (1966) [*Chem. Abstr.* **66**, 76000 (1967)].
[136] R. Raap, *Tetrahedron Lett.* 3493 (1969).

genation of the appropriate 3-methylisothiazoles with N-bromosuccinimide.[71,137] 3-Acylisothiazoles have been obtained from 3-cyanoisothiazoles and alkylmagnesium halide.[71,138]

4-Formylisothiazole and 4-formyl-3-methylisothiazole have been prepared by hydrolysis of Reissert compounds.[70,137] An alternative

route to 4-formyl-3-methylisothiazole is provided by reduction of ethyl 3-methylisothiazole-4-carboxylate to the hydroxymethylisothiazole with lithium aluminum hydride followed by oxidation with manganese dioxide.[67,68] 4-Acetyl-3-methylisothiazole is obtainable by direct ring synthesis,[69,139] and has also been prepared by treatment of 3-methylisothiazole-4-carboxylic acid chloride with methylmagnesium iodide in the presence of ferric chloride at $-20°$.[70] This technique was unsuccessful with some aryl Grignard reagents, but p-chlorobenzoyl-3-methylisothiazole could be prepared from the more reactive carbonylimidazole and p-chlorophenylmagnesium bromide.[101]

5-Acetyl- and 5-formylisothiazoles are readily available from 5-lithioisothiazoles.[71,102] However, 3-methyl-4-nitroisothiazole does not form a lithium derivative,[72] and 4-formyl-3-methyl-4-nitroisothiazole was prepared by reduction of the appropriate acid chloride with lithium tri-t-butoxyaluminum hydride.[140] A 5-formyl-4-hydroxyisothiazole has been prepared by direct ring synthesis [Eq. (12)].[29]

$$\text{Me—CH}_2\text{—CO—CH—Ph} \xrightarrow[\text{heat}]{\text{S}_2\text{Cl}_2} \quad (12)$$
$$\hspace{3em} | \hspace{6em}$$
$$\hspace{2em} \text{NH}_2$$

[137] H. P. Benschop, A. M. van Oosten, D. H. J. M. Platenburg, and C. van Hooidonk, *J. Med. Chem.* **13**, 1208 (1970).
[138] D. G. Jones and G. Jones, *J. Chem. Soc. C* 707 (1969).
[139] S. Rajappa, K. Nagarajan, and A. S. Akerkar, *Indian J. Chem.* **8**, 499 (1970).
[140] M. P. L. Caton, D. H. Jones, R. Slack, S. Squires, and K. R. H. Wooldridge, *J. Med. Chem.* **8**, 680 (1965).

5-Acetylisothiazoles have also been obtained from 5-cyanoisothiazoles and methylmagnesium iodide and by hydrolysis of β-ketoesters derived from isothiazole-5-carboxylic esters.[70]

The isothiazolecarbonyl derivatives behave normally and form derivatives with oximes,[69, 137] hydrazines,[102] semicarbazide, and thiosemicarbazide.[70, 140] The aldehydes condense with malonate and nitromethane,[32] form cyanhydrins[141] and acetals,[137] undergo the Wittig[67, 68] and Cannizzaro[70] reactions, and may be oxidized to acids with silver oxide[71] or reduced with borohydride to hydroxymethyl compounds.[71, 101, 119] 4-Bromo-5-formyl-3-methylisothiazole reacted anomalously with 2-aminopyridine [Eq. (13)].[101] The acetylisothiazoles form ketals[69] and bromacyl derivatives,[139] and the oximes undergo Beckmann rearrangement.[69]

(13)

I. Isothiazole Thiols, Alkylthiols, and Sulfides

Many isothiazoles with thiol or alkylthiol groups in the 3- or 5- (or both) positions are available by direct ring synthesis (see Sections I, B, E, and F). A 3-mercaptoisothiazole[8] and several 5-mercapto- and alkylmercaptoisothiazoles[9, 45, 51, 99, 121] have been prepared by nucleophilic displacement of halogen, and Lee and Volpp[85] obtained a disulfide from diazotized 5-amino-3-methylisothiazole and thiourea at 55°. Fanghänel[142] reported a thermal elimination of sulfur from 1,4,2-dithiazines to give 3-methylmercaptoisothiazoles (Scheme 38).

The sodium salts of mercaptoisothiazoles are readily alkylated by alkyl halides to form alkylmercapto derivatives,[43–46, 49–52, 143] which form sulfoxides[9, 99, 122, 143] and sulfones[9, 45, 99, 122] on oxidation.

---

[141] Chimetron, French Patent 1,452,171 (1966) [*Chem. Abstr.* **66**, 76015 (1967)].
[142] E. Fanghänel, *Z. Chem.* **5**, 386 (1965).
[143] G. A. Hoyer and M. Kless, *Tetrahedron Lett.* 4265 (1969).

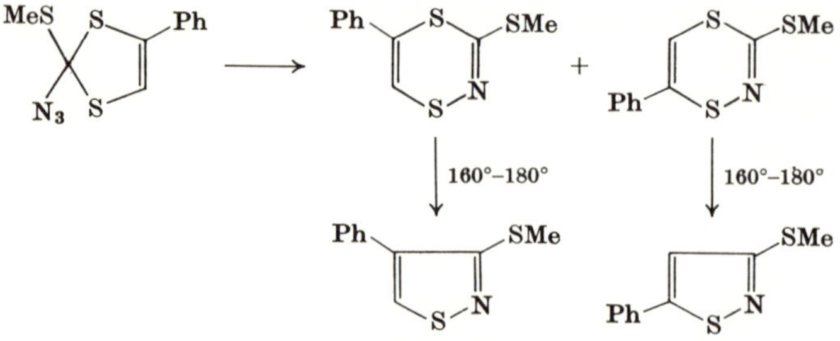

SCHEME 38

3,5-Dimercapto-4-cyanoisothiazole is alkylated primarily on the 3-mercapto group in agreement with molecular orbital calculations.[143] A 5-methylmercapto group may be displaced by other nucleophiles such as hydrazine[122] or alkoxide.[51] 3-Mercapto-5-phenylisothiazole has been oxidized to the 3-sulfonic acid and sulfonamide.[65]

Sulfides have been prepared from 5-halogenoisothiazoles and sodium sulfide[86, 12, 123] or thiourea[121] and also from a further variant of Naito's synthesis [Eq. (14)].[29] The sulfides have been oxidized to

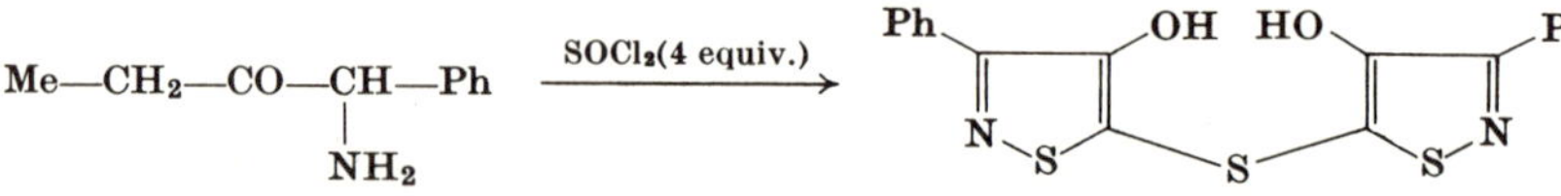

$$\text{Me—CH}_2\text{—CO—CH—Ph} \quad \xrightarrow{\text{SOCl}_2 (4 \text{ equiv.})} \quad (14)$$
$$|$$
$$\text{NH}_2$$

sulfoxides and sulfones,[121] and the latter compounds have also been prepared from 5-bromoisothiazoles and the sodium salts of sulfinic acids.[121] Incidentally these reactions demonstrate the marked stability of the isothiazole ring to powerful oxidizing reagents such as hot peracetic acid or permanganate at 100°. A nuclear S-oxide has been reported by treatment of 3-morpholino-5-phenylisothiazole with nitric acid and further treatment of this compound with hydrogen peroxide gave a product which was formulated as a nuclear sulfone on the basis of infrared spectroscopic evidence.[19] However, there must be an element of doubt about this structure, since it would involve the loss of the aromatic nature of the isothiazole ring, and it is possible that the second oxygen might be located on the ring or side-chain nitrogen.

J. Quaternary Isothiazoles

Simple isothiazoles undergo decomposition on heating with methyl iodide and form quaternary derivatives only slowly at lower temperatures.[144] However, methyl tosylate at higher temperatures,[137, 144] dimethyl sulfate,[63] and triethyloxonium fluoroborate[145] are more satisfactory, and benzyl halides react normally.[144] An intramolecular quaternization has been reported, but the product could not be aromatized to an isothiazolo[2,3-*a*]pyridinium system (Scheme 39).[138]

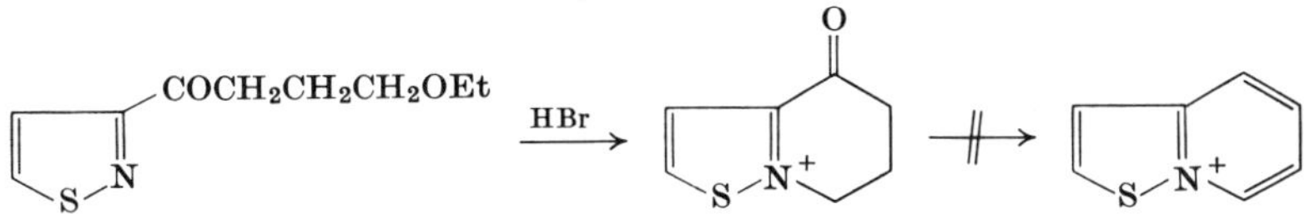

SCHEME 39

N-Alkylisothiazol-3-ones have been converted into 3-chloroisothiazolium chlorides[27] and *N*-substituted 3- and 5-thiones have been methylated to give the corresponding methylmercaptoisothiazolium products [Eq. (15)].[26, 63, 64]

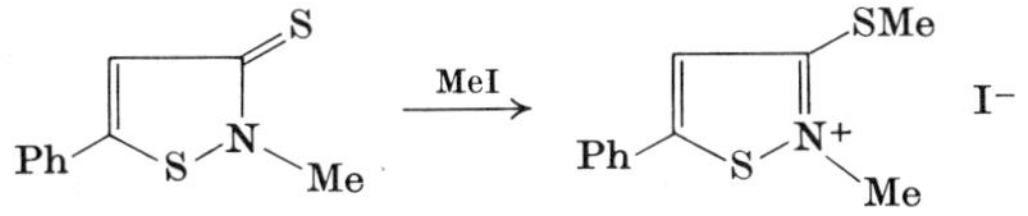

(15)

Isothiazolium salts have been reconverted to isothiazoles by dry distillation[27] and by treatment with ammonia.[145] The latter reaction incidentally confirms that isothiazoles are alkylated on the nitrogen and not the sulfur. Isothiazolium salts with a vacant 3-position have given 3-thiones on treatment with sulfur in pyridine [Eq. (16)].[63, 64]

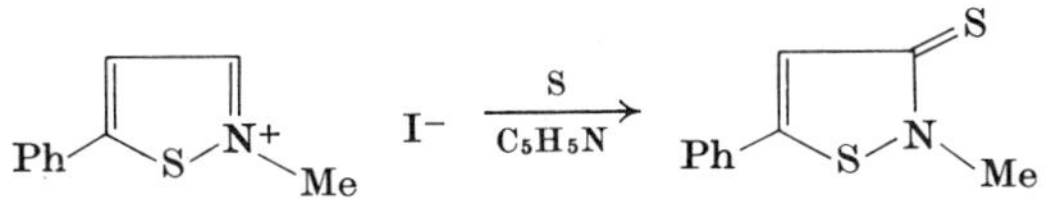

(16)

Finally some compounds related to the thiathiophthene "no-bond resonance" system may be regarded as containing contributions from isothiazolium forms (8 ↔ 9).[146]

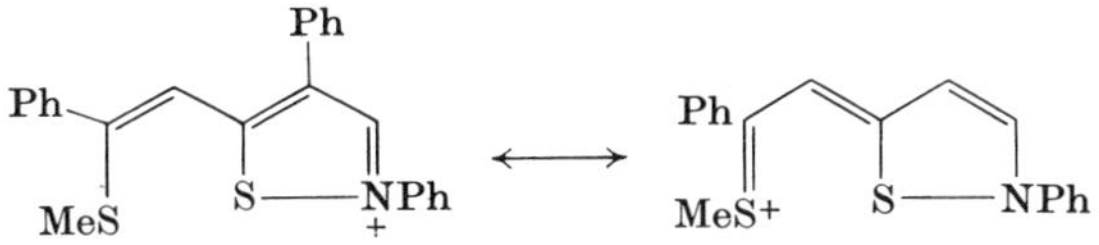

(8)   (9)

[144] P. Chaplen, R. Slack, and K. R. H. Wooldridge, *J. Chem. Soc.* 4577 (1965).
[145] J. M. Landesberg and R. A. Olofson, *Tetrahedron* **22**, 2135 (1966).
[146] E. Klingsburg, *J. Org. Chem.* **33**, 2915 (1968); D. H. Reid and J. D. Symon, *Chem. Commun.* 1314 (1969).

IV. Reactions Involving Cleavage of the Isothiazole Ring

As will have become clear from the previous section, the isothiazole nucleus is remarkably stable to chemical attack enabling interconversion of substituent groups by standard procedures. However, there are instances in which ring cleavage occurs or can be inferred from the nature of the products. Thus, 3-chloroisothiazoles with blocked 4- and 5-positions undergo attack on the ring sulfur by nucleophiles leading to ring-opened products (Scheme 40).[8, 51]

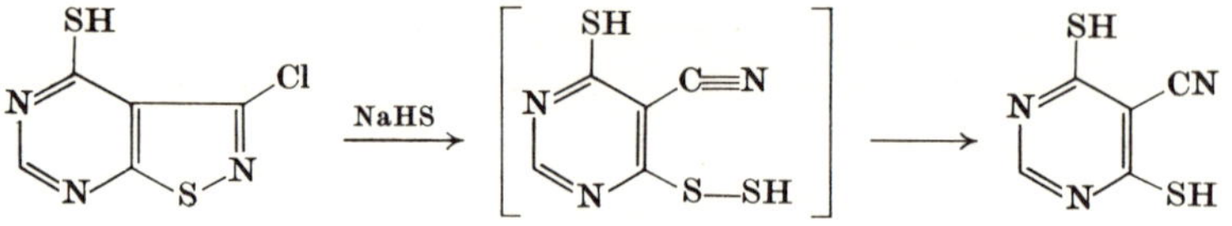

SCHEME 40

Cleavage of isothiazoles during lithiation was discussed in Section III,B, and a related example is provided by the fission of the isothiazole ring of an isothiazolo[5,4-b][1,4]benzothiazine during alkylation with sodium hydride and alkyl halide.[123]

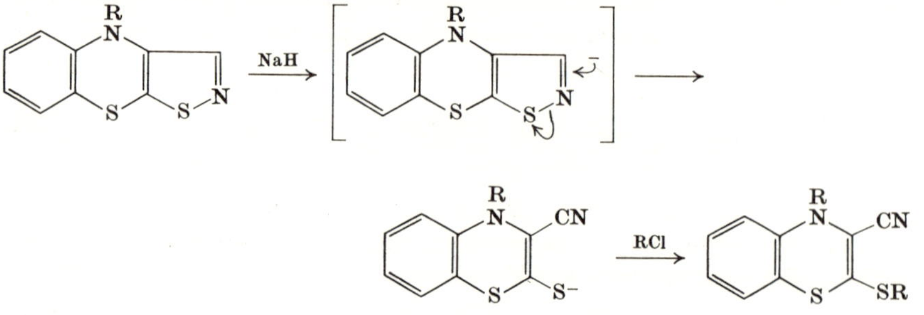

The synthesis of 3-hydroxyisothiazoles by cyclization of 3-thiocyanatoacrylamide is reversible (see Section I,D), and Crow and Gosney have demonstrated that 2-alkyl-3-isothiazolones are particularly vulnerable to nucleophilic attack [Eq. (17)].[132, 134] In the

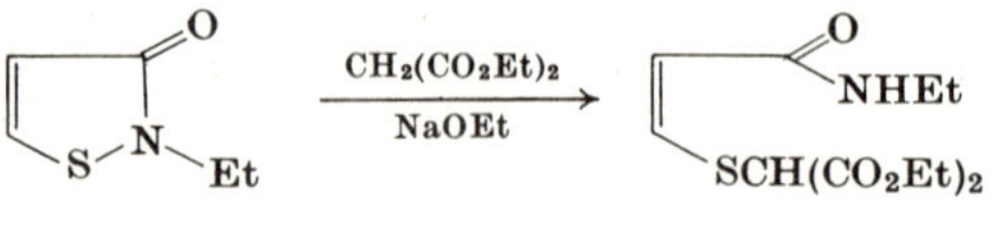

(17)

absence of an added nucleophile, 2-alkyl-3-isothiazolones dimerize under basic conditions to give 2,4-bismethylene-1,3-dithietanes (desaurins, **10**).[133]

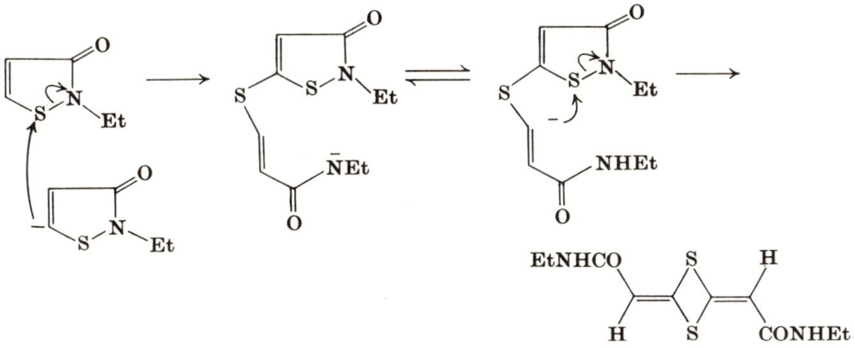

3,5-Dimercaptoisothiazoles on standing in the dark in ethanolic solution for several days[49, 50] or by acid treatment[47] give high yields of cyclic disulfides [Eq. (18)].

Treatment of quaternary isothiazoles with hydrazine or phenylhydrazine gives pyrazoles [Eq. (19)].[145]

Lee and Volpp[85] observed that although diazotized 5-aminoisothiazoles gave good yields of disulfides on treatment with thiourea, 4-aminoisothiazoles under the same conditions gave 1,2,3-thiadiazoles [Eq. (20)]. During discussion of the mechanism of this reaction, Lee and Volpp mentioned ring cleavage of quaternary isothiazoles by hydroxide [Eq. (21)], although no details were given.

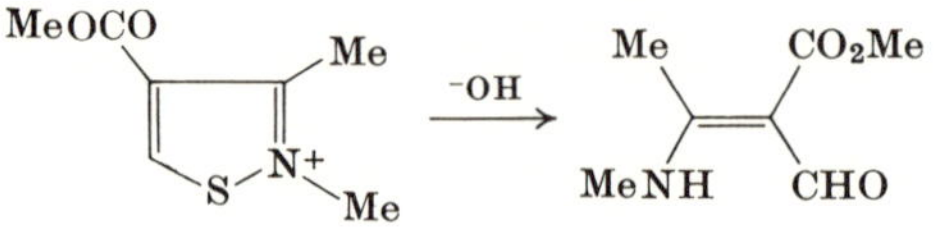

$$\text{(21)}$$

An isothiazolo[5,4-b]triazole prepared from a 4,5-diaminoisothiazole and nitrous acid suffered reductive cleavage on treatment with hypophosphorous acid (Scheme 41).[147]

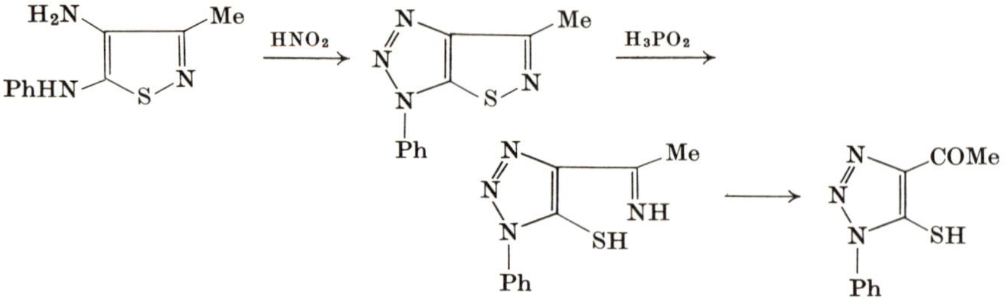

SCHEME 41

Isothiazoles are reductively desulfurized by Raney nickel, and this method has been used to confirm structures.[3, 51] In his remarkable synthesis of colchicine, Woodward employed the isothiazole nucleus as a template on which the various rings were constructed, and finally removed the sulfur with Raney nickel to leave a nitrogen function in the correct position [Eq. (22)].[67, 68]

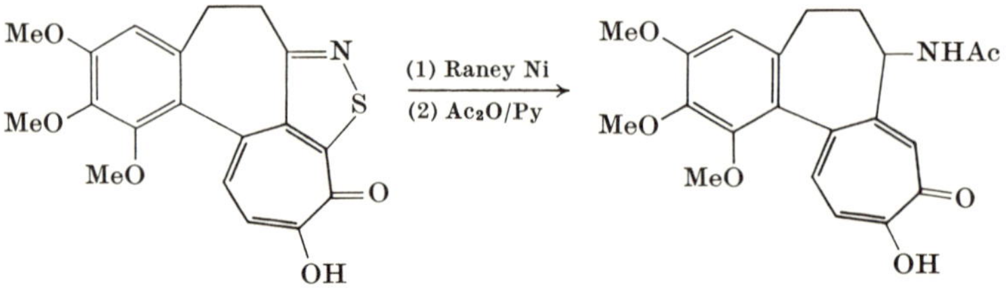

$$\text{(22)}$$

Behringer et al.[148] observed the formation of a thiazole on treatment of an iminoisothiazole with an acetylenic diester [Eq. (23)].

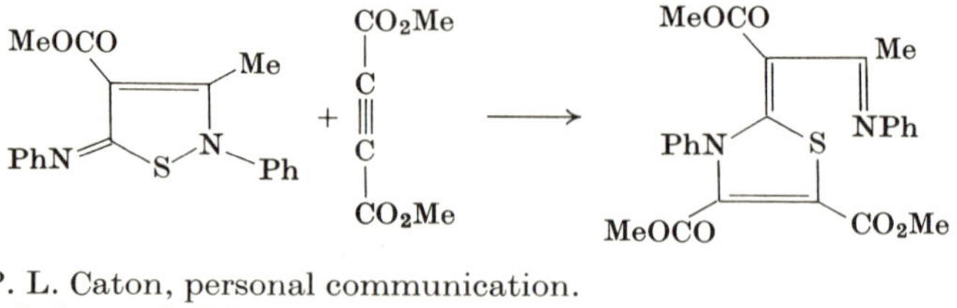

$$\text{(23)}$$

[147] M. P. L. Caton, personal communication.
[148] H. Behringer, J. Kilger, and R. Wiedenmann, *Tetrahedron Lett.* 1185 (1968).

V. Photochemical Reactions of Isothiazoles

Isothiazole gave a low yield of thiazole on irradiation in propylamine, but the reverse reaction was not observed.[149] Phenyl- and diphenyl-isothiazoles gave thiazoles on irradiation,[150] and the reverse reactions have also been reported.[84, 151] It appears likely that isothiazole-thiazole photorearrangements proceed through a common intermediate, possibly a tricyclic species in which the negative charge is stabilized by resonance with the phenyl group (Scheme 42).[80, 150]

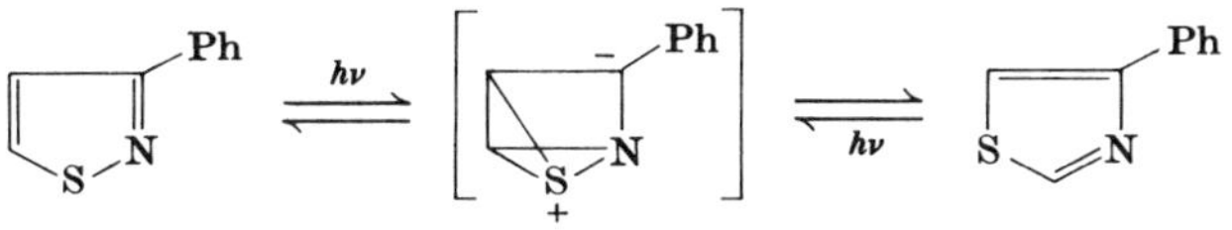

SCHEME 42

VI. Biologically Active Isothiazoles

The increasing understanding of the fundamental chemistry of the isothiazole system has enabled chemists to incorporate the ring into a wide variety of compounds with potential biological activity.

Heterocyclic or benzene rings of natural products have been replaced to give, for example, the isothiazole analogs of histidine,[119] histamine,[119] and thiamine,[144] but perhaps the greatest effort has been devoted to the synthesis of semisynthetic penicillins and cephalosporins. In particular, penicillins derived from isothiazole-4-carboxylic acids with substituents in the 3- and 5-positions have attracted attention[7, 23, 93, 94, 99, 107, 135, 152] because of their similarity to Celbenin and other penicillins active against penicillinase-producing organisms. Penicillins and cephalosporins have been prepared from isothiazole-3-, 4-, and 5- acetic acids,[95, 100, 153] and the 4- and 5-isothiazolylacetamido-

[149] J. P. Catteau, A. Lablanche-Combier, and A. Pollet, *Chem. Commun.* 1018 (1969).
[150] M. Ohashi, A. Iio, and T. Yonezawa, *Chem. Commun.* 1148 (1970).
[151] G. Vernin, H. J. M. Dou, and J. Metzger, *C. R. Acad. Sci. C* **271**, 1616 (1970).
[152] Bristol Banyu Co. Ltd., Japanese Patent 9389/69 (1969).
[153] R & L Molecular Research, U.S. Patent 3,268,523 (1966) [*Chem. Abstr.* **65**, 18597 (1967)].

penicillinates were reported to possess comparable activity to ampicillin against Gram-positive and Gram-negative organisms.[95]

Examples of the four possible modes of fusion of isothiazole to pyrimidine have been described, namely, isothiazolo[3,4-*d*]-,[53, 54] -[4,3-*d*],[112] -[4,5-*d*]-,[112] and -[5,4-*d*]pyrimidine.[8, 10] A derivative of an isothiazolo[5,4-*d*]pyrimidine has been reported to possess antitumor activity,[154] and others to have sedative properties.[155]

Isothiazoles have also been incorporated into a wide range of established drugs including phenothiazines,[123] sulfones,[121] sulfonamides,[3, 66, 156] thiosemicarbazones,[140, 157] amidines,[158] nitroheterocycles,[159] and benzimidazoles,[160] but this approach does not appear to have resulted in any major improvements. However, many isothiazoles not directly related to known drugs have biological activity, and representatives are listed in Table I. No attempt has been made to include the numerous patents which mention isothiazole derivatives as part of a series, but for which no special claims are made.

---

[154] Z. Machón, *Diss. Pharm. Pharmacol.* **21**, 325 (1969).
[155] Ciba Ltd., German Patent 1,950,990 (1970) [*Chem. Abstr.* **73**, 56118 (1970)].
[156] A. Adams, W. A. Freeman, A. Holland, D. Hossack, J. Inglis, J. Parkinson, H. W. Reading, R. Slack, R. Sutherland, and R. Wien, *Nature (London)* **186**, 221 (1960); J. W. Bridges and R. T. Williams, *J. Pharm. Pharmacol.* **15**, 565 (1963).
[157] R. Slack, K. R. H. Wooldridge, J. A. McFadzean, and S. Squires, *Nature (London)* **204**, 4958 (1964); A. R. Rao, J. A. McFadzean, and S. Squires, *Ann. N.Y. Acad. Sci.* **130**, 118 (1965).
[158] D. H. Jones and K. R. H. Wooldridge, *J. Chem. Soc.* 550 (1968).
[159] M. Robba and R. C. Moreau, *Ann. Pharm. Fr.* **22**, 201 (1964).
[160] Merck Inc., U.S. Patent 3,017,415 (1962) [*Chem. Abstr.* **56**, 15517 (1962)]; Merck Inc., U.S. Patent 3,055,907 (1962) [*Chem. Abstr.* **58**, 2456 (1963)]; Chemetron, French Patent 1,450,548 (1966); Chemetron, French Patent 1,472,101 (1967)[*Chem. Abstr.* **68**, 49606 (1968)]; Chemetron, French Patent 1,476,558 (1967); Chemetron, French Patent 1,476,560 (1967) [*Chem. Abstr.* **68**, 95827 (1968); Chemetron, French Patent 1,488,281 (1967) [*Chem. Abstr.* **69**, 43924 (1968)].

TABLE I
Biologically Active Isothiazoles

Compounds	Activity	References
 R_1R_2NCONH-[isothiazole] (and 3-isomer)	Herbicides	116, 161, 162
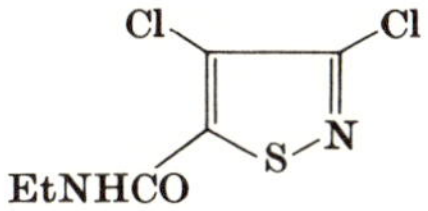 4,5-diCl-3-EtNHCO-isothiazole	Herbicide	4
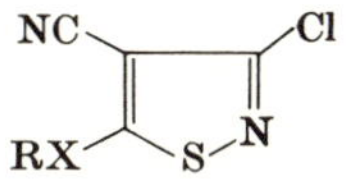 4-NC-3-Cl-5-RX-isothiazole, X = O, NH, or S	Herbicide and insecticide	55, 163
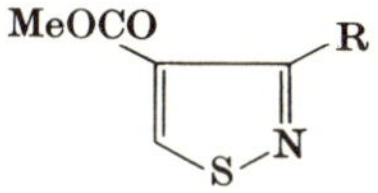 4-MeOCO-3-R-isothiazole (and 5-isomer)	Plant fungicide and bactericide	104
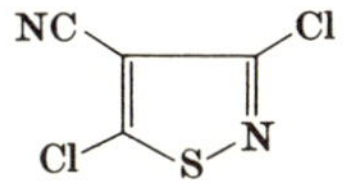 4-NC-3-Cl-5-Cl-isothiazole	Fungicide and slimicide	164
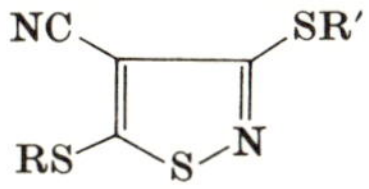 4-NC-3-SR'-5-RS-isothiazole	Fungicides	165

(continued)

[161] F.M.C. Corp., Belgian Patent 724,537 (1969) [*Chem. Abstr.* **71**, 112926 (1969)].
[162] Schering A.G., Netherlands Patent 6,605,902 (1966) [*Chem. Abstr.* **66**, 76001 (1967)].
[163] Du Pont, U.S. Patent 3,155,678 (1964) [*Chem. Abstr.* **62**, 2778 (1965)]; Merck A.G., British Patent 1,125,872 (1968) [*Chem. Abstr.* **70**, 11695 (1970)].
[164] Takeda Chem. Ind. Ltd., Japanese Patent 6840/67 (1967); F.M.C. Corp. U.S. Patent 3,375,161 (1968) [*Chem. Abstr.* **68**, 113581 (1968)].
[165] Merck A.G., Netherlands Patent 6,703,832 (1967) [*Chem. Abstr.* **69**, 96708 (1968)]; Merck, A.G., Netherlands Patent 6,808,400 (1968) [*Chem. Abstr.* **71**, 124429 (1969)].

TABLE I—continued

Compounds	Activity	References
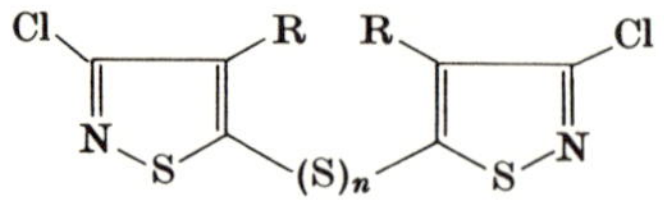 [structure: bis-isothiazole with Cl, R substituents linked by $(S)_n$]	Fungicides	166
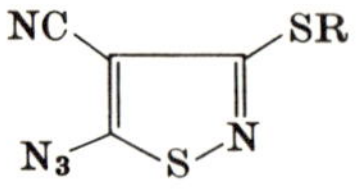 [structure: isothiazole with NC, SR, N_3 substituents] (also sulfoxides and sulfones)	Insect repellent	123
N- and O-derivatives of 3-hydroxyisothiazole	Bacteriocide and fungicide	167
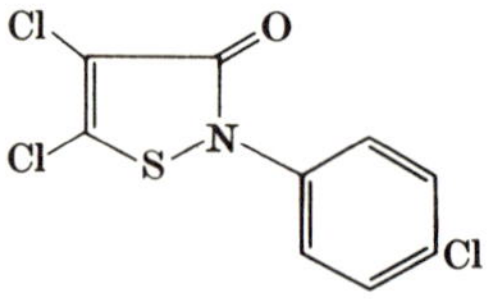 [structure: 4,5-dichloro-2-(chlorophenyl)isothiazol-3(2H)-one] (also variously substituted phenyl)	Bacteriocide and fungicide (textile preservative)	28
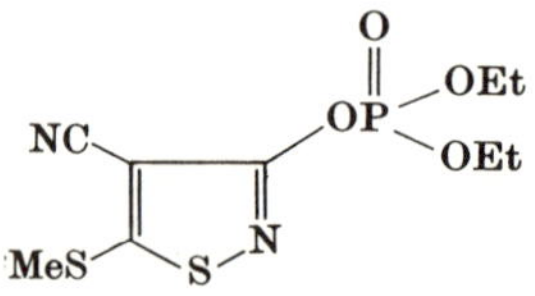 [structure: isothiazole with NC, MeS, and $OP(O)(OEt)_2$ substituents]	Insecticide and fungicide	52
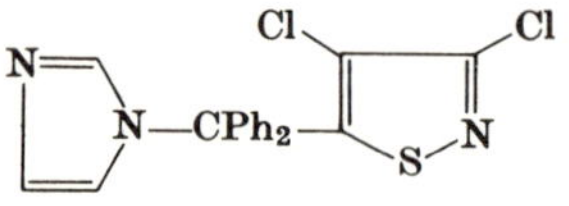 [structure: dichloroisothiazole linked via CPh_2 to imidazole]	Orally active fungicide	168

[166] Du Pont, U.S. Patent 3,155,679 (1964) [*Chem. Abstr.* **62**, 9141 (1965)].
[167] Rohm and Haas, Netherlands Patent 6,803,117 (1968) [*Chem. Abstr.* **72**, 43651 (1970)]; Rohm and Hass, Netherlands Patent 6,803,118 (1968) [*Chem. Abstr.* **72**, 111459 (1970)]; Rohm and Haas, Netherlands Patent 6,803,119 (1968) [*Chem. Abstr.* **72**, 55438 (1970); Rohm and Haas, Netherlands Patent 6,808,712 (1968).
[168] Farbenfabriken Bayer, A.G., Belgian Patent 746,983 (1970).

TABLE I—*continued*

Compounds	Activity	References
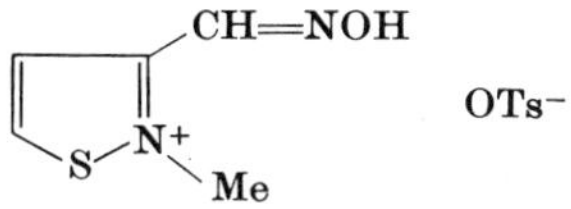 3-(CH=NOH)-2-Me-isothiazolium OTs⁻	Therapeutic agent in poisoning by organophosphorus compounds	137
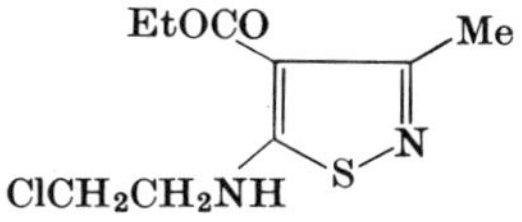 4-EtOCO-3-Me-5-(ClCH₂CH₂NH)-isothiazole	Cytostatic	117
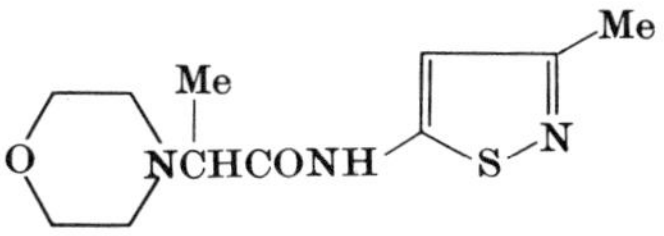 5-(morpholino-N-CH(Me)-CONH)-3-Me-isothiazole	Analgesic	169

[169] S. Uyeo, H. Fujimura, and A. Asai, *J. Pharm. Soc. Jap.* **83**, 195 (1963).

Benzisothiazoles

MICHAEL DAVIS

Department of Chemistry, La Trobe University, Bundoora, Victoria, Australia

I. Introduction	43
II. 1,2-Benzisothiazoles	44
A. Formation	44
B. Physical Properties	50
C. Chemical Properties	51
D. Reduced Derivatives	55
E. 1,2-Benzisothiazolinones	58
F. Applications	60
III. 2,1-Benzisothiazoles	63
A. Formation	63
B. Physical Properties and Structure	68
C. Chemical Properties	69
D. Reduced Derivatives	73
E. Applications	75
IV. Selenium and Tellurium Analogs	76
V. Tables	76

I. Introduction

There are two series of fused bicyclic compounds incorporating both an isothiazole and a benzene ring. These two ring systems are now generally known as 1,2-benzisothiazole (**1**) and 2,1-benzisothiazole (**2**).

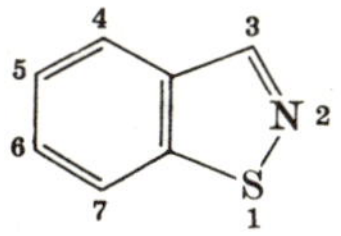

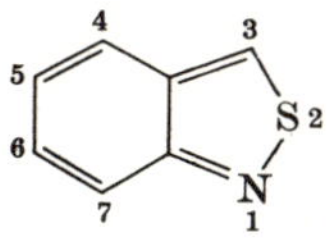

Although these names are the ones preferred by *Chemical Abstracts* and the *Ring Index*, other names have also been used; thus, **1** has been described as benz[*d*]isothiazole or thioindoxazene, and **2** as benz[*c*]isothiazole or thioanthranil.

Both series of compounds have been known for 50 years or more. The 2,1-isomer was first prepared nearly 80 years ago, but until recently attracted very little attention. Both ring systems were reviewed in 1952 by Bambas in the Interscience series of monographs on "The Chemistry of Heterocyclic Compounds."[1] In this review, the section on the 2,1-isomer occupied slightly more than one page, whereas the 1,2-isomer section was 50 pages long—an interesting example of the often erratic way in which chemical knowledge increases.

In the present review the earlier work will be mentioned only briefly, the main emphasis being placed on developments within the last 20 years. Reduced and oxidized derivatives will also be mentioned, but discussion of the latter will be restricted to N-oxides and sulfoxides; the benzisosulfonazoles (S-dioxides) include the very large group of saccharin and related compounds, and will not be discussed.

The literature was surveyed up to the end of 1970; some additional material on 2,1-benzisothiazoles originating from the author's laboratory and presently in course of publication has also been included. A few references to work published in 1971 have been added while this article was in proof.

II. 1,2-Benzisothiazoles

A. Formation

Almost all the syntheses presently known require as a precursor an aromatic sulfur-containing compound with a functionalized carbon atom ortho to the sulfur. The general pattern of these syntheses is thus either from 3 or 4 to 1,2-benzisothiazole (1). This

requirement limits the general availability of substituted 1,2-benzisothiazoles as the sulfur-containing precursors are often prepared only with some difficulty.

[1] L. L. Bambas, *in* "The Chemistry of Heterocyclic Compounds" (A. Weissburger, ed.), Vol. 4, pp. 225–277. Wiley (Interscience), New York, 1952.

1. From Thianaphthenequinones

The parent compound (1) was first prepared in 1923 by Stollé and co-workers.[2,3] They found that treatment of thianaphthenequinone (5) with aqueous ammonia and hydrogen peroxide afforded 1,2-benzisothiazole-3-carboxamide (6), which was hydrolyzed to the acid (7) and the latter decarboxylated.

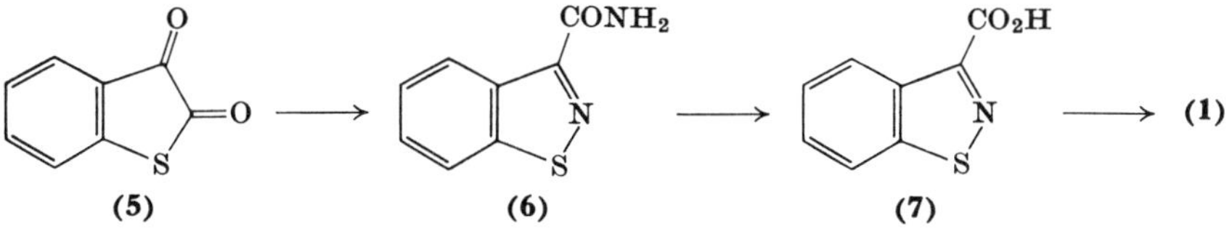

(5) → (6) → (7) → (1)

The formation of an isothiazole ring from the thiophene system (5) is of interest. Probably ammonia breaks the sulfur–carbonyl bond and then adds to the second carbonyl group, giving an imine (8), which is then oxidized by the hydrogen peroxide. A number of similar rearrangements and interconversions of isothiazole, thiophene, and dithiole systems are known and will be discussed later (Section II, C, 3).

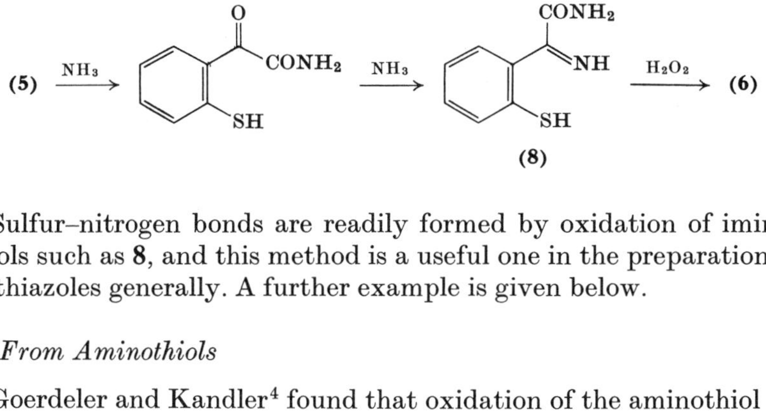

(5) $\xrightarrow{NH_3}$ [structure] $\xrightarrow{NH_3}$ (8) $\xrightarrow{H_2O_2}$ (6)

Sulfur–nitrogen bonds are readily formed by oxidation of iminothiols such as 8, and this method is a useful one in the preparation of isothiazoles generally. A further example is given below.

2. From Aminothiols

Goerdeler and Kandler[4] found that oxidation of the aminothiol (9) with iodine or bromine afforded an excellent yield of 5-methyl-1,2-

---

[2] R. Stollé and W. Geisel, *Angew. Chem.* **36**, 159 (1923).
[3] R. Stollé, W. Geisel, and W. Badstübner, *Chem. Ber.* **58**, 2095 (1925).
[4] J. Goerdeler and J. Kandler, *Chem. Ber.* **92**, 1679 (1959).

benzisothiazole (**10**). Later workers showed that alkaline ferricyanide is an equally effective oxidizing agent.[5]

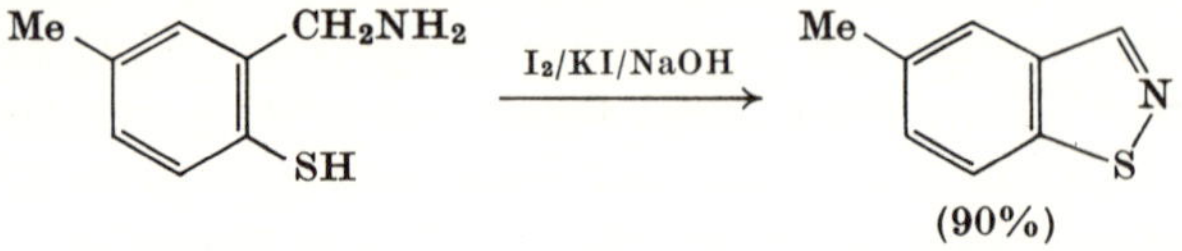

(**9**) →[I_2/KI/NaOH] (**10**) (90%)

3. *From Sulfenyl Halides*

Sulfenyl halides with an aldehyde function ortho to the sulfur react with ammonia with the formation of 1,2-benzisothiazoles[6] [Eq. (1)].

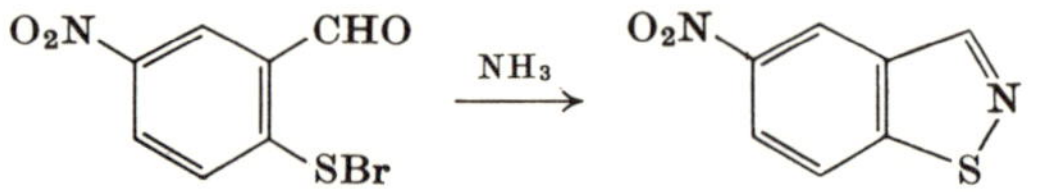

$$\text{(O}_2\text{N-Ar(CHO)(SBr))} \xrightarrow{NH_3} \text{1,2-benzisothiazole} \quad (1)$$

Ketones similarly afford 3-substituted 1,2-benzisothiazoles[7] [Eq. (2)].

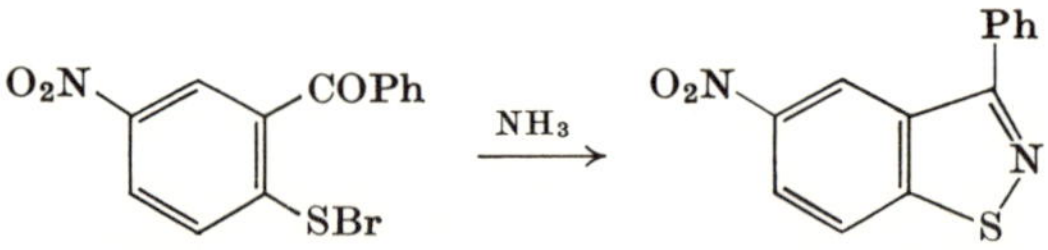

$$\text{(O}_2\text{N-Ar(COPh)(SBr))} \xrightarrow{NH_3} \text{3-Ph-1,2-benzisothiazole} \quad (2)$$

4. *From Mercaptophenyl Aldoximes or Ketoximes*

Ricci and Martani[8] found that *o*-mercaptobenzaldoxime (**11**) could be cyclized by treatment with hot polyphosphoric acid (PPA).

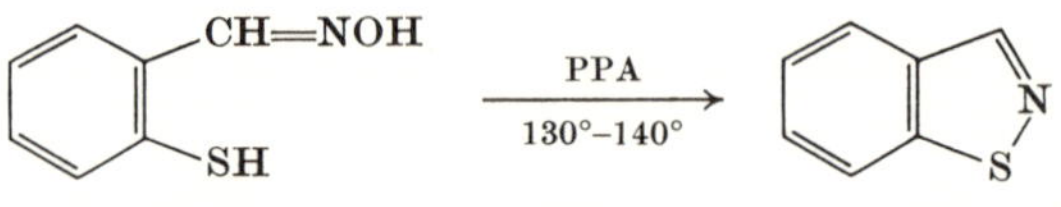

(**11**) →[PPA, 130°–140°] (**1**)

Ketoximes give 3-substituted derivatives in the same manner.[9] Thioethers undergo a similar reaction, the extra alkyl group on the sulfur

---

[5] R. Boudet and D. Bourgoin-Legay, *C. R. Acad. Sci. C* **262**, 596 (1966).
[6] K. Fries and G. Brothuhn, *Chem. Ber.* **56**, 1630 (1923).
[7] K. Fries, K. Eishold, and B. Vahlberg, *Ann.* **454**, 264 (1927).
[8] A. Ricci and A. Martani, *Ann. Chim. (Rome)* **53**, 577 (1963).
[9] A. Ricci and A. Martani, *Ric. Sci. Parte 2, Sez. B.* 117 (1962).

being eliminated during the reaction. Thus, the sodium salt of o-methylthiophenylmethyl ketoxime (12) on treatment with p-toluenesulfonyl chloride affords 3-methyl-1,2-benzisothiazole (14), presumably through the participation of the sulfur atom (13) in the reaction.[10] This reaction might well be developed into a general

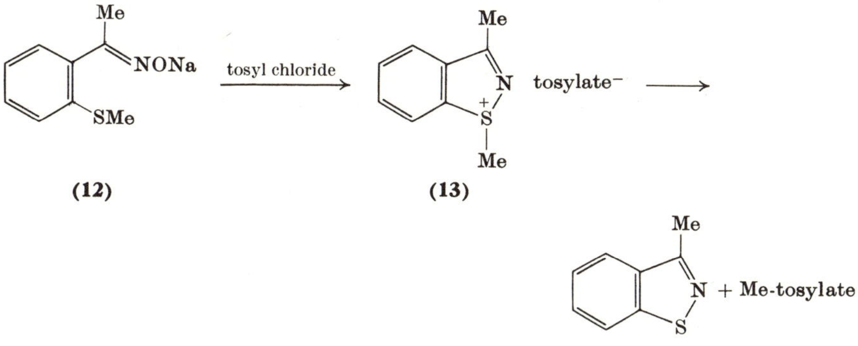

synthesis of 1,2-benzisothiazoles, as the alkyl-substituted thioethers such as 12 are much easier to handle than the corresponding air-sensitive thiols.

5. *From Benzylidene Dichlorides*

The heating together of 2,6-dichlorobenzylidene dichloride (15), sulfur, ammonia, and benzene in an autoclave affords a good yield of 4-chloro-1,2-benzisothiazole (16).[11,12] This is one of the few syntheses

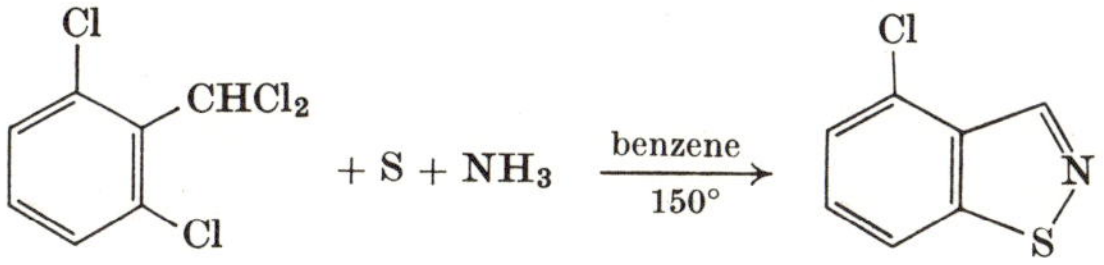

of 1,2-benzisothiazoles in which the sulfur atom is introduced by means of a simple inorganic sulfur compound or, as in this case, by

---

[10] R. J. Crawford and C. Woo, *J. Org. Chem.* **31**, 1655 (1966).
[11] F. Becke and H. Hagen, *Ann.* **729**, 146 (1969).
[12] Badische Anilin und Soda-Fabrik A.G., French Patent 1,558,071; *Chem. Abstr.* **72**, 43,656 (1970).

elemental sulfur itself. The second flanking chlorine atom is apparently necessary; only 4-chloro-substituted products can be prepared by this method, which somewhat limits its general utility. The mechanism of the reaction is uncertain.

6. From Benzothiophene Derivatives

An unusual synthesis reported recently[13] is the reaction of 3-acetamido-2-nitrobenzothiophene (**17**) with ferrous oxalate. The major products are 3-cyano-1,2-benzisothiazole (**18**) and the corresponding amide (**19**). Once again the mechanism of this curious

reaction remains in doubt. The method is equally useful for the preparation of the 5-chloro and 5-bromo derivatives.

7. From Dithiosalicylamides

A reaction of rather less general application, but one which produces valuable intermediates, is the treatment of dithiosalicylamide and its N-substituted derivatives with phosphorus pentachloride. Thus, N,N-diethyldithiosalicylamide (**20**) yields 3-chloro-2-ethyl-1,2-benzisothiazolium chloride (**21**).[14–16] This salt (**21**) is a versatile compound; for example, heating it with diethylamine gives 3-diethylamino-1,2-benzisothiazole (**22**)[16], treatment with base yields 2-ethyl-1,2-benzisothiazolinone (**23**), and reaction with ammonia, followed by hydrochloric acid, affords 3-ethylamino-1,2-benzisothiazole (**24**).[14] This final reaction involves an interesting rearrangement which will be discussed in more detail in Sections II,C,2 and II,C,3.

---

[13] M. S. El Shanta, R. M. Scrowston, and M. V. Twigg, *J. Chem. Soc. C* 2364 (1967).

[14] H. Böshagen, *Chem. Ber.* **99**, 2566 (1966).

[15] Farbenfabriken Bayer A.G., Netherlands Patent Application 6,610,677; *Chem. Abstr.* **68**, 59,572 (1968).

[16] H. Böshagen (Farbenfabriken Bayer A.G.), South African Patent 684,111; *Chem. Abstr.* **71**, 38,954 (1969).

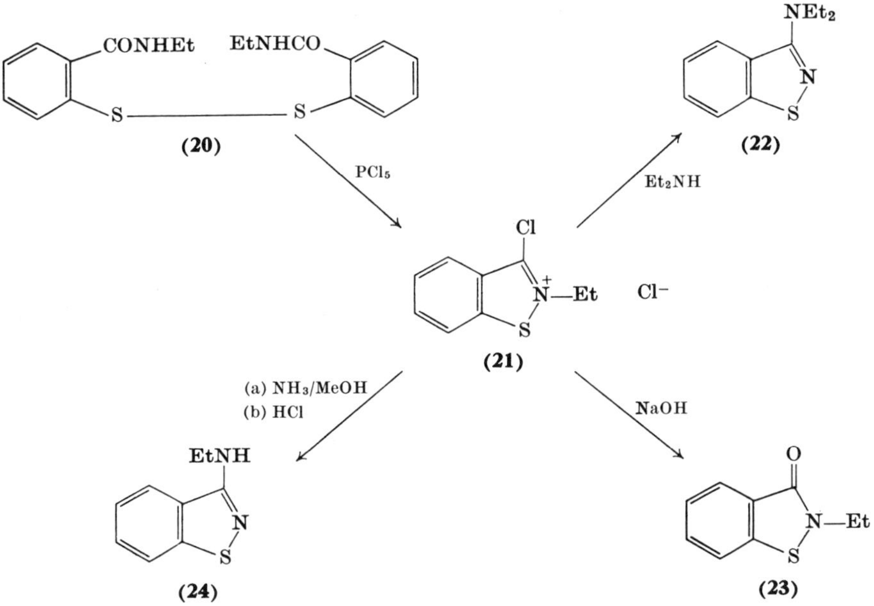

Sec. II. A.8] BENZISOTHIAZOLES 49

8. From 1,2-Benzisothiazolinones

These compounds and their reactions will be dealt with in Section II,E, but it may be mentioned here that treatment of 1,2-benzisothiazolinone (**25**) with phosphorus pentachloride,[17] phosphoryl chloride,[18,19] phosgene, or oxalyl chloride[20,21] affords 3-chloro-1,2-benzisothiazole (**26**). If the 1,2-benzisothiazoline is substituted at the nitrogen atom, then a salt similar to **21** is the product.

The chlorine atom in **26** is readily replaced by nucleophiles and

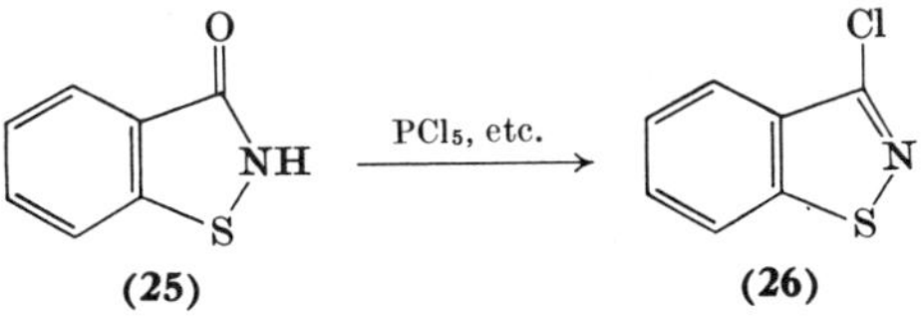

---

[17] A. Reissert, *Chem. Ber.* **61**, 1680 (1928).
[18] T. Vitali, E. Gaetani, P. Mantovani, and A. Agosti, *Farmaco, Ed. Sci.* **24**, 440 (1969); *Chem. Abstr.* **71**, 49,831 (1969).
[19] T. Vitali, R. Ponci, and F. Berteccini, German Patent 1,915,644; *Chem. Abstr.* **72**, 31,781 (1970).
[20] J. Faust and R. Meyer, *Z. Naturforsch.* **20B**, 712 (1965).
[21] J. Faust, *Z. Chem.* **8**, 170 (1968).

a variety of 3-substituted 1,2-benzisothiazoles can thereby be obtained.[18, 19, 22–24] Attempts to obtain the parent compound, 1,2-benzisothiazole, by zinc dust reduction of **26**, gave only 3,3-bis-1,2-benzisothiazolyl (**27**).[17] See Note below.*

B. Physical Properties

1,2-Benzisothiazole is a very pale yellow solid, m.p. 37°, with an odor of bitter almonds. It is slightly soluble in water, volatile in steam, and very soluble in concentrated acids and in almost all organic solvents. It boils at 220° without appreciable decomposition.[3] Few other physical properties or spectroscopic data appear to have been recorded, but the 60 MHz NMR spectrum of a solution in carbon tetrachloride shows a singlet at 8.73δ, ascribed to the proton on the heterocyclic ring, and a complex band between 7.12 and 8.00δ resulting from the benzenoid ring protons.[25]

Substituted 1,2-benzisothiazoles are, generally, solids of higher melting point; most compounds reported in the literature in the last 20 years are listed in the tables at the end of this chapter.

Some ultraviolet absorption data are available for certain substituted 1,2-benzisothiazoles,[13, 21, 27] as are limited NMR data.[26, 27]

---

* *Note Added in Proof.* It has recently been shown that **27** is actually bis(*o*-cyanophenyl)disulfide, and that in many cases nucleophilic attack on **26** leads to opening of the heterocyclic ring. See S. Huenig, G. Kiesslich, and H. Quast, *Ann.* **748**, 201 (1971); also D. E. L. Carrington, K. Clarke, and R. M. Scrowston, *Tetrahedron Lett.* 1075 (1971) and *J. Chem. Soc.* (*C*) 3262 (1971).

[22] O. Bup (Knoll A.G. Chemische Fabriken), German Patent 1,174,783; *Chem. Abstr.* **61**, 12,008 (1964).
[23] Farbenfabriken Bayer A.G., French Patent Application 2,007,380; *Chem. Abstr.* **73**, 25,447 (1970).
[24] T. Vitali, F. Mossini, G. Bertaccini, and M. Impicciatore, *Farmaco, Ed. Sci.* **23**, 1081 (1968); *Chem. Abstr.* **70**, 37,702 (1969).
[25] M. Davis, unpublished results (1968).
[26] H. Böshagen, W. Geiger, and H. Medenwald, *Chem. Ber.* **103**, 3166 (1970).
[27] W. Geiger, H. Böshagen, and H. Medenwald, *Chem. Ber.* **102**, 1961 (1969).

C. Chemical Properties

A good account of the earlier work on this series of compounds has been given by Bambas.[1] In more recent work, attention has been directed toward three main aspects—electrophilic substitution, reactions and rearrangements of 3-amino derivatives and related compounds, and formation of N-oxides and sulfoxides.

1. *Electrophilic Substitution*

Ricci and co-workers[28] found that nitrogen of 1,2-benzisothiazole (**1**) with potassium nitrate and sulfuric acid at 0° afforded a mixture of the 5-nitro compound (**28**) and the 7-nitro derivative (**29**). The

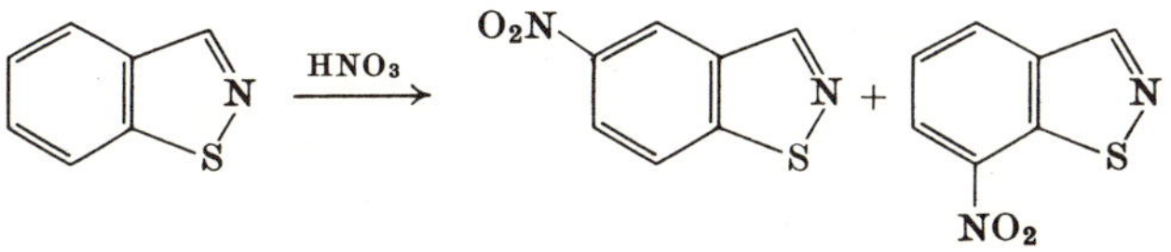

corresponding amino compounds are readily prepared by reduction, under mild conditions, of these nitro derivatives.[28, 29]

2. *Reactions of 3-Amino-1,2-benzisothiazole and Derivatives*

a. *Rearrangements.* Böshagen and his colleagues have discovered a number of novel reactions of this group of compounds. For example, the hydrochloride salt (**30**) of 3-ethylamino-1,2-benzisothiazole (**24**) mentioned earlier, is in equilibrium in aqueous solution with 2-ethyl-3-imino-1,2-benzisothiazolinyl hydrochloride (**31**).[27, 30] The free base,

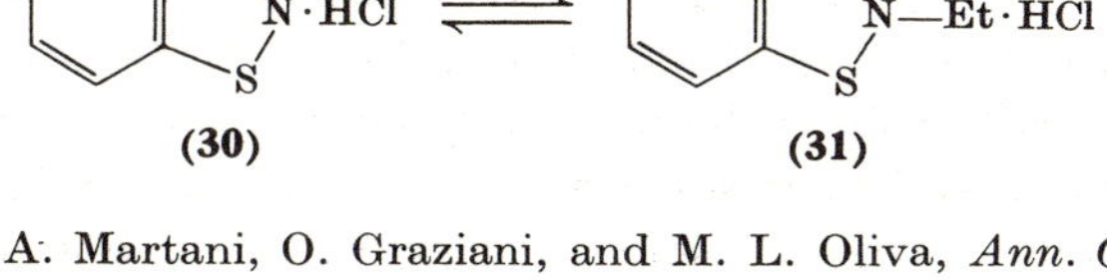

---

[28] A. Ricci, A. Martani, O. Graziani, and M. L. Oliva, *Ann. Chim.* (*Rome*) **53**, 1860 (1963).
[29] A. Adams and R. Slack, *J. Chem. Soc.* 3061 (1959).
[30] H. Böshagen, W. Geiger, and H. Medenwald (Farbenfabriken Bayer A.G.), South African Patent 07,624 (1969); *Chem. Abstr.* **72**, 90,444 (1970).

on the other hand, exists wholly in the 3-ethylamino form (**24**), and this structure is preserved when the base is N-alkylated or N-acylated.[27]

b. *Sulfoxides and Sulfones.* Oxidation of the base (**24**) with nitric acid affords a sulfoxide (**32**), whereas treatment with peracetic acid yields the sulfone (**33**).[26]

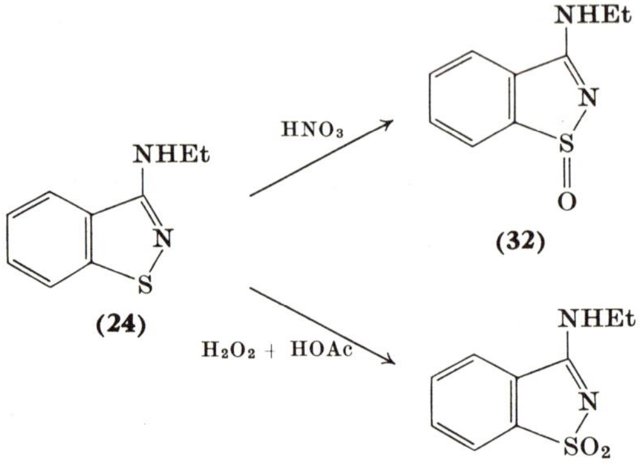

The sulfoxide (**35**) of the isomeric 2-ethyl-3-imino-1,2-benzisothiazoline system (**31**) is obtained by reaction of **31** with chlorine water, and dechlorination of the intermediate N-chloro compound (**34**) with hydrogen chloride in carbon tetrachloride.[26] Other sulfoxides have been reviewed by Bambas.[1]

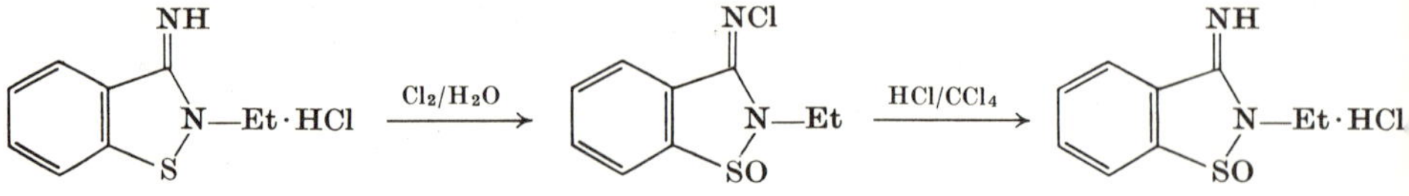

c. *N-Oxides.* If the salt (**21**) is heated with hydroxylamine, the product is an N-oxide, which can be isolated in the form of its hydrochloride. This salt (**36**) can be reversibly converted into the isomeric 2-ethyl-3-oximino-1,2-benzisothiazolium compound (**37**); treatment with formic acid, however, brings about an irreversible nitrogen-to-sulfur shift and the hydrochloride salt (**38**) of the sulfoxide (**32**) is produced.[26]

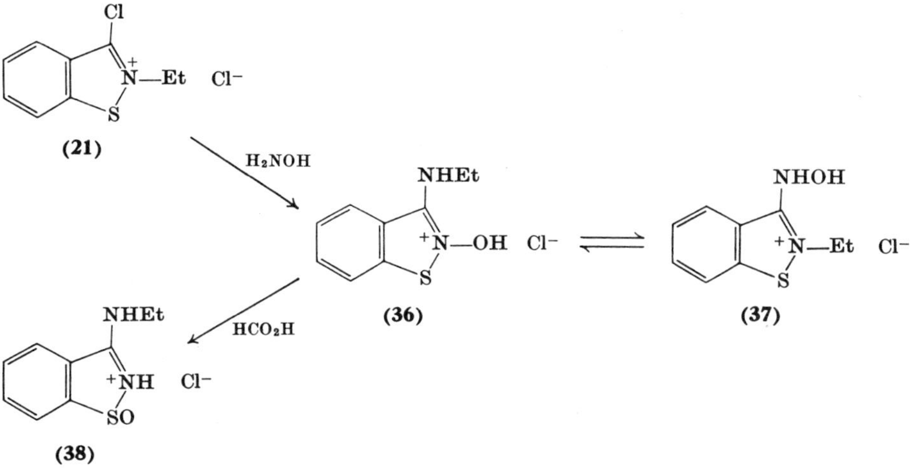

3. Heterocyclic Rearrangements

Displacement reactions of 3-chloro-1,2-benzisothiazole (**26**) have already been mentioned in Section II, A, 8.[18, 19, 22–24] In some cases this displacement is accompanied by a rearrangement. Böshagen has shown that treatment of the above-mentioned compound (**26**) with thioacetic acid yields an *N*-acyl-3*H*-1,2-benzodithiole (**39**).[31] This rearrangement is analogous to that described in Section II, C, 2 above.

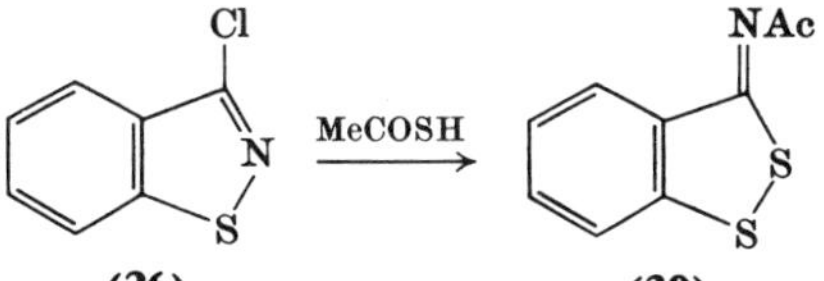

A further example of the same type of rearrangement is afforded by the 1,2-benzisothiazoline-3-thiones. If the benzisothiazolium salt (**40**) is treated with thioacetic acid, the product is the 3-arylimino-3*H*-1,2-benzodithiole (**41**), provided R is an aryl group; when R is alkyl, the product is the benzisothiazoline-3-thione (**42**).[32] Structures of types **41** and **42** are, however, in equilibrium with each other at 150°, presumably via a dipolar intermediate [Eq. (3)].[32] A diradical

---

[31] H. Böshagen and W. Geiger, *Chem. Ber.* **101**, 2472 (1968).
[32] H. Böshagen, H. Feltkamp, and W. Geiger, *Chem. Ber.* **100**, 2435 (1967).

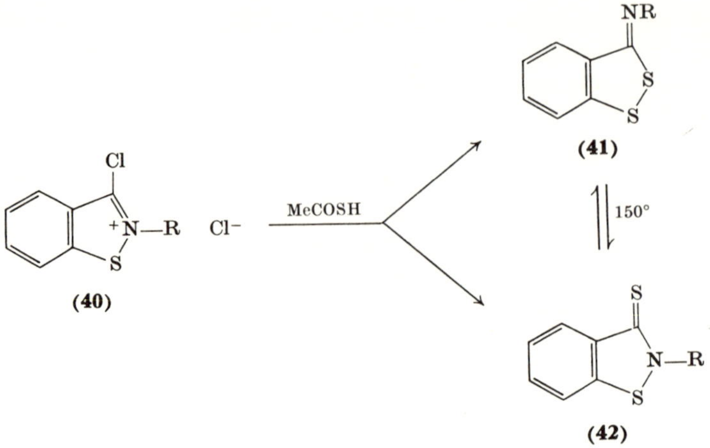

intermediate is also a possibility, but McClelland, who carried out much of the earlier work on compounds of this type, felt this to be unlikely. The arguments for and against radical dissociation of sulfur–sulfur and sulfur–nitrogen bonds have been summarized.[1]

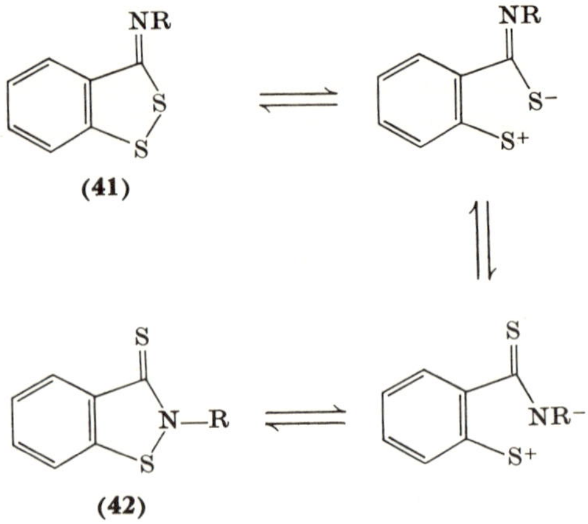

(3)

4. *Other Reactions of 1,2-Benzisothiazoles*

Few other reactions of simple 1,2-benzisothiazoles have been described in the literature. There is, for example, no report of the formation of a quaternary salt directly from the parent compound. The isothiazole ring is generally quite stable to oxidation or reduction;

Sec. II. D.1] BENZISOTHIAZOLES

thus, Adams and Slack were able to oxidize 5-amino-1,2-benziso-thiazole (itself prepared by reduction of the 5-nitro compound) to isothiazole-4,5-dicarboxylic acid using potassium permanganate solution.[29]

D. Reduced Derivatives

The 1,2-benzisothiazolinones, which can be considered as either oxidized or reduced derivatives of 1,2-benzisothiazole, will be discussed in Section II, E.

1. 3H-1,2-Benzisothiazolines

Goerdeler and Kandler, in their studies on the oxidation of aminothiols,[4] produced 2,5-dimethyl-3H-1,2-benzisothiazoline (**44**) by gentle iodine oxidation of the thiol (**43**). The product formed a picrate and a mercuric chloride adduct.

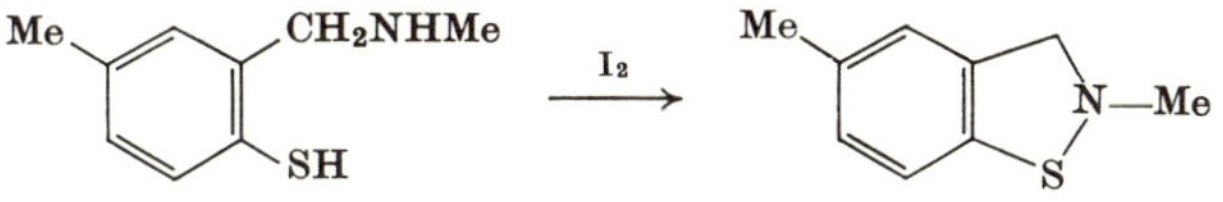

A claim in the patent literature[33] that a benzisothiazoline can be produced by the reaction between an N-benzyl-substituted sulfonamide and trichloromethanesulfenyl chloride [Eq. (4)], if confirmed, would be of great importance in opening the way to the preparation of many benzisothiazolines, as the benzylamines required for the reaction are readily available.

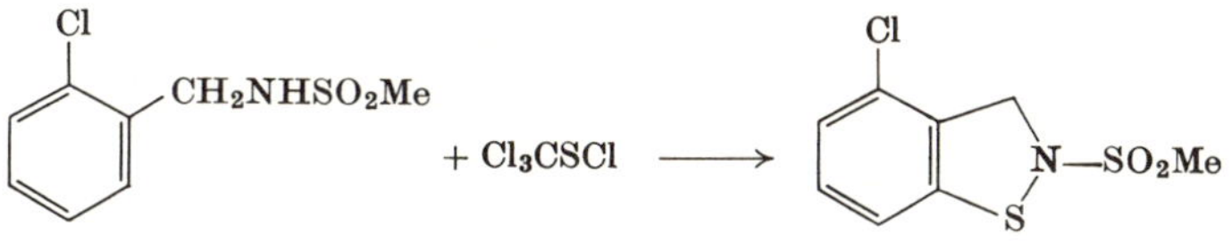

3,3-Bis-1,2-benzisothiazoline was prepared by Stollé and Merkle[34] by treatment of 3-amino-1,2-benzisothiazole with nitrous acid.

---

[33] P. F. Epstein and G. E. Lukes (Stauffer Chemical Co.), U.S. Patent 3,166,563; Chem. Abstr. **62**, 7764 (1965).
[34] R. Stollé and M. Merkle, J. Prakt. Chem. **138**, 221 (1933).

2. 1,2-Benzisothiazoline-3-thiones

The parent compound in this series is formally tautomeric with 1,2-benzisothiazole-3-thiol [Eq. (5)]. All the known compounds of this type are substituted at the nitrogen atom.

$$\text{(structure with S, NH, S)} \rightleftharpoons \text{(structure with SH, N, S)} \qquad (5)$$

With the exception of the compounds mentioned in Section II, C, 3, all 1,2-benzisothiazoline-3-thiones (46) have been prepared by the reaction between a benzodithiole-3-thione ("trithione") (45) and a

$$\text{(45)} + RNH_2 \longrightarrow \text{(46)} + H_2S$$

(45)   (46)

primary amine. As the products possess useful fungicidal activity, a large number have been prepared and tested.[35–38] The crystal structure of 2-methyl-1,2-benzisothiazoline-3-thione has been determined by X-ray diffraction,[39] and there has been a study of the ultraviolet spectra and chromatographic behavior of a number of thiazolinethiones.[40]

Oxidation of 1,2-benzisothiazoline-3-thiones with mercuric acetate affords the corresponding benzisothiazolinones.[32]

[35] J. T. Braunholtz (Imperial Chemical Industries Ltd.), British Patent 1,113,634; *Chem. Abstr.* **69**, 58,590 (1968).
[36] Montecatini Edison S.p.A., Italian Patent 800,015; *Chem. Abstr.* **70**, 37,807 (1969).
[37] A. Baruffini, P. Borgna, F. Gialdi, and R. Ponci, *Farmaco, Ed. Sci.* **23**, 3 (1968); *Chem. Abstr.* **69**, 10,423 (1968).
[38] L. Amoretti, F. Mossini, and V. Plazzi, *Farmaco, Ed. Sci.* **23**, 583 (1968); *Chem. Abstr.* **69**, 86,865 (1968).
[39] G. D. Andreetti, A. Corradi, P. Domiano, and A. Musatti, *Ric. Sci.* **38**, 1104 (1968); *Chem. Abstr.* **71**, 25,490 (1969).
[40] R. Mayer, P. Rosmus, and J. Fabian, *J. Chromatogr.* **15**, 153 (1964).

3. 1-Imino- and 3-Imino-1,2-benzisothiazolines

Klyuev and Snegireva[41,42] have prepared imino derivatives of 1,2-benzisothiazolines, including 1-imino-1,2-benzisothiazolinone (**49**) and 1,3-diimino-1,2-benzisothiazoline (**50**). These compounds are formed by treating dithiosalicylic acid (**47**) with excess chlorine and reacting the product (**48**) with ammonia or urea. The same workers have also prepared a number of derivatives of these imino compounds.[43,44]

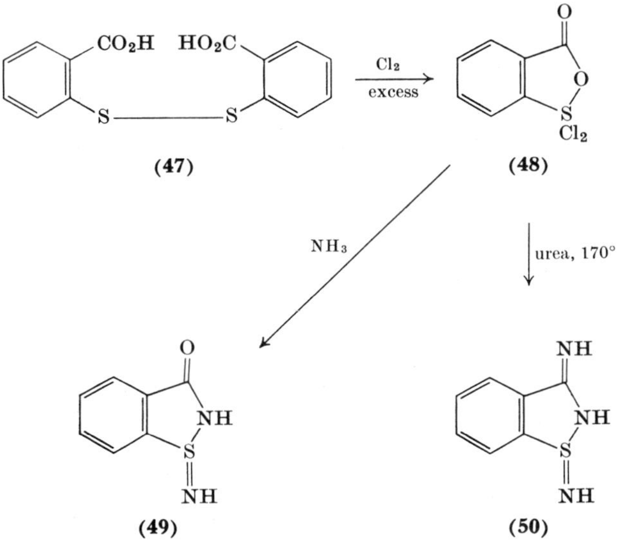

4. 4,5,6,7-Tetrahydro-1,2-benzisothiazoles

A remarkable sulfur extrusion reaction was discovered by Fanghaenel.[45] The dithiazine (**51**), if heated to about 180°, yields the

---

[41] V. N. Klyuev and F. P. Snegireva (Ivanovo Chemical-Engineering Institute), USSR Patent 188,977; *Chem. Abstr.* **67**, 90,837 (1967).
[42] V. N. Klyuev and F. P. Snegireva, *Izv. Vyssh. Ucheb. Zaved., Khim. Khim. Tekhnol.* **10**, 891 (1967); *Chem. Abstr.* **68**, 105,077 (1968).
[43] V. N. Klyuev and F. P. Snegireva (Ivanovo Chemical-Engineering Institute), USSR Patent 196,877; *Chem. Abstr.* **68**, 95,850 (1968).
[44] V. N. Klyuev and F. P. Snegireva (Ivanovo Chemical-Engineering Institute), USSR Patent 196,878; *Chem. Abstr.* **68**, 104,975 (1968).
[45] E. Fanghaenel, *Z. Chem.* **5**, 386 (1965).

tetrahydrobenzisothiazole (**52**). One might, perhaps, have expected a benzothiazole to result.

(**51**) →(Δ, 180°) (**52**)

E. 1,2-BENZISOTHIAZOLINONES

Since the discovery a few years ago that this group of compounds possessed useful bactericidal and, especially, fungicidal activity, intensive investigation has taken place and a very large number of compounds have been prepared. However, since the earlier review by Bambas[1] gives an excellent, detailed account of the general chemistry of these compounds, including a good discussion of McClelland's early work, they will be mentioned only briefly in the present review.

Benzisothiazolinones (**53**) are readily prepared by treatment of dithiosalicylic acid (**47**) with chlorine, followed by an amine.[46-55] Variations from this basic reaction include the reaction of dithio-

---

[46] C. F. H. Allen and J. H. Sagura (Eastman Kodak Company), U.S. Patent 2,870,015; *Chem. Abstr.* **53**, 7844 (1959).

[47] Kodak Soc. Anon., Belgian Patent 565,380; *Chem. Abstr.* **53**, 14,799 (1959).

[48] R. Fischer (Dr. A. Wander A.G.), German Patent 1,135,468; *Chem. Abstr.* **58**, 529 (1963).

[49] Knoll, A. G. Chemische Fabriken, Belgian Patent 617,384; *Chem. Abstr.* **58**, 12,570 (1963).

[50] R. Ponci, A. Baruffini, M. Croci, and F. Gialdi *Farmaco, Ed. Sci.* **18**, 732 (1963); *Chem. Abstr.* **60**, 2920 (1964).

[51] R. Ponci, F. Gialdi, and A. Baruffini, *Farmaco, Ed. Sci.* **19**, 254 (1964); *Chem. Abstr.* **61**, 3088 (1964).

[52] R. Ponci, T. Vitali, L. Amoretti, and F. Mossini, *Farmaco, Ed. Sci.* **22**, 935 (1967); *Chem. Abstr.* **68**, 105,078 (1968).

[53] R. Ponci, T. Vitali, F. Mossini, and L. Amoretti, *Farmaco, Ed. Sci.* **22**, 999 (1967); *Chem. Abstr.* **68**, 114,493 (1968).

[54] R. Ponci, T. Vitali, F. Mossini, and L. Amoretti, *Farmaco, Ed. Sci.* **22**, 989 (1967); *Chem. Abstr.* **69**, 27,322 (1968).

[55] T. Vitali, L. Amoretti, and F. Mossini, *Farmaco, Ed. Sci.* **23**, 468 (1968); *Chem. Abstr.* **69**, 59,147 (1968).

salicyamides,[56] o-mercaptobenzamides,[57] or hydrazides[58] with thionyl chloride, and the reactions of 1,2,3-benzodithiolones[59] or o-carboxybenzenesulfinic acid [60] with amines.

N-Alkyl-,[61] N-acyl-,[62,63] and N-sulfonyl-1,2-benzisothiazolinones[64] are readily obtained from the parent compound [(53) R = H] by treatment with alkylating, acylating, or sulfonylating agents or, in the case of N-acyl and N-sulfonyl derivatives, by using an amide or sulfonamide, instead of a primary amine, in the reaction above [Eq. (6)].

The possible tautomerism [Eq. (7)] of the parent compound has

[56] T. Vitali, L. Amoretti, and V. Plazzi, *Farmaco. Ed, Sci.* **23**, 1075 (1968); *Chem. Abstr.* **70**, 37,701 (1969).
[57] F. Gialdi, R. Ponci, and A. Baruffini, *Farmaco, Ed. Sci.* **16**, 509 (1961); *Chem. Abstr.* **60**, 512 (1964).
[58] L. Katz and W. Schroeder, *J. Org. Chem.* **19**, 103 (1954).
[59] J. S. Morley (Imperial Chemical Industries Ltd)., British Patent 848,130; *Chem. Abstr.* **55**, 9430 (1961).
[60] V. N. Klyuev and V. F. Borodkin, *Izv. Vyssh. Ucheb. Zaved., Khim. Khim. Tekhnol.* **3**, 1079 (1960); *Chem. Abstr.* **55**, 13,358 (1961).
[61] M. Fishman and P. A. Cruickshank, *J. Heterocycl. Chem.* **5**, 467 (1968).
[62] F. Gialdi, R. Ponci, A. Baruffini, and P. Borgna, *Farmaco, Ed. Sci.* **19**, 76 (1964); *Chem. Abstr.* **60**, 9266 (1964).
[63] A. Baruffini, F. Gialdi, and R. Ponci ("Montecatini" Societa Generale per l'Industria Mineraria e Chimica), French Patent 1,364,176; *Chem. Abstr.* **61**, 14,681 (1964).
[64] G. Pagani, L. Amoretti, and A. Baruffini, *Farmaco, Ed. Sci.* **22**, 801 (1967); *Chem. Abstr.* **68**, 105,067 (1968).

not been studied in detail, but O-alkyl[18, 19] and O-acyl[65] derivatives are known.

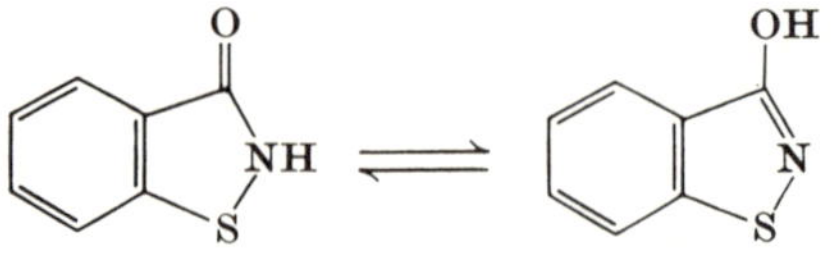

$$\text{(structure: benzisothiazolinone)} \rightleftharpoons \text{(structure: 3-hydroxybenzisothiazole)} \qquad (7)$$

Mustafa and co-workers investigated the reaction of N-sulfonyl-1,2-benzisothiazolinones with Grignard reagents[66] and organolithium compounds.[67] In either case the heterocyclic ring is broken and a tertiary alcohol is formed [Eq. (8)].

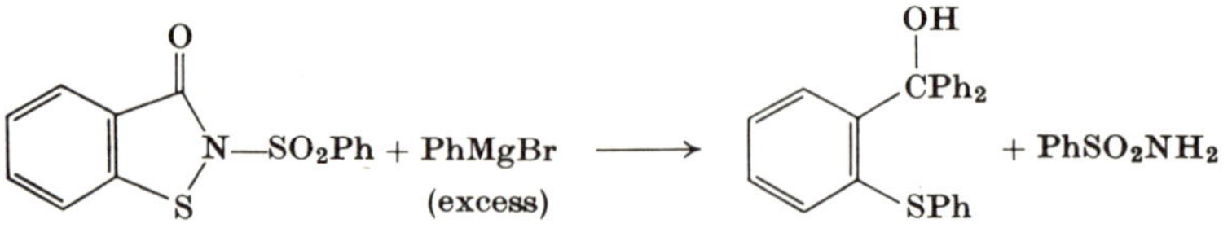

$$\text{(N-SO}_2\text{Ph benzisothiazolinone)} + \text{PhMgBr (excess)} \longrightarrow \text{(o-CPh}_2\text{OH, SPh arene)} + \text{PhSO}_2\text{NH}_2 \qquad (8)$$

F. APPLICATIONS

3-Amino-1,2-benzisothiazoles such as **54** have been claimed[22] to possess antiinflammatory, analgesic, and hypotensive properties, while 3-dialkylamino-1,2-benzisothiazoles[16] and 3-aryloxy- or 3-arylthio-1,2-benzisothiazoles (**55**)[23] have useful antimycotic activity.

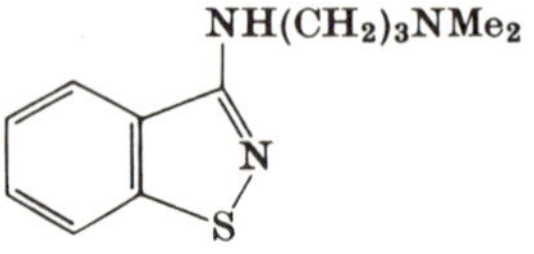

(**54**) NH(CH$_2$)$_3$NMe$_2$-benzisothiazole

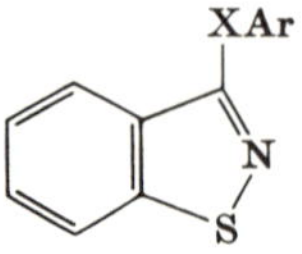

(**55**) XAr-benzisothiazole, X = O, S

Vitali and his colleagues have investigated the local anesthetic and antihistaminic activity of a series of alkyloxy derivatives such as the

---

[65] W. Lorenz (Farbenfabriken Bayer A.G.), German Patent 1,160,440; *Chem. Abstr.* **60**, 10,718 (1964).
[66] A. Mustafa and O. H. Hishmat, *J. Amer. Chem. Soc.* **75**, 4647 (1953).
[67] A. Mustafa, W. Asker, O. H. Hishmat, A. F. A. Shalaby, and M. Kamel, *J. Amer. Chem. Soc.* **76**, 5447 (1954).

salt (56);[18, 19, 24, 68] analogous compounds such as the phosphorus ester (57) are useful insecticides.[65] Azo compounds, prepared by

(56) (57)

diazotizing 4- or 5-amino-1,2-benzisothiazole and coupling the diazonium salt with tertiary aromatic amines have been proposed as dyes[69, 70] and schistosomicides.[71]

A purely chemical application has been suggested for 3-bromomethyl-1,2-benzisothiazole (58) and related compounds. This halide (58) can function as an effective alkylating agent by virtue of the benzylic nature of the compound; subsequent removal of the hetero-

(58)

atoms (in the alkylated product) yields a phenylethyl moiety which is often rather difficult to introduce directly on account of the competing elimination reaction.[72]

As mentioned earlier, 1,2-benzisothiazolinones have received considerable attention recently. Katz and Schroeder[73] were apparently the first to note the antifungal and antibacterial properties of these

[68] R. Ponci, T. Vitali, and G. Bartaccini, *Farmaco, Ed. Sci.* **23**, 620 (1968); *Chem. Abstr.* **69**, 42,630 (1968).
[69] A. Algerino, D. Dal Monte Casoni, and L. D'Aria, *Chim. Ind. (Milan)* **42**, 1339 (1960); *Chem. Abstr.* **55**, 22,827 (1961).
[70] A. Algerino, D. Dal Monte, and L. D'Aria, *Fibre Colori* **11**, 75 (1961); *Chem. Abstr.* **56**, 14,433 (1962).
[71] E. F. Elslager, D. B. Capps, L. M. Werbel, D. F. Worth, J. E. Meisenhelder, H. Najarian, and P. E. Thompson, *J. Med. Chem.* **6**, 217 (1963).
[72] R. A. Gillham, Thesis, California Institute of Technology, 1969; *Diss. Abstr. B.* **30**, 3095 (1970).
[73] L. Katz and W. Schroeder, U.S. Patent 2,767,172; *Chem. Abstr.* **51**, 6703 (1957); and U.S. Patent 2,767,174; *Chem. Abstr.* **51**, 6704 (1967).

compounds. Since this observation, numerous papers and patents have appeared dealing with these aspects.[48, 51–55, 57, 59, 62, 63, 74–91] Of the compounds of relatively simple structure, 5-chloro- **(59)**[75, 76, 79] and 6-chloro-1,2-benzisothiazolinone **(60)**[48, 83, 84] appear to be the most effective.

(59) (60)

[74] G. Ambrosoli, M. P. Ciuti, M. G. Menozzi, and M. R. Mingiardi, *Boll. Chim. Farm.* **109**, 251 (1970); *Chem. Abstr.* **73**, 95,725 (1970).
[75] J. S. Morley (Imperial Chemical Industries Limited), British Patent 861,379; *Chem. Abstr.* **55**, 22,723 (1961).
[76] A. J. Hinton, J. N. Turner, and J. S. Morley (Imperial Chemical Industries Limited), British Patent 844,541; *Chem. Abstr.* **56**, 10,637 (1962).
[77] B. R. Baker, W. W. Lee, E. Tong, L. O. Ross, and A. P. Martinez, *J. Theor. Biol.* **3**, 446 (1962).
[78] A. J. Hinton, G. A. Thomas, J. S. Morley, and J. N. Turner (Imperial Chemical Industries Limited), British Patent 918,869; *Chem. Abstr.* **58**, 13074 (1963).
[79] A. J. Hinton, J. S. Morley, and J. N. Turner (Imperial Chemical Industries Limited), U.S. Patent 3,065,123; *Chem. Abstr.* **60**, 16,447 (1964).
[80] R. Ponci, A. Baruffini, and F. Gialdi, *Farmaco, Ed. Sci.* **19**, 121 (1964); *Chem. Abstr.* **61**, 6300 (1964).
[81] F. Gialdi, R. Ponci, and A. Baruffini, *Farmaco, Ed. Sci.* **19**, 474 (1964); *Chem. Abstr.* **61**, 6300 (1964).
[82] F. Gialdi, R. Ponci, and P. Caccialanza, *Mycopathol. Mycol. Appl.* **24**, 163 (1964); *Chem. Abstr.* **63**, 429 (1965).
[83] R. Fischer and H. Hurni, *Arzneim-Forsch.* **14**, 1301 (1964).
[84] H. Hurni, H. U. Gubler, and R. Fischer, *Arzneim-Forsch.* **14**, 1306 (1964).
[85] A. Baruffini, F. Gialdi, and R. Ponci ("Montecatini" Societa Generale per l'Industria Mineraria e Chimia), French Patent 1,411,720; *Chem. Abstr.* **64**, 2092 (1966).
[86] A. Baruffini, F. Gialdi, and R. Ponci ("Montecatini" Societa per l'Industria Mineraria e Chimica), Italian Patent 670,877; *Chem. Abstr.* **64**, 3548 (1966).
[87] H. Klesper, F. Steinfatt, W. Lorenz, and K. Langheinrich (Farbenfabriken Bayer A.G.), Belgian Patent 671,490; *Chem. Abstr.* **65**, 8921 (1966).
[88] R. Fischer, *Int. Congr. Chemother. Proc., 3rd, Stuttgart, 1963* **2**, 1393 (1964); *Chem. Abstr.* **66**, 1364 (1967).
[89] R. Ponci, A. Baruffini, and P. Borgna, *Farmaco, Ed. Sci.* **21**, 249 (1966); *Chem. Abstr.* **66**, 37,816 (1967).
[90] R. J. Lukens and J. G. Horsfall, *Phytopathology* **57**, 876 (1967).
[91] Imperial Chemical Industries Limited, French Patent Application 2,002,353; *Chem. Abstr.* **72**, 120,490 (1970).

There have also appeared reports of the insecticidal,[92] schistosomicidal,[93,94] and antiinflammatory properties[95] of substituted 1,2-benzisothiazolinones; certain N-substituted derivatives have antifogging properties and can be used to stabilize photographic silver halide emulsions.[46,47]

As might be expected, antifungal and antibacterial activity is also possessed by 1,2-benzisothiazoline-3-thiones[35-38,74] and 3-imino-1,2-benzisothiazolines.[15,30]

III. 2,1-Benzisothiazoles

A. Formation

As was the case with 1,2-benzisothiazoles, two basic methods are known for the construction of the 2,1-benzisothiazole ring. The first requires a precursor of type **61**, in which cyclization can be brought

about by oxidation or reduction. The second route, from **62**, is less demanding, for the sulfur atom is introduced during the reaction.

1. *From o-Nitro-α-toluenethiol and Related Compounds*

Gabriel and co-workers[96,97] first prepared 2,1-benzisothiazole ("thioanthranil") (**2**) in the 1890's. Reduction of *o*-nitro-α-tolenethiolcarbamate (**63**) or *o*-nitro-α-toluenethiol (**64**) with stannous chloride and hydrochloric acid yielded a crystalline solid which, on

---

[92] W. Lorenz and G. Schrader (Farbenfabriken Bayer A.G.), Belgian Patent 631,729; *Chem. Abstr.* **61**, 8315 (1964).
[93] J. N. Ashley, R. F. Collins, and M. Davis (May & Baker Limited), British Patent 810,304; *Chem. Abstr.* **54**, 2369 (1960).
[94] J. N. Ashley, R. F. Collins, M. Davis, and N. E. Sirett, *J. Chem. Soc.* 3880 (1959).
[95] C. Hanna and H. J. Thomason, *Arch. Int. Pharmacodyn. Ther.* **173**, 262 (1968); *Chem. Abstr.* **69**, 75,405 (1968).
[96] S. Gabriel and T. Posner, *Chem. Ber.* **28**, 1025 (1985).
[97] S. Gabriel and R. Stelzner, *Chem. Ber.* **29**, 160 (1896).

addition of water and steam distillation, afforded 2,1-benzisothiazole (**2**). This crystalline solid is undoubtedly the stannic chloride complex of the free base; it is decomposed by steam and the base is liberated.

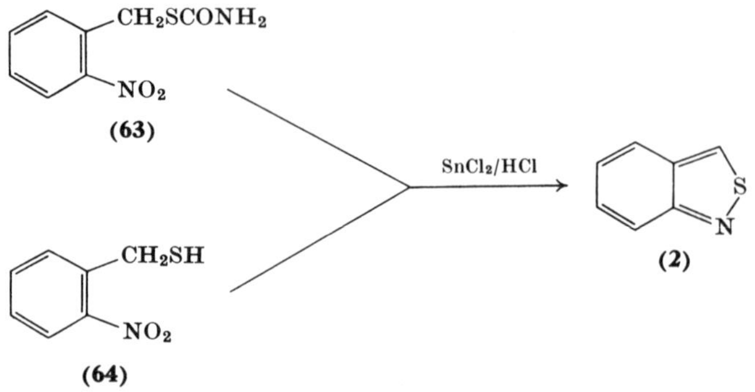

The thiol (**64**) also yields 2,1-benzisothiazole on treatment with concentrated potassium hydroxide solution;[2] in this case most of the thiol is oxidized to the disulfide (**65**) and the yield of product (**2**) is low.[98] It was later found that o-nitro-α-toluenethiolacetic acid (**66**) undergoes a similar reaction with sodium hydroxide solution; the yield of product (**2**) is however very small (less than 10%).[99]

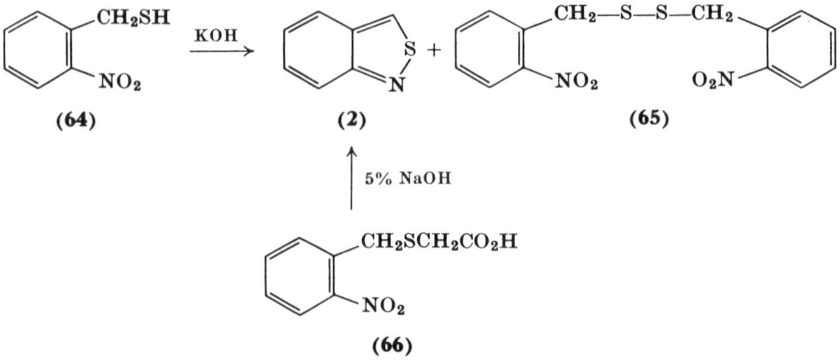

2. From o-Amino-α-toluenethiol

A generally useful synthesis of nitrogen–sulfur heterocycles, already referred to in Section II, A, is oxidative cyclization of amino-

[98] M. Davis, unpublished work (1967).
[99] Y. Iskander and Y. Riad, *J. Chem. Soc.* 2054 (1951).

thiols.[4] o-Amino-α-toluenethiol (67), treated with iodine and dilute base, affords 2,1-benzisothiazole (2); careful control of the pH of this reaction is essential, otherwise the yield of product (2) is low and that of the disulfide (68) correspondingly high.[4] The yield of product is also lowered by further substitution in the benzene ring of the thiol (67).[25]

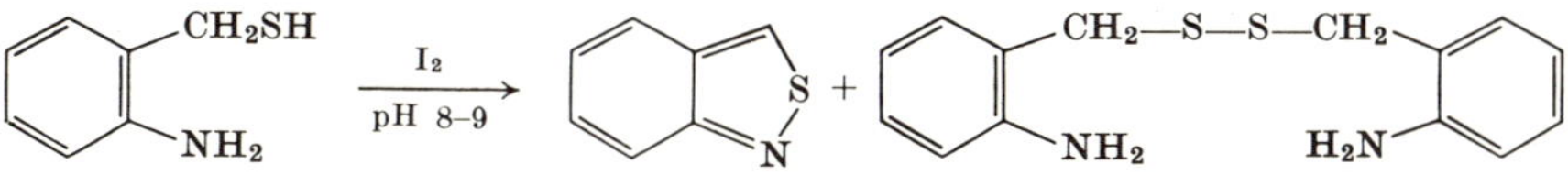

3. From o-Aminothiobenzamides and o-Nitrothiobenzamides

Although any of the foregoing methods could, in principle, have been applied to the synthesis of substituted 2,1-benzisothiazoles, no such attempt was made and until 1963 the parent compound (2) was the sole representative of this heterocyclic system. In that year a group from the Parke, Davis Company filed a patent application[100] in which the preparation of a number of 3-substituted 2,1-benzisothiazoles was described. This application, shortly followed by a definitive paper,[101] indicated how o-aminothiobenzamides (69) could be oxidized, and o-nitrothiobenzamides (70) reduced, with concomitant ring closure to yield 3-amino-2,1-benzisothiazoles (71).

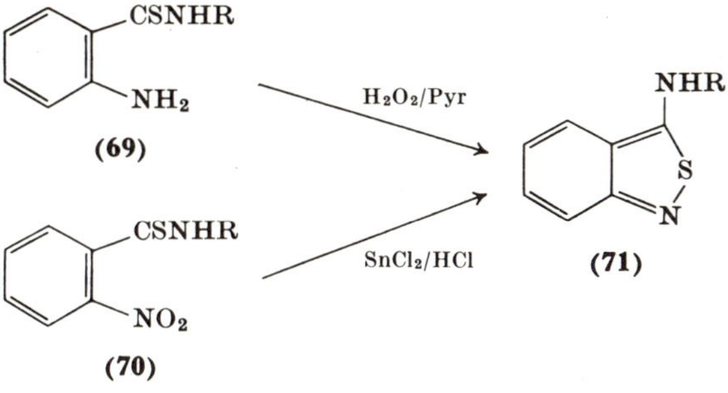

[100] Parke, Davis Company, Netherlands Patent Application 6,408,290; *Chem. Abstr.* **63**, 1768 (1965).
[101] R. F. Meyer, B. L. Cummings, P. Bass, and H. O. J. Collier, *J. Med. Chem.* **8**, 515 (1965).

Independently and almost simultaneously two other groups filed patent applications[102, 103] describing an essentially similar oxidative ring closure. Since the starting thiobenzamides are readily prepared, either by treatment of the corresponding benzamide with phosphorus pentasulfide, or by the action of hydrogen sulfide on the nitrile, a wide variety of substituted 2,1-benzisothiazoles (71) can be obtained.

The mechanism of the oxidative cyclization, which is the more useful of the two cyclizations described above, is not known with certainty, but it is probable that an S-oxide intermediate (72) is involved. Such oxides are known to be formed by the action of hydrogen peroxide on thioamides,[104] and the sulfur atom in the S-oxide (72) is then likely to be more susceptible to intramolecular nucleophilic attack by the adjacent nitrogen atom.

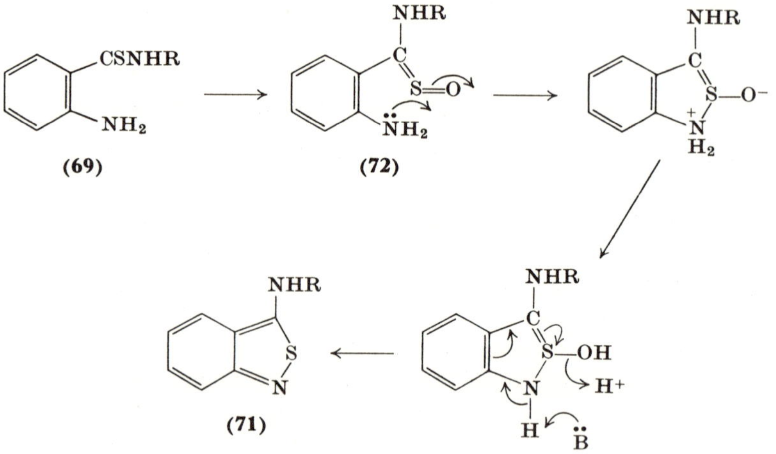

4. From o-Toluidines and Thionyl Chloride

Reaction between an o-toluidine (73) and thionyl chloride in a high-boiling inert solvent (such as xylene or bromobenzene) affords 2,1-benzisothiazoles (74) in moderate yields.[105, 106] Long reaction

[102] M. Seefelder and H. Armbrust (Badische Anilin und Soda-Fabrik A.G.), Belgian Patent 670,652; *Chem. Abstr.* 66, 65,467 (1967).
[103] S. T. Ross (Smith Kline and French Company), U.S. Patent 3,254,094; *Chem. Abstr.* 65, 5464 (1966).
[104] W. Walter and K. D. Bode, *Ann.* 660, 74 (1962).
[105] M. Davis and A. W. White, *Chem. Commun.* 1547 (1968).
[106] M. Davis and A. W. White, *J. Org. Chem.* 34, 2985 (1969).

Sec. III. A.4] BENZISOTHIAZOLES 67

times (24 hours or more) and excess of thionyl chloride ($2\frac{1}{2}$ to 3 moles) are required for successful reaction. The yields are markedly dependent on the nature of the R group; when R is methoxy para to the amino group, a 70% yield is obtained, whereas nitro groups ortho or para to the amino group inhibit the reaction completely.[106]

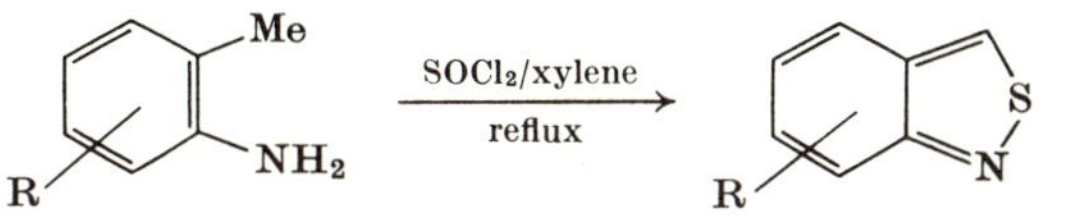

$$\underset{(73)}{\text{R}\text{-}{C_6H_3}(Me)(NH_2)} \xrightarrow[\text{reflux}]{SOCl_2/xylene} \underset{(74)}{\text{R}\text{-benzisothiazole}}$$

Once again, the mechanism of this somewhat unusual reaction is not known with certainty. The N-sulfinylamine (75) is undoubtedly an intermediate, but exactly how the methyl group is functionalized is difficult to ascertain. A second equivalent of thionyl chloride is required before cyclization of the sulfinylamine (75) occurs; other dehydrating agents, or heat alone, are not effective. It has been suggested[106] that the N-sulfinylamino group in 75 is an electron acceptor and that this renders the methyl group sufficiently acidic (cf. o-nitrotoluene) to react with a second molecule of thionyl chloride. If this is the case, then one might expect a sulfine (76) or related compound to be a further intermediate. An analogous sulfine intermediate has been suggested in a related synthesis of thiazoles.[107]

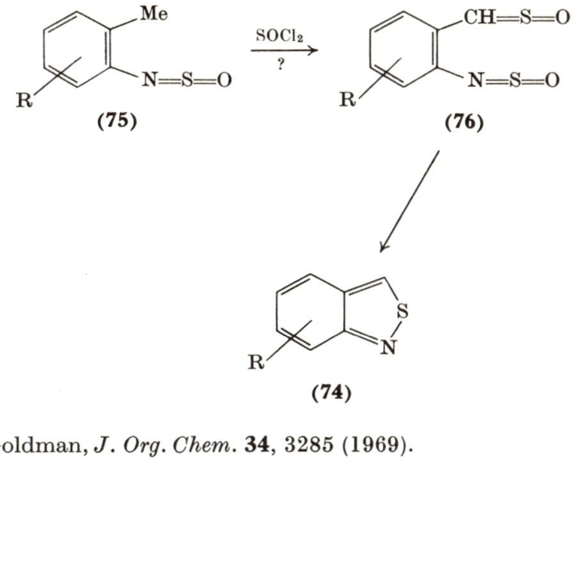

$$\underset{(75)}{\text{R}\text{-}{C_6H_3}(Me)(N=S=O)} \xrightarrow[?]{SOCl_2} \underset{(76)}{\text{R}\text{-}{C_6H_3}(CH=S=O)(N=S=O)} \longrightarrow \underset{(74)}{\text{R}\text{-benzisothiazole}}$$

[107] I. M. Goldman, *J. Org. Chem.* **34**, 3285 (1969).

5. From o-Azidoaryl Thioketones

Recently Ashby and Suschitzky have described the preparation of 3-substituted 2,1-benzisothiazoles by the pyrolysis of o-azidoaryl thioketones.[107a]

B. Physical Properties and Structure

2,1-Benzisothiazole is a very pale yellow oil, b.p. 242°/748 mm, 70°/0.5 mm, with a characteristic odor similar to that of quinoline.[96, 97, 105] The methyl-substituted derivatives so far examined are also liquids, but other substituted 2,1-benzisothiazoles are generally colorless crystalline solids. A full list of 2,1-benzisothiazoles is given in the tables at the end of this chapter. The nitrogen atom in 2,1-benzisothiazoles is weakly basic, and these compounds dissolve in concentrated hydrochloric acid, but are reprecipitated on dilution.

Early workers favored a tricyclic structure (77) for the parent compound, but by analogy with the corresponding benzisoxazoles (anthranils)[108] the o-quinonoid structure is now generally accepted, as in the conventional formula (2). This may not be quite the end of

(77)   (2)   (78)

the story, as a quadrivalent sulfur form (78) can be written, and this may make an important contribution to the ground state of the molecule. The chemical properties of 2,1-benzisothiazole, especially when compared to those of the corresponding isoxazole in which, of course, no corresponding canonical form is possible, indicate a substantial contribution from 78.

The proton NMR spectrum of 2,1-benzisothiazole has been analyzed[109] and the parameters are listed in Table I. The variation in the chemical shift of H-3 has been reported for a number of mono- and disubstituted 2,1-benzisothiazoles;[106, 110] this H-3 proton is

[107a] J. Ashby and H. Suschitzky, *Tetrahedron Lett.* 1315 (1971).
[108] K.-H. Wünsch and A. J. Boulton, *Advan. Heterocycl. Chem.* 8, 277 (1967).
[109] M. Davis, H. Hügel, and B. Ternai, unpublished work (1969).
[110] M. Davis and A. W. White, *J. Chem. Soc. C* 2189 (1969).

TABLE I

NMR PARAMETERS FOR 2,1-BENZISOTHIAZOLE[a]

H atom	δ (ppm)	Protons coupled	J (Hz)	Protons coupled	J (Hz)
3	9.06	3,4	0.38	4,6	1.23
4	7.63	3,5	0.18	4,7	0.85
5	7.09	3,6	0.17	5,6	6.02
6	7.31	3,7	0.94	5,7	1.21
7	7.78	4,5	7.92	6,7	8.40

[a] From Davis et al.[109] Spectrum obtained at 60 MHz of a sample containing tetramethylsilane, without solvent.

coupled with H-7 ($J \simeq 1$ Hz) and appears as a narrow doublet unless position 7 is substituted.

The ultraviolet spectra of a few 2,1-benzisothiazoles have been reported. In the unsubstituted compound absorption maxima are at 203 nm ($\epsilon = 14,300$), 221 (16,400), 288 sh (7600), 298 (9100), and 315 sh (4000).[106] Simple monosubstituted 2,1-benzisothiazoles possess similar spectra,[111] but in 3-amino derivatives the maxima are at ca. 230 and 380 nm, and the reported extinction coefficients are much smaller.[101]

Few details of the infrared spectra of 2,1-benzisothiazoles have been reported. Some infrared and NMR spectra are available from commercial reference collections.[112] The molecular refraction of 2,1-benzisothiazoles is similar to that of 2,1-benzisoxazole.[113]

C. CHEMICAL PROPERTIES

1. Stability

2,1-Benzisothiazole is quite stable, but in the presence of light and air it gradually darkens. The extent of decomposition seems to be small, for redistillation affords an almost complete recovery of pure compound. The compound displays none of the reactivity or instability generally associated with o-quinonoid systems.

[111] R. K. Buckley, M. Davis, and K. S. L. Srivastava, Aust. J. Chem. 24, 2405 (1971).

[112] E.g., from the Sadtler Research Laboratories, Inc., 3316 Spring Garden St., Philadelphia, Pennsylvania 19104.

[113] K. v. Auwers and W. Ernst, Z. Phys. Chem. 122, 231 (1926).

2. Salts

No simple salts of 2,1-benzisothiazole have been described in the literature, other than those of amino-substituted derivatives,[101] in which the site of protonation is probably the cyclic nitrogen atom, at least in the 3-substituted compounds. The parent compound does, however, form a simple crystalline nitrate [114] which is not very stable. A number of complex salts with platinum, gold, lead, and mercury salts and with picric acid have been described.[96, 97, 115] The picrate is highly crystalline and stable, and is useful for the isolation and characterization of 2,1-benzisothiazole. Many substituted 2,1-benzisothiazoles form similarly useful picrates.[106]

3. Quaternary Salts

2,1-Benzisothiazole forms stable, crystalline quaternary salts, such as **79**, quite readily. Quaternary methiodides such as the salt (**80**)

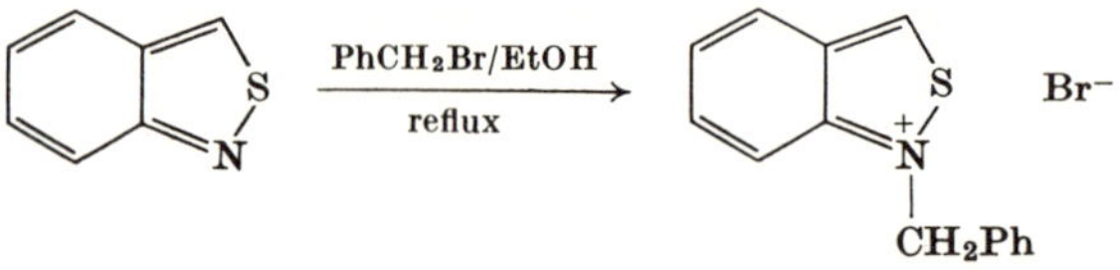

have also been prepared from various 3-amino-2,1-benzisothiazoles,[101] and methosulfates have been obtained from a number of 3-arylazo-2,1-benzisothiazoles.[116, 117] In the latter case the products, such as

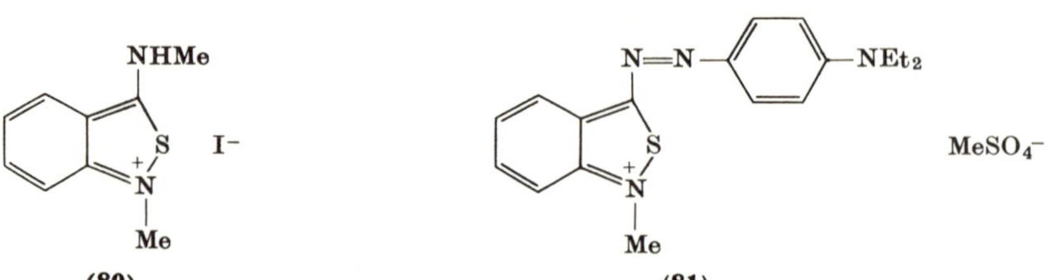

[114] M. Davis, unpublished work (1969).
[115] E. Bamberger, *Chem. Ber.* **42**, 1667 (1909).
[116] Badische Anilin und Soda-Fabrik A.G., Netherlands Patent Application 6,608,032; *Chem. Abstr.* **66**, 96,214 (1967).
[117] Farbenfabriken Bayer A.G., French Patent 1,540,834; *Chem. Abstr.* **72**, 80,338 (1970).

the salt (**81**), are highly colored and are claimed to be effective blue-green disperse dyes for polyacrylonitrile fibers.[117]

4. Electrophilic Substitution

a. *Bromination.* The action of bromine and silver sulfate (an effective source of Br^+) on 2,1-benzisothiazole dissolved in sulfuric acid gives a mixture of equal quantities of 5-bromo- and 7-bromo-2,1-benzisothiazole, together with a lesser amount of 4,7-dibromo-2,1-benzisothiazole and a trace of 4-bromo-2,1-benzisothiazole. With excess of brominating reagents the main product is 4,5,7-tribromo-2,1-benzisothiazole.[110]

b. *Nitration.* The parent compound, treated with a mixture of nitric and sulfuric acids, affords mainly 5,-nitro-2,1-benzisothiazole, with smaller quantities of 7-nitro and 4-nitro isomers.[110] In general, the direction of nitration in 2,1-benzisothiazoles already substituted in the benzenoid ring is controlled by the nature of that substituent rather than by the residual effect of the heterocyclic ring. In other words, the perturbing effect of the heterocyclic ring on the benzenoid ring is relatively slight.

c. *Acylation.* Attempts at Friedel–Crafts or Vilsmeier–Haack acylation reactions on the parent compound have been unsuccessful.[110]

5. Ring-Opening Reactions

It was reported by Gabriel and Leupold[118] that hydrazine opens the heterocyclic ring. Hydrogen sulfide is evolved and the hydrazone (**82**) is the product. Phenylhydrazine reacts in a similar fashion.[118] Hot aqueous alkali, on the other hand, has no effect,[97] although

2,1-benzisoxazole is rapidly cleaved by cold dilute base.[108] The quaternary salts derived from 3-amino-2,1-benzisothiazole are much more sensitive to acid and alkali. Thus, 1-methyl-3-methylamino-

---

[118] S. Gabriel and E. Leupold, *Chem. Ber.* **31**, 2185 (1898).

2,1-benzisothiazolium iodide (**83**), boiled with concentrated hydrochloric acid, gives *N*-methylanthranilic acid (**84**); the analogous 3-amino salt (**85**), similarly treated with ammonia, yields *o*-methylaminobenzonitrile (**86**).[101]

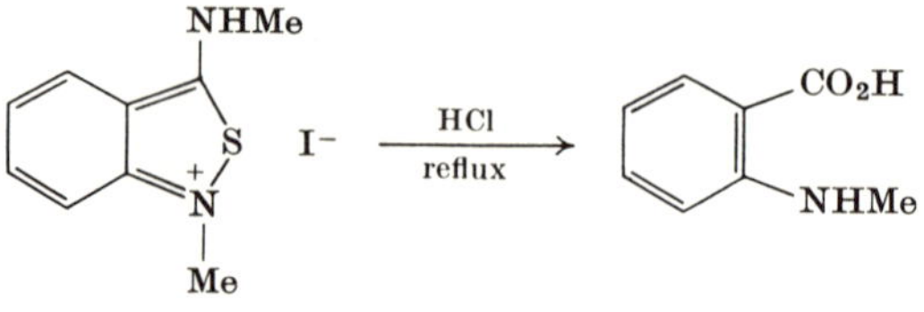

6. Reactions of Substituted 2,1-Benzisothiazoles

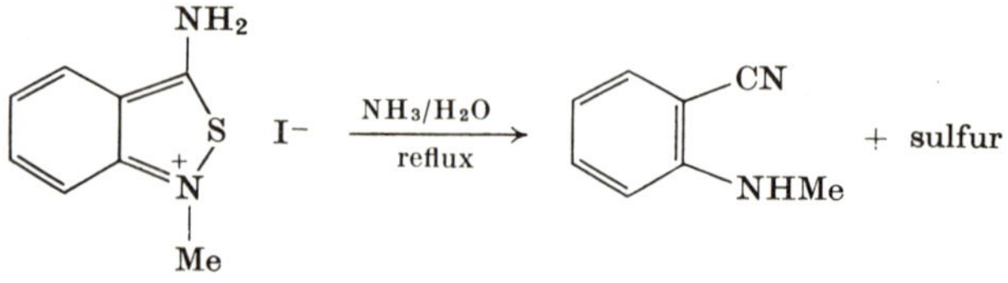

Electrophilic substitution reactions of substituted 2,1-benzisothiazoles have already been discussed. The only other functionally substituted compound so far studied in any detail is the 3-amino derivative (**87**). This compound is readily diazotized and the diazonium salt formed (**88**) couples with tertiary aromatic amines yielding azo dyes.[117, 119–121] The salt (**88**) also undergoes the usual Sandmeyer reactions, and other 3-substituted derivatives, such as the cyano compound (**89**) may be prepared.[111] Acylation of the amine (**87**) yields diacyl derivatives which appear to possess the 3-acylimino structure (**90**).[111]

---

[119] K. L. Moritz and C. Taube (Farbenfabriken Bayer A.G.), British Patent 1,134,579; *Chem. Abstr.* **71**, 14,164 (1969).
[120] Badische Anilin und Soda-Fabrik A.G., French Patent 1,551,777; *Chem. Abstr.* **72**, 112,792 (1970).
[121] Badische Anilin und Soda-Fabrik A.G., French Patent 1,568,501; *Chem. Abstr.* **73**, 67,671 (1970).

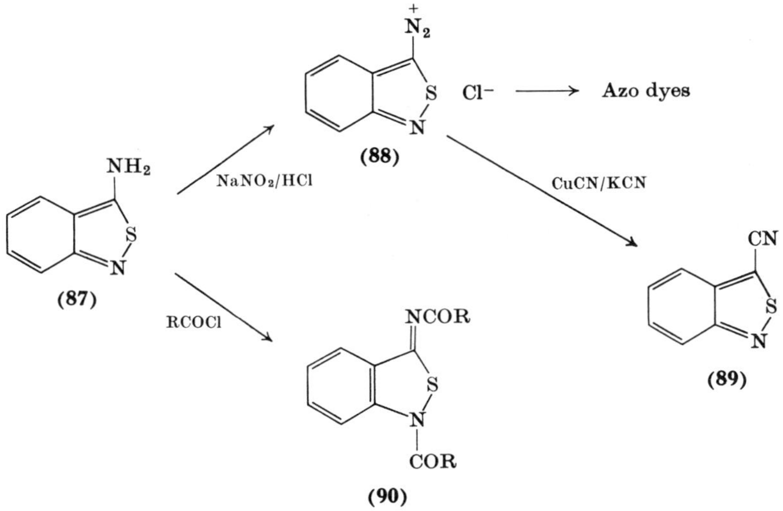

D. REDUCED DERIVATIVES

1. 4,5,6,7-Tetrahydro-2,1-benzisothiazoles

Dimedone monoimide (**91**) reacts with *p*-chlorophenyl isothiacyanate to give an adduct (**92**) which may be oxidized to a 4-keto-

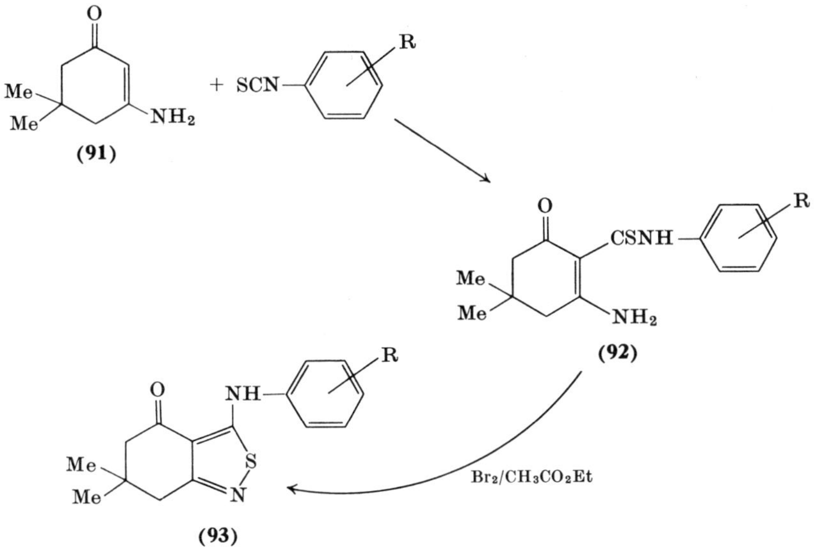

4,5,6,7-tetrahydro-2,1-benzisothiazole (**93**).[122] If the N-methyl derivative of the monoimide (**91**) is used, the product is a tetrahydrobenzisothiazoline.[122]

2. *2,1-Benzisothiazoline-3-thiones and 2,1-Benzisothiazolinones*

Reaction between an N-alkyl or N-aryl substituted anthranilic acid (**94**) and phosphorus pentasulfide, in refluxing xylene, affords a mixture of a 2,1-benzisothiazoline-3-thione (**95**) and a Δ-3,3'-bis-2,1-benzisothiazoline (**96**).[123, 124] Oxidation of the thione (**95**) with

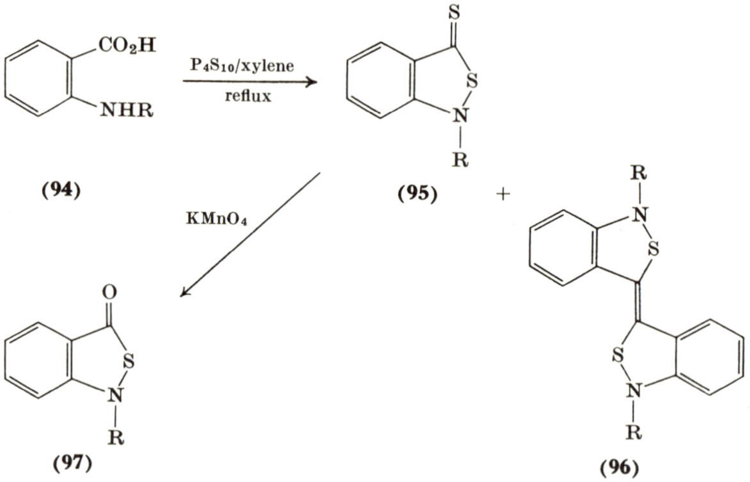

potassium permanganate yields the corresponding benzisothiazolinone (**97**).[123]

3. *4,5,6,7-Tetrahydro-2,1-benzisothiazoline-3-thiones*

Mayer and co-workers have shown[125–127] that the combined action of carbon disulfide and sulfur on ketimines (Schiff bases) (**98**) derived

[122] J. Goerdeler and U. Keuser, *Chem. Ber.* **97**, 2209 (1964).
[123] L. Legrand and N. Lozac'h, *Bull. Soc. Chim. Fr.* 1170 (1969).
[124] L. Legrand and N. Lozac'h, *Bull. Soc. Chim. Fr.* 1173 (1969).
[125] R. Mayer and J. Jentzsch, *J. Prakt. Chem.* **23**, 113 (1964).
[126] R. Mayer, P. Wittig, J. Fabian, and R. Heitmueller, *Chem. Ber.* **97**, 654 (1964).
[127] R. Mayer, H. J. Hartmann, and J. Jentzsch, *J. Prakt. Chem.* **31**, 312 (1966).

from cyclohexanone gives rise to 4,5,6,7-tetrahydro-2,1-benzisothiazoline-3-thiones (**99**). The reaction is a particular example of a general addition reaction of carbon disulfide and sulfur to enamines and Schiff bases.[128]

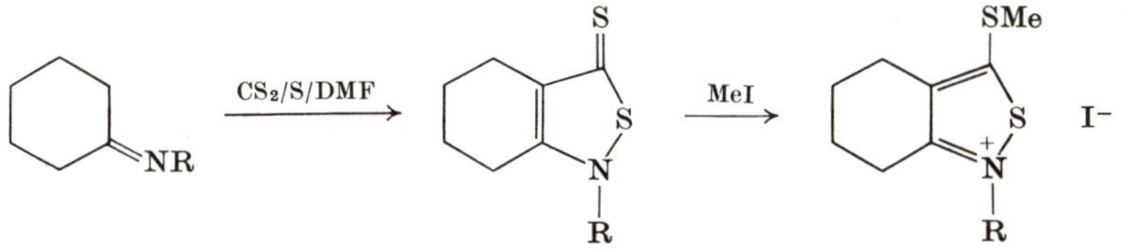

(**98**)   (**99**)   (**100**)

Treatment of the products (**99**) with methyl iodide affords S-methylated salts (**100**).[125]

4. *3-Imino-2,1-benzisothiazolines*

As mentioned earlier in Section III,C,6, 3-amino-2,1-benzisothiazole displays reactions which yield "fixed" products of the tautomeric 3-imino form. Ross has prepared a number of 3-alkylimino and 3-arylimino derivatives.[103]

E. Applications

3-Amino-2,1-benzisothiazoles possess analgesic, antipyretic, tranquilizing, and muscle-relaxant properties.[100–103] The 3-methylamino derivative, in particular, displays potent gastric antisecretory activity.[101, 129] Diazotization of 3-amino-2,1-benzisothiazoles and coupling with tertiary aromatic amines yield a variety of azo dyes especially suitable for disperse dyeing of synthetic polymer fibers.[116, 117, 119–121, 130, 131]

---

[128] R. Mayer and K. Gewald, *Angew. Chem. Int. Ed.* **6**, 294 (1967).
[129] P. Bass, R. A. Purdon, and M. A. Patterson, *J. Pharmacol. Exp. Ther.* **153**, 292 (1966); *Chem. Abstr.* **65**, 9573 (1966).
[130] H. G. Wippel, *Melliand Textilber.* 1090 (1969).
[131] Badische Anilin und Soda-Fabrik A.G., French Patent Application 2,010,017; *Chem. Abstr.* **73**, 99,998 (1970).

IV. Selenium and Tellurium Analogs

1,2-Benzisoselenazolinones were discussed in the earlier review.[1] Very little additional work has been reported; indeed, the sole reference appears to be one in which 1,2-benzisoselenazolinone was prepared by reaction between 2-selenocyanatobenzoic acid and ammonia [Eq. (9)].[132]

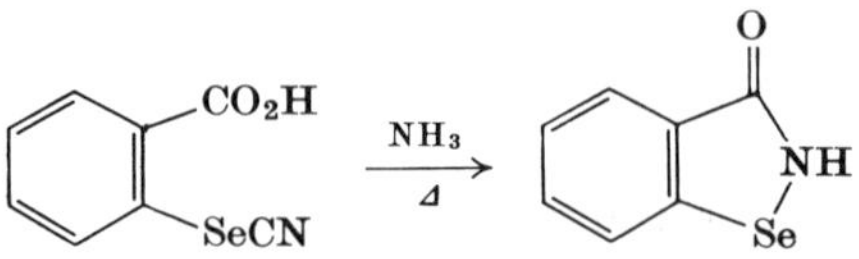

$$\underset{\text{SeCN}}{\text{C}_6\text{H}_4\text{CO}_2\text{H}} \xrightarrow[\Delta]{\text{NH}_3} \text{benzisoselenazolinone} \tag{9}$$

No tellurium analogs have, as yet, been described.

V. Tables

Tables II–XI list 1,2-benzisothiazoles, Tables XII–XV 2,1-benzisothiazoles. There are a number of minor groups in these series which, because they appear in only a small number of references, are comparatively easy for the reader to locate, and so they are not tabulated here. These are 2-arylsulfonyl-[64] and 2-arylideneamino-1,2-benzisothiazoles[58] [(**101**) and (**102**), respectively], certain 2,1-benzisothiazole quaternary azo dyes (**103**),[117] 2,1-benzisothiazolin-3-ones (**97**)[123] and -3-thiones (**95**),[123, 124] 3,3'-bis-2,1-benzisothiazolylidenes (**96**),[123] and 4,5,6,7-tetrahydro derivatives of types **93**,[122] **99**,[125, 127] and **100**.[125, 127]

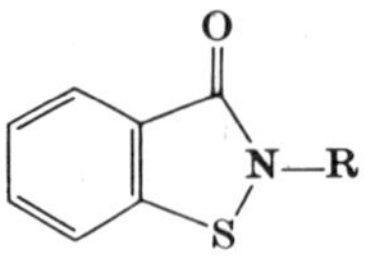

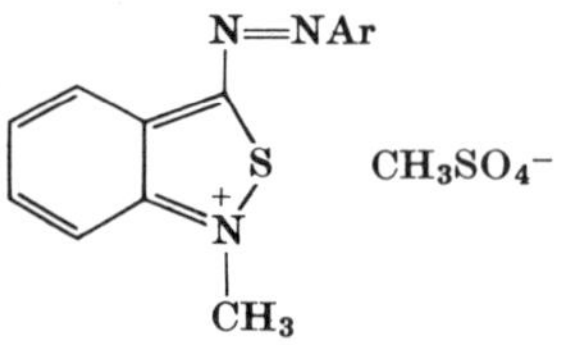

(**101**) R = SO₂Ar
(**102**) R = N=CHAr

(**103**)

[132] G. Wagner and P. Nuhn, *Pharm. Zentralh. Deut.* **104**, 328 (1965); *Chem. Abstr.* **63**, 7093 (1965).

TABLE II: 1,2-BENZISOTHIAZOLES[a]

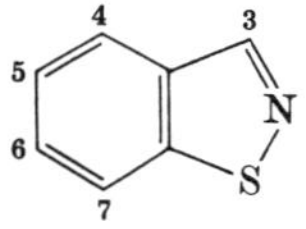

Substituent					M.p. or b.p. (°C)	Ref.
3	4	5	6	7		
—	—	—	—	—	38–39	5, 8, 9
Me	—	—	—	—	116 (HCl salt)	8, 9
—	—	Me	—	—	84	4
Cl	—	—	—	—	40	14, 21, 24, 31
—	Cl	—	—	—	41–42[28]; 45[11]	11, 12, 28
—	—	—	Cl	—	37–38	8
—	—	—	—	Cl	49–50	28
Br	—	—	—	—	59	21
NH_2	—	—	—	—	115	2, 3
—	—	NH_2	—	—	130	1, 29
—	—	Azoaryl[b]	—	—	147	69
—	—	Azoaryl[c]	—	—	147	69
—	—	Azoaryl[d]	—	—	125	69
—	—	Azoaryl[e]	—	—	161–165	71
—	—	—	—	NH_2	131–132	28
—	—	NO_2	—	—	149–150	9, 28, 29
—	—	—	—	NO_2	161–162	28
CH_2Br	—	—	—	—	f	72

(continued)

TABLE II (Continued)

Substituent						M.p. or b.p. (°C)	Ref.
3	4	5	6	7			
Ph	—	—	—	—		71	1, 8, 9
OCH_2CH_2OH	—	—	—	—		46–47	24
OCH_2CH_2NHMe	—	—	—	—		174–175 (HCl salt)	24
OCH_2CH_2NHEt	—	—	—	—		169–171 (HCl salt)	24
OCH_2CH_2NHBu	—	—	—	—		168–170 (HCl salt)	24
$OCH_2CH_2NMe_2$	—	—	—	—		177–178 (HCl salt)	24
$OCH_2CH_2NEt_2$	—	—	—	—		110–112/0.15 mm	24
				Hydrochloride		165–167	24
$OCH_2CH_2NPr_2$	—	—	—	—		198–199 (HCl salt)	24
				Methiodide		139–140	18
$OCH_2CH_2NisoPr_2$	—	—	—	—		168–169 (HCl salt)	24
$OCH_2CH_2NBu_2$	—	—	—	—		108–110 (HCl salt)	19, 24
$OCH_2CH_2pyrr^g$	—	—	—	—		172–173 (HCl salt)	24
$OCH_2CH_2pip^h$	—	—	—	—		218–219 (HCl salt)	24
$OCH_2CH_2morph^i$	—	—	—	—		201–202 (HCl salt)	24
$OCH_2CH_2CH_2NMe_2$	—	—	—	—		215–216 (HCl salt)	24
				Methiodide		201–202	18
$OCH_2CH_2CH_2NEt_2$	—	—	—	—		143–144 (HCl salt)	24
$O(CH_2)_5OC_6H_4NO_2(p)$	—	—	—	—		97–99	94
$OP(=S)(OMe)_2$	—	—	—	—		f	65
$OP(=O)(OEt)_2$	—	—	—	—		120–121/0.001 mm	65
$OP(=S)(OEt)_2$	—	—	—	—		110/0.01 mm	65
$OP(=S)(OEt)Et$	—	—	—	—		f	65

Sec. V.] BENZISOTHIAZOLES

OP(=S)Me₂	—	—	—	106	65
OP(=O)(NMe₂)₂	—	—	—	66–67	65
Me	—	NH₂	—	133–134	8, 9
Me	—	NO₂	—	95–96	8
CH₂Br	—	OMe	OMe	f	72
CH₂Br	—	—	—	f	72
Cl	Cl	—	—	130^{11}; 121^{24}; 126^{31}	11, 24, 31
Cl	—	Cl	—	$190–191^{24}$; 93^{31}	24, 31
Cl	—	—	Cl	$101–102^{24}$; 106.5^{31}	24, 31
Cl	—	—	—	70–72	24
Cl	Me	—	Cl	82–83	24
Cl	—	Me	—	$60–62^{24}$; 65^{31}	24, 31
Cl	—	—	Me	37–38	24
Cl	OMe	Me	Me	$42–43^{24}$; 45^{31}	24, 31
Cl	—	NO₂	—	73.5–74	18
Cl	Cl	—	—	144	31
—	Cl	—	Cl	117	11
—	—	—	NO₂	87	11
OMe	Cl	—	—	149–150	11
OEt	Cl	—	—	112	11
OPr	Cl	—	—	93.5	11
OCHMe₂	Cl	—	—	48–50	11
OBu	Cl	—	—	115	11
OCH₂CHMe₂	Cl	—	—	37–38	11
OCH₂CH₂NEt₂	—	Me	—	84–85	11
OCH₂CH₂NEt₂	—	—	—	213–214 (HCl salt)	24
OCH₂CH₂NEt₂	—	Me	—	143–145 (HCl salt)	24
OCH₂CH₂NEt₂	—	—	Me	178–179 (HCl salt)	24

(*continued*)

TABLE II (Continued)

Substituent					M.p. or b.p. (°C)	Ref.
3	4	5	6	7		
OCH$_2$CH$_2$NEt$_2$	Cl	—	—	—	218–220 (HCl salt)	24
OCH$_2$CH$_2$NEt$_2$	—	Cl	—	—	216–217 (HCl salt)	24
OCH$_2$CH$_2$NEt$_2$	—	—	Cl	—	176–178 (HCl salt)	24
OCH$_2$CH$_2$NEt$_2$	—	—	—	Cl	208–210 (HCl salt)	24
OCH$_2$CH$_2$NEt$_2$	—	OMe	—	—	186 (HCl salt)	18
OCH$_2$CH$_2$NEt$_2$	Me	—	—	—	203–204 (HCl salt)	24
Cl	—	Me	—	Me	82	31
Cl	—	Cl	—	Cl	126	31
Cl	Cl	Cl	—	—	157–158	11
OMe	Cl	Cl	—	—	92–93	11
Cl	Me	—	Me	Me	105	31
Cl	—	Me	Cl	Me	118	31
Cl	Me	Cl	Me	Cl	215	31
Cl	Me	Cl	—	OMe	157	31

[a] An earlier compilation is given in Bambas[1]; compounds included therein are not, in general, repeated here.
[b] —N=N—C$_6$H$_4$N(CH$_2$CH$_2$OH)$_2$(p).
[c] —N=N—C$_6$H$_4$N(Me)CH$_2$CH$_2$OH(p).
[d] —N=N—C$_6$H$_4$H(Et)CH$_2$CH$_2$OH(p).
[f] No data available.
[g] pyrr, 1-Pyrrolidinyl.
[h] pip, 1-Piperidinyl.
[i] morph, 1-Morpholinyl.
[e] —N=N-naphthyl(1,4)-NHCH$_2$CH$_2$N⟨pyrrolidine⟩

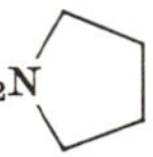

TABLE III: 3-AMINO-1,2-BENZISOTHIAZOLES

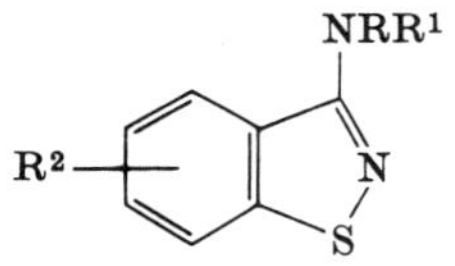

	Substituent			M.p. or b.p. (°C)	Ref.
R	R¹	R²			
H	H	—		115	2, 3
Me	H	—		93	14
Et	H	Hydrochloride		220	14
				78	14
Et	H	Hydrochloride		171	14
Pr	H	—		111 (HCl salt)	14
IsoPr	H	—		114 (HCl salt)	14
Bu	H	—		212 (HCl salt)	14
Isopentyl	H	—		216 (HCl salt)	14
Ph	H	—		99	14
Me	Me	—		76/0.01 mm	16
Et	Me	—		67–73/0.01 mm	16
Et	Et	—		110/0.3 mm	16
Et	CH₂Ph	—		165/0.01 mm	16
Et	Cyclohexyl	Perchlorate		156	16
Allyl	Allyl	—		151/0.3 mm	16
Pr	Pr	—		125/0.01 mm	16
Bu	Bu	—		108/0.01 mm	16
				128/0.01 mm	16

(continued)

TABLE III (Continued)

Substituent R	R¹	R²	M.p. or b.p. (°C)	Ref.
IsoBu	IsoBu	—	108–110/0.01 mm	16
2-Ethylhexyl	2-Ethylhexyl	—	124–128/0.01 mm	16
CH_2CH_2OH	H		132–133	24
		Hydrochloride	215–216	24
CH_2CH_2OH	Me	—	191–192	24
$CH_2CH_2CH_2NMe_2$	H		146–150/0.1 mm	22
		Dihydrochloride	227–229	22
$CH_2CH_2NEt_2$	H		152–155/0.1 mm	22
		Dihydrochloride	118–119	22
		Oxalate	154–155	22
$CH_2CH_2N\!\bigcirc\!\!O$ (morpholinyl) 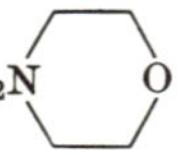	H	—	102–104	22
H	H	4-Cl	162	11
Me	H	4-Cl	128	11
Me	Me	4-Cl	120–121	11
CHO	H	4-Cl	135	11
Cyclohexyl	H	4-Cl	119	11
$CH_2CH_2CH_2NMe_2$	H	4-Cl	53–54	11
H	H	4-Cl, 5-Cl	173	11
$CH_2CH_2CH_2NMe_2$	H	4-Cl, 5-Cl	60	11
NRR^1 = 1-Morpholinyl	—		105/0.01 mm	16
NRR^1 = 1-Piperidinyl	—	—	140–142/0.5 mm	16
NRR^1 = Hexamethylenimino	—		165/0.01 mm	16
NRR^1 = 1-Morpholinyl	—	4-Cl	121	11

TABLE IV

3-Chloro-1,2-benzisothiazolium Salts

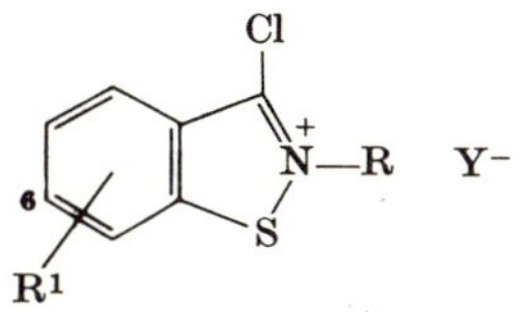

	Substituent			
R	R¹	Y	M.p. (°C)	Ref.
Me	—	Cl	164 (dec.)[14]; 116–119[20]	14, 20
Me	—	ClO$_4$	159–160	20
Me	—	ReO$_4$	134–135 (dec.)	20
Et	—	Cl	194 (dec.)	14
Et	—	BF$_4$	139	21
(3-Bromo derivative)	—	BF$_4$	165–172	21
Pr	—	Cl	171 (dec.)	14
IsoPr	—	Cl	152 (dec.)	14
Allyl	—	Cl	175 (dec.)	14
Bu	—	Cl	162 (dec.)	14
Isopentyl	—	Cl	162 (dec.)	14
CH$_2$CH$_2$Cl	—	Cl	174 (dec.)	14
Ph	—	Cl	202[14]; 239–240[20]	14, 20
Ph	—	ClO$_4$	161	20
p-Chlorophenyl	—	Cl	179	14
p-Methoxyphenyl	—	Cl	208–209	20
p-Methoxyphenyl	—	ClO$_4$	146–147 (dec.)	20
Et	6-OMe	Cl	167 (dec.)	14
Et	6-Cl	Cl	153 (dec.)	14

TABLE V

N-OXIDES AND SULFOXIDES OF 1,2-BENZISOTHIAZOLINES AND RELATED COMPOUNDS

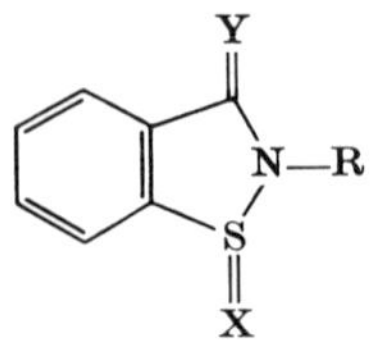

Substituent				
X	R	Y	M.p. (°C)	Ref.
NH	H	O	269–272	42
NH	H	NH	198–200	41, 42
O	H	O	159	1, 90
O	H	NMe	186	26
O	H	NEt	196	26
O	H	NPr	116	26
O	Me	NCl	122	26
O	Et	NCl	90–91	26
O	Me	NH	206 (HCl salt)	26
O	Et	NH	166 (HCl salt)	26
—	OH	NMe	157 (HCl salt)	26
—	OH	NEt	176 (HCl salt)	26
—	OH	NPr	147 (HCl salt)	26
—	Et	NOH	155 (HCl salt)	26

TABLE VI

REDUCED 1,2-BENZISOTHIAZOLES

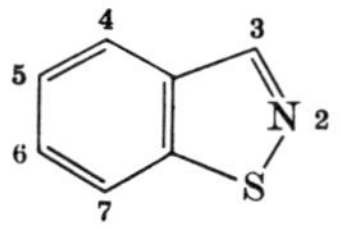

Position of extra hydrogen atoms	Substituent				M.p. (°C)	Ref.
	2	3	4	5		
2, 3	Me	—	—	Me	53	4
				HgCl$_2$ adduct	200–210 (dec.)	4
				Picrate	160–175 (dec.)	4
2, 3	SO$_2$Me	—	Cl	—	137.5–139.5	33
4, 5, 6, 7	—	SMe	—	—	23	45

TABLE VII

1,2-BENZISOTHIAZOLINE-3-THIONES

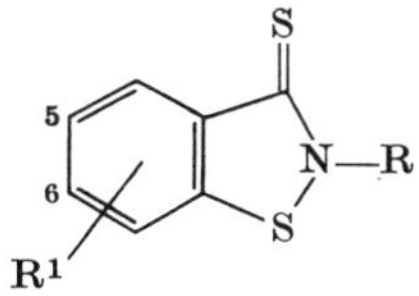

Substituent		M.p. or b.p. (°C)	Ref.
R	R^1		
Me	—	138	32, 36, 37, 39
Et	—	66[32]; 63–64[36]	32, 36
Pr	—	76[32]; 72–73[36, 37]	32, 36, 37
IsoPr	—	63[32]; 61–62[36, 37]	32, 36, 37
Bu	—	70[32]; 67[36, 37]	32, 36, 37
IsoBu	—	71–72	36, 37
Pentyl	—	74–75	36, 37
Isopentyl	—	100	36, 37
Hexyl	—	95	36, 37
Cyclohexyl	—	189–190	36, 37

(*continued*)

TABLE VII (Continued)

Substituent			
R	R¹	M.p. or b.p. (°C)	Ref.
Heptyl	—	77	36, 37
Decyl	—	49	36, 37
Dodecyl	—	60	36
Hexadecyl	—	80–81	36
CH_2CH_2OH	—	110	36, 37
$CH_2CH_2CH_2OH$	—	74–75	36
$CH_2CH_2CH_2OMe$	—	70	36, 37
Ph	—	77	36
NH_2	—	125	36
CH_2Ph	—	122–123	32, 36, 37
CH_2CH_2Ph	—	137–138	36
$CH_2C_6H_4CH_3(p)$	—	138–139	36
$CH_2C_6H_4Cl(p)$	—	137–139	36
$CH_2C_6H_3Cl_2(2,4)$	—	123–125	36
3-Picolyl	—	129–130	36
$CH_2CH_2SC_6Cl_5$	—	150–151	36
Et	5-Cl	139	32
Et	6-OMe	142	32
Pr	5-Me	50/0.005 mm	37
Bu	5-Me	a	36
Bu	5-Cl	77–78	36–38
Bu	5-Br	88	36, 37
Bu	5-NO_2	118–120	36, 37
Bu	6-Cl	123.5–125.5	38
$CH_2CH_2NEt_2$	6-Cl	88–90	38
Cyclohexyl	5-Me	159–160	36, 37
Cyclohexyl	5-Cl	154–155	36, 37
Cyclohexyl	5-Br	163	36, 37
Cyclohexyl	5-NO_2	225–226	36, 37
CH_2Ph	5-Me	125	36
CH_2Ph	5-Cl	123–124	36
CH_2Ph	5-Br	127	36
CH_2Ph	5-NO_2	145–147	36
CH_2CH_2Ph	5-Me	109–110	36
CH_2CH_2Ph	5-Cl	142–143	36
CH_2CH_2Ph	5-Br	155	36
CH_2CH_2Ph	5-NO_2	124–126	36

[a] No data available.

TABLE VIII

3-Imino-1,2-benzisothiazolines

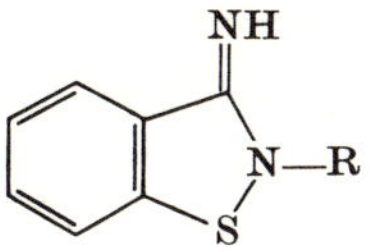

R	Salt	M.p. (°C)	Ref.
Me	HCl	220[15]; 232[30]	15, 30
Et	HCl	171[15]; 169[30]	15, 30
Pr	HCl	111	15
Bu	HCl	155–157, then 212	15
IsoBu	HCl	187, then 216	15
CH_2CH_2Cl	HCl	152[15]; 165–166 (dec.)[30]	15, 30
Cyclohexyl	HCl	166	15
4-Methylcyclohexyl	HCl	160	15
3-Methylcyclohexyl	HCl	197	15
2-Methylcyclohexyl	HCl	210	15
Ph	Free base	99	15
p-Tolyl	Free base	144	15
CH_2Ph	Free base	85	15

TABLE IX

2-ALKYL- AND 2-ARYL-SUBSTITUTED 1,2-BENZISOTHIAZOLINONES

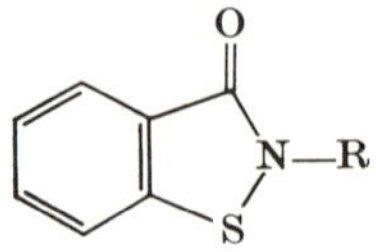

R	M.p. or b.p. (°C)	Ref.
H	158	1, 47, 56, 60
Me	54	1, 14, 17, 56
Et	118–120	1, 47
Allyl	154–157/0.2 mm	1, 83
Pr	170–172/0.8 mm	47, 83
Bu	115/0.15 mm	56
	144–147/0.1 mm	83
t-Bu	145/0.5 mm	59
CH_2CH_2Cl	92–93	56
Pentyl	152–156/0.2 mm	83
Dodecyl	Waxy solid	47, 59
Cetyl	Waxy solid	59
Hexadecyl	45–47	46, 47
Heptadecyl	Waxy solid	59
Octadecyl	Waxy solid	59
CH_2CH_2OH	104–106	83
CH_2CH_2OEt	164–167/0.2 mm	83
$C(Me)_2CH_2OH$	84–85	83
$CMe(CH_2OH)_2$	136	83
$CH_2CH_2CH_2OH$	74–75	83
$CH_2CH(Me)OH$	190–194/0.1 mm	83
$CH_2CH_2CH_2OC_8H_{17}$	190–194/0.1 mm	83
$CH_2CH_2NEt_2$	128–130/0.08 mm	24
Hydrochloride	179–180	24, 49
$CH_2CH_2NHC_6H_{11}$	216–218 (HCl salt)	49
$CH_2CH_2pyrr^a$	87–88	49
$CH_2CH_2pipMe^b$	268–269 (Di-HCl salt)	49
$CH_2CH_2NHCH_2CH_2CH_2OMe$	126–127 (HCl salt)	49
CH_2CH_2NHBu	168–169 (HCl salt)	49
$CH_2CH_2pip^c$	215–216 (HCl salt)	49
$CH_2CH_2morph^d$	225–227 (HCl salt)	49
$CH_2CH_2CH_2NHMe$	133–134 (HCl salt)	49
$CH_2CH_2CH_2NHCH_2CH=CH_2$	159–160 (HCl salt)	49
$CH_2CH_2CH_2NHCHMeEt$	194–196 (HCl salt)	49
$CH_2CH_2CH_2pyrr^a$	153–155 (HCl salt)	49

TABLE IX (Continued)

R	M.p. or b.p. (°C)	Ref.
$CH_2CH_2CH_2NHCH_2CH_2OH$	163–164 (Oxalate salt)	49
$CH_2CH_2CH_2NH_2$	204–205 (HCl salt)	49
$CH_2CH_2CH_2NMe_2$	114–115 (Maleate salt)	49
$CH_2CH_2CH_2CH_2CH_2OC_6H_4NO_2(p)$	109–111	94
$CH_2CH_2CH_2CH_2CH_2OC_6H_4NH_2(p)$	117–119	93, 94
$CH_2CH_2CH_2CH_2CH_2OC_6H_4NHAc(p)$	169–170	93
$CH_2CH_2CH_2OC_{12}H_{25}$	210–215/0.1 mm	83
$CH_2CH_2NH_2$	130	83
CH_2CO_2Et	89	83
$CH_2CH_2CH_2N\!\!\diagup\!\!\diagdown\!O$ (morpholino)	82	83
Cyclohexyl	89–90	56, 57, 83
Ph	140–141[56]; 145[83]	56, 83
o-Tolyl	124	83
m-Tolyl	126	57, 83
p-Tolyl	137	56, 83
o-Chlorophenyl	135–136[59]; 128–130[83]	59, 83
m-Chlorophenyl	140–142	56, 57, 59, 83
p-Chlorophenyl	128	57, 59, 83
p-Bromophenyl	133–135	59
o-Methoxyphenyl	136	59, 83
m-Methoxyphenyl	116–118	59
p-Methoxyphenyl	148–149	57, 59, 83
o-Ethoxyphenyl	104	83
p-Ethoxyphenyl	140	59, 83
2,5-Dimethoxyphenyl	173–174	59
p-Butoxyphenyl	90–91	59
o-Hydroxyphenyl	196–198	83
p-Hydroxyphenyl	247–251	82
$p\text{-}C_6H_4CO_2Me$	187–189	83
$o\text{-}C_6H_4CO_2Et$	152	57
$p\text{-}C_6H_4CO_2Et$	138–139	57
$o\text{-}C_6H_4CO_2H$	198–200	83
2-Hydroxy-6-carboxyphenyl	193–195	83
3-Hydroxy-4-carboxyphenyl	191–193	83
$p\text{-}C_6H_4COMe$	186–188	83
$p\text{-}C_6H_4SO_3Na$	> 300	83
$p\text{-}C_6H_4SO_2NH_2$	289–291	83
o-Nitrophenyl	238–240	83
p-Nitrophenyl	242	83

(continued)

TABLE IX (Continued)

R	M.p. or b.p. (°C)	Ref.
o-C$_6$H$_4$SH	146	83
p-Fluorophenyl	110	83
3,4-Dichlorophenyl	188	59, 83
2,4-Dichlorophenyl	142	83
2,5-Dichlorophenyl	163–164	59
2,3-Dimethylphenyl	171	83
2,4-Dimethylphenyl	138–139	59
2,5-Dimethylphenyl	117–118	59
2,6-Dimethylphenyl	150	59, 83
3,4-Dimethylphenyl	161–162[59]; 151–154[83]	59, 83
p-Butylphenyl	60–61	59
2,6-Diethylphenyl	118–119	59
p-Dodecylphenyl	85–86	59
Benzyl	88[56]; 83–85[83]	56, 83
CH$_2$CH$_2$Ph	97	56, 57, 83
2-(3,4-Dimethoxyphenyl)ethyl	102–103	83
2-(3,4-Methylenedioxyphenyl)ethyl	127	83
2-Chlorobenzyl	107–108	59
4-Chlorobenzyl	97–98[59]; 86–89[83]	59, 83
2,6-Dichlorobenzyl	163–164	59
2,4,5-Trichlorobenzyl	175–176	59
p-C$_6$H$_4$NH$_2$	230–231	59
p-C$_6$H$_4$NHAc	249–250	59
p-C$_6$H$_4$NMe$_2$	188–189	59
2-Pyridyl	198	59, 83
4-Pyridyl	183	59, 83
2-Chloro-3-pyridyl	155	83
5-Bromo-2-pyridyl	232–233	59
2-Pyrimidyl	236	59, 83
4,6-Dimethyl-2-pyrimidyl	252–253[59]; 258–266[83]	59, 83
2-Thiazolinyl	247	83
2-Benzothiazolyl	e	46, 47
2-Naphthyl	165	57

[a] pyrr, 1-Pyrrolidinyl.
[b] pipMe, 1-(4-Methyl)piperazinyl.
[c] pip, 1-Piperidinyl.
[d] morph, 1-Morpholinyl.
[e] No data available.

TABLE X

2-ACYL-1,2-BENZISOTHIAZOLINONES

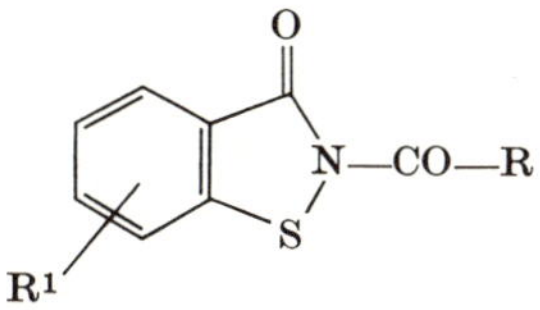

Substituent			
R	R^1	M.p. (°C)	Ref.
CHCl$_2$	—	157	62[a]
Pr	—	115	62
Pentyl	—	94–95	62
CHMeEt	—	104	62
CHEt$_2$	—	80–81	62
Heptyl	—	85–84	62
Pentadecyl	—	93–94	62
Heptadecyl	—	95–96	62
CH(Et)Ph	—	123	62
o-Chlorophenyl	—	125–126	62
m-Chlorophenyl	—	133–134	62
p-Chlorophenyl	—	205–206	62
m-Tolyl	—	137–138	62
p-Tolyl	—	196–197	62
p-Methoxyphenyl	—	155–156	62
p-Nitrophenyl	—	209–210	62
2,4-Dichlorophenyl	—	123–124	62
Me	5-NO$_2$	190–192	80
Me	6-NO$_2$	220–222	51
ClCH$_2$	5-NO$_2$	165–167	80
Et	5-NO$_2$	179–180	80
Et	6-NO$_2$	185–186	51
Pentyl	5-NO$_2$	108–109	80
CHEt$_2$	5-NO$_2$	107	80
Hexyl	5-NO$_2$	113	80
Ph	5-NO$_2$	180	80
Ph	6-NO$_2$	154–156	51
CH$_2$CH$_2$Ph	5-NO$_2$	138–140	80
p-Chlorophenyl	5-NO$_2$	233–234	80
p-Nitrophenyl	5-NO$_2$	>245 (dec.)	80

[a] Data given in Gialdi et al.[62] are duplicated in Baruffini et al.[63,86]

TABLE XI

1,2-Benzisothiazolinones Substituted in the Benzene Ring

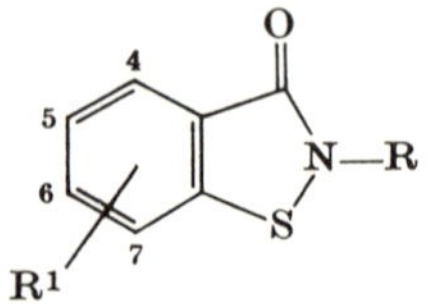

Substituent		M.p. or b.p. (°C)	Ref.
R	R^1		
H	4-Me	227–229	52
H	5-Me	197–199	52, 56, 83
H	6-Me	212–213	52, 83
H	7-Me	190–191	89
H	4-Cl	256–258	56
H	5-Cl	265–266	1, 54, 56, 89
H	6-Cl	271–273	48, 54, 56, 83, 88
H	4-OMe	172–173	55
H	5-OMe	189	55, 56, 83
H	6-OMe	208–212	55
H	7-OMe	190–191	55
H	5-NO$_2$	> 280 (dec.)	50, 56, 83, 89
H	6-NO$_2$	298–300	51, 83, 89
H	5-Bu	90	53
H	5-sec-Bu	79–80	53
H	5-t-Bu	171–172	53
H	6-CO$_2$H	280 (dec.)	83
H	6-CONH$_2$	~ 300 (dec.)	83
H	5-Me, 7-Me	185–188	83
H	5-OMe, 6-OMe	265	83
H	6-OMe, 7-OMe	183	83
H	5-Cl, 7-Cl	219	83
H	5-Cl, 6-Cl	272	83
H	5-SO$_2$NH$_2$, 6-Me	~ 290 (dec.)	83
Me	5-Bu	87–88	53
Me	5-NO$_2$	229–230	50
Et	5-Bu	142/0.13 mm	53
CH$_2$CH$_2$OH	6-Cl	140–142	83
Pr	6-Cl	76	83
Pr	5-Bu	180/0.9 mm	53
IsoPr	5-Bu	132/0.16 mm	53

TABLE XI (Continued)

Substituent			
R	R¹	M.p. or b.p. (°C)	Ref.
Allyl	6-Cl	102	83
Bu	4-Me	112/0.10 mm	52
Bu	5-Me	125/0.10 mm	52
Bu	6-Me	42–43	52
Bu	7-Me	117/0.09 mm	52
Bu	4-OMe	145/0.4 mm	55
Bu	5-OMe	145–147/0.15 mm	55
Bu	6-OMe	50–51	55
Bu	7-OMe	140/0.2 mm	55
Bu	4-Cl	82–83	54
Bu	5-Cl	76	54
Bu	6-Cl	102–103	54, 56
Bu	7-Cl	142–144/0.2 mm	54
Bu	5-NO$_2$	79	50, 56
Bu	6-NO$_2$	109–110	51
Bu	5-Bu	171/0.18 mm	53
Bu	5-sec-Bu	148/0.22 mm	53
Bu	5-t-Bu	150/0.25 mm	53
Pentyl	5-Bu	161/0.35 mm	53
Isopentyl	5-Bu	160/0.5 mm	53
Hexyl	5-Bu	173/0.15 mm	53
Heptyl	5-Bu	173/0.15 mm	53
Ph	5-NO$_2$	228	50
Ph	6-NO$_2$	229–231 (dec.)	51
p-C$_6$H$_4$Cl	5-NO$_2$	215–217	50
CH$_2$Ph	5-NO$_2$	142	50
CH$_2$Ph	6-Cl	140–142	83
CH$_2$Ph	6-Me	66–67	83
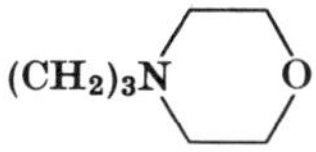 (CH$_2$)$_3$N⟨morpholine⟩	6-Cl	94–95	83
(NHAc[a]	—	145–147	58)
(SCCl$_3$[a]	—	95–97	85, 89)
SCCl$_3$	5-Cl	132–134	85, 89
SCCl$_3$	5-NO$_2$	154–156	85, 89
SCCl$_3$	6-NO$_2$	164–167	85, 89
SCCl$_3$	7-Me	130–132	89

[a] These are not "substituted in the benzene ring" but are placed here for convenience.

TABLE XII: 2,1-BENZISOTHIAZOLES

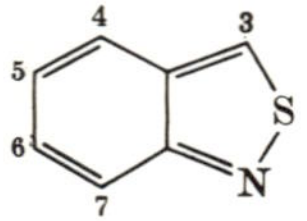

Substituent					M.p. or b.p. (°C)	Ref.
3	4	5	6	7		
—	—	—	—	—	242–242.5/748 mm	96, 97
—	—	—	—	—	116–118/18 mm	4
—	—	—	—	Picrate	70/0.5 mm	105
—	—	—	—	Picrate	123–124	96
Me	—	—	—	—	122–125 (dec.)	4
—	Me	—	—	—	55	107a
—	—	—	—	Picrate	116/2 mm	106
—	—	—	—	Me	135	106
—	—	—	—	Picrate	106/1 mm	106
Cl	—	—	—	—	98	106
—	Cl	—	—	—	ca. −10	111
—	—	Cl	—	—	41–42	106
—	—	—	Cl	—	99	106
—	—	—	—	Picrate	72	106
—	—	Br	—	—	71	106
Br	—	—	—	—	64–65	111
—	—	—	Br	—	85	106
—	—	—	—	—	81	106
—	—	—	—	Br	70–71.5	110
Ph	—	—	—	—	56	107a
CN	—	—	—	—	104–105	111
CO_2H	—	—	—	—	212 (dec.)	111
I	—	—	—	—	114–115	111
—	I	—	—	Br	144–145	110

Sec. V.]   BENZISOTHIAZOLES   95

3-Substituent	4	5	6	7	Derivative	m.p. (°C)	Refs.
—	—	—	—	Br	Br	216–217	110
NH₂	—	—	—	—	—	178–179	100, 101
NH₂	—	—	—	—	Hydrochloride	169.5–171.5	103
NH₂	—	—	—	—	Methiodide	112–113	101
NHMe	—	—	—	—	—	>200 (dec.)	100, 101
NHMe	—	—	—	—	Hydrochloride	206–207	101
NHMe	—	—	—	—	Methiodide[a]	>200	101
NHEt	—	—	—	—	—	211	101
NHPr	—	—	—	—	—	195–196	100, 101
NHisoPr	—	—	—	—	—	187–188	100, 101
NHisoBu	—	—	—	—	—	188–189	100, 101
NH-sec-Bu	—	—	—	—	—	179–181	100, 101
NMe₂	—	—	—	—	—	206–207	100, 101
3-(1-Pyrrolidinyl)	—	—	—	—	—	120–121	100, 101
NHEt	—	—	—	—	Hydrochloride	144–145	100, 101
NHEt	—	—	Cl	—	—	210–215	100, 101
NHEt	—	—	Me	—	—	231–232	100
NHEt	—	—	MeO	—	—	225–226	100, 101
NMe₂	—	—	Cl	—	—	181–182	100, 101
NMe₂	—	Cl	—	—	—	134–135	100, 101
NH₂	—	—	—	—	Methiodide[b]	254–260	100
NH₂	—	NH₂	—	—	—	140–141	101
NH₂	—	—	—	—	Dihydrochloride	125–127	100
NH₂	—	—	—	—	—	>300	102
NH₂	Cl	—	—	—	—	c	102
NH₂	—	Cl	—	—	—	191–192	103
NH₂	F	—	—	—	—	c	101
NH₂	Br	—	—	—	—	c	103
NH₂	NO₂	—	—	—	—	>300	103
NH₂	NO₂	—	—	—	—	>230 (dec.)	102
NHCH₂Ph	—	—	—	—	—	181–182	117
NHPh	—	—	—	—	—	>240	101

(continued)

TABLE XII (Continued)

3	4	5	6	7	M.p. or b.p. (°C)	Ref.
NHMe	—	Br	—	—	244–249	101
NHMe	—	—	Cl	—	250–251	101
3-(1-Morpholinyl)	—	—	Cl	—	125–126	101
NHEt	—	OMe	OMe	—	187–188	101
NHEt	—	—	—	Hydrochloride	240–242	101
NHEt	—	OMe	—	OMe	184–185	101
—	—	OMe	—	—	55	114
—	—	—	—	Picrate	152	114
NH_2	Cl	OMe	—	—	122–123	114
NH_2	—	CF_3	CF_3	—	202–203	103
NH_2	—	Br	—	Br	164.5–166	103
NH_2	—	NO_2	—	Br	207–209	102
NO_2	NO_2	—	—	—	259–260	102
—	—	NO_2	—	—	126–126.5	111
—	—	—	NO_2	—	92	110
—	—	NO_2	—	—	180	110
—	—	—	NO_2	NO_2	149	106
—	NO_2	NO_2	—	NO_2	149[d]	110
—	Me	—	—	—	220–221	110
—	NO_2	Cl	NO_2	NO_2	198–198.5	110
—	—	—	—	—	115	110
—	NO_2	—	Br	NO_2	140–140.5	110
—	—	—	—	—	150–151	110
—	—	—	—	Me	134–135	110

[a] Ring nitrogen quaternary salt.
[b] Exocyclic nitrogen quaternary salt.
[c] No data available.
[d] Earlier reported melting point of 127° is in error.

TABLE XIII

QUATERNARY SALTS OF 2,1-BENZISOTHIAZOLES
(OTHER THAN AZO DYES)

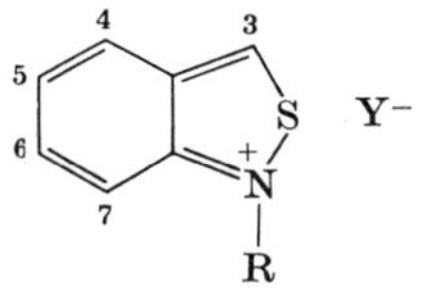

	Substituent							
R	3	4	5	6	7	Y	M.p. (°C)	Ref.
Me	NH_2	—	—	—	—	I	> 200 (dec.)	100
Me	NHMe	—	—	—	—	I	210 (dec.)	100
Me	—	—	—	—	—	I	177–178	114
Me	—	—	—	—	—	Picrate	165	114
CH_2Ph	—	—	—	—	—	Br	199–200	114

TABLE XIV

2,1-BENZISOTHIAZOLE AZO DYES

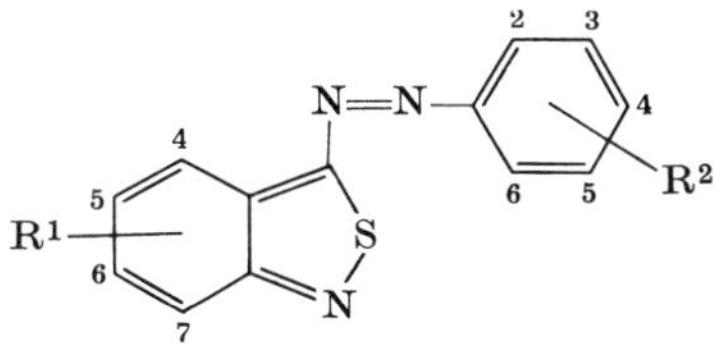

Substituent		λ max (nm)	
R^1	R^2	or color	Ref.
—	4-NMe_2	548	130
4-Cl	4-NMe_2	566	130
5-Cl	4-NMe_2	557	130
6-Cl	4-NMe_2	556	130
7-Cl	4-NMe_2	562	130
5-NO_2	4-NMe_2	598	130
5-NO_2	4-NMePh	Blue-violet	120

(*continued*)

TABLE XIV (Continued)

Substituent			
R^1	R^2	λ max (nm) or color	Ref.
—	4-N(Et)CH_2CH_2OH	Ruby red	116
5-NO_2	2-Me, 4-N$(CH_2CH_2OH)_2$	Gray-blue	116
5-Br	2-Me, 4-N$(CH_2CH_2OH)_2$	Blue	116
5-NO_2	2-Me, 4-N$(CH_2CH_2OH)CH_2CH_2Ph$	Blue	121
5-NO_2	2-Me, 4-N$(CH_2CH_2OAc)CH_2CH_2Ph$	Blue	121
5-Br	2-Me, 4-N$(CH_2CH_2OAc)CH_2CH_2Ph$	Red-violet	121
5-NO_2	2-Me, 4-N(Et)$CH_2CH_2NMe_3^+$, H_2PO_4	Navy blue	119
5-NO_2	2-Me, 4-N(Et)$CH_2CH_2NMe_3^+$, HSO_4	—	119
5-NO_2, 7-Cl	2-Me, 4-N(Et)$CH_2CH_2N(Me)_2$ CH_2Ph^+, HSO_4	Blue	119
6-Cl	4-N(Et)$CH_2CH_2NMe_3^+$, H_2PO_4	Red-violet	119
—	4-N(Et)$CH_2CH_2NMe_3^+$, H_2PO_4	Ruby red	119
5-NO_2	2-$NHCOCH_2NMe_3^+$, 4-NMe_2, H_2PO_4	Greenish blue	119
5-NO_2, 7-Br	2-$NHCOCH_2NMe_3^+$, 4-NMe_2, 5-OMe, OAc	Bluish green	119

TABLE XV

3-IMINO-2,1-BENZISOTHIAZOLINES

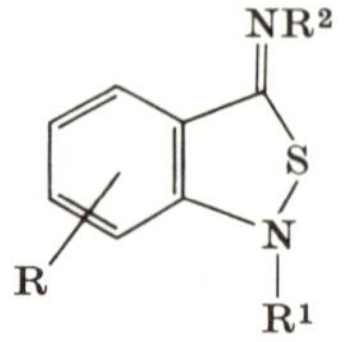

Substituent				
R	R^1	R^2	M.p. (°C)	Ref.
—	Ac	Ac	158–159	111
—	COPh	COPh	239–240	111
5-CF_3	Ph	H	161–163 (HI salt)	103
5-CF_3	Et	H	[a]	103
5-CF_3	Me	Me	[a]	103

[a] No data available.

Recent Advances in Pyrazine Chemistry

G. W. H. CHEESEMAN

Chemistry Department, Queen Elizabeth College, University of London, London, England

AND

E. S. G. WERSTIUK

Biochemistry Department, McMaster University, Hamilton, Ontario, Canada

I.	Introduction	99
II.	Physical and Spectroscopic Properties	105
III.	General Synthetic Methods	112
	A. From α,β-Dicarbonyl Compounds	113
	B. From α-Aminocarbonyl Compounds	114
	C. From α-Halogenoketones	115
	D. From Piperazines	117
	E. From Quinoxalines and Pteridines	119
	F. From Ring Transformation Reactions	120
	G. Miscellaneous Syntheses	121
IV.	General Chemical Properties	122
V.	Substituted Pyrazines	127
	A. Alkyl- and Arylpyrazines	127
	B. Pyrazinecarboxylic Acids and Their Derivatives	139
	C. Halopyrazines	153
	D. Aminopyrazines	165
	E. Hydroxypyrazines (Pyrazinones)	172
VI.	Reduced Pyrazines	182
VII.	Pyrazine N-Oxides	192
VIII.	Biological Activity	208

I. Introduction

The aim of the present review is to survey developments in the field of pyrazine chemistry since the appearance of the last major reviews by Pratt,[1] Ramage and Landquist,[2] and Novacek et al.[3] No attempt

---

[1] Y. T. Pratt, in "Heterocyclic Compounds" (R. C. Elderfield, ed.), Vol. 6, p. 377. Wiley, New York, 1957.

[2] G. R. Ramage and J. K. Landquist, in "Chemistry of Carbon Compounds" (E. H. Rodd, ed.), Vol. 4B, p. 1318. Elsevier, Amsterdam, 1959.

[3] L. Novacek, K. Palat, and M. Celadnik, *Chem. Listy* **57**, 298 (1963).

will be made to review the extensive chemistry of piperazines (hexahydropyrazines). Pyrazine, or 1,4-diazine (1), is a highly symmetrical molecule; only one monosubstituted derivative is possible and only one trisubstituted pyrazine, if a common substituent is involved. Nevertheless, monosubstituted pyrazines are frequently designated 2-substituted, and trisubstituted pyrazines 2,3,5-trisubstituted pyrazines. The terms pyrazyl and pyrazinyl are used synonymously in the literature for the $C_4H_3N_2$ radical (2).

Pyrazines occur naturally, although not in appreciable quantities. Fusel oil contains 2,5-dimethyl-, 2,5-diethyl-, trimethyl-, and tetramethylpyrazine. The last compound has been isolated from cultures of *Bacillus natto* and *B. subtilis* and from fermented soybeans.[4-6] Emimycin (pyrazine-2-one-4-oxide), a powerful antibiotic, has been obtained from *Streptomyces* No. 2020-I.[7,8]

The aspergillic acids are an important group of naturally occurring pyrazine derivatives. They are formed by cultures of the *Aspergillus* species and were the first natural products shown to be cyclic hydroxamic acids.[9] Aspergillic acid itself was isolated[10] in 1943 from *A. flavus* and subsequently shown to have structure 3. An isomer of deoxyaspergillic acid, also isolated from *A. flavus*, was shown to be 3,6-diisobutyl-2-pyrazinone (4) and named flavacol.[11]

Other antibiotics structurally analogous to aspergillic acid are

[4] T. Kosuge, T. Adachi, and H. Kamiya, *Nature (London)* **195**, 1103 (1962).
[5] T. Kosuge, H. Kamiya, and T. Adachi, *J. Pharm. Soc. Jap.* **82**, 190 (1962); *Chem. Abstr.* **56**, 15969 (1962).
[6] H. Kamiya, T. Adachi, and T. Kosuge, *J. Pharm. Soc. Jap.* **84**, 448 (1964); *Chem. Abstr.* **61**, 4729 (1964).
[7] M. Terao and K. Karasawa, *J. Antibiot. Ser. A* **13**, 401 (1960); *Chem. Abstr.* **56**, 5215 (1962).
[8] M. Terao, *J. Antibiot. Ser. A* **16**, 182 (1963); *Chem. Abstr.* **60**, 5495 (1964).
[9] J. B. Bapat, D.StC. Black, and R. F. C. Brown, *Advan. Heterocycl. Chem.* **10**, 199 (1969).
[10] E. C. White and J. H. Hill, *J. Bacteriol.* **45**, 433 (1943).
[11] G. Dunn, G. T. Newbold, and F. S. Spring, *J. Chem. Soc.* 2586 (1949).

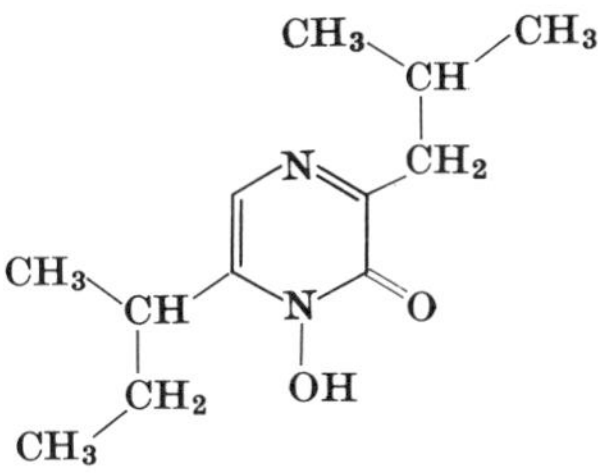

PYRAZINE CHEMISTRY

(3)

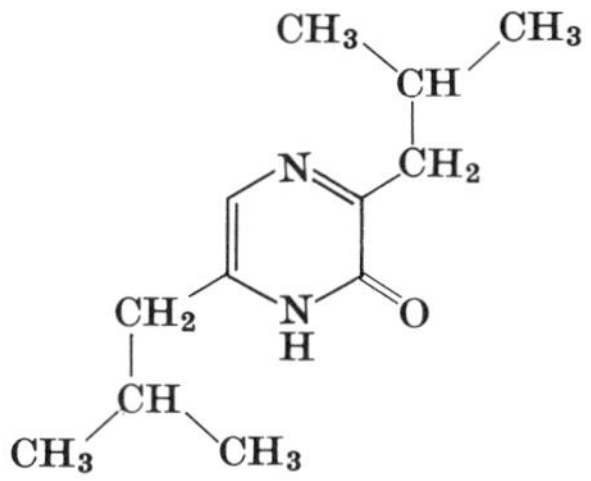

(4)

hydroxyaspergillic acid (5)[12,13] isolated from *A. flavus*, mutaaspergillic acid (6)[14] from *A. oryzae*, neohydroxyaspergillic acid (7)[15] and neoaspergillic acid (8)[16,17] from *A. sclerotiorum*.

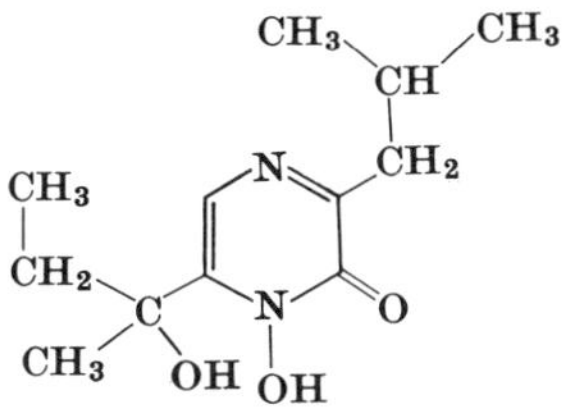

(5)

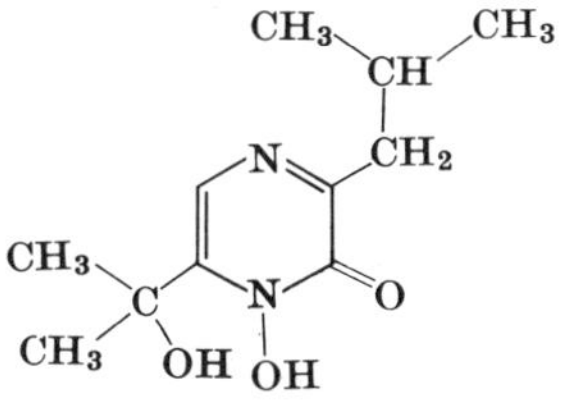

(6)

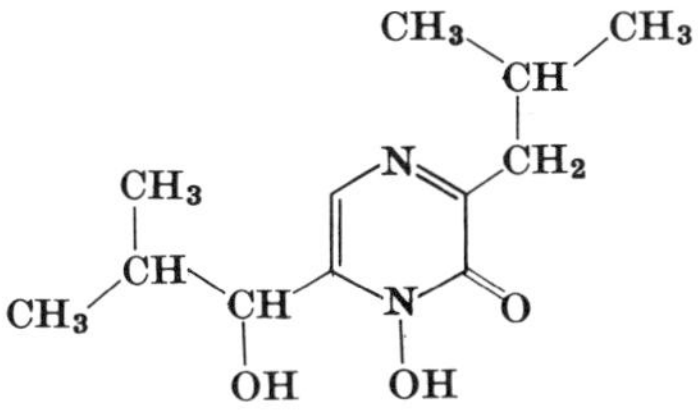

(7)

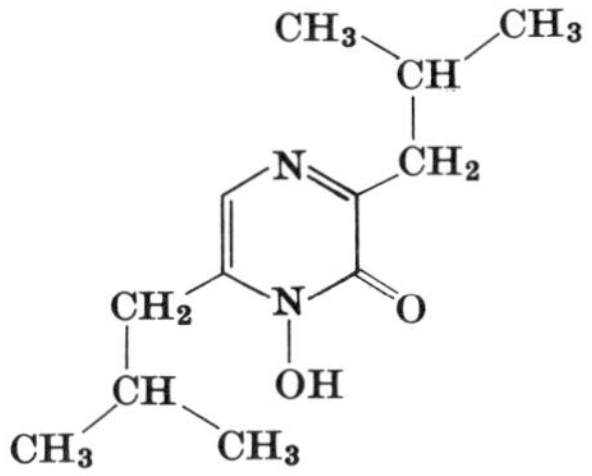

(8)

[12] A. E. O. Menzel, O. Wintersteiner, and G. Rake, *J. Bacteriol.* **46**, 109 (1943).
[13] J. D. Dutcher, *J. Biol. Chem.* **232**, 785 (1958).
[14] S. Nakamura, *Agr. Biol. Chem. (Tokyo)* **25**, 74,658 (1961); *Chem. Abstr.* **55**, 10456, 27329 (1961).
[15] U. Weiss, F. Strelitz, H. Flow, and I. N. Asheshov, *Arch. Biochem. Biophys.* **74**, 150 (1958).
[16] R. G. Micetich and J. C. MacDonald, *J. Chem. Soc.* 1507 (1964).
[17] J. C. MacDonald, R. G. Micetich, and R. H. Haskins, *Can. J. Microbiol.* **10**, 90 (1964).

The red pigment named pulcherrimin[18] isolated from *Candida pulcherrima* has structure 9. Structural assignment followed from its conversion into pulcherriminic acid (10). The structures originally proposed for pulcherrimin and pulcherriminic acid incorporated two additional ring hydrogen atoms.[19,20]

(9)   (10)

Synthetic pyrazines exhibit a wide range of physiological activity, in some cases superior to those of pyridine or pyrimidine analogs. Sulfalene (Kelfizina) (2-*p*-aminobenzenesulfonamido-3-methoxypyrazine) (11) is a long acting sulfonamide[21] and pryazinamide is a tuberculostatic drug.[22] *N*-Substituted pyrazinamides such as *N*-morpholinomethylpyrazinamide (12)[23] and *N*,*N*-dimethylpyrazinethioamide (13)[24] and pyrazinecarboxylic acid derivatives, especially 3-hydrazino-2-pyrazinecarboxylic acid hydrazide,[25] also exhibit antituberculous activity. 2-Methyl-3-piperidinopyrazine has antidepressant activity,[26] Amiloride (*N*-amidino 3,5-diamino-6-chloro-

---

[18] A. J. Kluyver, J. P. van der Walt, and A. J. van Triet, *Proc. Nat. Acad. Sci. U.S.* **39**, 583 (1953).
[19] A. H. Cook and C. A. Slater, *J. Inst. Brewing* **51**, 213 (1954); *J. Chem. Soc.* 4130, 4133 (1956).
[20] J. C. MacDonald, *Can. J. Chem.* **41**, 165 (1963).
[21] M. Ghione, C. Bertazzoli, A. Buogo, T. Chieli, and V. Zavaglio, *Chemotherapia* **6**, 344 (1963); *Chem. Abstr.* **60**, 2224 (1964).
[22] "British Pharmaceutical Codex," p. 692. Pharmaceutical Press, London, 1968.
[23] S. Yoshida, Japanese Patent 6843 (1962); *Chem. Abstr.* **58**, 13969 (1963).
[24] K. Yoshihira, Japanese Patent 20,580 (1963); *Chem. Abstr.* **60**, 2980 (1964).
[25] K. Palat, L. Novacek, M. Celadnik, and E. Kubala, *Sci. Pharm. Proc., 25th* 139 (1965); *Chem. Abstr.* **70**, 4037 (1969).
[26] J. A. Gylys, P. M. R. Muccia, and M. K. Taylor, *Ann. N.Y. Acad. Sci.* **107**, 899 (1963).

pyrazinecarboxamide) **(14)** is a diuretic,[27] and Thionazin (*O,O*-diethyl-*O*-2-pyrazinylphosphorothioate) **(15)** is a soil insecticide and nematicide.[28]

(11) pyrazine-OCH$_3$ with NHSO$_2$-C$_6$H$_4$-NH$_2$

(12) pyrazine-CONHCH$_2$N(morpholine)

(13) pyrazine-CSN(CH$_3$)$_2$

(14) 5-chloro-2,3-diamino pyrazine with CONHC(=NH)NH$_2$

(15) pyrazine-OP(=S)(OC$_2$H$_5$)$_2$

Alkylpyrazines are important flavor constituents of roasted food products such as coffee, cocoa, and peanuts; coffee, for instance, contains 2-isopropyl-5-methylpyrazine.[29, 29a] 2-Methoxy-3-isobutylpyrazine has been identified in the steam-volatile components of Californian green bell peppers and other 2-methoxy-3-alkylpyrazines have strong bell pepperlike odors.[30] 2-Methoxy-3-*sec*-butyl, -3-isobutyl-, and -3-isopropylpyrazine are present in galbanum oil[31] and in green peas.[31a] These three pyrazines are thought to be of major significance in the flavor of green peas.

In the last few years an increasing number of condensed pyrazine ring systems have been synthesized and their reactions studied. The stimulus for this work has often been the search for new drugs (see,

[27] C. O. Wilson and T. E. Jones, "American Drug Index." Pitman Medical Publ., London, 1969.
[28] "Pesticide Manual," p. 415. British Crop Protection Council, Droitwich, England (1968).
[29] H. G. Peer and A. van der Heijden, *Rec. Trav. Chim.* **88**, 1335 (1969).
[29a] H. G. Maier, *Angew. Chem. Int. Ed.* **9**, 917 (1970).
[30] R. M. Seifert, R. G. Buttery, D. G. Guadagni, D. R. Black, and J. G. Harris, *J. Agr. Food. Chem.* **18**, 246 (1970).
[31] A. F. Bramwell, J. W. K. Burrell, and G. Riezebos, *Tetrahedron Lett.* 3215 (1969); J. W. K. Burrell, R. A. Lucas, D. M. Michalkiewicz, and G. Riezebos, *Chem. Ind.* (*London*) 1409 (1970).
[31a] K. E. Murray, J. Shipton, and F. B. Whitfield, *Chem. Ind.* (*London*) 897 (1970).

e.g., references in footnotes 32–36). Examples of the preparation of condensed ring systems from pyrazine intermediates are given in subsequent sections of this review.

Pyrazine and its alkyl derivatives are potentially useful bidentate ligands and form complexes with the following common transition metals. Cu^I,[37] Co^{II},[38-40] Ni^{II},[41-43] Fe^{II},[44] Mo^{VI},[45] and Ti^{IV}.[46] Complexes of pyrazines with uranyl chloride,[47] ammonium, and potassium chloroiridates,[48] and the hexacarbonyls of chromium, molybdenum, and tungsten[49] have also been prepared. Vanadium(IV) chloride forms complexes with pyrazine and 2,6-dimethylpyrazine[50] and vanadyl sulfate forms a complex with pyrazine 2,3-dicarboxylic acid.[51] Methyl-, 2,5-dimethyl, and 2,6-dimethylpyrazines and some chloro-substituted methylpyrazines form 1:1 complexes with silver nitrate;[52] 1:1 complexes have also been prepared from pyrazine and

[32] Netherlands Patent Appl. 6,613,937 (1967); *Chem. Abstr.* **68**, 49650 (1968).
[33] C. Temple, J. D. Rose, R. D. Elliott, and J. A. Montgomery, *J. Med. Chem.* **11**, 1216 (1968).
[34] T. Okamoto, Y. Torigoe, M. Sato, and Y. Isogai, *Chem. Pharm. Bull.* **16**, 1154 (1968).
[35] G. R. Wendt and K. W. Ledig, U.S. Patent 3,299,064 (1967); *Chem. Abstr.* **68**, 21958 (1968).
[36] C. K. Cain, French Patent M6,196 (1968); *Chem. Abstr.* **72**, 12767 (1970).
[37] A. B. P. Lever, J. Lewis, and R. S. Nyholm, *Nature (London)* **189**, 58 (1961).
[38] A. B. P. Lever, J. Lewis, and R. S. Nyholm, *J. Chem. Soc.* 1235 (1962).
[39] E. Koros, S. M. Nelson, F. Paulik, L. Erdey, and F. Ruff, *Magy. Chem. Foly.* **70**, 468 (1964); *Chem. Abstr.* **62**, 8655 (1965).
[40] A. B. P. Lever and S. M. Nelson, *J. Chem. Soc., A* 859 (1966).
[41] F. D. Ayres, P. Pauling, and G. B. Robertson, *Inorg. Chem.* **3**, 1303 (1964).
[42] A. B. P. Lever, J. Lewis, and R. S. Nyholm, *J. Chem. Soc.* 4761 (1964).
[43] A. B. P. Lever, *Inorg. Chem.* **4**, 763 (1965).
[44] G. Beech and C. T. Mortimer, *J. Chem. Soc. A* 1115 (1967).
[45] W. M. Carmichael, D. A. Edwards, G. W. A. Fowles, and P. R. Marshall, *Inorg. Chim. Acta* **1**, 93 (1967).
[46] G. W. A. Fowles and R. A. Walton, *J. Less-common Metals* **5**, 510 (1963).
[47] P. V. Balakrishnan, S. D. Kamath, and H. V. Venkatasetty, *Proc. Nucl. Radiat. Chem. Symp., Waltair, India, 1966* p. 72; *Chem. Abstr.* **66**, 25,622 (1967).
[48] F. Larèze, *C.R. Acad. Sci.* **261**, 3420 (1965).
[49] M. C. Ganorkar and M. H. B. Stiddard, *J. Chem. Soc.* 5346 (1965).
[50] B. E. Bridgland, G. W. A. Fowles, and R. A. Walton, *J. Inorg. Nucl. Chem.* **27**, 383 (1965).
[51] R. L. Dutta and S. Ghosh, *J. Indian Chem. Soc.* **44**, 290 (1967).
[52] R. F. Trimble, L. Ling-Tse Yoh, and K. Ashley, *Naturwissenschaften* **51**, 37 (1964); *Chem. Abstr.* **61**, 6616 (1964).

Sec. II.]   PYRAZINE CHEMISTRY   105

zinc, cadmium, and mercury halides, and pyrazine-d_4 has been complexed with mercuric chloride.[53] 2,3,5,6-Tetrakis(α-pyridyl)pyrazine is an excellent tridentate ligand and gives a red-violet chelate with Fe^{II} salts.[54]

II. Physical and Spectroscopic Properties

The pyrazine ring may be represented as a resonance hybrid of the following canonical structures.

Interatomic distances as calculated from the analysis of the rotational fine structure of the ultraviolet spectrum are C–C, 1.395 Å; C–N, 1.341 Å; and C–H, 1.085 Å.[55] These are very similar to the bond lengths for pyridine which are C-2–C-3, 1.3945 Å; C-3–C-4, 1.3944 Å; and C-2–N, 1.3402 Å. The C–N–C bond angle in pyrazine is 115° and the C–C–N bond angle 122.5°.[56,57] A delocalization energy for pyrazine of ca. 18 kcal/mole is indicated from heats of combustion data.[58] The C=N bond energy in 2,2,5,5-tetramethyl-2,5-dihydropyrazine has been calculated to be 130.3 kcal.[58a]

The π-electron density distribution in the pyrazine ring has been calculated,[59-61] and compared with that of the pyridine ring.[3] These figures indicate an increase in π-electron density at the nitrogen atoms and a depletion of π-electron density at the carbon atoms. The values for charge distribution vary slightly with the method of calculation

[53] H. D. Stidham and J. A. Chandler, *J. Inorg. Nucl. Chem.* **27**, 397 (1965).
[54] H. A. Goodwin and F. Lions, *J. Amer. Chem. Soc.* **81**, 6415 (1959).
[55] J. A. Merritt and K. K. Innes, *Diss. Abstr.* **20**, 4291 (1960).
[56] C. A. Coulson and H. Looyenga, *J. Chem. Soc.* 6592 (1965).
[57] P. J. Wheatley, *Acta Cryst.* **10**, 182 (1957).
[58] M. H. Palmer, "The Structure and Reactions of Heterocyclic Compounds," p. 66. Arnold, London, 1967.
[58a] R. G. West, *Diss. Abstr.* B **30**, 2640 (1969).
[59] P. J. Black and C. A. McDowell, *Mol. Phys.* **12**, 233 (1967).
[60] S. Kwiatkowski and B. Zurawski, *Bull. Acad. Pol. Sci., Ser. Sci. Math. Astron. Phys.* **13**, 487 (1965); *Chem. Abstr.* **64**, 15719 (1966).
[61] P. J. Black, R. D. Brown, and M. L. Heffernan, *Austr. J. Chem.* **20**, 1305 (1967).

and the assumptions made. It is significant that the π-electron densities at the α-carbon atoms in both pyrazine and pyridine are very similar and this is reflected in the comparable chemical reactivities of α-substituted pyrazines and pyridines. The total electron densities $(\sigma + \pi)$ at the 2-position in pyrazine and its mono- and dication have been calculated by application of the extended Hückel theory (EHT).[62] In general, there is a good correlation between both the proton chemical shifts and carbon-13 chemical shifts of six-membered heterocyclic rings and EHT total electron densities.

Calculations by the self-consistent field LCAO–MO method for the ground state wave function of the pyrazine molecule indicate that the lone pairs are quite different. The lower lone pair is little delocalized (1.88 electrons on nitrogen), but the second lone pair is as delocalized as the lone pair in pyridine with 1.37 electrons on nitrogen, 0.22 electrons on hydrogen, and 0.40 electrons on carbon.[63]

Evidence for the delocalization of the "lone pairs" of pyrazine is deduced from the large differences in the magnetic susceptibilities of the three isomeric diazines. Thus the magnetic susceptibilities of pyrazine and pyridazine have been found to be smaller than that of pyrimidine.[63a] Further π-electron calculations have been carried out on the singlet excited states of pyrazine.[63b]

The lowest ionization potential of pyrazine and related nitrogen-containing heteroaromatic compounds has been measured. The experimental value of 9.29 eV is very similar to that of benzene (9.25 eV) and pyrimidine (9.35 eV) and is thought to be due to ionization from the highest occupied π orbital. Subsequent calculations have supported this assignment.[64,65] (For a discussion of the relative ordering of π, σ, and n orbitals, see Turner et al.[65a])

[62] W. Adam, A. Grimson, and G. Rodriguez, *Tetrahedron* **23**, 2513 (1967).
[63] E. Clementi, *J. Chem. Phys.* **46**, 4737 (1967).
[63a] J. D. Wilson, *J. Chem. Phys.* **53**, 467 (1970).
[63b] T. M. Bustard and H. H. Jaffe, *J. Chem. Phys.* **53**, 534 (1970).
[64] A. J. Yencha and M. A. El-Sayed, *J. Chem. Phys.* **48**, 3469 (1968).
[65] J. Del Bene and H. H. Jaffe, *J. Chem. Phys.* **50**, 563 (1969).
[65a] D. W. Turner, C. Baker, A. D. Baker, and C. R. Brundle, "Molecular Photoelectron Spectroscopy," Chapter 12. Wiley (Interscience), New York, 1970.

The ionization constants of pyrazines have been measured from potentiometric and spectroscopic data and from titrations with perchloric acid in glacial acetic acid.[66-68] The first pK_a of pyrazine is 0.65, the second is -5.8. Compared to pyridine with a pK_a of 5.2, it is clear that the second ring nitrogen has a marked base-weakening effect. Pyrazine is a weaker base than pyrimidine (pK_a 1.30) or pyridazine (pK_a 2.33), thus the basicity of the diazines varies inversely with the separation of the nitrogen atoms. The pK_a values of the methylpyrazines are proportional to the number of methyl groups present in the molecule. Methyl groups increase the basic strength by ca. 0.7 pK units, similar to their effect in pyridine.[67]

The total energies of the diazines and of their protonated forms have been calculated using the CNDO method. The data in the table show that differences between the total energies of the bases and their protonated forms correlate with the pK_a of the base except in the case of pyridazine.[67a]

	Pyridine	Pyridazine	Pyrimidine	Pyrazine
Total energy (a.u.)	−50.87418	−54.69181	−54.63493	−54.61563
Total energy of corresponding salts	−51.38771	−55.19645	−55.14151	−55.11446
ΔE	−0.51352	−0.50464	−0.50657	−0.49883
pK_a	5.17	2.33	1.30	0.65

Tetrafluoropyrazine is not completely protonated either in concentrated sulfuric acid or in fluorosulfuric acid (FSO_3H); it is estimated to have a pK_a value of approximately -13.[67b] The pK_a of pyrazine N-oxide is 0.05, it is thus a considerably weaker base than the parent heterocycle.[67c]

The infrared spectra of pyrazines and pyrazine 1-oxides show four stretching vibrations in the region 1600–1370 cm^{-1}. The absorption in the 1600–1575 cm^{-1} region is of variable intensity, the 1550–1520 cm^{-1} band is of weak-medium intensity, and the 1500–1465 and

[66] A. Albert, in "Physical Methods in Heterocyclic Chemistry" (A. R. Katritzky, ed.), Vol. I, Chapter 1. Academic Press, New York, 1963.
[67] A. Shih Chiuen Chia and R. F. Trimble, J. Phys. Chem. 65, 863 (1961).
[67a] M. H. Palmer and P. S. McIntyre, personal communication.
[67b] S. L. Bell, R. D. Chambers, W. K. R. Musgrave, and J. G. Thorpe, J. Fluorine Chem. 1, 51 (1971).
[67c] W. W. Paudler and S. A. Humphrey, J. Org. Chem. 35, 3467 (1970).
[68] D. A. Keyworth, J. Org. Chem. 24, 1355 (1959).

1420–1370 cm^{-1} bands are of medium-strong intensity.[69, 70] For pyrazine N-oxides a strong absorption in the region 1350–1250 cm^{-1} is assigned to the N–O stretching frequency.[71] The infrared and Raman spectra of deuterated pyrazines have also been reported.[72–75] The effects of hydrogen bond formation on pyrazine vibrational frequencies have been studied in a number of hydrogen-donor solvents. Shifts to higher frequency are observed indicating considerable changes of electron distribution on hydrogen-bond formation.[76]

A study of the infrared spectra of pyrazine monocarboxamides in the solid phase (potassium bromide and Nujol) and in solution (dioxane) gave no evidence of hydrogen bonding, as deduced from the carbonyl stretching frequency and the absence of concentration effects. In chloroform, however, the carbonyl stretching absorption moves to lower frequency indicating possible intramolecular hydrogen bonding between amide hydrogens and the ring nitrogen atom.[77]

The ultraviolet spectrum of pyrazine in cyclohexane shows maxima at 260 nm (corresponding to a π–π* transition) and 328 nm (corresponding to a n–π* transition) in each case with vibrational fine structure; the coefficients of molecular extinction are 5600 and 1040, respectively.[78, 79] Substitution of halogen has a bathochromic effect on the ultraviolet spectrum of pyrazine.[80] A useful index of the ultraviolet and visible spectra of pyrazine derivatives is available for the period from 1955 to 1963.[81] The far-ultraviolet spectrum of pyrazine

[69] A. R. Katritzky and A. P. Ambler, *in* "Physical Methods in Heterocyclic Chemistry" (A. R. Katritzky, ed.), Vol. II, Chapter 10. Academic Press, New York, 1963.
[70] A. D. Cross and R. A. Jones, "An Introduction to Practical Infrared Spectroscopy," p. 79. Butterworths, London, 1969.
[71] H. Shindo, *Chem. Pharm. Bull.* **8**, 33 (1960); *Chem. Abstr.* **55**, 6139 (1961).
[72] H. H. Perkampus and E. Baumgarten, *Spectrochim. Acta* **19**, 1473 (1963).
[73] S. Califano, G. Adembri, and G. Sbrana, *Spectrochim. Acta* **20**, 385 (1964).
[74] M. Scrocco, C. Di Lauro, and S. Califano, *Spectrochim. Acta* **21**, 571 (1965).
[75] J. D. Simmons, K. K. Innes, and G. M. Begun, *J. Mol. Spectrosc.* **14**, 190 (1964).
[76] H. Takahashi, K. Mamola, and E. K. Plyler, *J. Mol. Spectrosc.* **21**, 217 (1966).
[77] H. Negoro and Y. Morotomi, *Yakuzaigaku* **23**, 308 (1963); *Chem. Abstr.* **61**, 1390 (1964).
[78] S. F. Mason, *in* "Physical Methods in Heterocyclic Chemistry" (A. R. Katritzky, ed.), Vol. II, Chapter 7. Academic Press, New York, 1963.
[79] Y. Kanda, *Kagaku To Kogyo (Tokyo)* **17**, 1023 (1964); *Chem. Abstr.* **61**, 14042 (1964).
[80] B. Majee and B. K. Das, *J. Indian Chem. Soc.* **44**, 1086 (1967).
[81] H. M. Hershenson, "Ultraviolet and Visible Absorption Spectra, Indexes for 1955–1959 and 1960–1963." Academic Press, New York, 1961 and 1966.

in cyclohexane shows a maximum at 194 nm ($\epsilon = 6100$);[82] the spectrum of the vapor has been measured in the 153–195 nm region and shows a maximum at 165 nm.[83] The fluorescence[84] and phosphorescence[85, 86] spectra of pyrazine have also been measured; from similar measurements on pyrazine N-oxide it is deduced that the effect of the N-oxide group on the absorption spectrum resembles that of a substituent with π-donor action.[87] The electronic absorption and circular dichroism (c.d.) spectra of some optically active pyrazines have been measured and the observed Cotton effects assigned to the respective n–π* and π–π* transitions of the pyrazine chromophore.[88]

The PMR spectrum of pyrazine shows a singlet at 1.37τ in deuterochloroform, the coupling constants as calculated from the ^{13}C-satellite signals are $J_{23} = 1.8$ Hz, $J_{25} = 1.8$ Hz, and $J_{26} = 0.5$ Hz.[89] An analysis of the PMR spectra of eight monosubstituted pyrazines indicates that the effect of substituents on chemical shift is roughly parallel to that in monosubstituted benzenes. Thus, as the electron-withdrawing effect of the substituent increases, the resonance of all three ring protons moves to lower field. Spin tickling experiments indicate that the meta-coupling constant across nitrogen is negative in sign. The ortho-, para-, and meta-coupling constants for monosubstituted pyrazines are in the ranges 2.4–2.9, 1.3–1.6, and -0.5–0.0 Hz, respectively. Similar values for the ortho-, para-, and meta-coupled protons in 2,3-, 2,5-, and 2,6-dialkyl- and alkylalkoxypyrazines have been observed. In the PMR spectrum of 2-hydroxypyrazine the 5- and 6-protons are shifted upfield 0.4 and 0.5 ppm with respect to the corresponding protons of 2-aminopyrazine, and by 0.92 and 0.83 ppm when compared with the corresponding protons of 2-methoxypyrazine. This supports the conclusion obtained by other methods (e.g., infrared and ultraviolet spectroscopy) that the 2-hydroxy compound exists predominantly in the tautomeric 1,2-dihydro-2-oxo

---

[82] G. Favini and I. R. Bellobono, *Rend. Ist. Lomb. Sci. Lett. A* **99**, 381 (1965); *Chem. Abstr.* **64**, 18711 (1966).
[83] J. E. Parkin and K. K. Innes, *J. Mol. Spectrosc.* **15**, 407 (1965).
[84] H. Baba, L. Goodman, and P. C. Valenti, *J. Amer. Chem. Soc.* **88**, 5410 (1966).
[85] M. A. El-Sayed and R. G. Brewer, *J. Chem. Phys.* **39**, 1623 (1963).
[86] W. R. Moomaw and M. A. El-Sayed, *J. Chem. Phys.* **45**, 3890 (1966).
[87] B. Ziolkowsky and F. Doerr, *Ber. Bunsenges. Phys. Chem.* **69**, 448 (1965); *Chem. Abstr.* **63**, 7770 (1965).
[88] H. E. Smith and A. E. Hicks, *J. Chem. Soc.* D1112 (1970).
[89] K. Tori and M. Ogata, *Chem. Pharm. Bull.* **12**, 272 (1964).

form.[90] The benzylic coupling constants for methylpyrazine and the three isomeric methoxymethylpyrazines have been measured.[90a] The results show that $^6J_p \sim {}^4J_o > {}^5J_m$, whereas it had suggested earlier[90] that $^5J_m > {}^6J_p \sim {}^4J_o$. Since the determination of the relative position of substituents on the pyrazine ring has in the past often been a difficult problem, such correlations of NMR parameters and structure are particularly useful. The effect of solvent on the PMR spectrum of pyrazine has been reported; a singlet at 1.41τ is observed in carbon tetrachloride and 1.91τ in benzene.[91] A study of the protonation of pyrazine in sulfuric and trifluoroacetic acids has been carried out using this technique.[92] The ^{13}C NMR signal of pyrazine is 17.4 ppm deshielded with respect to benzene and the ^{13}C–H coupling constant is 184 Hz. In methylpyrazines, ring carbon atoms are *deshielded* by directly bound methyl groups.[93] The effect of protonation is also surprising, this causes *shielding* of the ring carbon atoms.[94]

PMR has been used to study hydrogen–deuterium exchange in pyridine, pyrimidine, pyridazine, and pyrazine in CH_3OD–CH_3ONa at 164.6°. Hydrogen–deuterium exchange in pyrazine occurs faster than in pyridine, but rate data indicate that ring nitrogen activates hydrogen–deuterium exchange preferentially at meta and para positions.[95] The hydrogen–deuterium exchange rates for H_2 and H_6 in 3-substituted pyrazine 1-oxides have been correlated with σ constants. There is a linear free-energy relationship between the H–D exchange rates for H_2 and the σ_I parameters. There is also a linear relationship between the log of the H_2 exchange rates and the pK_a values of these compounds, the exchange rates being faster for oxides of lower basic strength. The considerable exchange-rate enhancement caused by the replacement of the C_4–H function in a pyridine N-oxide by a ring nitrogen is illustrated by the fact that the half-life for H_2–D exchange in 3-chloropyrazine 1-oxide is 4 min at 31° in 0.0025 N NaOD, compared with a half-life of 40 min for the exchange of H_2 in 3-chloropyridine 1-oxide at 74° in 0.045 N NaOD.[67c]

[90] R. H. Cox and A. A. Bothner-By, *J. Phys. Chem.* **72**, 1642, 1646 (1968).
[90a] A. F. Bramwell, G. Riezebos, and R. D. Wells, *Tetrahedron Lett.* 2489 (1971).
[91] J. N. Murrell and V. M. S. Gil, *Trans. Faraday Soc.* **61**, 402 (1965).
[92] H. Kamei, *J. Phys. Chem.* **69**, 2791 (1965).
[93] P. C. Lauterbur, *J. Chem. Phys.* **43**, 360 (1965).
[94] A. Mathias and V. M. S. Gil, *Tetrahedron Lett.* 3163 (1965).
[95] J. A. Zoltewicz, G. Grahe, and C. L. Smith, *J. Amer. Chem. Soc.* **91**, 5501 (1969).

Reduction of pyrazine with potassium in 1,2-diethoxyethane at $-70°$ gives the paramagnetic radical anion. Anions have also been generated from the low-temperature reaction of pyrazine, methylpyrazine, 2,5- and 2,6-dimethylpyrazines with sodium or potassium in tetrahydrofuran or 1,2-dimethoxyethane. From measurements of their ESR spectra, ion-pair formation is proposed between the alkali metal cation and the pyrazine anion.[96-101]

Polarographic studies on pyrazine and methylpyrazines indicate that 1 4-dihydropyrazines are the products of reduction. The reduction of pyrazine itself at the dropping mercury electrode proceeds reversibly. The substitution of methyl groups makes the reduction more difficult; with an increased number of methyl groups an increased tendency toward irreversible reduction is noted.[102-104] The half-wave reduction potentials for pyrazine, methylpyrazine, 2,6-dimethylpyrazine, and tetramethylpyrazine are 2.17, 2.23, 2.28, and 2.50 eV, respectively. Pyrazine is thus more easily reduced than pyridine which has a half-wave potential of 2.76 eV, and less easily reduced than quinoxaline which has a half-wave potential of 1.80 eV.[105]

The mass spectra of pyrazine, methylpyrazine, 2,5- and 2,6-dimethylpyrazine have been reported.[106] The base peak in the mass spectrum of tetramethylpyrazine corresponds to the loss of two molecules of methyl cyanide.[107] The mass spectra of 2,5-bis(p-fluorophenyl)-3,6-diphenylpyrazine,[108] 2-hydroxy-3-alkylpyrazines, and their N- and O-methyl derivatives have also been reported.[30]

A detailed study has been made of the mass spectra of pyrazine 1-oxides and 4-oxides and 1,4-dioxides. The most prominent feature

[96] A. Carrington and J. Dos Santos-Veiga, *Mol. Phys.* **5**, 51 (1962).
[97] R. L. Ward, *J. Amer. Chem. Soc.* **84**, 332 (1962).
[98] C. A. McDowell and K. F. G. Paulus, *Mol. Phys.* **7**, 541 (1963–64).
[99] N. M. Atherton and A. E. Goggins, *Trans. Faraday Soc.* **61**, 1399 (1965).
[100] C. A. McDowell and K. F. G. Paulus, *Can. J. Chem.* **43**, 224 (1965).
[101] N. M. Atherton and A. E. Goggins, *Mol. Phys.* **8**, 99 (1964).
[102] J. Volke, D. Dumanovic, and V. Volkova, *Collect. Czech. Chem. Commun.* **30**, 246 (1965).
[103] L. F. Wiggins and W. S. Wise, *J. Chem. Soc.* 4780 (1956).
[104] J. M. Hale, *J. Electroanal. Chem.* **8**, 181 (1964); *Chem. Abstr.* **62**, 2510 (1965).
[105] B. J. Tabner and J. R. Yandle, *J. Chem. Soc. A* 381 (1968).
[106] A. L. Jennings and J. E. Boggs, *J. Org. Chem.* **29**, 2065 (1964).
[107] H. Budzikiewicz, C. Djerassi, and D. H. Williams, *in* "Mass Spectrometry of Organic Compounds," p. 582. Holden-Day, San Francisco, 1967; Q. N. Porter and J. Baldas, *in* "Mass Spectrometry of Heterocyclic Compounds," p. 491. Wiley (Interscience), New York, 1971.
[108] M. M. Bursey and T. A. Elwood, *J. Org. Chem.* **35**, 793 (1970).

is the appearance of M-16 ion peaks. In the case of the methyl and methoxy derivative of 1-oxides the intensity ratio of the M-17/M-16 ions is larger than in the 4-oxides.[108a] The negative-ion mass spectra of some pyrazines and their N-oxides have also been reported.[108b]

III. General Synthetic Methods

Methods of heterocyclic synthesis can be divided into two main types: those in which the ring system is built up from aliphatic components and those in which derivatives of other heterocyclic systems are used as starting materials. The essential step in most pyrazine syntheses from aliphatic components is C–N bond formation and various schemes for the synthesis of the required C_4N_2 ring system are illustrated below. The primary product is in many cases a reduced

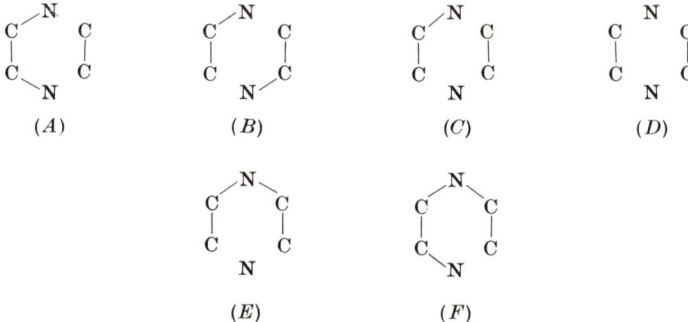

pyrazine and thus a final step of aromatization is required. It has been found convenient to classify pyrazine syntheses according to the starting material used, and the most important of these are α,β-dicarbonyl compounds, α-aminocarbonyl compounds, α-halogenoketones, piperazines, quinoxalines, and pteridines. Many pyrazine derivatives are normally prepared by transformations of other pyrazine derivatives; for example, halopyrazines are usually prepared from hydroxypyrazines or by halogenation. These reactions do not fall within the definition of heterocyclic synthesis outlined above and they are discussed in subsequent sections dealing with pyrazine derivatives having specific functional groups.

[108a] F. Uchimaru, S. Okada, A. Kosasayama, and T. Konno, *J. Heterocycl. Chem.* **8**, 99 (1971).
[108b] W. W. Paudler and S. A. Humphrey, *Org. Mass. Spectrometry Suppl.* **4**, 513 (1970).

A. From α,β-Dicarbonyl Compounds

The condensation of α,β-diketones with 1,2-diamines is a classical route for the synthesis of alkyl- and arylpyrazines. For example, good yields of dihydropyrazines are obtained from reaction of 2,3-dioxoalkanes and ethylenediamine: dehydrogenation over a copper chromite catalyst at 300° then gives 3-alkyl-2-methylpyrazines (Scheme 1). Attempts to carry out the dehydrogenation using a variety of milder and more convenient laboratory procedures were not successful.[109, 110]

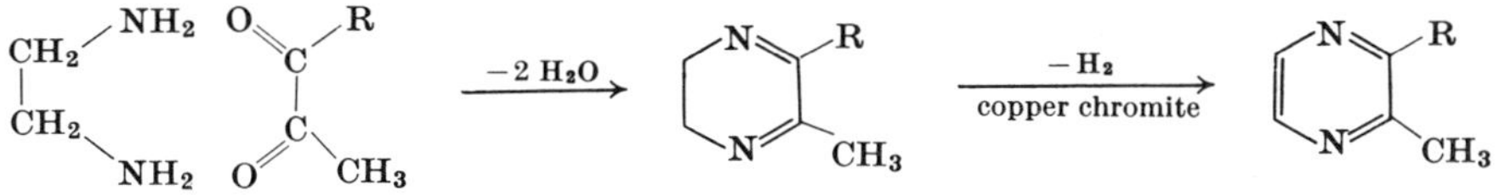

SCHEME 1. Type A synthesis.

Condensation of 1,2-diaminopropane and 2,3-dioxobutane similarly gives 5,6-dihydro-2,3,5-trimethylpyrazine which is oxidized to the corresponding pyrazine in 58% yield by treatment with potassium hydroxide pellets.[111] Hydroxypyrazines are very conveniently prepared from α,β-dicarbonyl compounds and α-amino acid amides [Eq. (1)],[30, 112] and pyrazinecarboxylic acids have been prepared by condensation of an α,β-diketone with an α,β-diaminocarboxylic acid, followed by oxidation (Scheme 2). Thus, condensation of benzil and

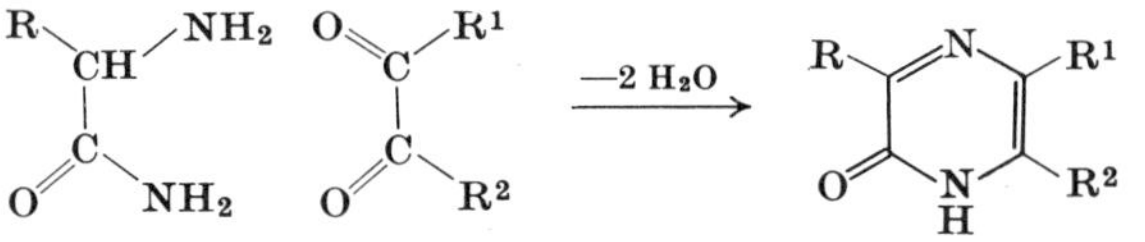

(1)

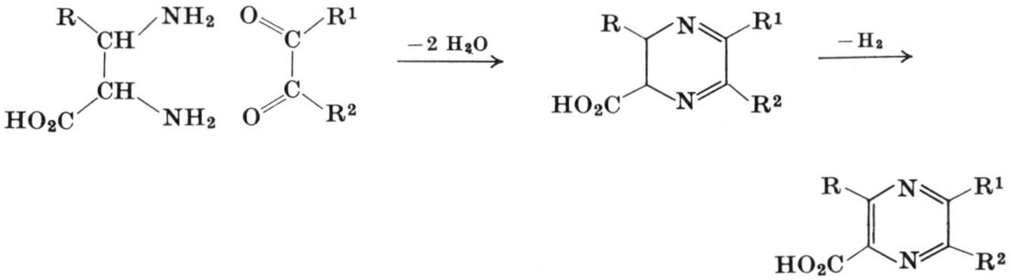

SCHEME 2. Type A synthesis.

[109] I. Flament and M. Stoll, *Helv. Chim. Acta* **50**, 1754 (1967).
[110] Netherlands Patent Appl. 6,812,899 (1969); *Chem. Abstr.* **71**, 61421 (1969).
[111] J. P. Marion, *Chimia* **21**, 510 (1967).
[112] R. G. Jones, *J. Amer. Chem. Soc.* **71**, 78 (1949).

2,3-dioxobutane and α,β-diaminopropionic acid gives, after aerial oxidation of the intermediate dihydropyrazine, 5,6-diphenyl- and 5,6-dimethylpyrazine-2-carboxylic acid, respectively.[113, 114] 2-Aminopyrazine and its alkyl and aryl derivatives have been obtained from the condensation of α,β-dicarbonyl compounds with aminoacetamidine [$NH_2CH_2C(=NH)NH_2$] [Eq. (2)][115] and 2-aminopyrazine-3-carboxamides are obtained from the analogous condensation of α,β-dicarbonyl compounds and aminomalonamidamidine [$NH_2CH(CONH_2)C(=NH)NH_2$].[116]

Condensation of α,β-carbonyl compounds and aminomalonamide [$NH_2CH(CONH_2)CONH_2$] similarly affords 2-hydroxy-3-carboxamidopyrazines.[117]

$$\begin{array}{c}\text{structure}\end{array} \xrightarrow{-2 H_2O} \begin{array}{c}\text{pyrazine}\end{array} \quad (2)$$

Type A synthesis.

B. From α-Aminocarbonyl Compounds

The self-condensation of two molecules of an α-aminoketone to a 2,5-dihydropyrazine and subsequent oxidation represents an important method for the preparation of 2,5-disubstituted and 2,3,5,6-tetrasubstituted pyrazines. The second step may proceed spontaneously in the presence of air, or be carried out with oxidizing agents such as hydrogen peroxide or mercuric chloride. The required α-aminoketones are by no means easily accessible intermediates and an alternative route to tetrasubstituted pyrazines has been developed which involves

---

[113] E. Felder, D. Pitre, S. Boveri, and E. B. Grabitz, *Chem. Ber.* **100**, 555 (1967).
[114] M. Litmanowitsch, E. Felder, D. Pitre, Swiss Patent 458,361 (1968); *Chem. Abstr.* **70**, 20083 (1969).
[115] D. Pitre and S. Boveri, *Chem. Ber.* **100**, 560 (1967).
[116] O. Vogl and E. C. Taylor, *J. Amer. Chem. Soc.* **81**, 2472 (1959).
[117] F. L. Muehlmann and A. R. Day, *J. Amer. Chem. Soc.* **78**, 242 (1956).

heating α-hydroxyketones with ammonium acetate.[118, 119] In this way α-aminoketones are probably formed *in situ* and the reaction then proceeds as illustrated (Scheme 3).

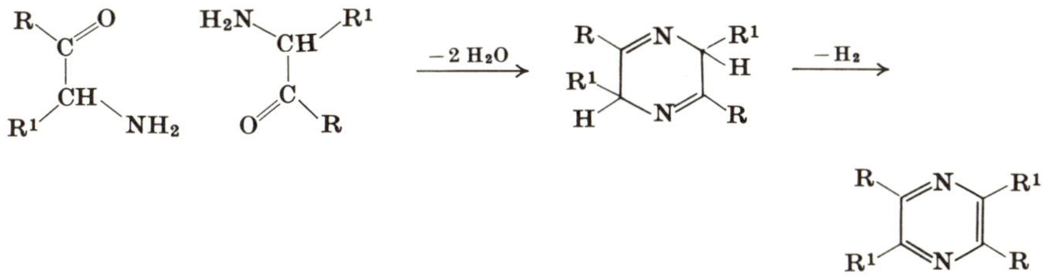

SCHEME 3. Type *B* synthesis.

Pyrazine-2,3-diones are conveniently prepared from aminoacetaldehyde diethyl- or dimethylacetal. The condensation product of the acetal and diethyl oxalate is treated with ammonia or a primary amine and the resulting intermediate cyclized in hot acetic acid containing a trace of hydrochloric acid (Scheme 4).[120]

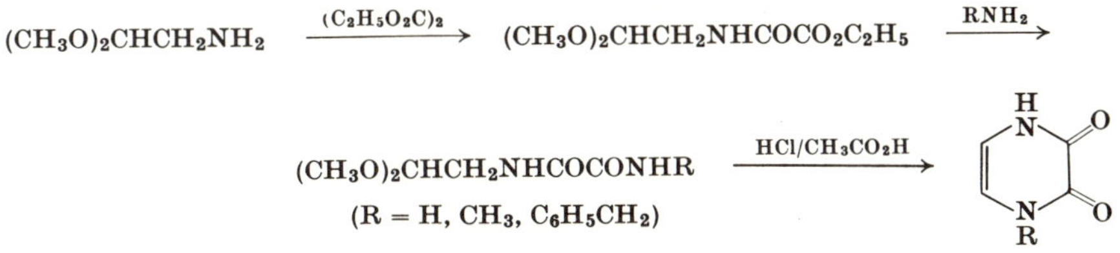

$(CH_3O)_2CHCH_2NH_2 \xrightarrow{(C_2H_5O_2C)_2} (CH_3O)_2CHCH_2NHCOCO_2C_2H_5 \xrightarrow{RNH_2}$

$(CH_3O)_2CHCH_2NHCOCONHR \xrightarrow{HCl/CH_3CO_2H}$
(R = H, CH$_3$, C$_6$H$_5$CH$_2$)

SCHEME 4. Type *F* synthesis.

C. FROM α-HALOGENOKETONES

The condensation of an α-halogenoketone with ammonia yields both 2,5- and 2,6-disubstituted pyrazines via the dihydropyrazines [Eq. (3)]. For example, when ω-chloroacetophenone is heated with alcoholic ammonia, 2,5- and 2,6-diphenylpyrazines are formed in about equal amounts.[121]

---

[118] J. Wiemann, N. Vinot, and M. Villadary, *Bull. Soc. Chim. Fr.* 3476 (1965).
[119] N. Vinot and J. Pinson, *Bull. Soc. Chim. Fr.* 4970 (1968).
[120] G. Palamidessi and L. Panizzi, French Patent 1,372,807 (1964); *Chem. Abstr.* **62**, 1674 (1965).
[121] F. Tutin, *J. Chem. Soc.* **97**, 2495 (1910).

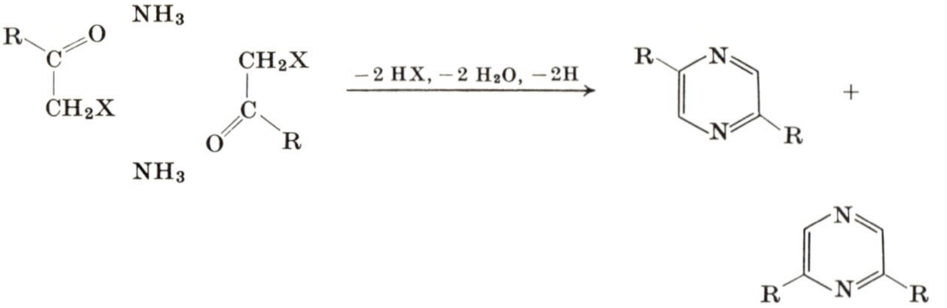

Type *D* synthesis.

Recently Zbiral and Stroh reported that treatment of α-azidoketones (obtained generally by reaction of the α-halogenoketone with sodium azide) with triphenylphosphine is a convenient route to 2,5-disubstituted pyrazines (Scheme 5). Thus, 2,5-diphenyl-, 2,5-bis(4-methoxyphenyl)-, and 2,5-diisopropylpyrazine have been obtained in 75, 45, and 44% yield, respectively. The intermediate *P–N*-ylide undergoes intermolecular condensation to give a 2,5-dihydropyrazine which is subsequently aromatized.[122]

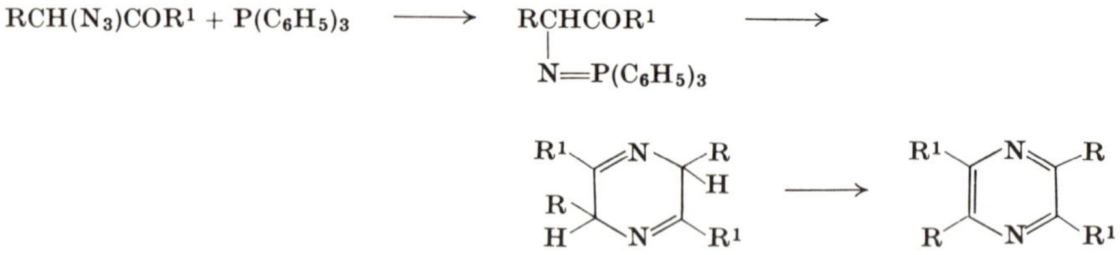

SCHEME 5. Type *B* synthesis.

A new and flexible route (Scheme 6) to compounds containing the aspergillic acid skeleton has been developed by Masaki and Ohta.[123] This has been successfully used for the synthesis of racemic aspergillic acid (3),[124] neoaspergillic acid (8),[124] and mutaaspergillic acid (6).[125] The initial step is the condensation of an α-chloromethyl ketone protected as the oxime with an α-aminohydroxamic acid protected as the *O*-benzyl ester. After removal of protective groups, the product is treated with ammonia to give a ketimine which undergoes ring

[122] E. Zbiral and J. Stroh, *Ann.* **727**, 231 (1969).
[123] M. Masaki and M. Ohta, *J. Org. Chem.* **29**, 3165 (1964).
[124] M. Masaki, Y. Chigira, and M. Ohta, *J. Org. Chem.* **31**, 4143 (1966).
[125] M. Sugujama, M. Masaki, and M. Ohta, *Tetrahedron Lett.* 845 (1967).

closure and finally spontaneous air oxidation to the aspergillic acid-type compound.

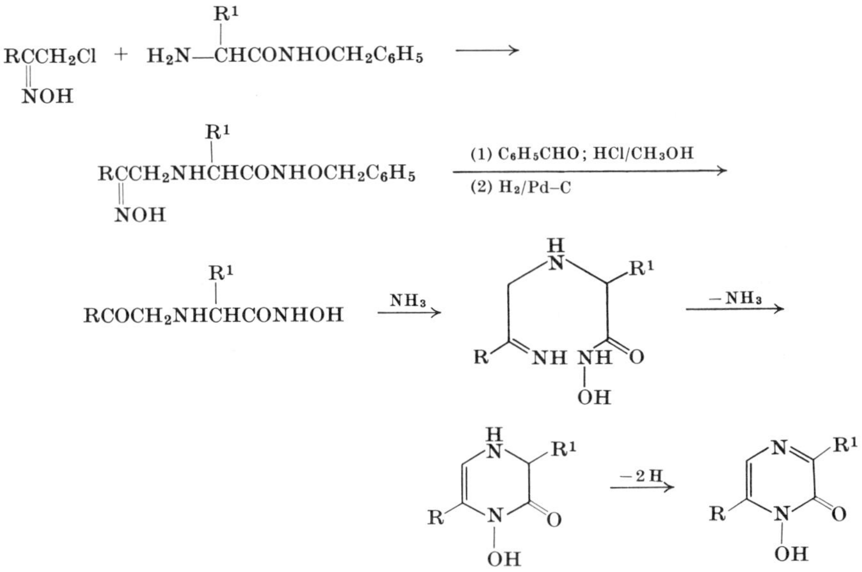

SCHEME 6. Type *F* synthesis.

D. FROM PIPERAZINES

The continuous vapor phase dehydrogenation of piperazines is effectively carried out over a prereduced 80% CuO–20% Cr_2O_3 catalyst at 350°. In this way 2-ethyl-, 2,6-dimethyl-, and 2,3,5,6-tetramethylpiperazines have been converted into the corresponding pyrazines in 79–89% yield.[126-128] Pyrazine itself is obtained from piperazine in 18% yield by the use of 5% Pt on alkali-washed firebrick catalyst.[129] The catalytic deamination of diethylenetriamine over alumina-based catalysts at temperatures of 300°–400° gives good

---

[126] M. Cenker, D. R. Jackson, W. K. Langdon, W. W. Levis, Jr., S. D. Tarailo, and G. E. Baxter, *Ind. Eng. Chem. Prod. Res. Develop.* **3**, 11 (1964).
[127] M. Cenker and G. E. Baxter, U.S. Patent 3,005,820 (1961); *Chem. Abstr.* **56**, 7335 (1962).
[128] S. D. Tarailo, U.S. Patent 2,945,858 (1960); *Chem. Abstr.* **55**, 1668 (1961).
[129] I. C. Nigam, *J. Chromatogr.* **24**, 188 (1966).

yields of pyrazine and piperazine (Scheme 7).[130] 2-Chloropyrazine is prepared in 53% yield by thermal decomposition of 2,3,5,6-tetra-

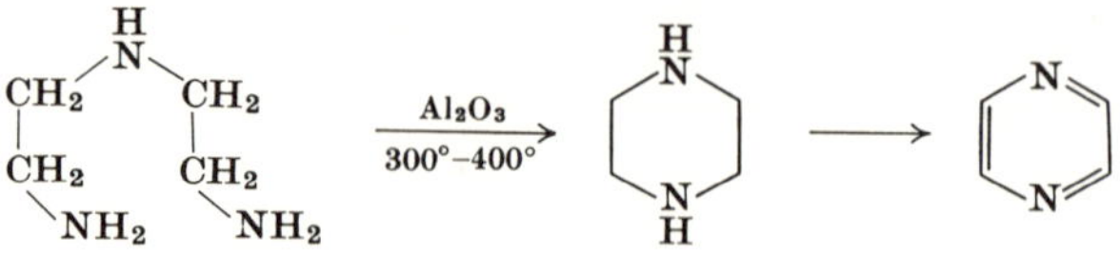

SCHEME 7. Type *F* synthesis.

chloro-1,4-diformylpiperazine at 185°–200°. The latter compound is obtained by chlorination of the condensation product of glyoxal and formamide (Scheme 8).[131, 132] 2,3-Dichloropyrazine has been obtained by treatment of 1,4-bischlorocarbonylpiperazine with chlorine in the presence of ultraviolet light and ferric chloride at 155°–160° (Scheme 9).[133]

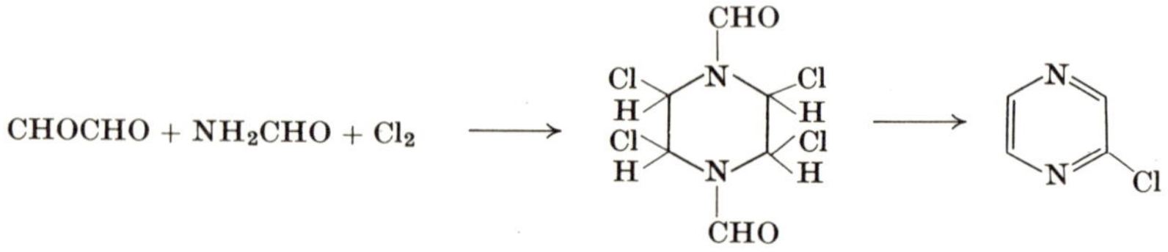

SCHEME 8

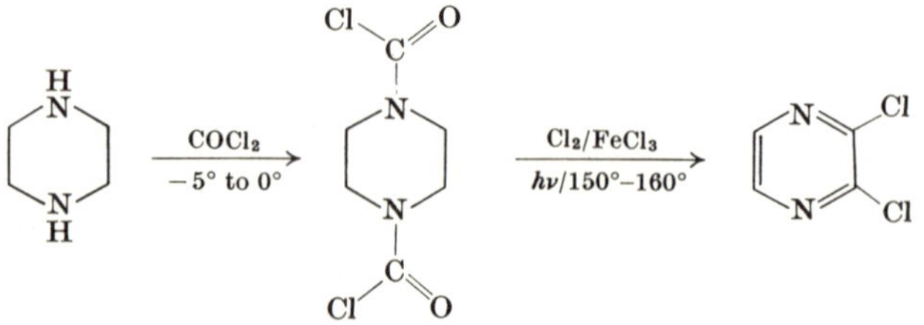

SCHEME 9

2,5-Diketopiperazines are among the most readily available piperazine derivatives and are therefore attractive starting materials

[130] A. A. Anderson, S. P. Yurel, and M. V. Shimanskaya, *Chem. Heterocycl. Compounds* **3**, 271 (1967); *Chem. Abstr.* **67**, 100107 (1967).
[131] G. Fort, U.S. Patent 3,356,679 (1967); *Chem. Abstr.* **69**, 10471 (1968).
[132] G. Fort, British Patent 1,100,890 (1968); *Chem. Abstr.* **69**, 19207 (1968).
[133] A. Carrara, A. Leone, and D. Fabris, French Patent 1,413,599 (1965); *Chem. Abstr.* **64**, 5114 (1966).

for pyrazine synthesis. Spring and Newbold and their co-workers found that mixtures of mono- and dichloropyrazines are formed on treatment of these compounds with phosphoryl chloride.[134] More recently it has been shown by Blake and Sammes that 2,5-diketopiperazines can be alkylated with triethyloxonium fluoroborate.[135] In the case of 2,5-diketopiperazine itself, the derived diethoxydihydropyrazine is conveniently oxidized with dichlorodicyanoquinone (DDQ) (Scheme 10).

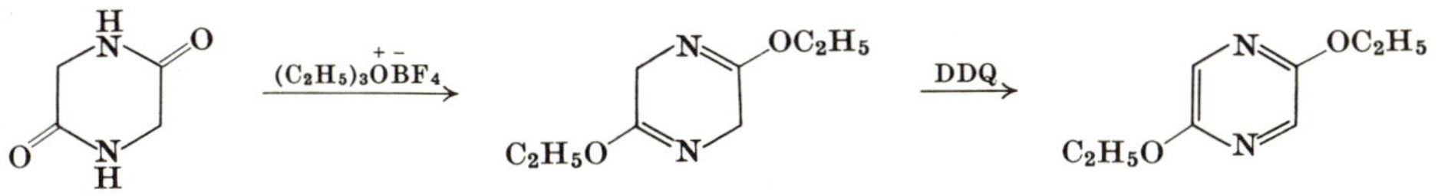

SCHEME 10

E. FROM QUINOXALINES AND PTERIDINES

The greater stability of the pyrazine ring to oxidation compared with that of benzene enables pyrazinecarboxylic acids to be prepared by permanganate oxidation of either quinoxalines or phenazines. The pyrazine ring is more stable than the pyrimidine ring to acid and alkaline hydrolysis. Thus, pteridine is converted into 2-amino-3-formylpyrazine on treatment with dilute sulfuric acid and N-(3-formyl-2-pyrazinyl)formamidine oxime on treatment with sodium carbonate and hydroxylamine (Scheme 11).[136] Aminopyrazines and

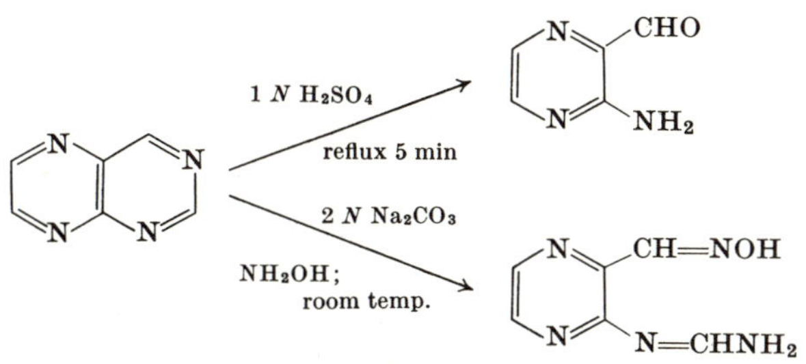

SCHEME 11

[134] R. A. Baxter and F. S. Spring, *J. Chem. Soc.* 1147 (1947).
[135] K. W. Blake and P. G. Sammes, *J. Chem. Soc. C* 1070 (1970).
[136] A. Albert, D. J. Brown, and H. C. S. Wood, *J. Chem. Soc.* 2066 (1956).

aminopyrazinecarboxylic acids are prepared by acid- or alkali-promoted hydrolysis of 2,4-dihydroxypteridines. Detailed investigations by Clark and his co-workers on the action of nucleophiles on 4-hydroxypteridines have revealed that although only pyrimidine ring cleavage occurs on treatment with alkali, hydrazine and hydroxylamine and their methylated derivatives are able to cleave either the pyrimidine or the pyrazine ring of suitably substituted 4-hydroxypteridines.[137, 138, 138a]

F. From Ring Transformation Reactions

Musgrave, Chambers, and their co-workers have reported that perfluoropyridazines are isomerized on ultraviolet irradiation almost quantitatively to perfluoropyrazines.[139] The earlier suggestion that prismanes are formed intermediately has been withdrawn and the revised mechanism shown in Scheme 12 is now postulated.[140]

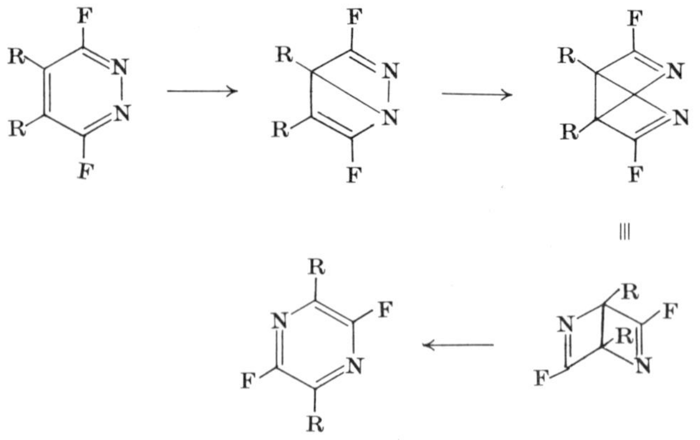

(R = F or iso-C_3F_7)

Scheme 12

[137] J. Clark and G. Neath, J. Chem. Soc. C 1112 (1966), C, 919 (1968).
[138] J. Clark, G. Neath, and C. Smith, J. Chem. Soc. C 1297 (1969).
[138a] J. Clark and C. Smith, J. Chem. Soc. C 1948 (1971).
[139] C. G. Allison, R. D. Chambers, Y. A. Cheburkov, J. A. H. MacBride, and W. K. R. Musgrave, Chem. Commun. 1200 (1969).
[140] R. D. Chambers, W. K. R. Musgrave, and K. C. Srivastava, J. Chem. Soc. D 264 (1971).

Ultraviolet irradiation of 4,5-dichloro-3,6-difluoropyridazine similarly gives 2,5-dichloro-3,6-difluoropyrazine. Lemal and his co-workers consider that this result also eliminates the prismane mechanism for pyridazine-pyrazine interconversion.[140a] Methyl-, 2,5-dimethyl-, 2,6-dimethyl-, and trimethylpyrazine are formed when cis-3,7-dihydroxy-octahydro-1,5-diazocine is passed over hot alumina [Eq. (4)].[141]

G. MISCELLANEOUS SYNTHESES

It has been reported earlier that ketones on treatment with nitrogen iodide and aqueous ammonia yield pyrazines.[142] Recently electrolysis of ketones in aqueous ammoniacal solution containing potassium iodide was also found to yield pyrazines. Thus, 2,5-dimethylpyrazine was isolated in 7% yield after electrolysis of an aqueous medium containing acetone, potassium iodide, and ammonia using a platinum anode and a nickel cathode. The mechanism proposed for this reaction is *in situ* formation of nitrogen iodide followed by reaction with the ketone in several steps to give dihydropyrazine and then pyrazine. In support of this proposal is the observation that electrolysis of potassium iodide in aqueous ammonia under similar conditions, but in the absence of ketones, produces nitrogen iodide.[143]

Recently 2,3,5,6-tetraphenylpyrazine has been prepared by a dimerization reaction [Eq. (5)]. This represents a case of pyrazine ring

[140a] D. W. Johnson, V. Austel, R. S. Feld, and D. M. Lemal, *J. Amer. Chem. Soc.* **92**, 7505 (1970).
[141] W. W. Paudler, A. G. Zeiler, and G. R. Gapski, *J. Org. Chem.* **34**, 1001 (1969).
[142] J. H. Fellman, S. H. Wilen, and C. A. VanderWerf, *J. Org. Chem.* **21**, 713 (1956).
[143] S. H. Wilen and A. W. Levine, *Chem. Ind. (London)* 237 (1969).

synthesis by C–C bond formation rather than the more usual C–N bond formation.[144]

$$\underset{\underset{C_6H_5}{|}}{\overset{\overset{C_6H_5}{|}}{N}}\overset{CH}{\underset{CHNa}{\diagdown}} \xrightarrow{CuBr_2} \begin{array}{c} C_6H_5 \diagdown N \diagup C_6H_5 \\ C_6H_5 \diagup N \diagdown C_6H_5 \end{array} \qquad (5)$$

IV. General Chemical Properties

Pyrazines have considerable aromatic character and therefore in the majority of their reactions tend to revert to type. The main features of their reactivity may be predicted by regarding them as pyridines into which a nitrogen has been inserted in the para position. Pyrazines also show close similarity in their reactions to the other diazines, pyrimidines, and pyridazines.

The pyrazine ring, like the pyridine ring is stable to *oxidizing agents*. It is more resistant to alkaline permanganate than the benzene ring itself; it also survives strong acid and alkaline treatments which will destroy the pyrimidine ring (the preparative significance of these observations is illustrated in Section III).

Reduction, carried out by the standard reagents such as sodium and ethanol, aluminum amalgam, sodium amalgam, tin and hydrochloric acid, or by catalytic hydrogenation normally gives piperazines (hexahydropyrazines). These are stable and react in the characteristic manner of secondary amines. The pK_a of piperazine is 9.8, whereas the pK_a of piperidine (hexahydropyridine) is 11.2, and is illustrative of the base-weakening effect of the additional ring nitrogen atom. Electrochemical reduction of 2,3-dihydro-5,6-diphenylpyrazine in aqueous ethanolic sodium hydroxide (pH 12.9) at −1.4 V gives 70% of 1,2,3,4-tetrahydro-5,6-diphenylpyrazine.[144a]

Pyrazines are more resistant to electrophilic substitution reactions at the ring carbon atoms than the corresponding pyridines. Electrophilic attack normally takes place on the ring nitrogen atoms; thus pyrazines form mono- and disalts with proton acids and mono- and

---

[144] Th. Kauffmann, G. Beissner, H. Berg, E. Koppelmann, J. Legler, and M. Schonfelder, *Angew. Chem. Int. Ed. Engl.* **7**, 540 (1968).

[144a] J. Pinson, J. P. Launay, and J. Armand, *C.R. Acad. Sci. Ser. C* **270**, 1881 (1970).

dicoordination compounds with Lewis acids (e.g., boron trichloride).[145] At room temperature pyrazine–alkyl halide interaction yields monoquaternary salts, and 2-substituted pyrazines react mainly at position 4.[146] 2-Aminopyrazine has been shown by both UV and PMR spectroscopy to form a 4-methiodide,[147, 148] and 2-methylpyrazine forms a methiodide of analogous structure.[148] The PMR spectra of these salts show well-defined ^{14}N–H splittings and this is attributed to the presence of an electron-attracting nitrogen atom para to the quaternized nitrogen.[148] 2,5-Dimethylpyrazine 1-oxide quaternizes at position 4 on treatment with methyl iodide and benzyl chloride; no alkylation at the exocyclic oxygen atom is observed [Eq. (6)].[149]

(RX = CH$_3$I or C$_6$H$_5$CH$_2$Cl)

Pyrazines form diquaternary salts on treatment with triethyloxonium fluoroborate. Using this reagent the 1,4-diethylpyrazinium difluoroborates of pyrazine, 2,5-dimethyl-, and 2,6-dimethylpyrazines have been obtained in 96, 97, and 46% yield, respectively. The reaction is subject to considerable steric hindrance since the yield of diquaternary salt from tetramethylpyrazine is only 6%. The diquaternary salts are extremely reactive substances and that of the parent compound is shown by ESR measurements to be readily reducible to the radical cation [Eq. (7)].[150]

[145] A. B. P. Lever, J. Lewis, and R. S. Nyholm, *J. Chem. Soc.* 3156 (1963).
[146] G. F. Duffin, *Advan. Heterocycl. Chem.* **3**, 1 (1964).
[147] G. W. H. Cheeseman, *J. Chem. Soc.* 242 (1960).
[148] T. Goto and M. Isobe, *Tetrahedron Lett.* 1511 (1968).
[149] C. F. Koelsch and W. H. Gumprecht, *J. Org. Chem.* **23** 1603 (1958).
[150] T. J. Curphey, *J. Amer. Chem. Soc.* **87**, 2063 (1965).

Pyrazine diquaternary salts of types **16**[151] and **17**[152] are known and also 6,7-dihydro derivatives of **16**.[152a] The diquaternary salt (**17**) was originally formulated as the monohydrate of **18**, but attempts to dehydrate **17** to **18** have not been successful.[152] Dipyrido[1,2-a; 2',1'-c]pyrazinium dibromide (**16**) has herbicidal properties; it is reduced in aqueous solution to a stable radical cation.[153] The related pyrido[1,2-a]pyrazinium salts of type **19** have also been prepared.[153, 154a]

The formation of N-oxides is analogous to the preparation of quaternary salts. Treatment with 30% hydrogen peroxide–glacial acetic acid yields pyrazine mono- and di-N-oxides depending on temperature and length of reaction.[155] Under these conditions pyridine also gives its N-oxide, but pyrimidine is oxidized only to its mono-N-oxide, and the pyridazine ring opens. Trifluoroperacetic acid is the reagent of choice for di-N-oxidation.[135]

Nucleophilic attack on the ring carbon atoms of pyrazine follows the pattern of pyridine and pyrimidine substitutions.[156, 157] Thus,

---

[151] D. H. Corr and E. E. Glover, *J. Chem. Soc.* 5816 (1966).
[152] A. L. Black and L. A. Summers, *J. Chem. Soc. D* 482 (1970).
[152a] A. L. Black and L. A. Summers, *J. Heterocycl. Chem.* **8**, 29 (1971).
[153] A. L. Black and L. A. Summers, *J. Chem. Soc. C* 610 (1969).
[154] E. E. Glover and M. J. R. Loadman, *J. Chem. Soc. C* 2391 (1967).
[154a] J. Adamson and E. E. Glover, *J. Chem. Soc. C* 1524 (1970).
[155] B. Klein and J. Berkowitz, *J. Amer. Chem. Soc.* **81**, 5160 (1959).
[156] G. Illuminati, *Advan. Heterocycl. Chem.* **3**, 285 (1964).
[157] R. G. Shepherd and J. L. Fedrick, *Advan. Heterocycl. Chem.* **4**, 146 (1965).

aminopyrazine is obtained by the reaction of pyrazine and sodamide. Organometallic reagents may give products of ring or side-chain substitution. In the case of 2,5- and 2,6-dimethylpyrazines ring substitution occurs on treatment with ethyllithium[158] (see Section V,A for further examples). Monosubstituted pyrazines and α-substituted pyridines have comparable reactivities. Kinetic studies by Barlin and Brown on the reaction of methylsulfonylpyridines, -pyridazines, and -pyrazine with methoxide ion have shown that the methylsulfonyl group is much more readily displaced than a chloro group. Methylsulfonylpyrazine is less reactive than 3-methylsulfonylpyridazine which is, in turn, less reactive than 4-methylsulfonylpyridazine.[159] Subsequent work has shown that the methylsulfinyl group is displaced with approximately the same ease as the methylsulfonyl group, but that the methylthio group is much more difficult to displace.[160,161] Pyrazine and other heteroaromatic bases undergo homolytic amidation. Thus, carboxamidopyrazine is obtained in greater than 80% yield when the base is treated in concentrated sulfuric acid with $CONH_2$ radicals generated from the action of hydrogen peroxide and ferrous sulfate on formamide.[162] Alkyl-, dialkyl-, amino-, and methoxypyrazines have been alkylated with aldehydes and ketones in the presence of an alkali metal and liquid ammonia. The mechanism of this reaction is envisaged as the attack of the pyrazine radical anion on the carbonyl group.[162a]

Recently some interesting ring isomerization reactions of pyrazines have been reported. Both under photolytic and pyrolytic conditions pyrimidines are formed. Lahmani and Ivanoff irradiated pyrazine in various organic solvents and pyrazine, methyl-, and 2,5- and 2,6-dimethylpyrazine in the vapor phase.[163,164] The products of these reactions are shown in Scheme 13. The presence of 2-methylpyrimidine in the photolysis products of 2-methylpyrazine was suspected, but not proved. The authors interpret the above data as indicative of the

[158] J. Gelas and R. Rambaud, *C.R. Acad. Sci. C* **266**, 625 (1968).
[159] G. B. Barlin and W. V. Brown, *J. Chem. Soc. B* 648 (1967).
[160] G. B. Barlin and W. V. Brown, *J. Chem. Soc. B* 1435 (1968).
[161] G. B. Barlin and W. V. Brown, *J. Chem. Soc. C* 921 (1969).
[162] F. Minisci, G. P. Gardini, R. Galli, and F. Bertini, *Tetrahedron Lett.* 15 (1970).
[162a] A. F. Bramwell, L. S. Payne, G. Riezebos, P. Ward, and R. D. Wells, *J. Chem. Soc. C* 1627 (1971).
[163] F. Lahmani, N. Ivanoff, and M. Magat, *C.R. Acad. Sci. C* **263**, 1005 (1966).
[164] F. Lahmani and N. Ivanoff, *Tetrahedron Lett.* 3913 (1967).

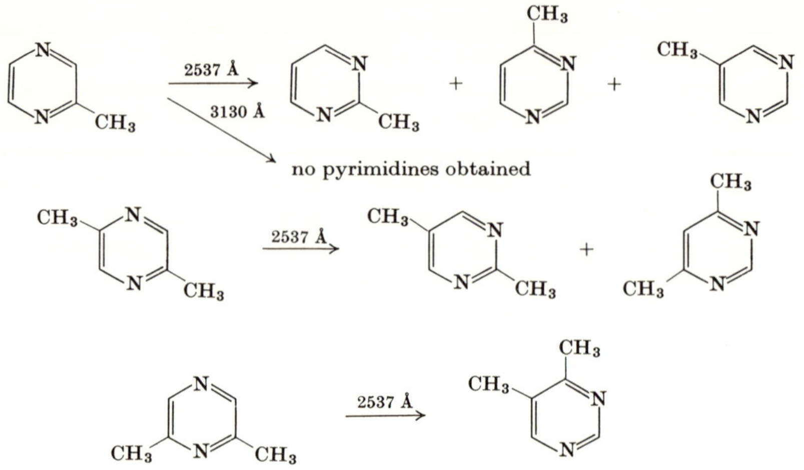

SCHEME 13

formation of diazabenzvalene-type intermediates arising from the lowest singlet $\pi\pi^*$ excited state as shown in Scheme 14. Crow and Wentrup's work on the thermal decomposition of pyrazine suggests

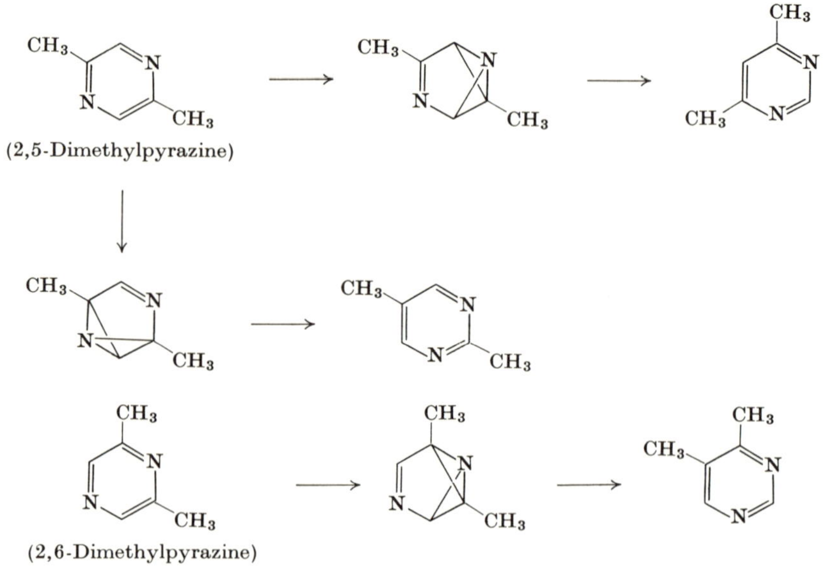

(2,5-Dimethylpyrazine)

(2,6-Dimethylpyrazine)

SCHEME 14

that, besides ring isomerism via diazabenzvalene, *cycloaddition* may also be an important route for pyrimidine ring formation.[165] Thus, vapor phase pyrolysis of pyrazine (carried out at 1270°K in a silica tube) yielded acetylene and hydrogen cyanide (36%), pyrimidine (3%), and pyridine (0.7%); 45% unreacted pyrazine was recovered and no other products were obtained. Pyridine must be formed from cycloaddition of one molecule of hydrogen cyanide and two of acetylene. Cycloaddition of two molecules of hydrogen cyanide and one of acetylene could, however, give pyrimidine or pyridazine, but since N–N orientation is unfavorable no pyridazine would be expected. The primary products of thermal decomposition, hydrogen cyanide and acetylene, are formed in a 2:1 molecular ratio, so pyrimidine formation is favored over pyridine formation in a 4:1 ratio. This is approximately the ratio in which pyrimidine and pyridine are found in the pyrolysis products. As in the ultraviolet-induced rearrangement,

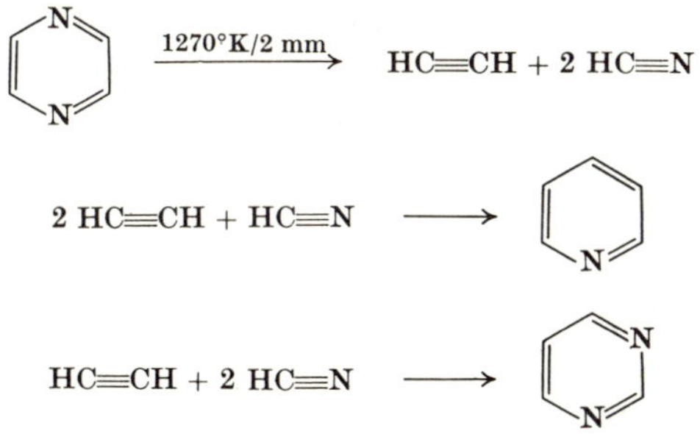

pyrazine $\xrightarrow{1270°K/2\,mm}$ HC≡CH + 2 HC≡N

2 HC≡CH + HC≡N ⟶ pyridine

HC≡CH + 2 HC≡N ⟶ pyrimidine

diazabenzvalene formation would also lead to pyrimidine. Because of N–N repulsions the once-formed pyrimidine is not likely to undergo further rearrangement by a diazabenzvalene-type intermediate to pyridazine.

V. Substituted Pyrazines

A. Alkyl- and Arylpyrazines

Alkyl- and arylpyrazines are the most readily accessible pyrazine derivatives and are obtained by a number of the general methods described in Section III.

[165] W. D. Crow and C. Wentrup, *Tetrahedron Lett.* 3115 (1968).

An improved procedure for the laboratory preparation of 2,5-dimethylpyrazine has been reported.[166] α-Amino alcohols are convenient precursors for the industrial preparation of alkylpyrazines. Thus when they are heated in the vapor phase with copper chromite catalysts, they are converted mainly into pyrazines [Eq. (8)]; with hydrogenation catalysts such as Raney nickel, piperazines are the

$$\underset{\underset{NH_2}{|}}{\overset{CH_3}{\underset{CH_2}{\overset{|}{CH}}}}\text{—OH} \xrightarrow{\text{copper chromite}} \underset{(65\%)}{\text{[2,5-dimethylpyrazine]}} + \underset{(10\%)}{\text{[2,5-dimethylpiperazine]}} \quad (8)$$

major products.[167-169] 2,3-Dihydropyrazines [Eq. (9)] and piperazines [Eq. (10)] may be dehydrogenated in the vapor phase to alkyl- or arylpyrazines (see Section III) and this method is particularly valuable for the preparation of methylpyrazines.[170]

$$\text{[2,3-dimethyl-5,6-dihydropyrazine]} \xrightarrow[300°]{\text{Fe}_2\text{O}_3/\text{acid clay}} \text{[2,3-dimethylpyrazine]} \quad (9)$$

$$\text{[2,5-dimethylpiperazine]} \xrightarrow[300°-375°]{\text{copper chromite}} \text{[2,5-dimethylpyrazine]} \quad (10)$$

The following synthesis (Scheme 15) of 2-ethylpyrazine, based on the reaction of ethylenediamine and 1,2-butylene oxide, is illustrative of the further application of these reactions.[171]

Pyrazines and piperazines are obtained by catalytic reduction of α-nitroketones; thus, reduction of 1-nitro-2-butanone in the presence

---

[166] H. I. X. Mager and W. Berends, *Rec. Trav. Chim.* **77**, 827 (1958).
[167] British Patent 791,050 (1958); *Chem. Abstr.* **52**, 15599 (1958).
[168] W. K. Langdon, W. W. Levis, Jr., D. R. Jackson, M. Cenker, and G. E. Baxter, *Ind. Eng. Chem. Prod. Res. Develop.* **3**, 8 (1964).
[169] W. K. Langdon, U.S. Patent 2,813,869 (1957); *Chem. Abstr.* **52**, 5489 (1958).
[170] T. Ishiguro, M. Matsumura, and H. Murai, *J. Pharm. Soc. Jap.* **80**, 314 (1960).
[171] H. Gainer, *J. Org. Chem.* **24**, 691 (1959).

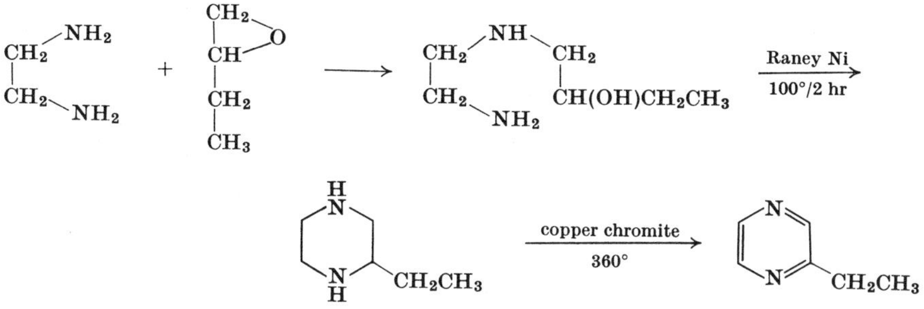

SCHEME 15

of Raney nickel gives 2,5-diethylpyrazine as the major product.[172] 2,6-Dialkylpyrazines have been obtained from acetylenic alcohols and ammonium formate, or formamide, or urea in the presence of formic acid.[172a] Reduction of 3,4-dimethyl-1,2,5-selenadiazole with ammonium hydrogen sulfide gives tetramethylpyrazine in excellent yield (Scheme 16)[173] and tetraphenylpyrazine has been prepared by the carbon monoxide and hydrogen reduction of benzil dioxime N,N'-dimethyl ether in benzene and in the presence of dicobalt octacarbonyl.[174]

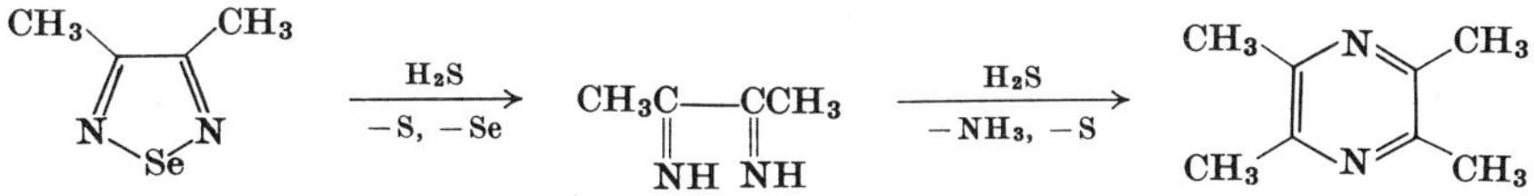

SCHEME 16

The methyl group of a methylpyrazine is activated similarly to the methyl group of 2- and 4-methylpyridines (the α- and γ-picolines). Thus treatment with strong base (e.g., sodamide in liquid ammonia) generates a reactive anion which may undergo aldol-type addition, acylation, alkylation, and nitrosation. These and other features of the reactivity of methylpyrazines are illustrated by reference particularly to the work of Levine and his co-workers.

[172] A. F. Ellis, U.S. Patent 3,453,278 (1969); *Chem. Abstr.* **71**, 91526 (1969).
[172a] A. Bonzom and B. Tramier, *Ger. Offen.* 1,958,619 (1970); *Chem. Abstr.* **73**, 35401 (1970).
[173] V. Bertini, *Gazz. Chim. Ital.* **97**, 1870 (1970).
[174] A. Rosenthal, *Can. J. Chem.* **38**, 2025 (1960).

The sodio derivative of methylpyrazine (20) reacts with aldehydes and ketones to give secondary and tertiary pyrazine carbinols, respectively.[175, 176] Yields in the range 21–99% are obtained when the molar ratio of methylpyrazine:sodamide:carbonyl compound used is 2:2:1. Oppenauer oxidation of 1-phenyl-2-pyrazylethanol (21), the aldol addition product of methylpyrazine and benzaldehyde, gives phenacylpyrazine (22) in 18% yield.[176]

Pyrazylmethylsodium (20) has been reacted with a wide range of esters. Thus, reaction with methyl benzoate gives phenacylpyrazine in 95% yield. When the ester used for acylation is ethyl formate, bis(pyrazylmethyl)carbinol (24) is obtained instead of the expected pyrazylacetaldehyde (23).[177]

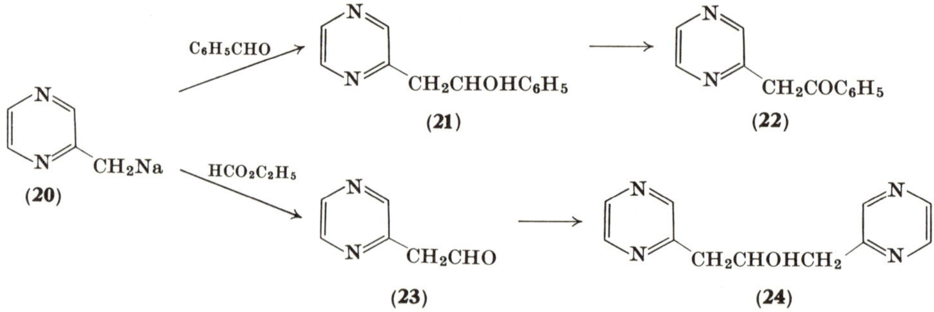

The reaction of the sodium derivative of methylpyrazine with alkyl halides in liquid ammonia gives good yields (44–81%) of mono-alkylated products; phenylation has been achieved by reaction with benzyne (generated from bromobenzene and sodamide in liquid ammonia) and gives 53% benzylpyrazine.[178] The alkylation of the sodio derivative of 2-methoxy-3-methylpyrazine with methyl iodide, and of the product with ethyl bromide, gives 2-methoxy-3-sec-butylpyrazine (25) a constituent of galbanum oil.[31]

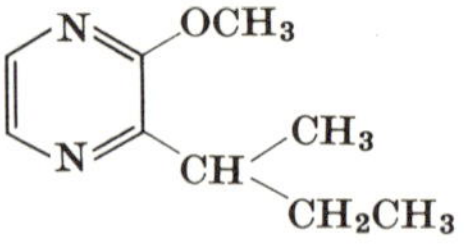

[175] H. E. Zaugg, R. W. deNet, and M. Freifelder, J. Amer. Chem. Soc. **80**, 2773 (1958).
[176] J. D. Behun and R. Levine, J. Amer. Chem. Soc. **81**, 5666 (1959).
[177] J. D. Behun and R. Levine, J. Amer. Chem. Soc. **81**, 5157 (1959).
[178] J. D. Behun and R. Levine, J. Org. Chem. **26**, 3379 (1961).

The reaction of pyrazylmethylsodium with styrene oxide gives the secondary carbinol (**26**) in 73% yield and oxidation of compound **26** with acidic potassium dichromate yields the corresponding ketone.[179] The sodio derivative reacts with butyl nitrite to give pyrazinaldoxime (**27**) in 39% yield.[180]

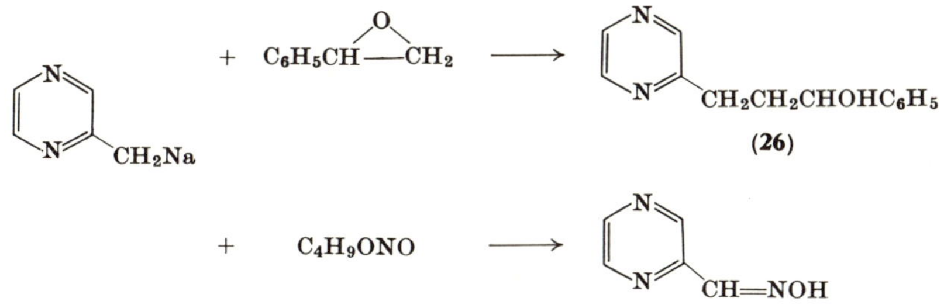

The alkylation of pyrazylmethylsodium with β-dimethylaminoethyl chloride gives 1-pyrazyl-3-dimethylaminopropane (**28**).[178] Compound **28** forms the expected metalated derivative on treatment with sodamide and liquid ammonia and this with ketones gives carbinols,[179] and with esters yields ketones (e.g., **29**).[181] Reaction of pyrazylmethyl sodium with N-methyl-N-phenylcyanamide gives pyrazylacetonitrile.[182]

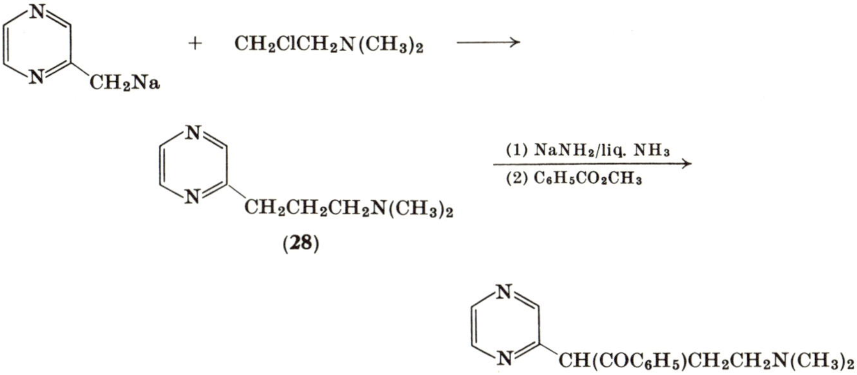

---

[179] M. R. Kamal and R. Levine, *J. Org. Chem.* **27**, 1360 (1962).
[180] S. E. Forman, U.S. Patent 3,150,135 (1964); *Chem. Abstr.* **62**, 2765 (1965).
[181] M. R. Kamal and R. Levine, *J. Org. Chem.* **27**, 1355 (1962).
[182] A. A. Akkerman, G. C. van Leeuwen, and J. F. Michels, French Patent 1,404,514 (1965); *Chem. Abstr.* **64**, 3572 (1966).

It is of interest to note that the sodio derivative of phenacylpyrazine undergoes *O*- rather than *C*-alkylation with β-dimethylaminoethyl chloride and thus compound **30** is formed.[181]

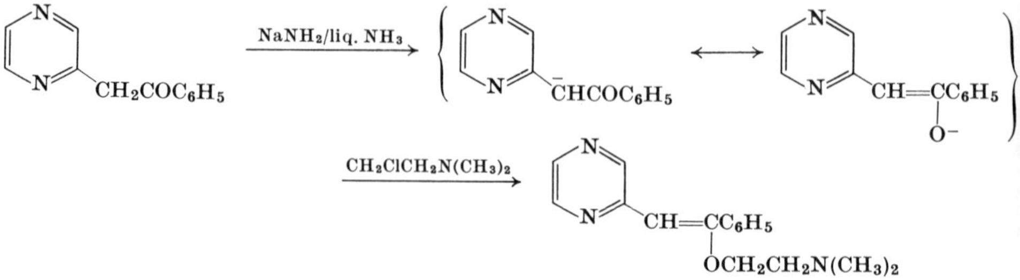

Oxidation of phenacylpyrazine with potassium hypochlorite gives a mixture of dichloromethylpyrazine and benzoic acid; the corresponding reaction with 2-acetonylpyrazine give dichloromethylpyrazine and acetic acid [Eq. (11)].[183]

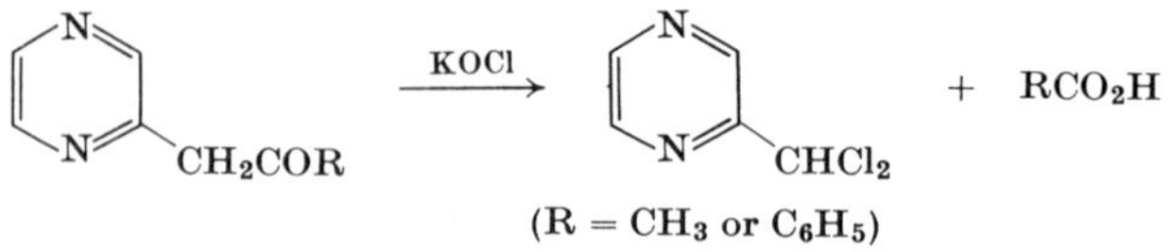

Acylation of 2-methylpyrazine with the Vilsmeier reagent gives 3-dimethylamino-2-(2-pyrazyl)acrolein (**31**) and this on hydrolysis with potassium hydroxide is converted into the malondialdehyde

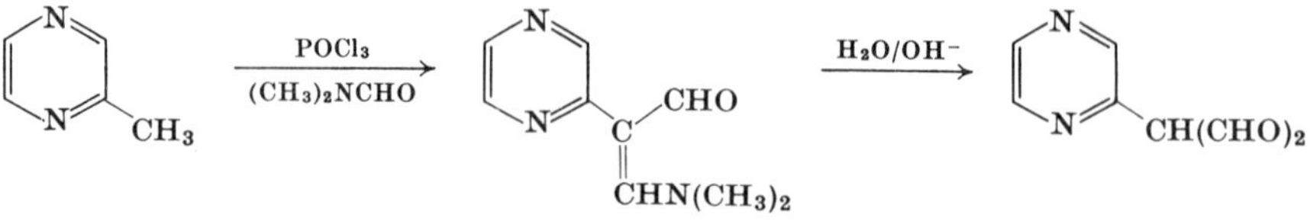

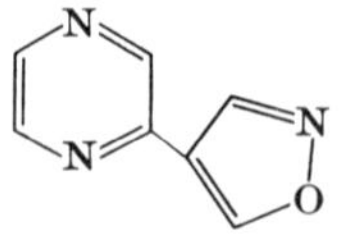

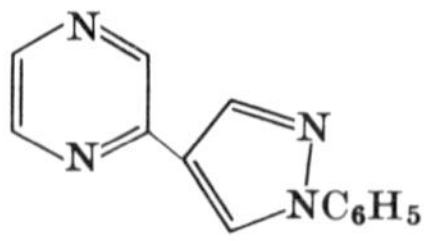

[183] J. D. Behun and R. Levine, *J. Org. Chem.* **23**, 406 (1958).

(32).[184] The Vilsmeier adduct is a versatile intermediate; it reacts with hydroxylamine and phenylhydrazine to give the isoxazole (33) and the pyrazole (34) and with ureas and amidines to give pyrimidine derivatives. 2-Methylpyrazine has been condensed with benzaldehyde in the presence of zinc chloride to give 2-styrylpyrazine in good yield.[185]

2,5-Dimethyl-[186] and 2,6-dimethylpyrazine react very similarly to methylpyrazine. Thus, 2,6-dimethylpyrazine can be converted into a monoanion with sodamide in liquid ammonia which can be condensed with aldehydes and ketones,[179] acylated with esters,[181] and alkylated with alkyl halides[181] to give the corresponding 2-methyl-6-substituted pyrazines. Acylation under suitable conditions also yields diacylated derivatives. Thus, when 2,6-dimethylpyrazine, sodamide, and ethyl benzoate are reacted in 1:3:2 molecular proportion, 38% 2,6-diphenacylpyrazine (35) and 25% 2-methyl-6-phenacylpyrazine (36) is obtained. From the preparative point of view it is better to form the diacyl derivative by the further acylation of the monoacyl derivative rather than by direct diacylation.[187]

$$\text{CH}_3\text{-pyrazine-CH}_3 \xrightarrow[\text{(2) C}_6\text{H}_5\text{CO}_2\text{CH}_3]{\text{(1) NaNH}_2/\text{liq. NH}_3} \text{C}_6\text{H}_5\text{COCH}_2\text{-pyrazine-CH}_2\text{COC}_6\text{H}_5 \quad + $$

(35)

$$\text{CH}_3\text{-pyrazine-CH}_2\text{COC}_6\text{H}_5$$

(36)

The reaction of acetonylpyrazine with phenyllithium in the absence, and in the presence, of methyl benzoate has been studied in some detail by Chakrabartty and Levine.[188] With phenyllithium alone, 2-phenyl-6-acetonylpyrazine (37) is formed, whereas in the presence of ester, a mixture of compound 37 and 3-(2-phenyl-6-pyrazyl)-4-phenylbutane-2,4-dione (38a) is obtained. Spectroscopic data indicate that compound 38a exists predominantly as the dihydropyrazine

[184] H. V. Hausen, J. A. Caputo, and R. I. Meltzer, *J. Org. Chem.* **31**, 3845 (1966).
[185] S. Yamada and T. Ueda, Japanese Patent 3367 (1961); *Chem. Abstr.* **57**, 3458 (1962).
[186] M. E. Strem, *Diss. Abstr.* **26**, 1355 (1965).
[187] M. R. Kamal and R. Levine, *J. Org. Chem.* **29**, 191 (1964).
[188] S. K. Chakrabartty and R. Levine, *J. Heterocycl. Chem.* **4**, 109 (1967).

(38b). The preference for acylation at the methylene rather than the terminal methyl group is explained by the stabilization resulting from cross-conjugation of both carbonyl groups in tautomer 38b with the dihydrophenylpyrazine ring, and the additional stabilization resulting from enolic tautomeric forms. 2-Acetonylpyridine, unlike 2-acetonylpyrazine, does not react at the heterocyclic ring with phenyllithium; the increased electrophilic character of the 6-carbon in the pyrazine is attributed to the additional ring nitrogen atom.

The reactions of the isomeric dimethylpyrazines and trimethylpyrazine with methyllithium have been studied in order to gain insight into the factors involved in the competition between ring methylation and side-chain metalation.[189] The major product from the reaction of 2,5-dimethylpyrazine and ethereal methyllithium was shown to be trimethylpyrazine, thus confirming Klein and Spoerri's earlier observation.[190, 191] Tetramethylpyrazine was also formed as a by-product. Proof of side-chain metalation was obtained by treatment of the reaction mixture with methyl benzoate and isolation of 2-methyl-5-phenacylpyrazine. Evidence for the presence of dihydro- and tetrahydropyrazine intermediates is derived from the infrared spectrum of the crude product obtained on hydrolysis of the reaction mixture which shows C=N and N–H absorptions (see Scheme 17).

The major product from the reaction of 2,6-dimethylpyrazine and an ethereal solution of methyllithium is trimethylpyrazine, but 2,3-dimethylpyrazine undergoes exclusive side-chain metalation under

[189] G. P. Rizzi, J. Org. Chem. 33, 1333 (1968).
[190] B. Klein and P. E. Spoerri, J. Amer. Chem. Soc. 72, 1844 (1950).
[191] B. Klein and P. E. Spoerri, J. Amer. Chem. Soc. 73, 2949 (1951).

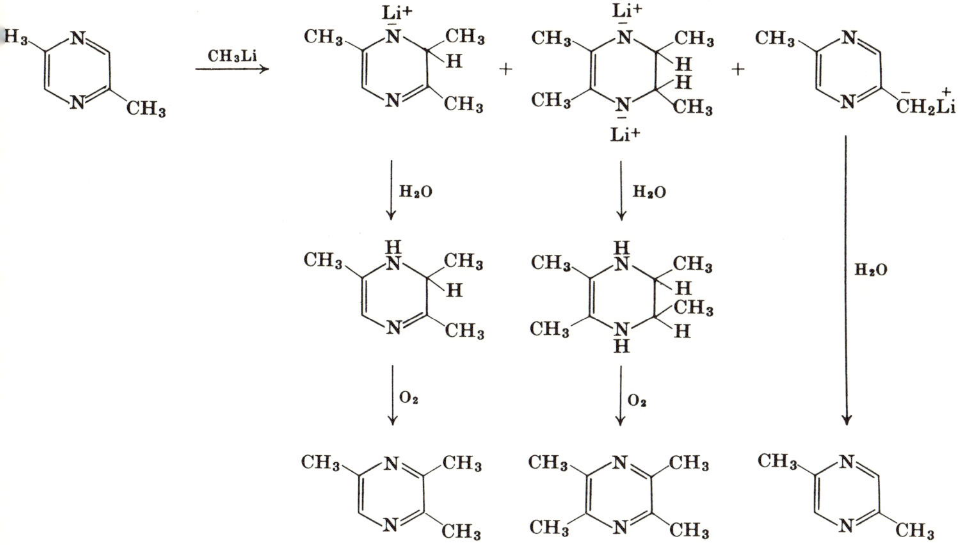

SCHEME 17

these conditions. Ring-alkylated products are obtained, however, when the 2,3-isomer is treated with either ethyllithium or *n*-butyllithium in hexane or benzene solution.

Tetramethylpyrazine can be converted into its monoanion by the use of sodamide in liquid ammonia or phenyllithium in ether.[192] Thus, when phenyllithium, tetramethylpyrazine, and methyl benzoate are reacted in 2:2:1 molar proportion, a 67% yield of the monoacylated product, phenacyltrimethylpyrazine, is formed. A lower yield of monoacylated product is obtained when sodamide and liquid ammonia is used as the condensing agent, but reaction of sodamide, tetramethylpyrazine, and methyl benzoate in 6:1:1 molar proportion gives the optimum yield of diacylated derivative, formulated as 2,6-diphenacyl-3,5-dimethylpyrazine (39), by UV spectral comparison with 2,6-dimethylpyrazine and by the formation of a dioxime.

The reaction of tetramethylpyrazine with alkyl halides using either sodamide in liquid ammonia or phenyllithium in ether as condensing agent gives mixtures of alkyltrimethylpyrazines and 2,5-dialkyl-3,6-dimethylpyrazines. A number of carbinols were also prepared from

[192] S. K. Chakrabartty and R. Levine, *J. Heterocycl. Chem.* **1**, 196 (1964).

$$\text{CH}_3 \underset{\text{CH}_3}{\overset{\text{N}}{\bigcirc}} \text{CH}_3 \xrightarrow[\text{(2) C}_6\text{H}_5\text{CO}_2\text{CH}_3]{\text{(1) NaNH}_2/\text{liq. NH}_3} \text{CH}_3 \underset{\text{C}_6\text{H}_5\text{COCH}_2}{\overset{\text{N}}{\bigcirc}} \text{CH}_3$$

(39)

tetramethylpyrazine and aldehydes and ketones using phenyllithium as condensing agent.[193, 194]

Methylpyrazine reacts with formaldehyde and dimethylamine to give the expected product (40) of Mannich reaction. This is converted in 63% overall yield to vinylpyrazine (41) by quaternization with methyl iodide followed by sodium hydroxide treatment.[195]

[Reaction scheme showing methylpyrazine → (40) CH₂CH₂N(CH₃)₂ via (CH₃)NH/HCHO, then CH₃I → CH₂CH₂N(CH₃)₃⁺ I⁻, then NaOH → (41) CH=CH₂]

Ozonolysis of vinylpyrazine in methanol at $-30°$ furnishes pyrazine aldehyde in 73% yield.[196] Vinylpyrazine undergoes a variety of addition reactions and pyrazylethyl derivatives of amines, ketones, ethyl phenyl acetate, phenylacetonitrile, and acetamide have been obtained.[197–199] 2-(2-Pyrazylethyl)cyclohexanone (42) has been prepared both by the condensation of vinylpyrazine with cyclohexanone in the presence of sodium metal and by interaction of vinylpyrazine with the pyrrolidine enamine of cyclohexanone followed by hydrolysis.[200]

The addition reactions of methyl-, 2,5-dimethyl-, 2,6-dimethyl-, and tetramethylpyrazine with dimethyl acetylenedicarboxylate have

[193] S. K. Chakrabartty and R. Levine, *J. Heterocycl. Chem.* **3**, 265 (1966).
[194] S. K. Chakrabartty, *Diss. Abstr.* **27**, 2646 (1967).
[195] M. R. Kamal, M. Neubert, and R. Levine, *J. Org. Chem.* **27**, 1363 (1962).
[196] A. Hirschberg and D. P. Smith, *J. Heterocycl. Chem.* **3**, 103 (1966).
[197] G. M. Singerman and R. Levine, *J. Heterocycl. Chem.* **1**, 151 (1964).
[198] G. M. Singerman and R. Levine, *J. Org. Chem.* **30**, 4379 (1965).
[199] G. M. Singerman, *Diss. Abstr.* **26**, 100 (1965).
[200] G. M. Singerman and S. Danishefsky, *Tetrahedron Lett.* 2249 (1964).

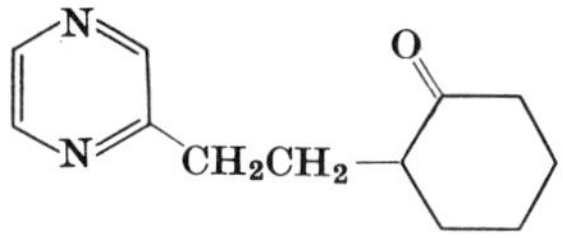

(42)

been studied by Acheson and his co-workers. From methyl- and 2,6-dimethylpyrazine low yields of pyrrolo[1,2-*a*]pyrazines (43) are obtained.[201]

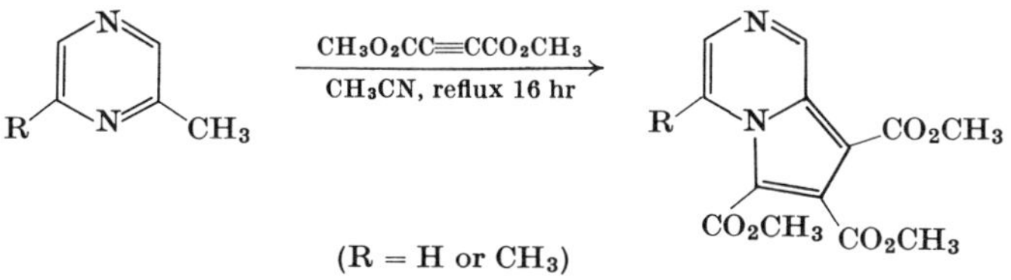

(R = H or CH$_3$)

(43)

The reaction of 2,5-dimethylpyrazine and dimethyl acetylenedicarboxylate is more complex: the initially formed 1:2-adduct (44) undergoes Diels–Alder addition with a third molecule of the ester. Subsequent elimination of methyl cyanide yields the pyridoazepine (45).[202] The reaction of 2-methylpyridine and dimethyl acetylenedicarboxylate also leads to azepine formation and similar mechanisms are proposed for both reactions.[203, 204]

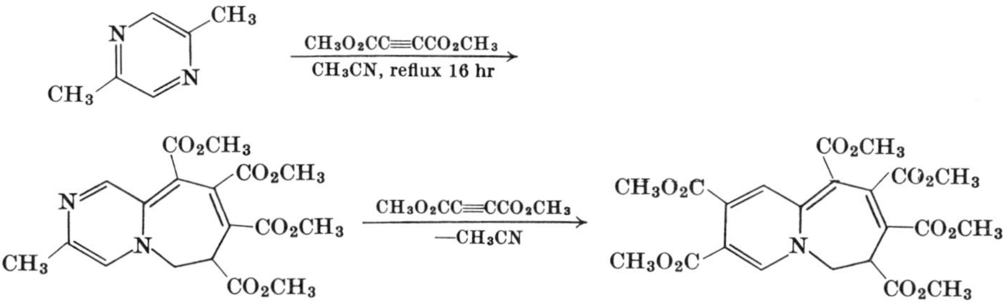

(44)   (45)

[201] R. M. Acheson and M. W. Foxton, *J. Chem. Soc. C* 2218 (1966).
[202] R. M. Acheson, M. W. Foxton, and J. K. Stubbs, *J. Chem. Soc. C* 926 (1968).
[203] R. M. Acheson, J. M. F. Gagan, and D. R. Harrison, *J. Chem. Soc. C* 362 (1968).
[204] R. M. Acheson, *Advan. Heterocycl. Chem.* **1**, 125 (1963).

Tetramethylpyrazine combines with three molar proportions of dimethyl acetylenedicarboxylate to give, after elimination of methanol, a dipyridopyrazine formulated as compound **46** or **47**. The spectroscopic data do not enable the relative positions of the carbonyl and ester functions at positions 9 and 11 to be determined.[202]

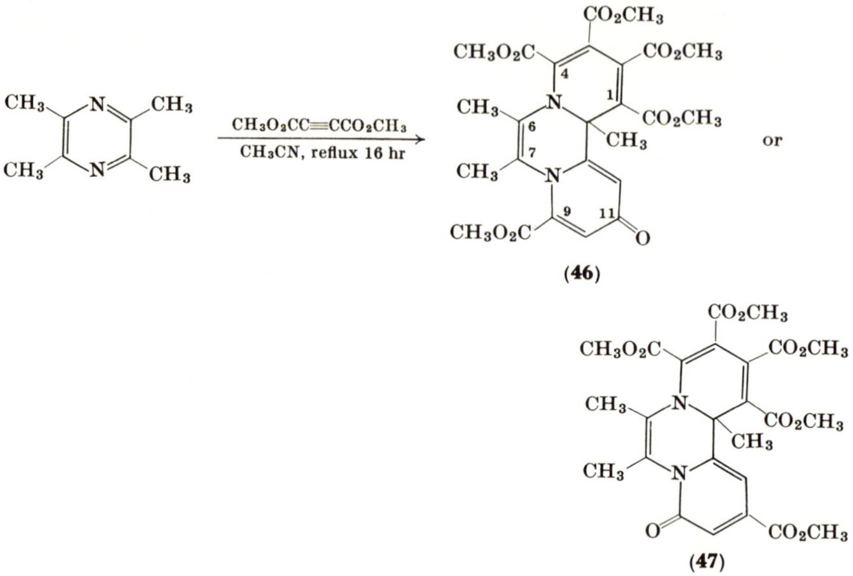

The reaction between dimethyl acetylenedicarboxylate and the zwitterion derived by the action of base on 1-phenacyl-2,5-dimethylpyrazinium bromide gives mainly 1,2-dimethoxycarbonyl-5,8-di-

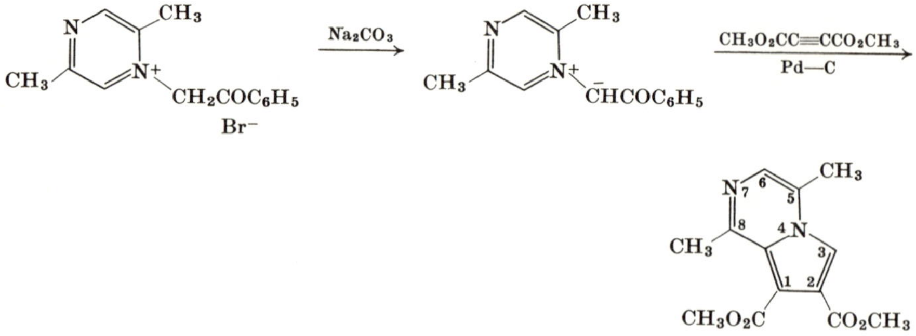

SCHEME 18

methyl-7-azaindolizine together with some of the corresponding 3-benzoyl derivative (Scheme 18).[204a]

B. Pyrazinecarboxylic Acids and Their Derivatives

Pyrazinecarboxylic acids are conveniently prepared by the permanganate oxidation of condensed ring systems containing the pyrazine ring (see Section III). Commercially the electrolytic oxidation of quinoxaline (**48**) is the preferred route to pyrazine-2,3-dicarboxylic acid (**49**). High yields (up to 92%) have been obtained by carrying out the electrolysis at 40°, in 5% sodium hydroxide solution containing 5% potassium permanganate as electrolyte, and supplying quinoxaline and 40% sodium hydroxide continuously.[205, 206]

(**48**)   →   (**49**)

5,6-Dimethylpyrazine-2,3-dicarboxylic acid is obtained in 67% yield by permanganate oxidation of 2,3-dimethylquinoxaline.[207] 2,3-Dialkoxyquinoxalines are very resistant to permanganate oxidation, but their 5- or 6-amino derivatives can be converted into pyrazine dicarboxylic acids under fairly mild conditions.[208] 2,3-Dihydroxyquinoxaline (**50**) has been converted into the corresponding dihydroxypyrazinedicarboxylic acid (**52**) indirectly through 2,3-dichloroquinoxaline (**51**).[209] Permanganate oxidation of 2-acetyl-

(**50**) $\xrightarrow{PCl_5}$ (**51**) $\xrightarrow[OH^-]{KMnO_4}$ (**52**)

[204a] V. Boekelheide and K. Fahrenholtz, *J. Amer. Chem. Soc.* **83**, 458 (1961).
[205] T. Kimura, S. Yamada, K. Yoshisue, and T. Nagoya, *J. Pharm. Soc. Jap.* **77**, 891 (1957).
[206] T. Kimura, S. Yamada, and K. Koshisue, Japanese Patents 7932 and 8125 (1959); *Chem. Abstr.* **54**, 13911 and 16472 (1960).
[207] R. A. Pages and P. E. Spoerri, *J. Org. Chem.* **28**, 1702 (1963).
[208] H. I. X. Mager and W. Berends, *Rec. Trav. Chim.* **78**, 5 (1959).
[209] H. I. X. Mager and W. Berends, *Rec. Trav. Chim.* **77**, 842 (1958).

aminoquinoxaline (53) gives 2-acetylaminopyrazine-5,6-dicarboxylic acid (54). This, on acid hydrolysis, followed by decarboxylation at 170°–175° *in vacuo* yields 2-aminopyrazine-5-carboxylic acid (55).[210]

The decarboxylation of pyrazine-2,3-dicarboxylic acids may be carried out by heating *in vacuo* or in glacial acetic acid at 130°–145°,[211] or by heating in a solvent of high dielectric constant such as nitrobenzene.[212] It is suggested that these solvents promote decarboxylation by zwitterion formation.

The rearrangement of the potassium salt of pyrazine-2,3-dicarboxylic acid at 320° has been reported to give the 2,5-diacid. The catalysts for this process are of the type $R_2(CdX_2Y_2)$ and $R(CdX_3)$, where R is an alkali metal, X is fluorine, and Y any halogen or "half" carbonate. This rearrangement is analogous to the rearrangement of potassium phthalate to potassium terephthalate on heating at 400° in an inert atmosphere.[213–215]

Oxidation of alkyl- or arylpyrazines with dichromate or permanganate yields pyrazinecarboxylic acids. Thus, Beck reported that on heating 2-methylpyrazine with aqueous dichromate and phosphoric acid in an autoclave at 225°–300°, pyrazinecarboxylic acid was obtained in 74% yield.[216] Pyrazine-2,5- and -2,6-dicarboxylic acids

[210] E. Felder, D. Pitre, and E. B. Grabitz, *Helv. Chim. Acta* **47**, 873 (1964).
[211] M. Asai and R. Takasaki, Japanese Patent 8187 (1958); *Chem. Abstr.* **54**, 4634 (1960).
[212] L. Bernardi and G. Larini, *Ann. Chim. (Rome)* **48**, 239 (1958); *Chem. Abstr.* **52**, 15539 (1958).
[213] H. Schutt, U.S. Patent, 2,919,273 (1959); *Chem. Abstr.* **54**, 19729 (1960).
[214] B. Raecke, *Angew. Chem.* **70**, 1 (1958).
[215] British Patent 827,467 (1960); *Chem. Abstr.* **54**, 18565 (1960).
[216] L. H. Beck, U.S. Patent 3,154,549 (1964); *Chem. Abstr.* **62**, 1673 (1965).

are also obtained in this manner from the corresponding dimethylpyrazines. Carefully controlled permanganate oxidation of 2,5- and 2,6-dimethylpyrazines gives 5- and 6-methylpyrazine-2-carboxylic acids, respectively. The former acid is prepared in much higher yield (76% compared to 6%) by permanganate oxidation of 2-hydroxymethyl 5-methylpyrazine.[217]

Pyrazinecarboxylic acid has been obtained by selenious acid oxidation in pyridine of methylpyrazine or aqueous permanganate oxidation of ethylpyrazine, in yields of 64 and 48%, respectively.[171, 218] It has also been obtained in 70% yield by partial decarboxylation of pyrazine-2,3-dicarboxylic acid on heating *in vacuo* at 210°.[219] Aqueous permanganate oxidation of 2,5-distyrylpyrazine gives the 2,5-dicarboxylic acid.[220] Pyrazine-2,5-dicarboxylic acid has also been prepared in 45% yield by direct carboxylation of pyrazine with carbon dioxide at 50 atm pressure at 250° for 3 hours in the presence of a potassium carbonate and calcium fluoride catalyst.[221] Pyrazinetricarboxylic acid (57), obtainable in only very poor yields by oxidation of 2,5-dimethyl-3-ethylpyrazine, is prepared in 87% yield by alkaline permanganate oxidation of 2-(D-arabo)tetrahydroxybutylquinoxaline (56).[222] Decarboxylation of the tricarboxylic acid by

[structure 56: quinoxaline with (CHOH)₃CH₂OH side chain] →KMnO₄→ [structure 57: pyrazine with HO₂C, HO₂C, and CO₂H groups]

(56)   (57)

heating *in vacuo* at 210° gives pyrazine-2,6-dicarboxylic acid, whereas on heating in boiling dimethylformamide the isomeric 2,5-dicarboxylic acid is formed.[222] Pyrazinetetracarboxylic acid has been prepared in 27% yield by alkaline permanganate oxidation of phenazine.[223]

Some alkyl- and aryl-substituted pyrazine carboxylic acids have been prepared by direct condensation (see Section, III, A). For

[217] D. Pitre, S. Boveri, and E. B. Grabitz, *Chem. Ber.* **99**, 364 (1966).
[218] H. Gainer and M. Cenker, U.S. Patent, 3,096,330 (1964); *Chem. Abstr.* **60**, 2975 (1964).
[219] S. F. Hall and P. E. Spoerri, *J. Amer. Chem. Soc.* **62**, 664 (1940).
[220] T. Kimura, S. Yamada, K. Kanzaki, and K. Kato, Japanese Patent 10,510 (1960); *Chem. Abstr.* **55**, 9439 (1961).
[221] British Patent 816,531 (1959); *Chem. Abstr.* **54**, 1552 (1960).
[222] H. I. X. Mager and W. Berends, *Rec. Trav. Chim.* **77**, 827 (1958).
[223] R. J. Light and C. R. Hauser, *J. Org. Chem.* **26**, 1296 (1961).

example, condensation of 2,3-diaminopropionic acid with phenylglyoxal gives 5-phenyl- (58) and 6-phenylpyrazine-2-carboxylic acids (59) in 20 and 31% yield, respectively.

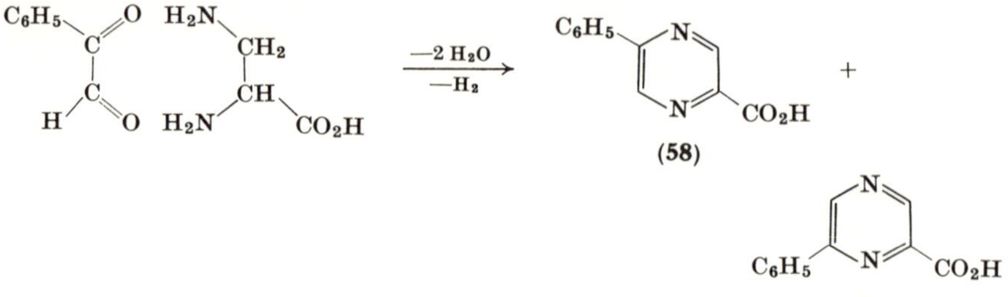

The amides of pyrazine carboxylic acids are generally prepared by conventional methods, i.e., by ester ammonolysis, or acid hydrolysis of the corresponding nitrile. Alternatively these derivatives are available from alkaline hydrolysis of 2,4-dihydroxypteridines or the ring-opening reactions of 4-hydroxypteridines with nucleophilic reagents. A direct synthesis of 2-aminopyrazine-3-carboxamides is available from the condensation of α,β-dicarbonyl compounds with aminomalonamidamidine dihydrochloride (60) in dilute ammonium hydroxide. For example, condensation with glyoxal gives 2-aminopyrazine-3-carboxamide (61) in 76% yield.[116] An alternative synthesis involves the reductive ring cleavage of the N–N bond in 3-hydroxy-1-pyrazolo-[b]pyrazine (62).[224]

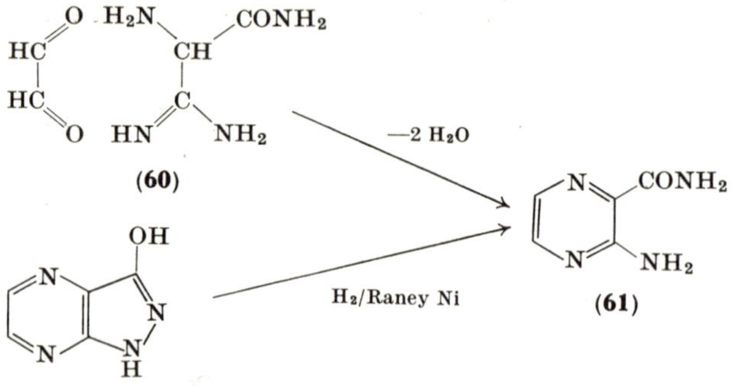

[224] E. C. Taylor, J. W. Barton, and T. S. Osdene, J. Amer. Chem. Soc. 80, 421 (1958).

The corresponding hydroxypyrazinecarboxamides (64) are synthesized by direct condensation of an α,β-dicarbonyl compound with aminomalondiamide (63).[112] A further example of this general synthetic method is the condensation of α,β-dicarbonyl compounds with the tetramer of hydrogen cyanide (65) to give 2,3-dicyanopyrazines (66).[225, 226]

The self-condensation of the bisulfite addition compound of an α-oximinoketone followed by reaction with potassium cyanide and hydrochloric acid yields 3,6-dicyano-2,5-dialkylpyrazines or, in the case of oximinoacetophenone, a mixture of 3-cyano-2,5-diphenylpyrazine and 3,6-dicyano-2,5-diphenylpyrazines.[227]

Cyanopyrazine is obtained in 90% yield by the reaction of phosphoryl chloride and carboxamidopyrazine. The reactants are first kept at room temperature for 5 hours and then heated under reflux for 40 minutes.[228] Commercially, cyanopyrazine is prepared by reaction of methylpyrazine with a mixture of ammonia and air in the presence of a catalyst. The catalysts used are Co, Fe, Sn, and Ag vanadates and vanadium phosphotungstate.[229] Bromopyrazines are readily converted into cyanopyrazines by treatment with cuprous cyanide in boiling γ-picoline.[230]

---

[225] L. E. Hinkel, G. O. Richards, and O. Thomas, *J. Chem. Soc.* 1432 (1937).
[226] H. Bredereck and G. Schmötzer, *Ann.* **600**, 95 (1956).
[227] E. Golombok and F. S. Spring, *J. Chem. Soc.* 1364 (1949).
[228] R. Delaby, R. Damiens, and M. Robba, *C.R. Acad. Sci.* **247**, 822 (1958).
[229] P. Davis, E. F. Schoenewaldt, A. Kaufman, G. T. Wildman, and D. E. Gagliostro, French Patent 1,334,102 (1963); *Chem. Abstr.* **60**, 5520 (1964).
[230] G. Karmas and P. E. Spoerri, *J. Amer. Chem. Soc.* **78**, 2141 (1956).

The pyrazinecarboxylic acids have properties similar to the pyridinecarboxylic acids and aromatic carboxylic acids in general. The pK_a of pyrazine-2-carboxylic acid is 2.92; it is thus considerably stronger than pyridine-2-carboxylic acid (pK_a 5.52), and comparable in acidic strength to pyridazine-3-carboxylic acid (pK_a 3.0). The pK_a values of pyrazine-2,3-dicarboxylic acid are 0.9 and 3.57.[231] Pyrazinecarboxylic acids form colored salts with Fe^{II} ions and they are readily esterified and decarboxylated.

Reference has already been made to the partial decarboxylation of pyrazinedicarboxylic acids.[219] Both pyrazine itself,[232] and 2,3-dimethylpyrazine,[207] are conveniently prepared by decarboxylation of the appropriate dicarboxylic acids. The decarboxylation of pyrazine-2,3-dicarboxylic acid is carried out by heating in di-n-butyl phthalate and gives pyrazine in 90% yield. The 2,3-dicarboxylic acid forms an anhydride in the normal way. Pyrolysis of the anhydride at 800°/0.05 mm through a silica tube gives in 80% yield an approximately 1:1 mixture of maleonitrile and fumaronitrile (Scheme 19). 2,3-Dehydropyrazine is thought to be an intermediate in this reaction and a strong peak of m/e 78, corresponding to the dehydropyrazine ion, is observed in the mass spectrum of the anhydride.[233]

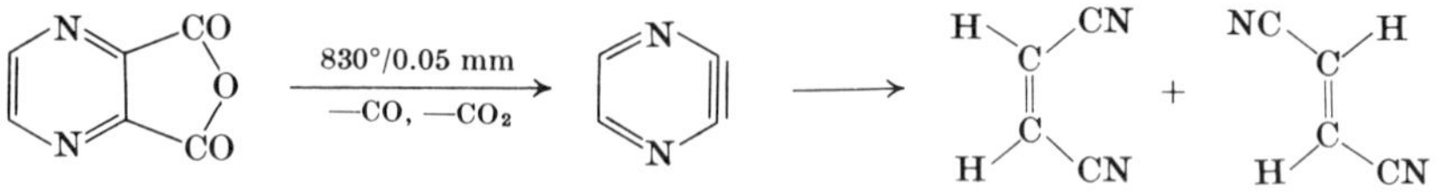

SCHEME 19

Pyrazinetetracarboxylic acid forms a dianhydride on treatment with acetic anhydride,[234] which is useful in the preparation of polyamides and polyimides.[235, 236] For example, polymeric material is obtained by reaction of the dianhydride with 4,4'-diaminodiphenyl ether.

---

[231] S. F. Mason, *J. Chem. Soc.* 1247 (1959).
[232] British Patent 560,965; *Chem. Abstr.* **40**, 5074 (1946).
[233] R. F. C. Brown, W. D. Crow, and R. K. Solly, *Chem. Ind. (London)* 343 (1966).
[234] S. S. Hirsch, French Patent, 1,545,420 (1968); *Chem. Abstr.* **71**, 124, 488 (1969).
[235] Netherlands Patent Appl. 298,294 (1965) and 298,295 (1965); *Chem. Abstr.* **64**, 14386a and d (1966).
[236] V. V. Korshak, G. M. Tseitlin, A. I. Pavlov, T. G. Pogorelova, and V. A. Smirnova, *Bull. Acad. Sci. USSR, Div. Chem. Sci.* 1811 (1968).

Cyanomethyl pyrazinecarboxylate has been prepared by treatment of the acid with α-chloromethyl cyanide in the presence of triethylamine [Eq. (12)].[237]

$$\underset{\text{CO}_2\text{H}}{\text{[pyrazine]}} \xrightarrow{\text{ClCH}_2\text{CN}/(\text{C}_2\text{H}_5)_3\text{N}} \underset{\text{CO}_2\text{CH}_2\text{CN}}{\text{[pyrazine]}} \quad (12)$$

N-(2-Pyrazinoyl)morpholine (**67**) and its 6-chloro (**68**) and 6-methoxy derivatives (**69**) are useful antidiabetic drugs.[238-240] Compound **67** is prepared from pyrazinoyl chloride and morpholine,[241] or alternatively by reaction of pyrazinoic acid with ethyl chloroformate in the presence of triethylamine, followed by reaction of the product with morpholine.[238]

(**67**) (R = H)
(**68**) (R = Cl)
(**69**) (R = OCH$_3$)

Pyrazinoyl chloride reacts with glycine under alkaline conditions to give N-pyrazinoylglycine; this and related pyrazinamides have antituberculosis activity.[242]

Reduction of pyrazinoyl chloride with lithium tri-t-butoxyaluminohydride gives 55% pyrazylmethyl pyrazinoate (**70**) and only 20% pyrazinealdehyde (isolated as the 2,4-dinitrophenylhydrazone). The

---

[237] S. Grudzinski, A. Kotelko, H. Mikolajewska, J. Strumillo, T. Kaczynski, and B. Zajaczkowska, *Acta Pol. Pharm.* **21**, 445 (1964); *Chem. Abstr.* **62**, 6430 (1965).

[238] Y. Abe, Y. Shigeta, F. Uchimaru, and S. Okada, Japanese Patent 12,739 (1969); *Chem. Abstr.* **71**, 112984 (1969).

[239] Y. Abe, Y. Shigeta, F. Uchimaru, S. Okada, and E. Ozasayama, Japanese Patent 12,899 (1969); *Chem. Abstr.* **71**, 91494 (1969).

[240] H. Abe, Y. Shigeta, F. Uchimaru, S. Okada, and E. Kosasayama, Japanese Patent 20,346 (1969) and 20,347 (1969); *Chem. Abstr.* **71**, 124501 and 124502 (1969).

[241] H. Abe, Y. Shigeta, F. Uchimaru, and S. Okada, Japanese Patent 15,777 (1969); *Chem. Abstr.* **71**, 91531 (1969).

[242] K. Kakemi, *J. Pharm. Soc. Jap.* **81**, 1609 (1961).

aldehyde is obtained in 46% yield by reduction of methyl pyrazinoate with lithium aluminum hydride at $-70°$ in tetrahydrofuran[243] (ozonolysis of 2-vinylpyrazine gives an improved yield of the aldehyde).[196]

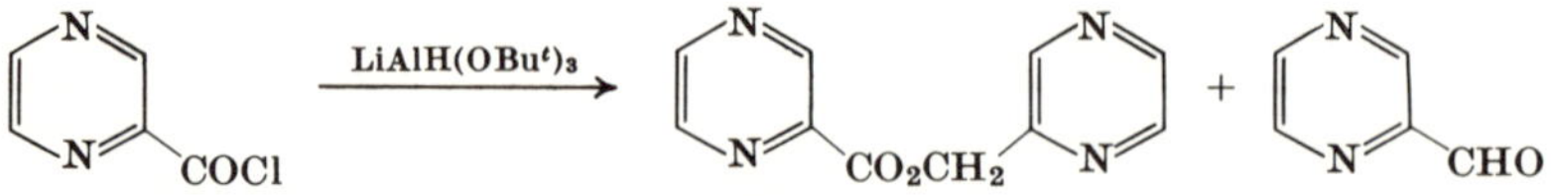

[Structure: pyrazine-COCl → LiAlH(OBu^t)$_3$ → pyrazine-CO$_2$CH$_2$-pyrazine + pyrazine-CHO]

(70)

Pyrazinealdehyde undergoes the normal reactions of aromatic aldehydes. Thus, the Cannizzaro reaction gives the expected products, pyrazylmethanol and pyrazinoic acid, and reaction with aqueous potassium cyanide gives 1,2-dipyrazylethene-1,2-diol (71) in 85% yield.[243]

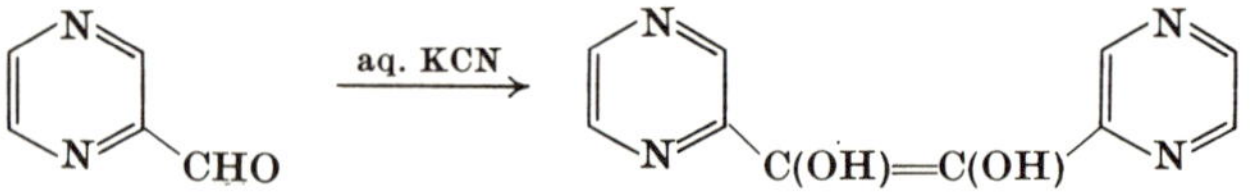

[Structure: pyrazine-CHO → aq. KCN → pyrazine-C(OH)=C(OH)-pyrazine]

(71)

Pyrazinetetracarboxylic acid is converted into tetra(trifluoromethyl)pyrazine by the action of sulfur tetrafluoride [Eq. (13)].[244]

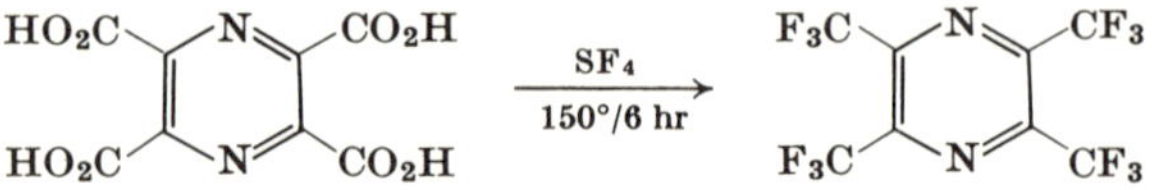

[Structure: 2,3,5,6-tetracarboxypyrazine → SF$_4$, 150°/6 hr → 2,3,5,6-tetrakis(trifluoromethyl)pyrazine] (13)

The amides of pyrazine-2-carboxylic acids and -2,3-dicarboxylic acids undergo the Hofmann reaction. In the case of the conversion of pyrazinamide to aminopyrazine (Scheme 20), the intermediate sodium carbamate can be isolated.[219] Pyrazinamides react with phosphoryl chloride to give the corresponding cyanopyra-

[243] H. Rutner and P. E. Spoerri, *J. Org. Chem.* **28**, 1898 (1963).
[244] W. R. Hasek, W. C. Smith, and V. A. Engelhardt, *J. Amer. Chem. Soc.* **82**, 543 (1960).

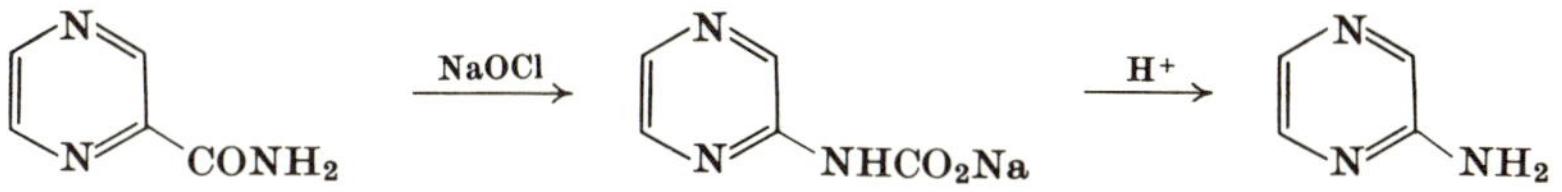

Scheme 20

zines.[228, 245–249] Thus, Cragoe and Jones converted 3-amino-6-chloropyrazinecarboxamide (72) into the 2-cyano derivative (73) by treatment with phosphoryl chloride and dimethylformamide followed by hydrolysis of the intermediate amidino compound.[247] Treatment of pyrazine 2,3-dicarboxamide (74) with thionyl chloride in boiling dimethylformamide gives 77% o-dinitrile (75).[250] The diamide (74) can be converted with potassium hypobromite either into 2-aminopyrazine-3-carboxylic acid or into 2,4-dihydroxypteridine(lumazine) (76).[251] Pyrazine-2,5-dicarboxamide does not undergo the Hofmann reaction; the 2,5-diamine is prepared from the 2,5-diacid azide by 2,5-dibenzylurethane formation followed by hydrolysis with cold concentrated sulfuric acid.[252]

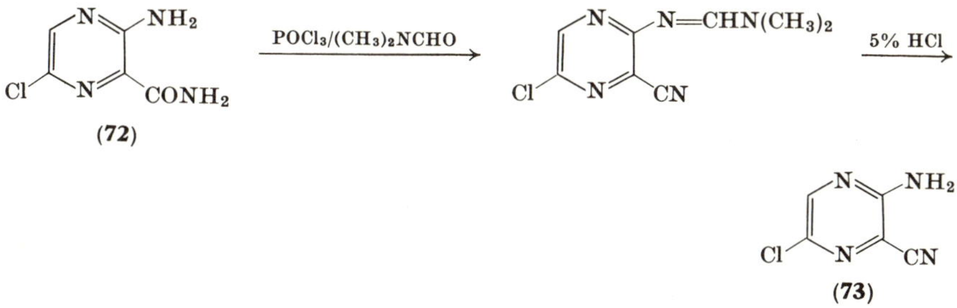

[245] P. I. Pollak and R. J. Tull, U.S. Patent 3,328,404 (1967); *Chem. Abstr.* **68**, 49653 (1968).
[246] Netherlands Patent Appl. 6,613,934 (1967); *Chem. Abstr.* **68**, 59614 (1968).
[247] E. J. Cragoe and J. H. Jones, U.S. Patent 3,341,540 (1967); *Chem. Abstr.* **68**, 105237 (1968).
[248] P. I. Pollak and R. J. Tull, French Patent 1,528,217 (1969); *Chem. Abstr.* **71**, 91530 (1969).
[249] E. J. Cragoe and J. H. Jones, French Patent 5,811 (1968); *Chem. Abstr.* **71**, 30504 (1969).
[250] M. G. Gal'pern and E. A. Luk'yanets, *Zh. Vses. Khim. Obshchest.* **12**, 474 (1967); *Chem. Abstr.* **68**, 2789 (1968).
[251] S. Gabriel and A. Sohn, *Ber.* **40**, 4857 (1907).
[252] D. M. Sharefkin and P. E. Spoerri, *J. Amer. Chem. Soc.* **73**, 1637 (1951).

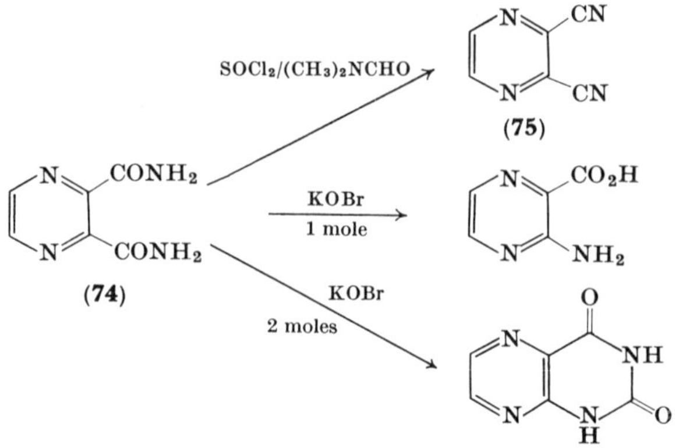

A modified Lossen rearrangement occurs when 2-aminopyrazine-carbohydroxamic acid (**77**) is treated with benzenesulfonyl chloride and the imidazopyrazine (**78**) is formed in good yield.[253]

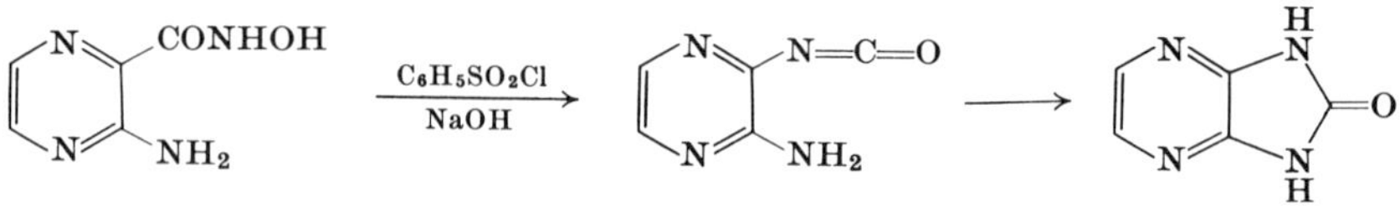

The *O*-carboxyhydroxamic acid, prepared by treatment of pyrazine 2,3-dicarboxylic acid anhydride with hydroxylamine, gives an *O*-benzoyl derivative. This forms a sodium salt which is isolated as a methanolate and which on heating in boiling toluene gives a mixture of 24% 3-aminopyrazine 2-carboxylic acid and 55% of the corresponding methyl ester (Scheme 20a).[253a]

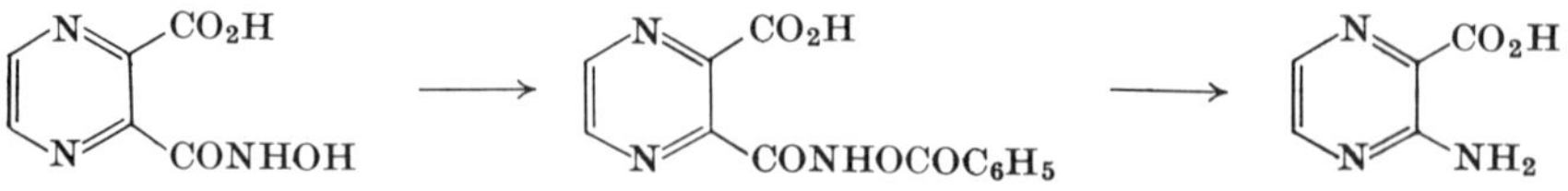

SCHEME 20a

Numerous patents and papers from the Merck, Sharp and Dohme Laboratories deal with the preparation of pyrazinecarboxamido-

[253] F. M. Hershenson, L. Bauer, and K. F. King, *J. Org. Chem.* **33**, 2543 (1968).
[253a] C. D. Hurd and V. G. Bethune, *J. Org. Chem.* **35**, 1471 (1970).

amidines (**80**) and pyrazinecarboxamidoguanidines (**81**).[254–260] These derivatives have useful diuretic activity, and one of them, *N*-amidino-3,5-diamino-6-chloropyrazinecarboxamide (**14**), is in clinical use, and is known as Amiloride. Compounds of this type promote sodium ion excretion, but do not affect the excretion of potassium ions. They are most simply prepared by the action of a guanidine or aminoguanidine on the appropriate methyl ester. In cases where the guanidine is not sufficiently nucleophilic, the indicated indirect route for carboxamidoamidine formation (Scheme 21) is used. Analogous ring opening of the 1,3-oxazinone (**79**) with an aminoguanidine leads after hydrolysis to compounds of type **81**. 3-Amino-2-(*N*-carbamoyl)pyrazinecarboxamides have also been tested as diuretics; they are prepared by

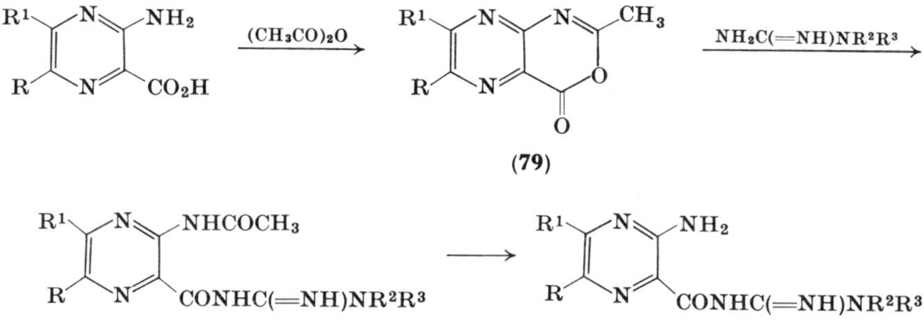

$$R^1\text{-pyrazine-}NH_2,\ CO_2H \xrightarrow{(CH_3CO)_2O} \textbf{(79)} \xrightarrow{NH_2C(=NH)NR^2R^3}$$

$$R^1\text{-pyrazine-}NHCOCH_3,\ CONHC(=NH)NR^2R^3 \longrightarrow R^1\text{-pyrazine-}NH_2,\ CONHC(=NH)NR^2R^3\ \textbf{(80)}$$

SCHEME 21

$$R^1\text{-pyrazine-}NH_2,\ CONHNHC(=NH)NR^2R^3$$

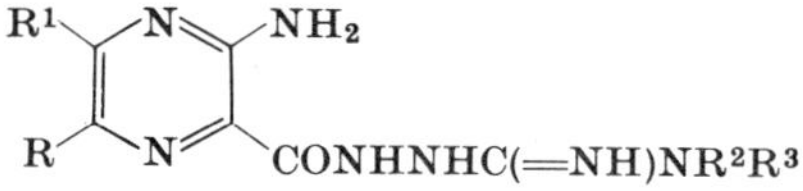

(81)

[254] J. B. Bicking, J. W. Mason, O. W. Woltersdorf, J. H. Jones, S. F. Kwong, C. M. Robb, and E. J. Cragoe, *J. Med. Chem.* **8**, 638 (1965).
[255] E. J. Cragoe, O. W. Woltersdorf, J. B. Bicking, S. F. Kwong, and J. H. Jones, *J. Med. Chem.* **10**, 66 (1967).
[256] J. B. Bicking, C. M. Robb, S. F. Kwong, and E. J. Cragoe, *J. Med. Chem.* **10**, 598 (1967).
[257] J. H. Jones, J. B. Bicking, and E. J. Cragoe, *J. Med. Chem.* **10**, 899 (1967).
[258] J. H. Jones and E. J. Cragoe, *J. Med. Chem.* **11**, 322 (1968).
[259] K. L. Shepard, J. W. Mason, O. W. Woltersdorf, J. H. Jones, and E. J. Cragoe, *J. Med. Chem.* **12**, 280 (1968).
[260] J. H. Jones, W. J. Holz, and E. J. Cragoe, *J. Med. Chem.* **12**, 285 (1968).

reaction of the methyl ester with the sodium salt of urea (NaNHCONH$_2$).[261] A series of 1H-imidazo[4,5-b]pyrazin-2-ones (e.g., **82**), prepared as illustrated below have antihypertensive and diuretic activity.[262]

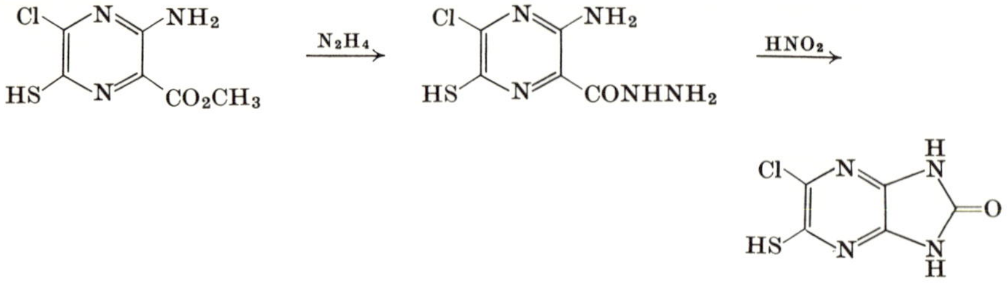

N-t-Butyl-3-(3,5-diamino-6-chloropyrazinecarbonyloxy)crotonamide and 3,5-diamino-6-chloropyrazinecarboxylic NN-diphenylcarbamic anhydride are prepared from the base-promoted reaction of 3,5-diamino-6-chloropyrazine carboxylic acid and N-t-butyl-5-methylisoxazolium perchlorate and diphenylcarbamoyl chloride, respectively. The amide and the anhydride are versatile acylating reagents and they react with a wide range of nucleophiles to give pyrazine derivatives of the general type shown.[262a]

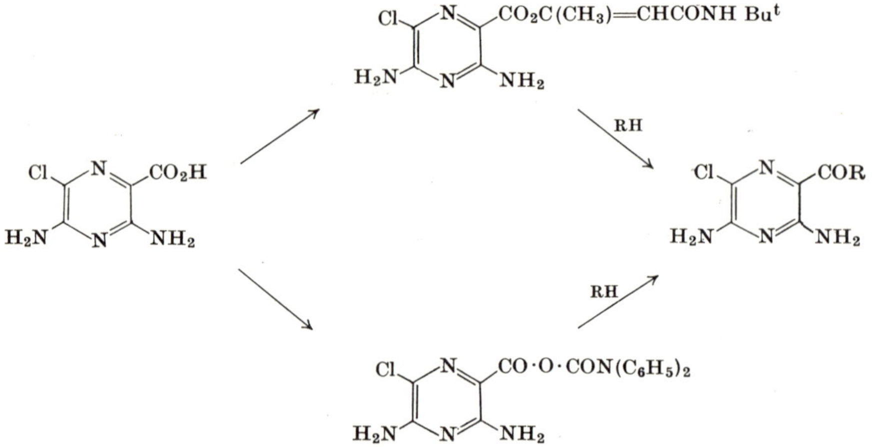

[261] J. W. Hanifin, R. Capuzzi, and E. Cohen, *J. Med. Chem.* **12**, 1102 (1969).
[262] J. H. Jones and E. J. Cragoe, U.S. Patent 3,461,123 (1969); *Chem. Abstr.* **71**, 101883 (1969).
[262a] K. L. Shepard, W. Halczenko, and E. J. Cragoe, Jr., *Tetrahedron Lett.* 4757 (1969).

Reaction of cyanopyrazine and its 6-chloro derivative with hydrazine gives the corresponding amidrazones, these compounds have been prepared for testing as antitubercular agents.[262b]

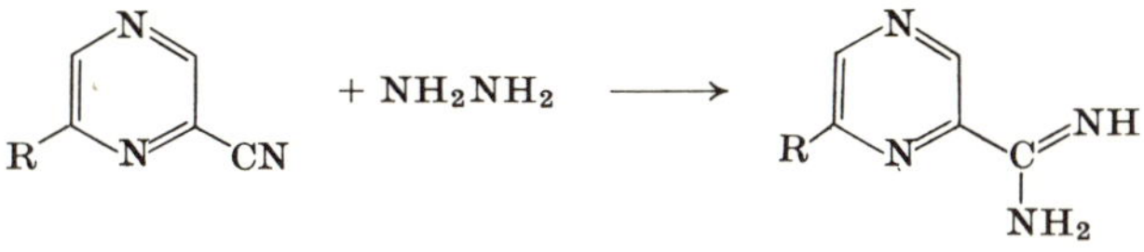

(R = H or Cl)

Cyanopyrazines react with aromatic amines in the presence of aluminum chloride as catalyst at temperatures below 200° and give amidine derivatives in good yield [Eq. (14)].[228] Hydrogenation of

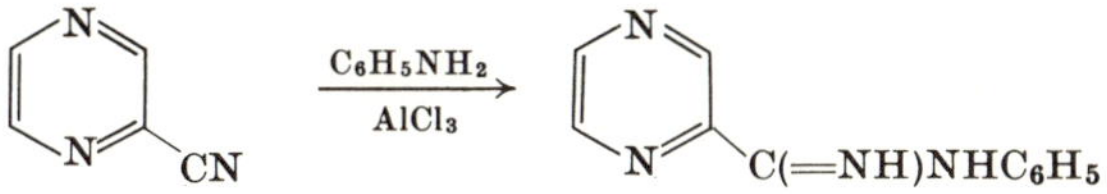

$$(14)$$

2-amino-3-cyanopyrazine over Raney nickel catalyst gives 2-amino-3-aminomethylpyrazine (**83**). The latter compound is a useful precursor for the preparation of 3,4-dihydropteridines (**84**) and imidazo[1,5-a]-pyrazines (**85**).[263] 2-Amino-3-cyano-5,6-diphenylpyrazine (**87**), pre-

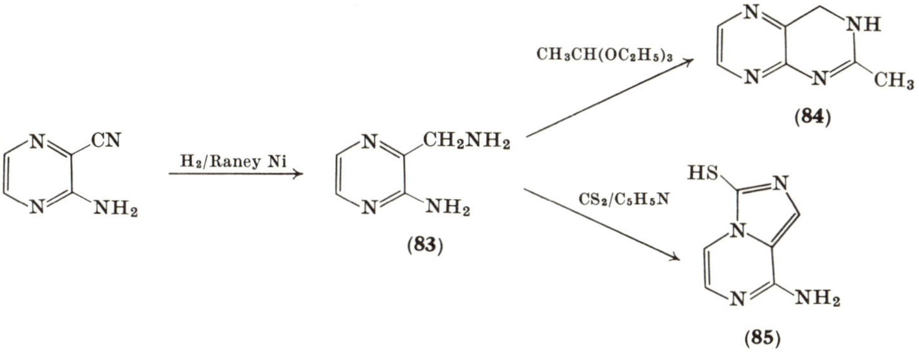

[262b] H. Foks, M. Buraczewska, W. Manowska, and J. Sawlewicz, *Diss. Pharm. Pharmacol.* **23**, 49 (1971).
[263] A. Albert and K. Ohta, *J. Chem. Soc. C* 1540 (1970).

pared by ring-opening of 4-mercapto-6,7-diphenylpteridine (86), is converted into the corresponding carboxamide (88) by the action of 30% hydrogen peroxide and alkali and into the thiocarboxamide (89) by reaction with hydrogen sulfide in triethanolamine.[264] Methyl

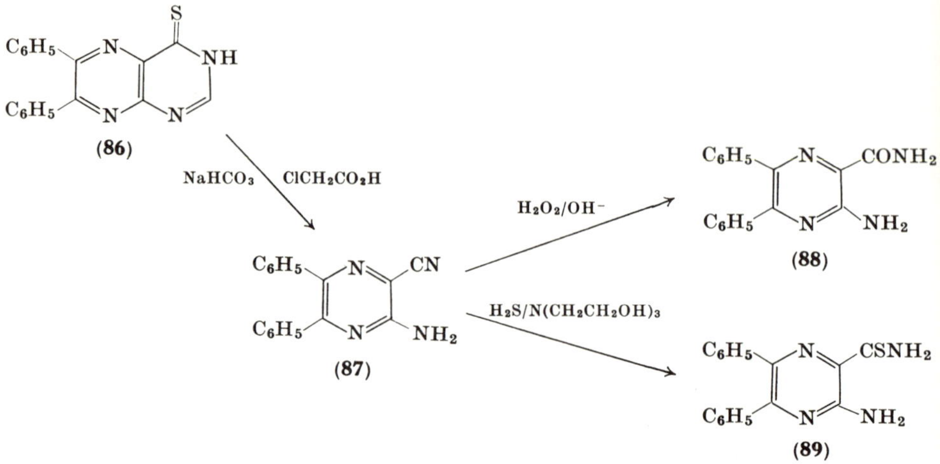

3-amino-6-chloropyrazine-2-carbothioximidate (90) has been prepared by the addition of methylmercaptan to the corresponding nitrile in the presence of base. Reaction of compound 90 with 2-amino-2-imidazoline (91) gives a derivative of the type (92) shown.[265]

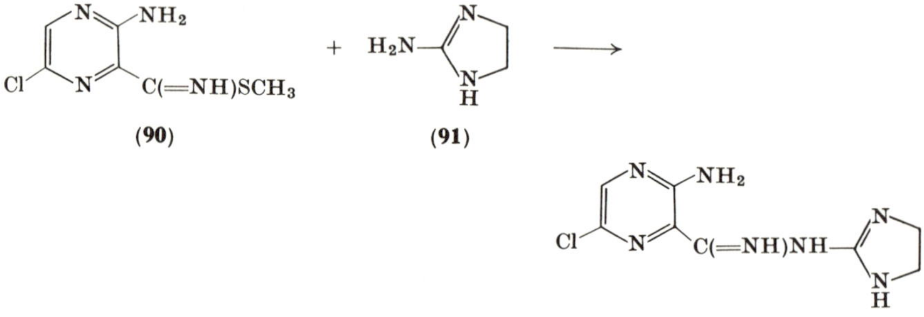

[264] E. C. Taylor, R. J. Knopf, J. A. Cogliano, J. W. Barton, and W. Pfleiderer, J. Amer. Chem. Soc. 82, 6058 (1960).
[265] E. J. Cragoe and J. H. Jones, U.S. Patent 3,299,063; Chem. Abstr. 68, 78309 (1968).

C. Halopyrazines

The classical route to chloropyrazines is by treatment of a hydroxypyrazine with phosphoryl chloride; bromopyrazines are similarly prepared by using phosphoryl bromide, phosphorus tribromide, or a mixture of both. Thus, treatment of hydroxypyrazine with phosphoryl chloride gives chloropyrazine in 92% yield,[147] and treatment of the hydroxy compound with a mixture of phosphoryl bromide and phosphorus tribromide gives bromopyrazine in 58% yield.[266] The use of phosphorus pentahalides frequently leads to substitution products (Scheme 22); for example, when hydroxypyrazine is treated with a mixture of phosphoryl bromide and phosphorus pentabromide both monobromo- and 2,6-dibromopyrazines are formed.[267, 268] Bromination of hydroxypyrazine with bromine in the presence of small

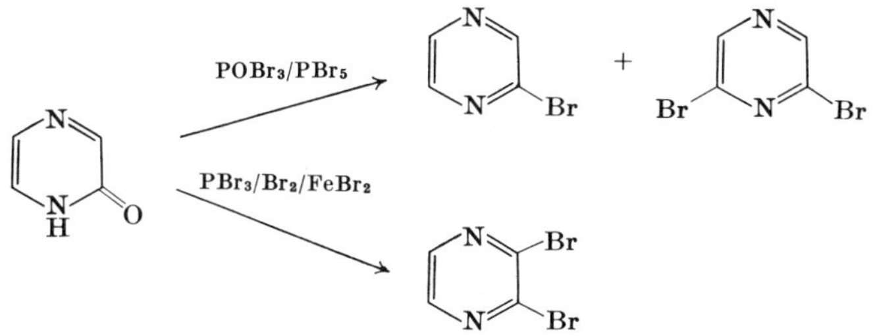

Scheme 22

quantities of phosphorus tribromide and ferrous bromide affords 2,3-dibromopyrazine in 32% yield.[269] 2,3-,[270] 2,5-,[271] and 2,6-.[272] Dichloropyrazines have been obtained by the reaction of 2,3-dihydroxy-, 5-chloro-2-hydroxy-, and 6-chloro-2-hydroxypyrazines with phosphoryl chloride, respectively.

---

[266] G. Karmas and P. E. Spoerri, *J. Amer. Chem. Soc.* **78**, 2141 (1956).
[267] A. E. Erickson and P. E. Spoerri, *J. Amer. Chem. Soc.* **68**, 400 (1946).
[268] K. H. Schaaf and P. E. Spoerri, *J. Amer. Chem. Soc.* **71**, 2043 (1949).
[269] G. Karmas and P. E. Spoerri, *J. Amer. Chem. Soc.* **78**, 680 (1957).
[270] G. Palamidessi and M. Bonanomi, *Farmaco, Ed. Sci.* **21**, 799 (1966); *Chem. Abstr.* **66**, 37884 (1967).
[271] G. Palamidessi and L. Bernardi, *J. Org. Chem.* **29**, 2491 (1964).
[272] R. A. Godwin, Ph.D. Thesis, London University, 1970.

2,5-Diketopiperazines (the bimolecular self-condensation products of α-amino acids) also react with phosphoryl and phosphorus halides to give halopyrazines. Treatment of 2,5-diketo-3,6-dimethylpiperazine with phosphoryl chloride yields a mixture of 2-chloro- and 2,5-dichloro-3,6-dimethylpyrazines; the monochloro compound is the predominant product [Eq. (14a)].[134, 135] Trichloropyrazine has been isolated from the reaction of triketopiperazine with phosphoryl chloride or phosphorus pentachloride [Eq. (15)].[273]

Halogenation of the parent base requires the use of elevated temperatures. Thus, bromopyrazine is prepared by heating pyrazine perbromide (obtained by treating pyrazine hydrochloride or hydrobromide with bromine in an inert solvent) at 200°–250°. Chloropyrazine is obtained by vapor phase chlorination in a steel tube at 400° or at lower temperatures in the presence of a catalyst such as activated carbon or copper chloride. By increasing the chlorine-to-pyrazine ratio tetrachloropyrazine can be prepared. The gas phase chlorination of pyrazine at 580° is reported to give good yields of tetrachloropyrazine.[273a]

Taft reported the conversion of chloropyrazine into 2,6-dichloropyrazine by chlorination with sulfuryl chloride or liquid chlorine.[274] This reaction is conveniently carried out in the laboratory by heating 2-chloropyrazine and sulfuryl chloride in a sealed tube at 120° for 3 hours. More recently 2,6-dichloropyrazine has been prepared in 90% yield from chloropyrazine by reaction with chlorine in dimethylformamide at 70°–75°; 2,3-dichloropyrazine is obtained in 84% yield

[273] L. Bernardi, G. Larini, and A. Leone, German Patent 1,178,436 (1964); *Chem. Abstr.* **62**, 4039 (1965).

[273a] H. Johnston and S. H. Ruetman, Ger. Offen. 1,911,023 (1970); *Chem. Abstr.* **74**, 31692 (1971).

[274] E. Taft, U.S. Patent 2,797,219 (1957); *Chem. Abstr.* **52**, 460 (1958).

when chlorine and phosphoryl chloride are used as the chlorinating agent.[275,276] The different orientation of substitution is attributed to the presence of small amounts of water in the reaction medium when chlorine is used alone. It is noteworthy that no significant amounts of 2,5-dichloropyrazine are produced in any of these reactions. Reaction of chloropyrazine with phosphorus pentachloride in an autoclave at 320° gives tetrachloropyrazine (Scheme 23), which can also be obtained by similar treatment of hydroxypyrazine or 2,5-diketopiperazine.[277,278] The preferred substrate for chlorination is pyrazine-2,3-dicarboxylic acid;[279] tetrachloropyrazine is also obtained by chlorination of 2,3-dichloropyrazine at 200°.[280] Preparations of

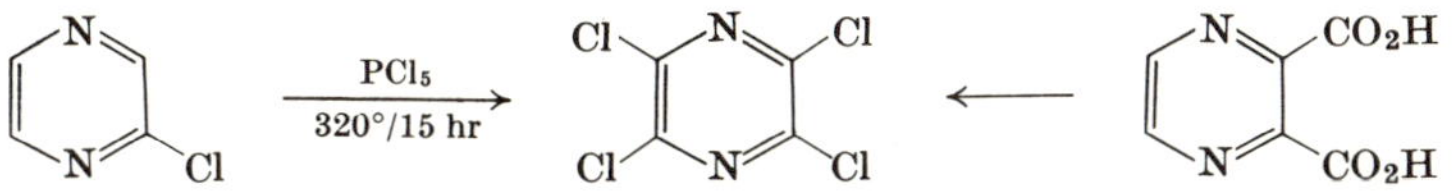

SCHEME 23

chloropyrazine and 2,3-dichloropyrazine from piperazine intermediates are mentioned in Section III.[131-133]

Treatment of methylpyrazine with chlorine in carbon tetrachloride at 40° gives two products and the total yield of monochloro derivatives is 65%.[281-283] The major product is 2-chloro-3-methylpyrazine and the minor product, originally thought to be the isomeric 2-chloro-5-methyl derivative, was later shown to be 2-chloro-6-methylpyrazine [Eq. (16)].[284] 2,5-Dimethylpyrazine also undergoes ring substitution

[275] French Patent 1,457,963 (1966); *Chem. Abstr.* **68**, 2917 (1968).
[276] K. H. Collins, U.S. Patent, 3,291,802 (1966); *Chem. Abstr.* **66**, 95086 (1967).
[277] R. D. Chambers, J. A. H. MacBride, and W. K. R. Musgrave, *Chem. Ind. (London)* 1721 (1966).
[278] R. D. Chambers, J. A. H. MacBride, and W. K. R. Musgrave, British Patent 1,163,582 (1969); *Chem. Abstr.* **71**, 124495 (1969).
[279] C. G. Allison, R. D. Chambers, J. A. H. MacBride, and W. K. R. Musgrave, *J. Chem. Soc. C* 1023 (1970).
[280] G. Palamidessi and F. Luini, *Farmaco, Ed. Sci.* **21**, 811 (1966); *Chem. Abstr.* **66**, 37886 (1967).
[281] H. Gainer, M. Kokorudz, and W. K. Langdon, *J. Org. Chem.* **26**, 2360 (1961).
[282] A. Hirschberg and P. E. Spoerri, *J. Org. Chem.* **26**, 2356 (1961).
[283] G. E. Baxter and W. W. Levis, U.S. Patent 3,113,132 (1963); *Chem. Abstr.* **60**, 5521 (1964).
[284] W. B. Lutz, S. Lazarus, S. Klutchko, and R. I. Meltzer, *J. Org. Chem.* **29**, 415 (1964).

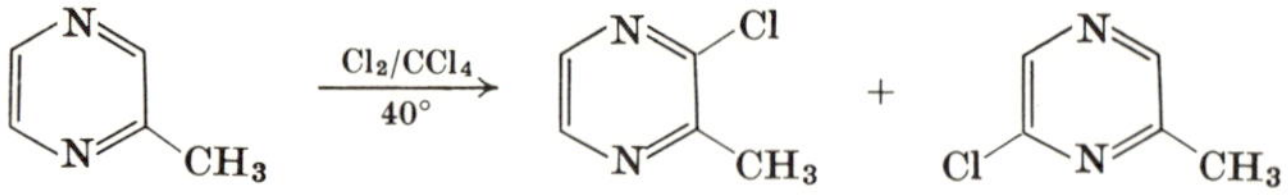

under the above conditions to give 2-chloro-3,6-dimethylpyrazine.[281-283] An addition–elimination mechanism (Scheme 24) is suggested for the formation of this compound. Chlorination via

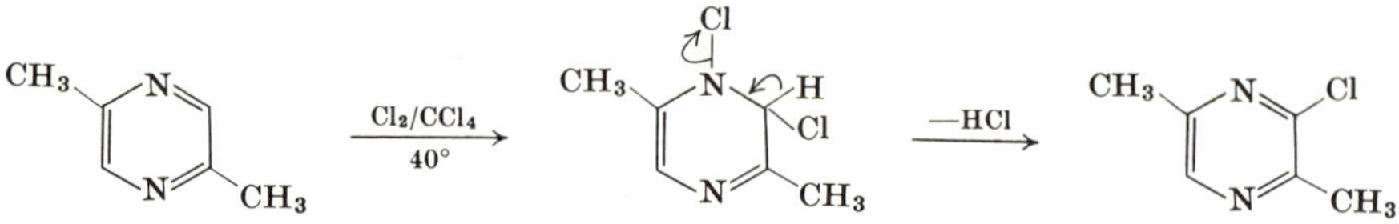

SCHEME 24

perchloride formation is an alternative possibility; a free radical mechanism is unlikely since chlorination of both methyl- and 2,5-dimethylpyrazines with N-chlorosuccinimide in the presence of catalytic amounts of benzoyl peroxide leads to side-chain substitution [Eq. (17)].[282] Chlorination of methylpyrazine in acetic acid at 100°

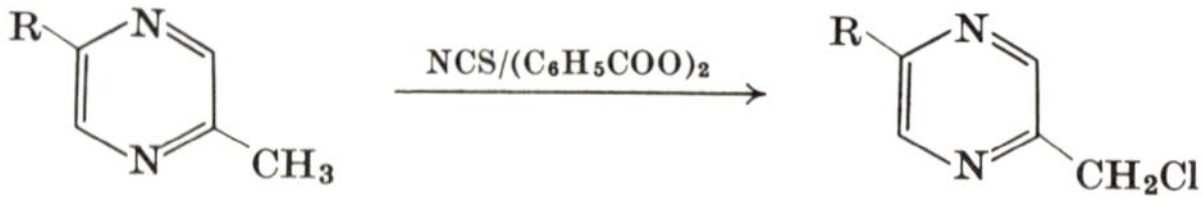

(R = H, CH$_3$)

gives trichloromethyl pyrazine and similar treatment of 2-chloro-3-methylpyrazine gives 2-chloro-3-dichloromethylpyrazine.[285, 286] The chlorination of 2,6-[282] and 2,3-dimethylpyrazines[207] with chlorine in carbon tetrachloride will only proceed in the presence of ultraviolet light. The products isolated are the bischloromethyl derivatives resulting from side-chain chlorination (Scheme 25). Chlorination with N-chlorosuccinimide in the presence of benzoyl peroxide yields the same products. The bischloromethylpyrazines are reactive and are readily converted into the bisalkoxymethyl ethers. The chlorination of 2,5-diethylpyrazine with chlorine in an inert solvent (e.g.,

[285] E. J. J. Grabowski, E. W. Tristram, R. J. Tull, and P. I. Pollak, *Tetrahedron Lett.* 5931 (1968).
[286] R. D. Wilcox, D. J. Horne, and H. Johnston, U.S. Patent 3,501,472; *Chem. Abstr.* **72**, 111508 (1970).

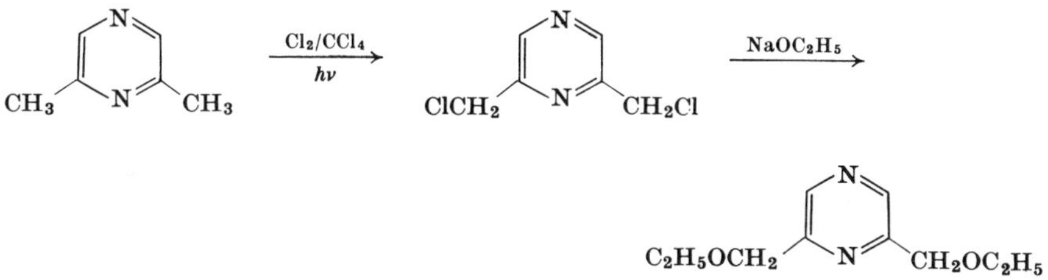

SCHEME 25

chloroform or carbon tetrachloride) gives 2-chloro-3,6-diethylpyrazine.[281, 283] Halogenated alkylpyrazines are described as useful starting materials for the preparation of pesticides and polymers.[283, 286, 287]

There are many examples from the work of Cragoe and his colleagues on the activating effect of a *para*-amino group on the halogenation of pyrazines. Thus, treatment of methyl 3-aminopyrazine 2-carboxylate (**93**) with chlorine in aqueous acetic acid gives the 6-chloro derivative (**94**);[288, 289] reaction with iodine and mercuric acetate in dioxane yields the 6-iodo derivative (**95**).[290] The 5,6-dichloro compound (**96**) is obtained from methyl 3-aminopyrazine 2-carboxylate (**93**) by sulfuryl chloride or chlorine in acetonitrile (Scheme 26).[290]

5-Bromo derivatives of 2-amino-3-chloro- and 2-amino-3-bromopyrazines are obtained by bromination with bromine in 20% hydrobromic acid at 5°. Similar treatment of 2-amino-5-bromopyrazine-3-carboxylic acid gives 2-amino-3,5-dibromopyrazine.[291, 292]

Tetrabromopyrazine, which shows bactericidal activity, is obtained from tetrachloropyrazine (HBr/HOAc)[293] or from tetrafluoropyrazine (AlBr$_3$/HBr).[279]

Spoerri and Rutner have prepared fluoropyrazine from aminopyrazine by thermal decomposition of the diazonium fluoroborate in

---

[287] W. K. Langdon and M. Kokorudz, U.S. Patent, 3,096,331 (1963); *Chem. Abstr.* **60**, 2982 (1964).
[288] E. J. Cragoe, Belgian Patent 639,393 (1964); *Chem. Abstr.* **62**, 7778 (1965).
[289] E. J. Cragoe, R. J. Tull, and J. T. Broeke, British Patent 1,082,060 (1967); *Chem. Abstr.* **68**, 105248 (1968).
[290] Netherlands Patent Appl. 6,615,406 (1967); *Chem. Abstr.* **68**, 21953 (1968).
[291] H. Brachwitz, German (East) Patent 66,877 (1969); *Chem. Abstr.* **71**, 124498 (1969).
[292] H. Brachwitz, *J. Prakt. Chem.* **311**, 40 (1969).
[293] A. H. Gulbenk, U.S. Patent 3,471,496 (1969); *Chem. Abstr.* **71**, 124489 (1969).

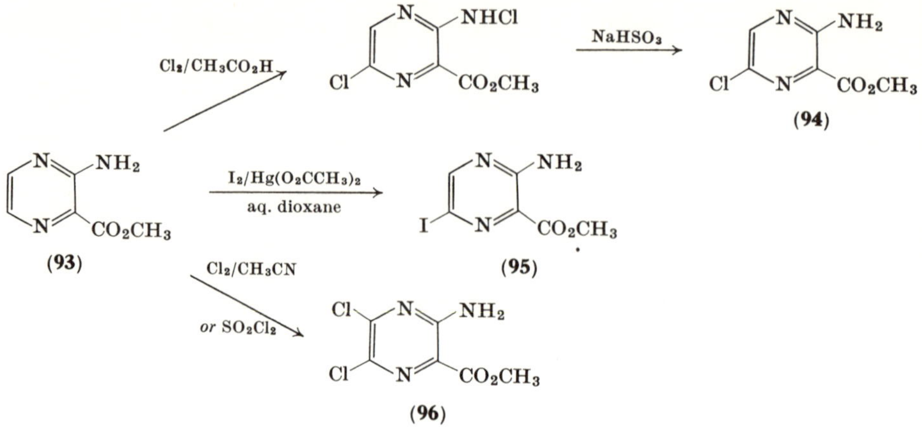

SCHEME 26

the presence of copper powder.[294] It has also been prepared by treatment of chloropyrazine with potassium fluoride in N-methylpyrrolidone at 185°.[272] The displacement reaction between tetrachloropyrazine and potassium fluoride at 310°–320° for 15 hours yields the tetrafluoro compound in 95% yield. At 280°, a mixture of 2-chloro-3,5,6-trifluoro- and 2,5-dichloro-3,6-difluoropyrazines is obtained.[279]

Iodopyrazines are prepared from the corresponding chloropyrazines by treatment with a saturated solution of sodium iodide in methyl ethyl ketone containing a little hydriodic acid.[295]

Phosphoryl chloride rearrangement of suitably substituted pyrazine N-oxides is a convenient route to some substituted chloropyrazines. Thus, Palamidessi, Bernardi, and Leone obtained trichloropyrazine from 2,6-dichloropyrazine 4-oxide (Scheme 27).[296]

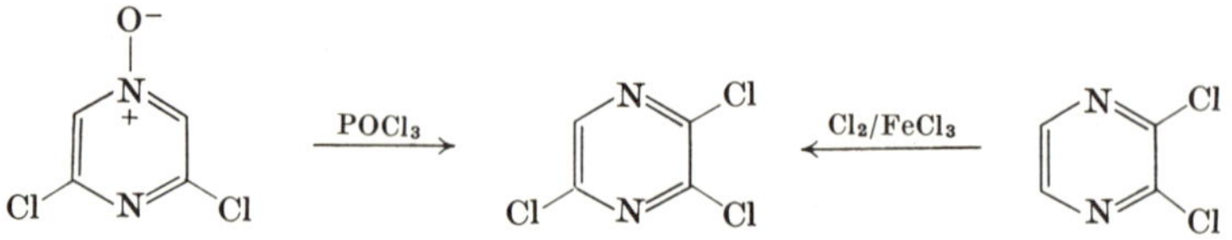

SCHEME 27

[294] H. Rutner and P. E. Spoerri, *J. Heterocycl. Chem.* **2**, 492 (1965).
[295] A. Hirschberg and P. E. Spoerri, *J. Org. Chem.* **26**, 1907 (1961).
[296] G. Palamidessi, L. Bernardi, and A. Leone, *Farmaco, Ed. Sci.* **21**, 805 (1966); *Chem. Abstr.* **66**, 37885 (1967).

Trichloropyrazine may also be prepared by chlorination of 2,3-dichloropyrazine[296] or, as previously mentioned, from triketopiperazine,[273, 296] but N-oxide rearrangement is probably the laboratory method of choice. A further application of N-oxide rearrangement for chloropyrazine preparation is taken from the work of Cragoe and his colleagues (Scheme 28).[265]

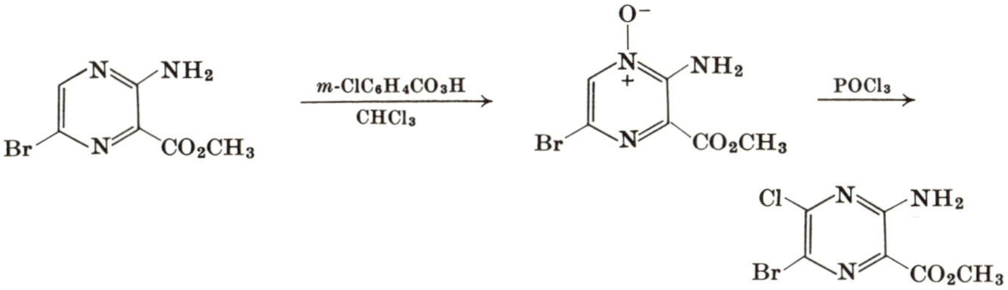

SCHEME 28

Halopyrazines react with nucleophilic reagents under similar conditions to the halopyridines. Kinetic data obtained by Chan and Miller[297] on the reaction of monochlorodiazines with p-nitrophenoxide ion has established the following order of reactivity: 2- > 4-chloropyrimidine > 4-chloropyridazine $\simeq$ 3-chloropyridazine $\simeq$ chloropyrazine. The unusual feature of this reactivity sequence is the greater reactivity of 2-chloropyrimidine compared to 4-chloropyrimidine; with most anionic reagents, 4-chloropyrimidine is the more reactive halide. Chloropyrazine undergoes the expected displacement reaction with ammonia, methylamine, dimethylamine, sodium methoxide, and sodium hydrogen sulfide to give the corresponding monosubstituted pyrazines.[147] 2-Chloropyrazine is converted into a mixture of imidazole, 2-cyanoimidazole, and 2-aminopyrazine on treatment with potassium amide in liquid ammonia [Eq. (18)].[298] Fluoropyrazine is

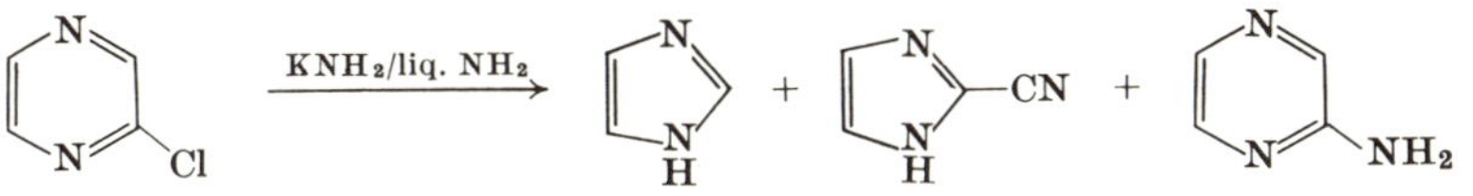

significantly more reactive to nucleophilic reagents than chloropyrazine; it reacts with anhydrous sodium sulfite to give the sodium

[297] T. L. Chan and J. Miller, *Austr. J. Chem.* **20**, 1595 (1967).
[298] P. J. Lont, H. C. van der Plas, and A. Koudijs, *Rec. Trav. Chim.* **90**, 207 (1971).

salt of pyrazinesulfonic acid and with sodium azide to give tetrazolo[1,5-a]pyrazine (**97**).[299] The equilibrium between 5,6-diphenyltetrazolo[1,5-a]pyrazine and the corresponding azidopyrazine has been investigated by PMR measurements. In DMSO-d_6 solution, only the tetrazolo form can be detected, and in CF_3CO_2H only the azido form is present. In $CDCl_3$ the composition of the equilibrium mixture is approximately 10 parts of tetrazole to 3 parts of azide. In boiling acetic acid, thermal rearrangement takes place to give a mixture of 56% 4,5-diphenylimidazole and 13% 1-acetyl-4,5-diphenylimidazole. The reaction is thought to involve the formation of a nitrene which undergoes valence-bond isomerization to a thermally unstable triazepine which finally rearranges to an imidazole.[299a] Treatment of iodopyrazine with n-butyllithium gives 2-pyrazinyllithium which is converted into pyrazinecarboxylic acid on reaction with carbon dioxide.[300]

(**97**)

Alkyl substituents considerably reduce the reactivity of halogen attached to a pyrazine ring. Thus, 2-chloro-3,6-dimethylpyrazine has been converted into the corresponding 2-hydroxy- and 2-ethoxypyrazines, but it does not undergo ammonolysis; 2-chloro-3,6-di-sec-butylpyrazine is recovered unchanged after prolonged treatment with boiling 20% potassium hydroxide solution, conditions which converted 2-chloro-3,6-dimethylpyrazine into 2-hydroxy-3,6-dimethylpyrazine.[301] 2-Dimethylamino-3,5,6-trimethylpyrazine is prepared by heating 2-chloro-3,5,6-trimethylpyrazine with dimethyl amine at 180° for 3 days[302] and 2-mercapto-3,5,6-trimethylpyrazine is prepared by treatment with sodium hydrogen sulfide at 120°–130° for 8 days.[303] These compounds have useful antibacterial properties.

[299] H. Rutner and P. E. Spoerri, *J. Heterocycl. Chem.* **3**, 435 (1966).
[299a] T. Sasaki, K. Kanematsu, and M. Murata, *J. Org. Chem.* **36**, 446 (1971).
[300] H. S. Hertz, F. F. Kabacinski, and P. E. Spoerri, *J. Heterocycl. Chem.* **6**, 239 (1969).
[301] R. A. Baxter and F. S. Spring, *J. Chem. Soc.* 1179 (1947).
[302] British Patent 1,031,915 (1966); *Chem. Abstr.* **65**, 5471 (1966).
[303] Y. Yamanishi and N. Kawasaki, *J. Pharm. Soc. Jap.* **87**, 105 (1967).

2-Chloro-3-methylpyrazine undergoes the expected reactions with aniline, methylaniline, sodium phenoxide, and sodium thiophenoxide.[304] Normal products of substitution are also obtained with piperidine and other heterocyclic amines such as pyrrolidine.[305] However, when 2-chloro-3-methylpyrazine (98) is treated with sodamide in liquid ammonia 2-chloro-3-(3'-methyl-2'-pyrazylmethyl)-pyrazine (99) is produced as indicated (Scheme 29) in 70% yield.[304]

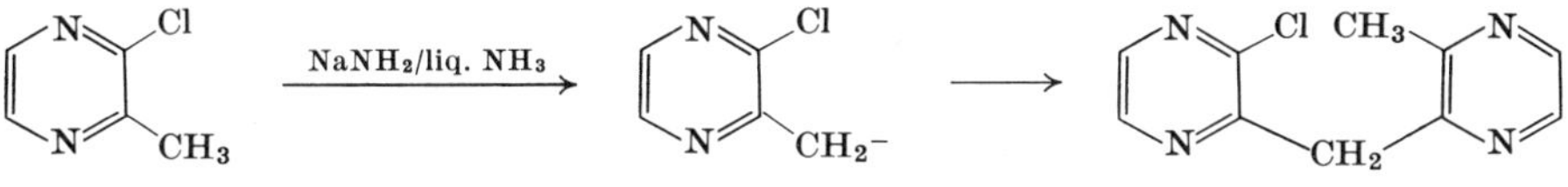

SCHEME 29

2-Iodo-3,6-dimethylpyrazine is converted into the corresponding cyanopyrazine in 70% yield on reaction with cuprous cyanide in boiling anhydrous β-picoline.[306] Halogen–metal exchange is achieved by reaction with ethereal n-butyllithium; the resulting lithio derivative has been condensed with a number of aromatic aldehydes, and with 2-acetylpyridine, to give pyrazinylcarbinols (Scheme 30).[300,306]

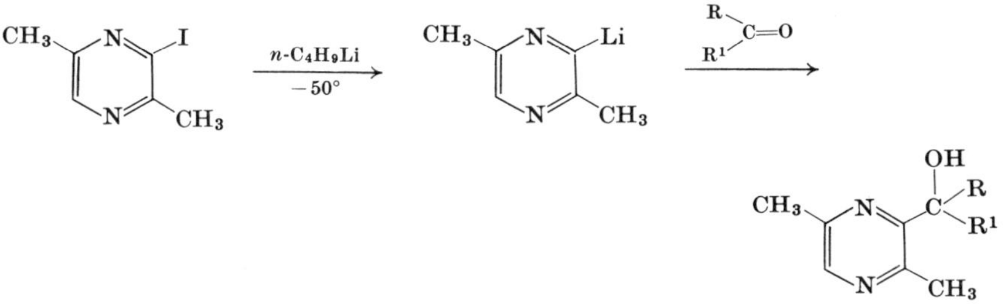

SCHEME 30

Palamidessi and Bernardi report the displacement of chloride ion from 2-carboxamido- and 2-amino-6-chloropyrazine with methanolic sodium methoxide.[307] In the former case, reaction is carried out for

---

[304] J. D. Behun, P. T. Kan, P. A. Gibson, C. T. Lenk, and E. J. Fujiwara, *J. Org. Chem.* **26**, 4981 (1961).

[305] W. B. Lutz, French Patent M 2229 (1964); *Chem. Abstr.* **60**, 15890 (1964).

[306] A. Hirschberg, A. Peterkofsky, and P. E. Spoerri, *J. Heterocycl. Chem.* **2**, 209 (1965).

[307] G. Palamidessi and L. Bernardi, *Gazz. Chim. Ital.* **91**, 1438 (1961).

45 minutes in an open vessel, and in the latter case, in a sealed tube at 140° for 15 hours. 2-Amino-6-methylthiopyrazine is prepared from reaction of 2-amino-6-chloropyrazine with sodium methylmercaptide in boiling ethanol for 5 hours.[308] These compounds are also useful bactericides.

Only one halogen is frequently displaced in the reactions of dihalogenopyrazines with nucleophiles. Thus, 2-amino-3-chloropyrazine is isolated in 85% yield from the reaction of 2,3-dichloropyrazine and aqueous ammonia in an autoclave at 130° for 14 hours.[270] Similarly Cragoe and his colleagues report a series of displacements in which the chlorine para to the electron-attracting carbomethoxy group in the dichloropyrazine (96) is preferentially displaced.[309,310] These are illustrated in Scheme 31. The reaction of 2,6-dichloropyrazine with

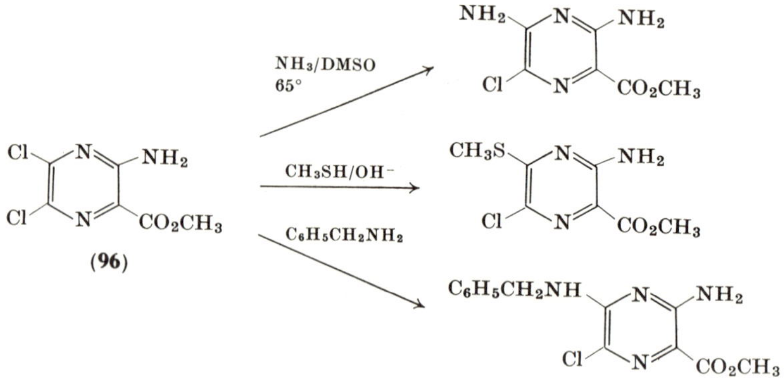

SCHEME 31

ammonia or sulfanilamide gives 2-amino-6-chloropyrazine and 2-chloro-6-sulfanilamidopyrazine, respectively.[307]

When trichloropyrazine is treated with ammonia under pressure, 2-amino-3,5-dichloropyrazine (101) is formed.[273] This, on reaction with sodium methoxide, gives the 3-methoxy derivative (104). When trichloropyrazine is allowed to react with aqueous ammonia at 80° for 15 hours, the 3-chlorine is displaced to give compound 100. The latter

[308] N. Okuda, Y. Fukuda, I. Kuniyoshi, and H. Shinoda, Japanese Patent 12,712 (1965); *Chem. Abstr.* **63**, 11589 (1965).
[309] E. J. Cragoe, Belgian Patent 639,386 (1964); *Chem. Abstr.* **62**, 14698 (1965).
[310] E. J. Cragoe and J. H. Jones, British Patent 1,083,901 (1967); *Chem. Abstr.* **69**, 19214 (1968).

compound on reaction with sodium methoxide gives both possible monomethoxy derivatives (**102**) and (**103**). Reductive dechlorination of compounds **102** and **103** gives 2-amino-6-methoxy- and 2-amino-3-methoxypyrazine, respectively.[296]

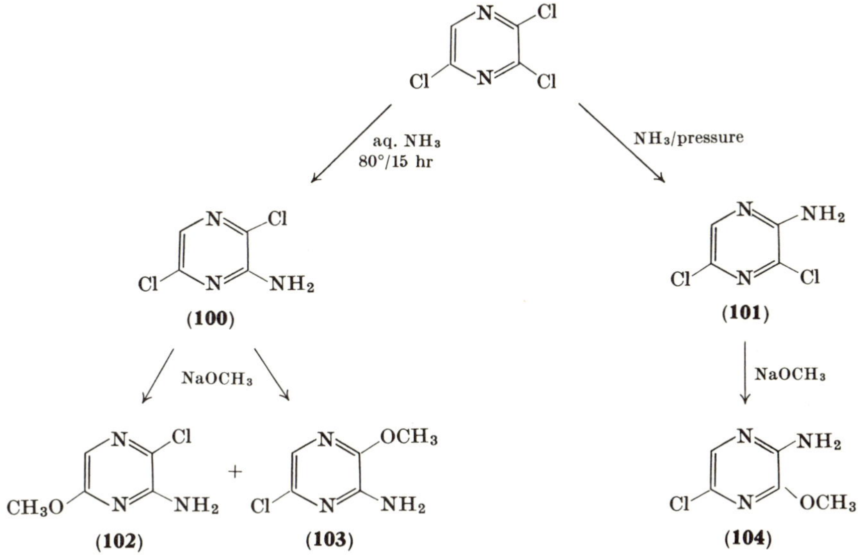

Under mild conditions, only *one* halogen of tetrahalopyrazine is displaced by sodium methylmercaptide, and 2-methylthio-3,5,6-trichloro-, -bromo-, and -iodopyrazines are obtained.[311] Tetrafluoropyrazine reacts with aqueous ammonia at room temperature to give 2-amino-3,5,6-trifluoropyrazine. A number of analogous displacement reactions have been carried out using nucleophiles such as lithium alkyls, potassium hydroxide, and sodium alkoxides.[279] Alkyl and chloro substituents in a trifluoropyrazine direct nucleophilic attack to the para position, and alkoxy substituents to the ortho position. The prolonged reaction of tetrafluoropyrazine and concentrated aqueous ammonia at 25° gives 2,6-diamino-3,5-difluoropyrazine.[310a] Treatment of 2-amino-3,5,6-trichloropyrazine with oxalyl chloride in boiling benzene gives the corresponding 2-isocyanato compound.[310b]

[310a] C. G. Allison, R. D. Chambers, J. A. H. MacBride, and W. K. R. Musgrave, *J. Fluorine Chem.* **1**, 59 (1971).
[310b] U. von Gizycki, *Angew. Chem. Int. Ed.* **10**, 402 (1971).
[311] D. H. Horne, U.S. Patent 3,452,016 (1969); *Chem. Abstr.* **71**, 81415 (1969).

Chlorination of 2-methoxy-3,5,6-trichloropyrazine (**105**) (prepared from tetrachloropyrazine and sodium methoxide) gives a mixture of 2-trichloromethyloxy- and 2-dichloromethyloxy-3,5,6-trichloropyrazines. The former compound (**106**) is the major product; compounds of this type are useful pesticides.[311]

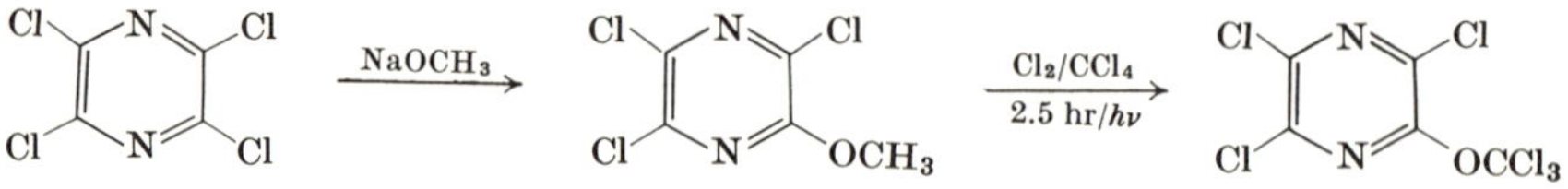

Treatment of 2-chloro-3-dichloromethylpyrazine (**107**) with three equivalents of methoxide ion in refluxing methanol gives a quantitative yield of 3,5-dimethoxy-2-methoxymethylpyrazine (**108**). From experiments in which one and two equivalents of the ethoxide ion were used, the reaction sequence shown in the accompanying diagram was deduced. It is noteworthy that initial attack of the nucleophile is at C-6 of the ring.[285]

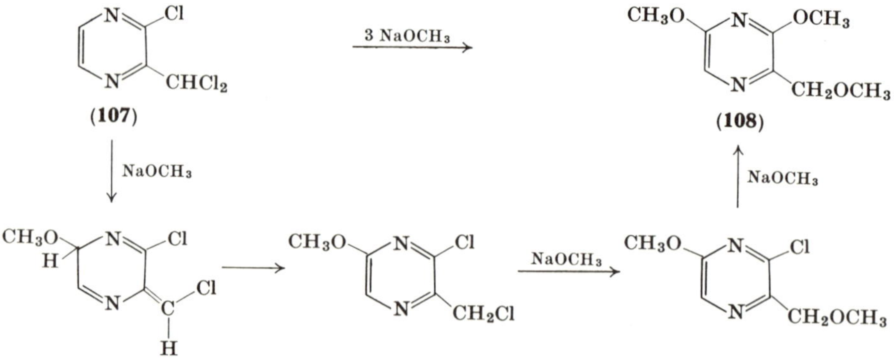

Trichloromethylpyrazine (**109**) reacts in a similar manner; thus, with one equivalent of methoxide ion in methanol at 5°, 2-dichloromethyl-5-methoxypyrazine (**110**) is formed; treatment with three equivalents of methoxide ion in boiling methanol gives a mixture of compounds **111**, **112**, and **113**, in 75, 15, and 10% yield, respectively. The mechanisms by which these products of abnormal substitution are formed is being further investigated.[285, 286]

Sec. V. D.]   PYRAZINE CHEMISTRY   165

[Structures: (109) pyrazine-CCl₃ → NaOCH₃ → (110) CH₃O-pyrazine-CHCl₂ → 2 NaOCH₃ →

(111) CH₃O-pyrazine-CH(OCH₃)₂ +

(112) 3,6-di(CH₃O)-2-CH₃-5-OCH₃ pyrazine + (113) 3,6-di(CH₃O)-pyrazine-CH₂OCH₃]

D. AMINOPYRAZINES

Aminopyrazines are conveniently prepared from carboxamidopyrazines by application of the Hofmann reaction (see Section V, B). Thus, Camerino and Palamidessi prepared aminopyrazine in 80% yield from carboxamidopyrazine.[312] Aminopyrazine may also be prepared from the reaction of pyrazine with sodamide in liquid ammonia,[313] and 3-amino-2,5-dimethylpyrazine is the product of amination of 2,5-dimethylpyrazine with sodamide in dimethylaniline.[314] The ammonolysis of halopyrazines also represents a useful preparative procedure for aminopyrazines (see Section V, C). This reaction proceeds most easily in the case of fluoro compounds; for example, fluoropyrazine is converted into aminopyrazine in 70% yield by treatment with concentrated aqueous ammonia at room temperature for 3 days,[299] whereas the corresponding reaction with chloropyrazine has been carried out in a sealed tube at 150°.[147] Alkaline hydrolysis of 2,4-dihydroxypteridines followed by decarboxylation yields aminopyrazines;[315] thus, high-temperature alkaline hydrolysis of 7-methyl-2,4-dihydroxypteridine (7-methyllumazine) gives, after decarboxylation of the intermediate pyrazinecarboxylic

[312] B. Camerino and G. Palamidessi, *Gazz. Chim. Ital.* **90**, 1807 (1960).
[313] M. L. Crossley and J. P. English, U.S. Patent 2,394,963 (1946); *Chem. Abstr.* **40**, 3143 (1946).
[314] R. R. Joiner and P. E. Spoerri, *J. Amer. Chem. Soc.* **63**, 1929 (1943).
[315] H. Saikachi and J. Matsuo, *J. Pharm. Soc. Jap.* **89**, 1071 (1969).

acid, 78% 2-amino-6-methylpyrazine (Scheme 32)[316] Aminopyrazine 1-oxides may be directly synthesized by condensation of an α-amino-

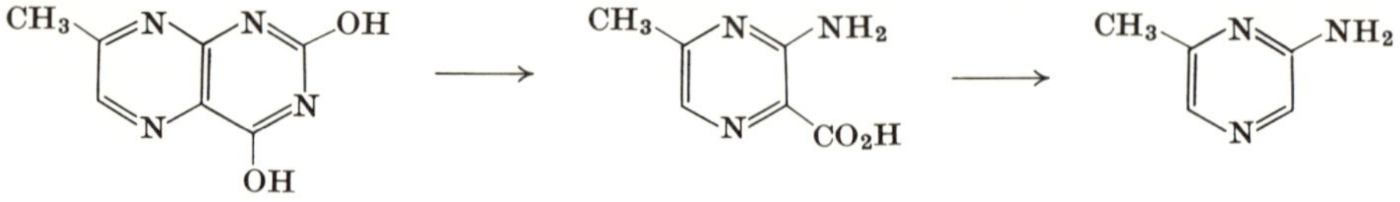

SCHEME 32

nitrile and an α-oximino ketone (Scheme 33). These, on reduction with sodium dithionite, yield aminopyrazines.[317] The formation of aminopyrazines by condensation of α,β-dicarbonyl compounds and aminoacetamidine has been mentioned in Section III.[115]

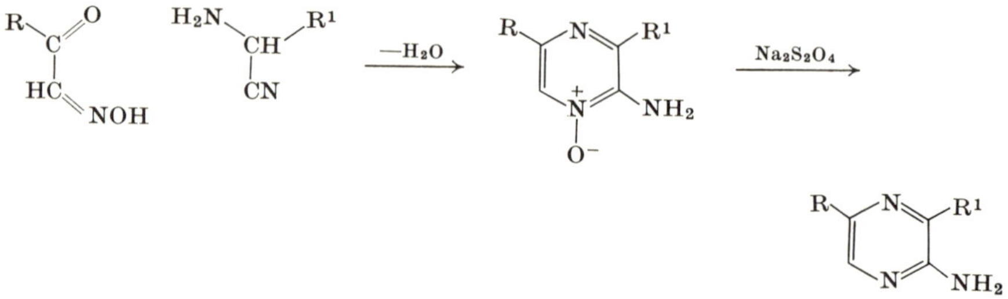

SCHEME 33

Comparison of the ultraviolet absorption and ionization constants of aminopyrazine, methylaminopyrazine, and dimethylaminopyrazine indicates that aminopyrazine exists as such and not in the tautomeric imino form.[147] Aminopyrazines, like their pyridine analogs, form diazonium salts, which readily decompose to the corresponding pyrazinones. For example, nitrous acid treatment of aminopyrazine[318] and 2-aminopyrazine-5-carboxylic acid[210] gives the corresponding pyrazinones in 30 and 59% yield, respectively. The diazonium salt from aminopyrazine cannot be converted into bromopyrazine under the conditions of the Sandmeyer reaction.

The amino substituent attached to the pyrazine ring facilitates electrophilic attack at the ring carbon atoms in the ortho and para positions. Thus, bromination of aminopyrazine in glacial acetic acid

[316] D. M. Sharefkin, J. Org. Chem. 24, 345 (1959).
[317] E. C. Taylor and K. Lenard, J. Amer. Chem. Soc. 90, 2424 (1968).
[318] R. A. Baxter, G. T. Newbold, and F. S. Spring, J. Chem. Soc. 370 (1947).

gives 2-amino-3,5-dibromopyrazine (Scheme 34);[319] as already mentioned this product is also obtained by bromination of 2-aminopyrazine-3-carboxylic acid and 2-amino-3-bromopyrazine.[292] In 2-amino-3,5-dibromopyrazine the bromine atom ortho to the amino group is the more readily displaced by nucleophilic reagents. For example, displacement with ammonia or dimethylamine at 130° under pressure gives 2,3-diamino-5-bromo- and 2-amino-5-bromo-3-dimethylaminopyrazine, respectively. Debromination of the bromodiamine with hydrogen over a palladium on charcoal catalyst gives 2,3-diaminopyrazine, which can also be prepared by ammonolysis of 2-amino-3-chloropyrazine. 2-Amino-3,5-dibromopyrazine is converted into 2-amino-3-methoxypyrazine by reaction with sodium methoxide followed by catalytic removal of bromine (Scheme 34).[312]

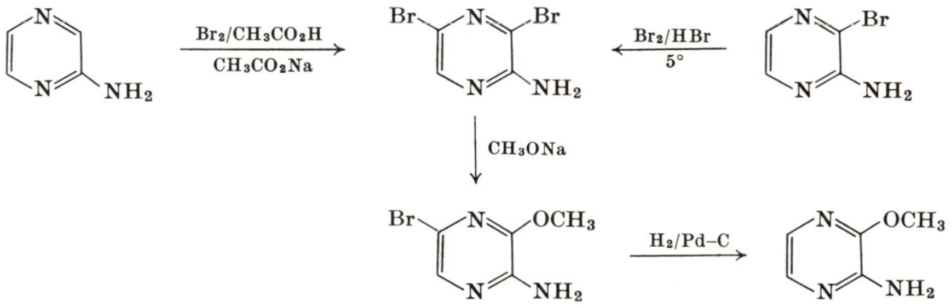

SCHEME 34

Aminopyrazines, like aminopyridines and aminopyrimidines, form p-aminobenzenesulfonyl derivatives, and one of these, 2-sulfanilamido-3-methoxypyrazine (**11**), is clinically used as an antibacterial agent.[320] Bromination of compound **11** in methanol gives a product (**114**)[321,322] which was originally incorrectly formulated as a hydrate.[323] The correct structure follows from spectroscopic evidence and alkaline

[319] B. Camerino and G. Palamidessi, British Patent 928,152 (1963); *Chem. Abstr.* **60**, 2971 (1964).
[320] B. Camerino and G. Palamidessi, *Gazz. Chim. Ital.* **90**, 1815 (1960).
[321] W. Barbieri, L. Bernardi, G. Palamidessi, and M. Tacchi Venturi, *Farmaco, Ed. Sci.* **23**, 821 (1968); *Chem. Abstr.* **70**, 4044 (1969).
[322] W. Barbieri, L. Bernardi, G. Palamidessi, and M. Tacchi Venturi, *Tetrahedron Lett.* 2931 (1968).
[323] J. Esche and H. Wojahn, *Arch. Pharm. (Weinheim)* **299**, 147 (1966); *Chem. Abstr.* **64**, 15880 (1966).

cleavage to compounds **115** and **116**. The structure of compound **115** was proved by the formation of a *O*-methyl derivative, catalytic debromination, and comparison with an authentic specimen of 2-sulfanilamido-3,6-dimethoxypyrazine (**117**).

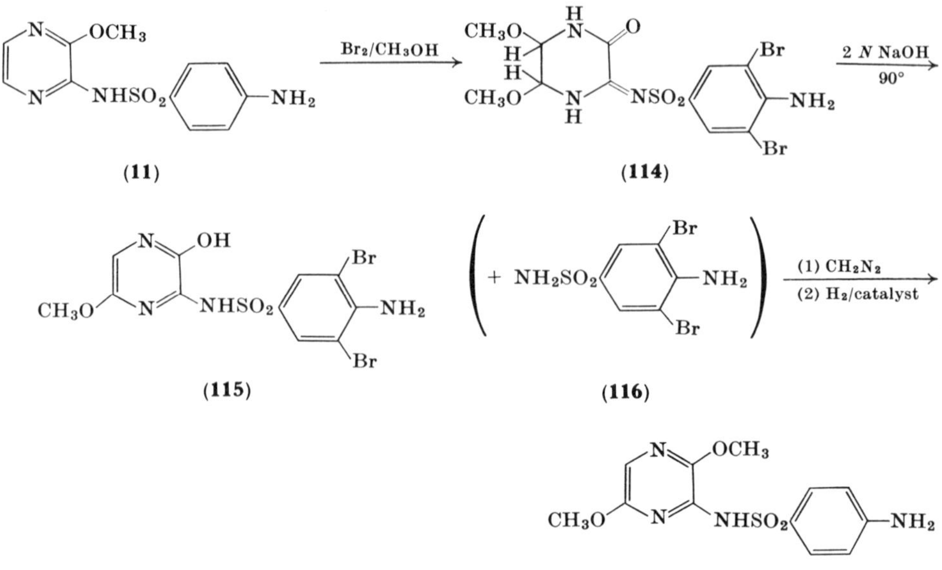

The isomeric 2-amino-5-methyl- and 2-amino-6-methylpyrazines are obtained by Hofmann degradation of 2-carboxamido-3-hydroxy-5- and 6-methylpyrazines, respectively, followed by phosphoryl chloride treatment and catalytic dechlorination (Scheme 35).[324] Two methods for the preparation of 2-aminopyrazine-5-carboxylic acid have been reported. One of these is based on the permanganate oxidation of 2-acetylaminoquinoxaline (see Section V, B), the other on pyrazine 2,5-dicarboxylic acid.[325,326] The required transformations are illustrated in Scheme 36.

[324] G. Palamidessi, *Farmaco, Ed. Sci.* **18**, 557 (1963); *Chem. Abstr.* **59**, 13975 (1963).
[325] W. J. Schut, H. I. X. Mager, and W. Berends, *Rec. Trav. Chim.* **80**, 391 (1961).
[326] W. J. Schut, H. I. X. Mager, and W. Berends, *Rec. Trav. Chim.* **82**, 282 (1963).

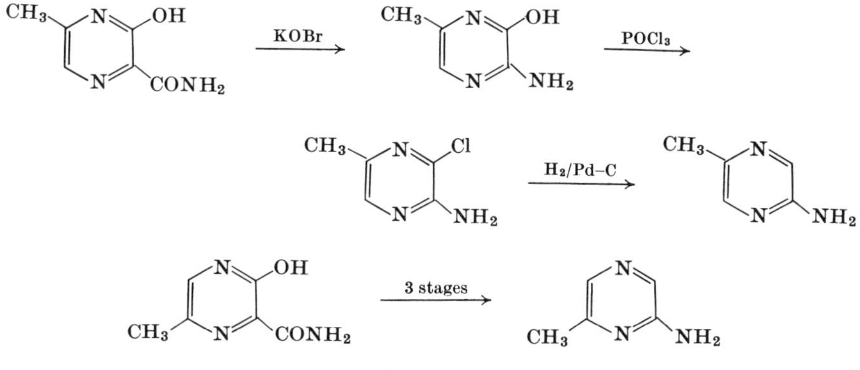

SCHEME 35

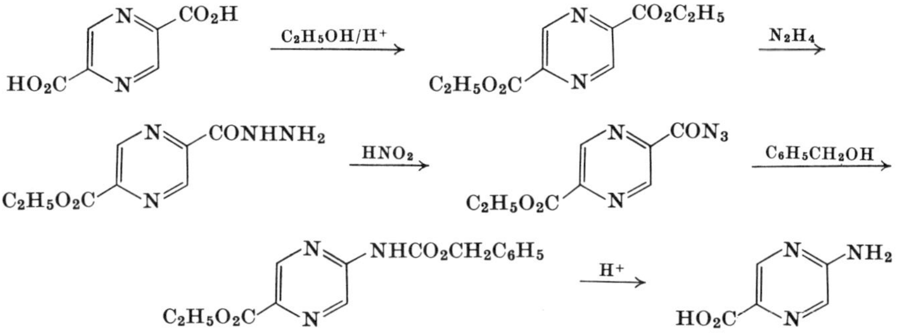

SCHEME 36

2-Aminopyrazines condense with α-ketoaldehydes in dilute hydrochloric acid to give good yields of dihydroimidazopyrazinone derivatives; for example, compound **118** is obtained in 73% yield by condensation of aminopyrazine with methylglyoxal. This ring-closure reaction is the key step in a synthesis of *Cypridina hilgendorfii* luciferin (**119**).[327] Simple analogs of the luciferin have been prepared[327a] and *Cypridina* oxyluciferin, a product of *Cypridina* bioluminescence has been shown by synthesis to be an acylaminopyrazine, rather than an imidazopyrazine.[327b] A further synthesis of *Cypridina*

---

[327] S. Inoue, S. Sugiura, H. Kakoi, and T. Goto, *Tetrahedron Lett.* 1609 (1969).
[327a] S. Sugiura, S. Inoue, and T. Goto, *Yakugaku Zasshi* **90**, 707 (1970); *Chem. Abstr.* **73**, 98904 (1970) and preceding papers in this series.
[327b] S. Sugiura, S. Inoue, and T. Goto, *Yakugaku Zasshi* **90**, 711 (1970); *Chem. Abstr.* **73**, 98906 (1970).

luciferin has been announced which removes the previous uncertainty whether the indolyl residue is attached in the luciferin to carbon-5 or to carbon-6 of the pyrazine ring. The key intermediate in this synthesis is a suitably substituted aminopyrazine N-oxide prepared by the condensation of an α-aminonitrile and an α-oximinoketone (see Scheme 33). Improved yields are obtained in the presence of titanium tetrachloride.[327c] Amidines have been prepared in good yield from aminopyrazine by aluminum chloride-promoted reaction with alkyl, aryl,

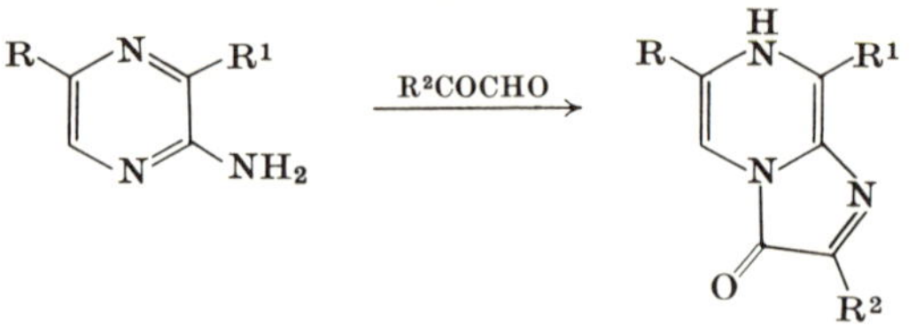

(118) (R = R¹ = H, R² = CH₃)
(119) [R = 3-indolyl, R¹ = CH₂CH₂CH₂NHC(=NH)NH₂, R² = CH(CH₃)CH₂CH₃]

or alkaryl cyanides.[328] Oxidation of N-2-pyrazinylbenzamidine with lead tetraacetate gives 2-phenyltriazolo[1,5-a]pyrazine [Eq. (19)].[329]

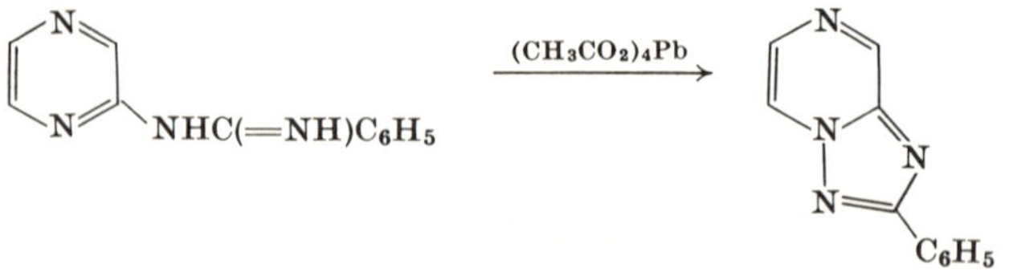

(19)

Brief reference has already been made to the work of Cragoe and his colleagues on the preparation of amino-N-amidinopyrazinecarboxamides which has led to the development of clinically useful diuretics (see Section V, B[254–260]). Aminocarboxamides of type **120** give 4-pteridones (**121**) on ring closure with triethyl orthoformate in boiling acetic anhydride.[330]

[327c] T. P. Karpetsky and E. H. White, *J. Amer. Chem. Soc.* **93**, 2333 (1971).
[328] T. Okamoto, Y. Torikoshi, Japanese Patent 12,149 (1969); *Chem. Abstr.* **71**, 81408 (1969).
[329] G. M. Badger, P. J. Nelson, and K. T. Potts, *J. Org. Chem.* **29**, 2542 (1964).
[330] E. J. Cragoe and J. B. Bicking, U.S. Patent 3,361,748 (1968); *Chem. Abstr.* **69**, 19022 (1968).

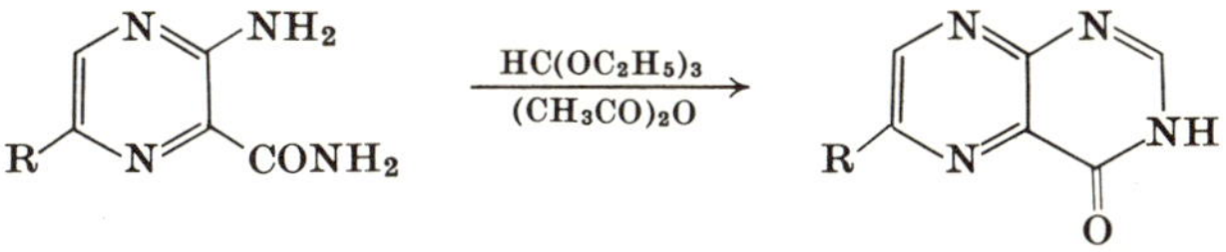

Sec. V. D.]   PYRAZINE CHEMISTRY   171

(120) → (121)
reagents: HC(OC$_2$H$_5$)$_3$, (CH$_3$CO)$_2$O

A number of 3-amino-s-triazolo[4,3-a]pyrazines (122) have been prepared by the action of cyanogen chloride on hydrazinopyrazines. The latter compounds also undergo ring closure to s-triazolo[4,3-a]-pyrazines on treatment with phosgene, carbon disulfide, and triethyl orthoformate.[331–335]

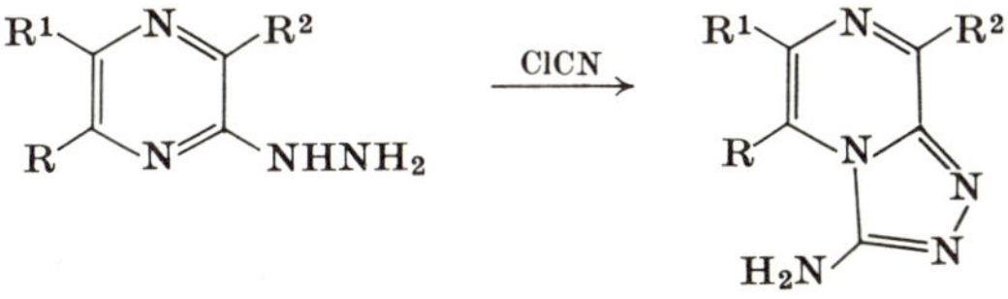

(122)

2,3-Diaminopyrazines undergo the expected condensation reactions with α,β-dicarbonyl compounds. For example, condensation of 2,3-diamino-5,6-dimethylpyrazine with α,β-diketones gives pyrazino-[2,3-b]pyrazines (124).[36] 2,3-Diaminopyrazines are also starting materials for the preparation of imidazo[4,5-b]pyrazines (123).[117]

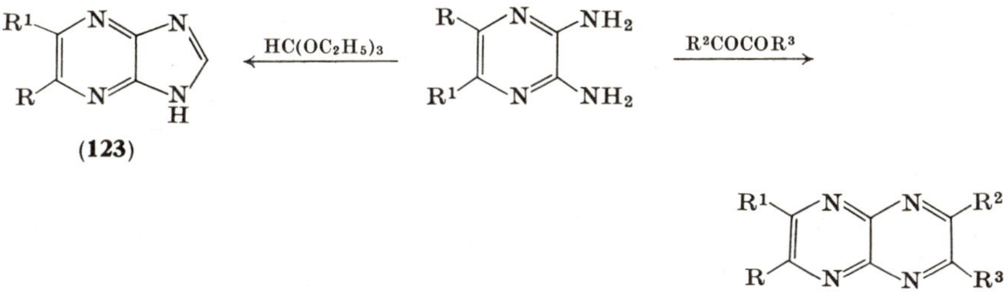

(123) ← HC(OC$_2$H$_5$)$_3$ — 2,3-diaminopyrazine — R^{2}COCOR^3 → (124)

[331] S. E. Mallett and F. L. Rose. *J. Chem. Soc. C* 2038 (1966).
[332] F. L. Rose, *Proc. Int. Symp. Drug. Res., 1967*, 93; *Chem. Abstr.* **72**, 43506 (1970).
[333] J. Maguire, D. Paton, and F. L. Rose, *J. Chem. Soc. C* 1593 (1969).
[334] J. Maguire and F. L. Rose, British Patent 1,146,770 (1969); *Chem. Abstr.* **71**, 39005 (1969).
[335] K. T. Potts and S. W. Schneller, *J. Heterocycl. Chem.* **5**, 485 (1968).

E. Hydroxypyrazines (Pyrazinones)

Pyrazines with hydroxyl groups are generally in the oxo form; however, substituents like chlorine may profoundly influence the position of the tautomeric equilibria. Ultraviolet measurements indicate that in ethanol solution 2-chloro-6-hydroxypyrazine (**125a**) exists predominantly in the hydroxy rather than in the oxo form (**125b**);[272] trifluorohydroxypyrazine also does not tautomerize appreciably to a pyrazinone.[279]

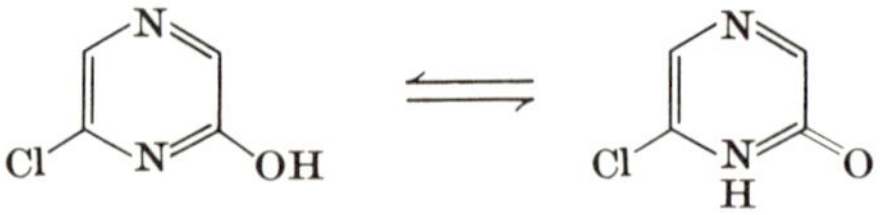

(125a)   (125b)

Hydroxypyrazines are prepared from aminopyrazines, hydroxypyrazinecarboxylic acids, and halopyrazines, and also by direct ring synthesis. The conversion of aminopyrazines into hydroxypyrazines is carried out by treatment with nitrous acid under a variety of conditions (see, e.g., references in footnotes 271, 284, 336, and also Section V,D). Hydroxypyrazinecarboxylic acids are convenient starting materials for hydroxypyrazine preparation as these are readily decarboxylated. 2-Hydroxy-3-carboxamidopyrazines are available by direct synthesis from α,β-dicarbonyl compounds and aminomalonamide,[117] and these on alkaline hydrolysis give the corresponding hydroxycarboxylic acid. Alternatively, alkaline hydrolysis of 2,5-dicyanopyrazines gives 2-hydroxypyrazine 5-carboxylic acids (Scheme 37); the dimethyl derivative gives **126** on decarboxylation.[337]

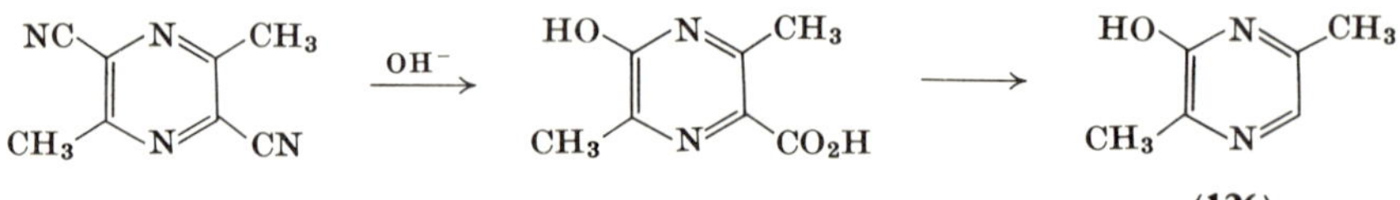

Scheme 37

Hydroxypyrazines are also prepared from halopyrazines either directly by hydroxide ion treatment, e.g., the conversion of fluoro-

---

[336] G. W. H. Cheeseman and E. S. G. Törzs, *J. Chem. Soc.* 6681 (1965).
[337] W. Sharp and F. S. Spring, *J. Chem. Soc.* 1862 (1948).

pyrazine into hydroxypyrazine,[299] or indirectly by intermediate ether formation. Thus, Lutz and Meltzer prepared the hydroxypyrazine (**128**) from the corresponding chloropyrazine by reaction with methoxide ion followed by acid cleavage of the methoxy compound (**127**).[338] 3,6-Dimethyl- and 3,6-diphenyl-2,5-dihydroxypyrazines are obtained from the corresponding dichloro compounds by formation of dimethoxy derivatives and high-temperature methoxide ion demethylation.[269] 2-Hydroxy-6-methoxy- and 2-benzyloxy-6-hydroxypyrazine have been prepared from 2,6-dichloropyrazine by prior

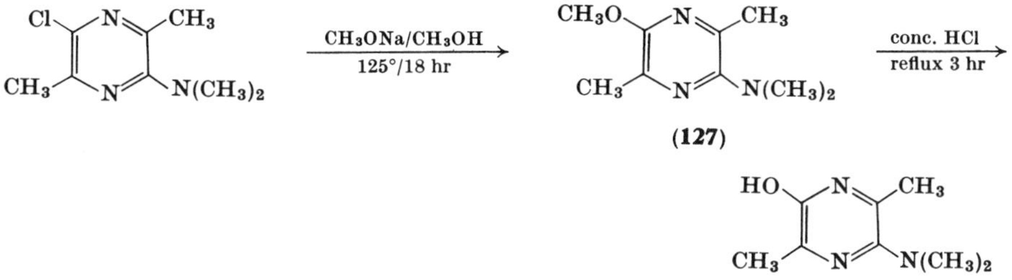

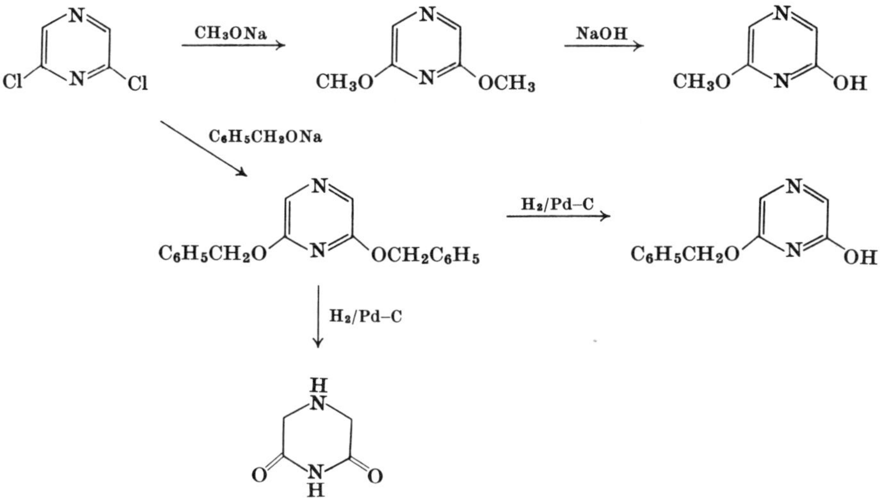

SCHEME 38

[338] W. B. Lutz and R. I. Meltzer, U.S. Patent 3,155,663 (1964); *Chem. Abstr.* **62**, 1674 (1965).

formation of the appropriate 2,6-dialkoxy compounds. Mild sodium hydroxide treatment of the dimethoxy compound gives 2-hydroxy-6-methoxypyrazine; catalytic hydrogenation of the dibenzyloxy compound gives 2-benzyloxy-6-hydroxypyrazine, but attempts to prepare 2,6-dihydroxypyrazine by further hydrogenation failed. Thus, when 2.0 M proportions of hydrogen are used 2,6-dioxopiperazine is formed (Scheme 38).[336] 2,6-Dihydroxy-3,5-diphenylpyrazine (**131**) is prepared by mild base treatment of the corresponding diacetoxy compound (**130**). This is obtained by reaction of the hydroxamic acid (**129**) with acetic anhydride and acetic acid.[338a]

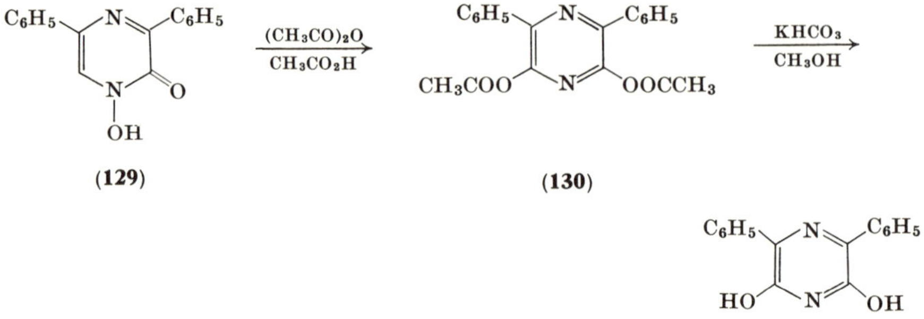

Direct ring syntheses are also available for the preparation of hydroxypyrazines. Thus, haloacylation of an α-aminoketone, followed by reaction with ammonia and oxidation represents a general synthesis of 5,6-disubstituted and 3,5,6-trisubstituted 2-hydroxypyrazines.[339] This is illustrated by the preparation of 5,6-dimethyl-2-hydroxypyrazine (Scheme 39). Hydroxypyrazines are very conveniently

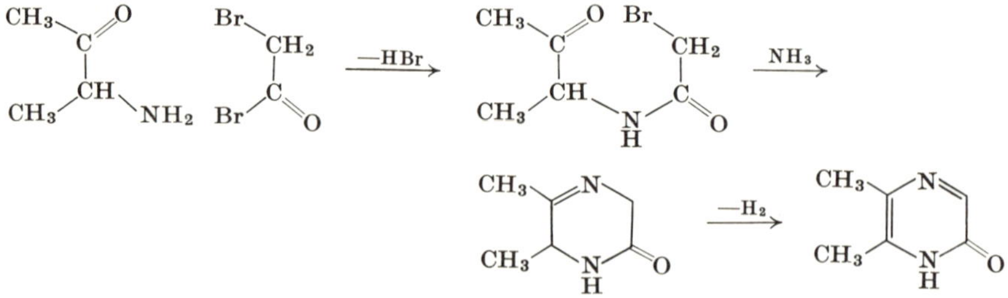

SCHEME 39

[338a] G. W. H. Cheeseman and R. A. Godwin, *J. Chem. Soc. C* 2977 (1971).
[339] Y. A. Tota and R. C. Elderfield, *J. Org. Chem.* **7**, 313 (1942).

prepared by condensation of α,β-dicarbonyl compounds with α-amino acid amides (see Section III, A). An extension of this method involving α-amino acid nitriles may be exemplified by the condensation of aminoacetonitrile with glyoxal in the presence of sodium hydroxide to give the sodium salt of 2-hydroxypyrazine.[340] 2-Hydroxy-3,6-diphenylpyrazine is obtained from an unusual condensation reaction of benzoylformoin and ammonia (Scheme 40).[341] 2,3-Dihydroxypyrazine(pyrazine-2,3-dione) is synthesized by either of the two

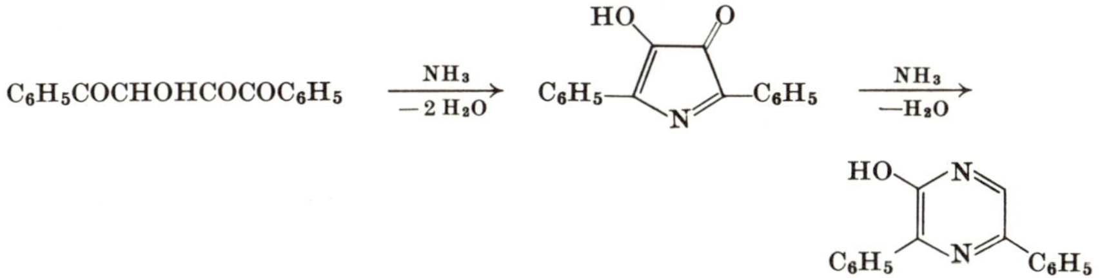

SCHEME 40

sequences shown in Scheme 41.[120, 270] The second of these routes is the method of choice.

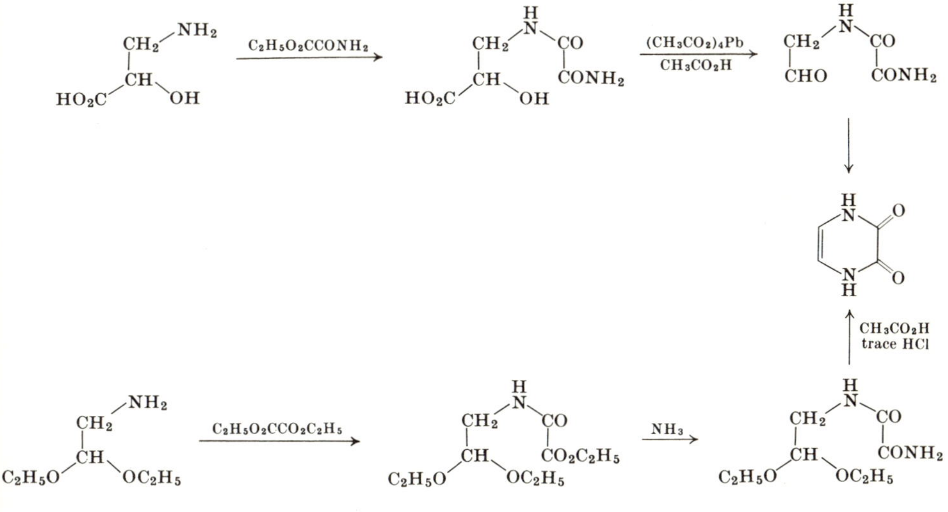

SCHEME 41

[340] M. E. Hultquist, U.S. Patent 2,805,223 (1958); *Chem. Abstr.* **52**, 2935 (1958).
[341] H. von Euler and H. Hasselquist, *Ark. Kemi* **11**, 407 and 481 (1957); *Chem. Abstr.* **52**, 3488 and 5298 (1958).

As already mentioned, hydroxypyrazines exist in tautomeric equilibria with the corresponding pyrazinones which are normally the predominant species in the equilibria. Some of the reactions of hydroxypyrazines are reminiscent of those of phenols; they can, for example, be coupled with diazonium salts and brominated and nitrated in either the ortho or para position to the hydroxyl group. Coupling with diazonium salts occurs in neutral or weakly alkaline solution, but if the reaction is carried out in 1 M sodium hydroxide solution, arylation of the pyrazine ring takes place. From hydroxypyrazine and benzenediazonium chloride 47% 2-hydroxy-3-phenyl- and 4% 2-hydroxy-3,6-diphenylpyrazine are obtained.

Various 2-hydroxy-3-alkylpyrazines give 5-bromo derivatives on bromination with bromine in chloroform in the presence of pyridine as hydrogen bromide acceptor.[342] Nitration of 2-hydroxy-3-phenylpyrazine in acetic acid gives a 5-nitro derivative and nitration of 2-hydroxy-5,6-diphenylpyrazine in sulfuric acid forms a 3-nitro derivative (132).[342] The nitro group in compound 132 is readily displaced, for example, by hydrazine, to give the 3-hydrazino derivative (133),[343] and by a variety of chlorinating and brominating reagents.[344] Reaction with thionyl chloride gives the chlorohydroxy compound (134) and with a mixture of thionyl chloride and pyridine,

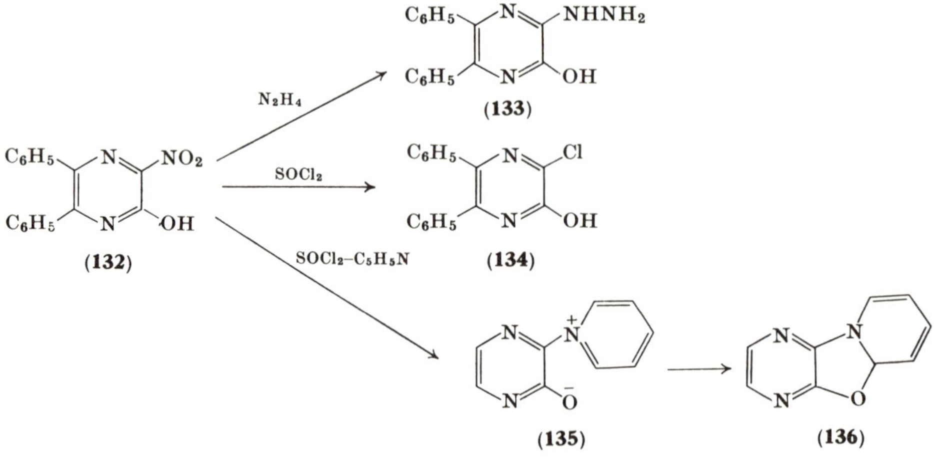

[342] G. Karmas and P. E. Spoerri, *J. Amer. Chem. Soc.* **78**, 4071 (1956); **75**, 5517 (1953).
[343] G. W. H. Cheeseman and M. Rafiq, *J. Chem. Soc. C* 452 (1971).
[344] A. Hirschberg and P. E. Spoerri, *J. Heterocycl. Chem.* **6**, 975 (1969).

compound **135** is formed. Acid hydrolysis of **135**, which may well exist in a cyclic (**136**) rather than in a zwitterionic form, gives 2,3-dihydroxy-5,6-diphenylpyrazine.[345] Methylation of hydroxypyrazine with dimethyl sulfate and alkali gives the *N*-methyl derivative (**137**)[147] and similar treatment of 2,3-dihydroxy- and 2-amino-3-hydroxypyrazine also gives *N*-methylated products. Diazomethane methylation of 2,3-dihydroxypyrazine gives a mixture of *O,O*-, *O,N*-, and *N,N*-dimethyl derivatives, and a mixture of *O*- and *N*(4)-monomethyl derivatives (**138**) and (**139**) is obtained from 2-amino-3-hydroxypyrazine.[336]

Reaction of the silver salt of hydroxypyrazine with acetobromoglucose gives the *O*-glucosyl derivative (**140**) in 30% yield; the corresponding reaction of the sodium salt gives 16% *O*-glucosyl derivative and 5% *N*-glucosyl derivative (**142**). Structural assignment is made by comparison of the ultraviolet absorption properties of the *O*- and *N*-deacetylated glucosides (**141**) and (**143**) with those of the *O*- and *N*-methyl derivatives of hydroxypyrazine.[346–351]

3-Benzyl-6-methyl-2,5-dihydroxypyrazine (**144**) undergoes Diels–Alder addition with dimethyl acetylenedicarboxylate; on heating the

---

[345] J. D. Ratajczyk and J. A. Carbon, *J. Org. Chem.* **27**, 2644 (1962).
[346] W. Pfleiderer, R. Lohrmann, F. Reisser, and D. Soell, *Pteridine Chem., Proc. 3rd Int. Symp., Stuttgart, 1962* 87; *Chem. Abstr.* **62**, 9224 (1965).
[347] F. Reisser and W. Pfleiderer, *Chem. Ber.* **99**, 542 (1966).
[348] G. Wagner and H. Frenzel, *Z. Chem.* **5**, 24 (1965).
[349] G. Wagner and R. Metzner, *Pharmazie* **20**, 752 (1965); *Chem. Abstr.* **64**, 19748 (1966).
[350] G. Wagner and R. Metzner, *Naturwissenschaften* **52**, 83 (1965).
[351] G. Wagner and H. Frenzel, *Arch. Pharm. (Weinheim)* **300**, 421 and 591 (1967).

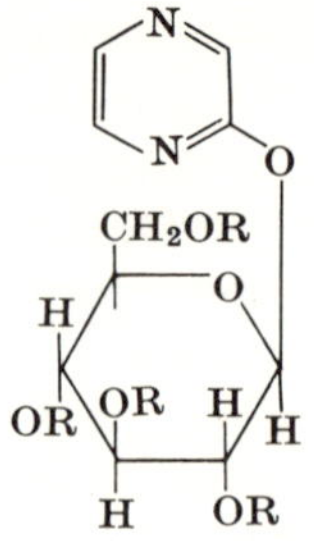

(140) (R = COCH₃)
(141) (R = H)

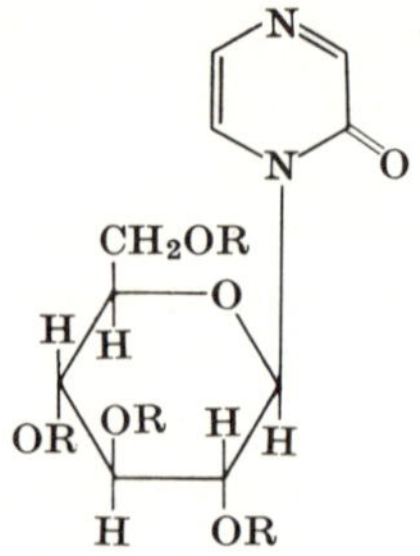

(142) (R = COCH₃)
(143) (R = H)

bicyclic adduct (145) at 100°, the isomeric pyridones (146) and (147) are formed in approximately equal quantities by the elimination of isocyanic acid from either of the two amide bridges. This interesting reaction has possible biosynthetic significance.[352] Irradiation of the

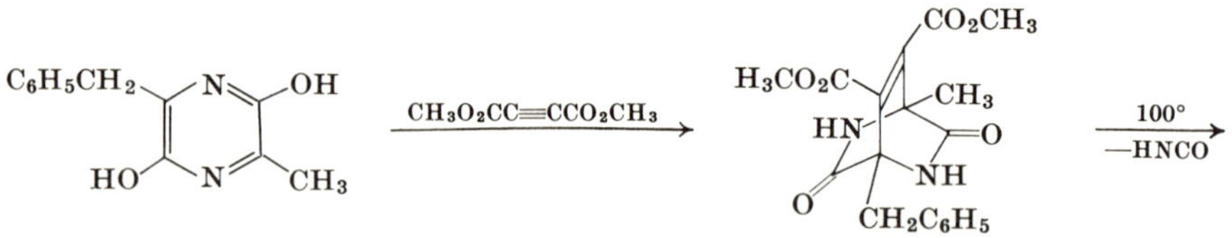

(144)

(145)

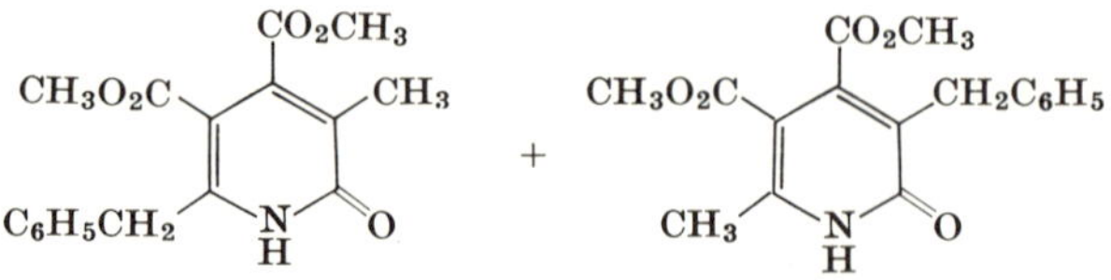

(146)

(147)

bicyclic aziridine (148) gives a dipolar species (149) which undergoes 1,3-cycloaddition with dimethyl acetylenedicarboxylate to give compound (150). The intermediate (149) is related in structure to 2,6-dihydroxypyrazine.[353]

[352] A. E. A. Porter and P. G. Sammes, *J. Chem. Soc. D* 1103 (1970).
[353] R. Huisgen and H. Maeder, *Angew. Chem. Int. Ed. Engl.* **8**, 604 (1969).

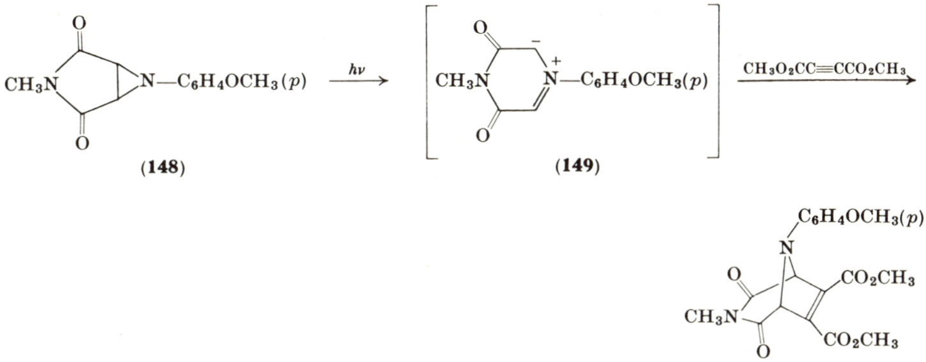

Mercaptopyrazine (**151a**) is best prepared by the action of sodium hydrogen sulfide on chloropyrazine in dimethylformamide.[147] It decomposes on heating with evolution of hydrogen sulfide and formation of di-2-pyrazinyl sulfide. The latter compound is also obtained together with mercaptopyrazine on treatment of chloropyrazine with aqueous potassium hydrogen sulfide solution. Comparison of the

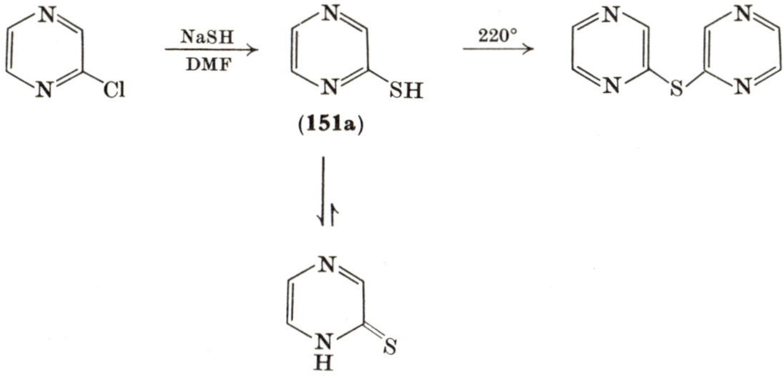

ultraviolet absorption and ionization properties of mercaptopyrazine (**151a**) with those of its N- and S-methyl derivatives (**152**) and (**153**) indicates that it exists predominantly in the 2-thiono form (**151b**). Compound **152** is prepared by the action of phosphorus pentasulfide in pyridine on the corresponding oxo compound, and compound **153** is prepared by methylation of mercaptopyrazine with methyl iodide and alkali.[147] Treatment of the sodium salt of mercaptopyrazine with acetobromoglucose gives mainly the *S*-glucoside, but a little of the *N*-glucoside is also formed.[351]

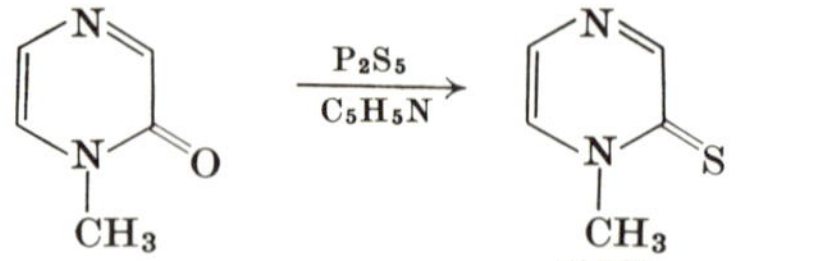

Some substituted 2-mercaptopyrazines have also been prepared. 2-Mercapto-3,5,6-trimethylpyrazine has been obtained from the 2-chloro derivative by treatment with hydrogen sulfide and sodium ethoxide[303] and 2,3,5-tribromo and triodo-6-methylthiopyrazines from the corresponding tetrahalo compounds by nucleophilic displacement with sodium methylmercaptide (CH_3SNa).[311] Treatment of 2-amino-6-chloropyrazine with this reagent similarly gives 2-amino-6-methylthiopyrazine;[308] these compounds have antibacterial properties. 2,3-Dimercapto-[354] and 2,5-dimercaptopyrazines[355] have also

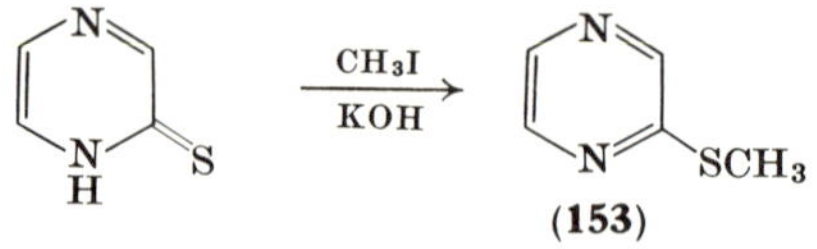

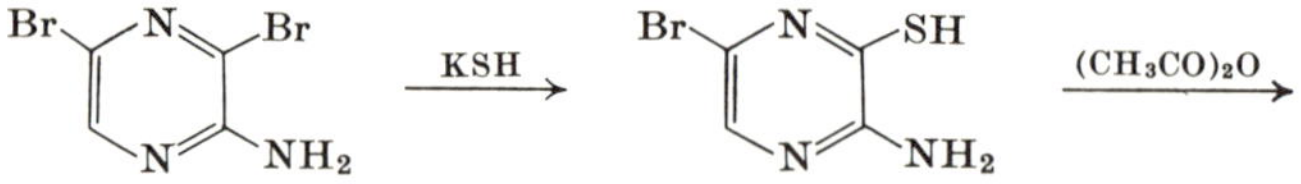

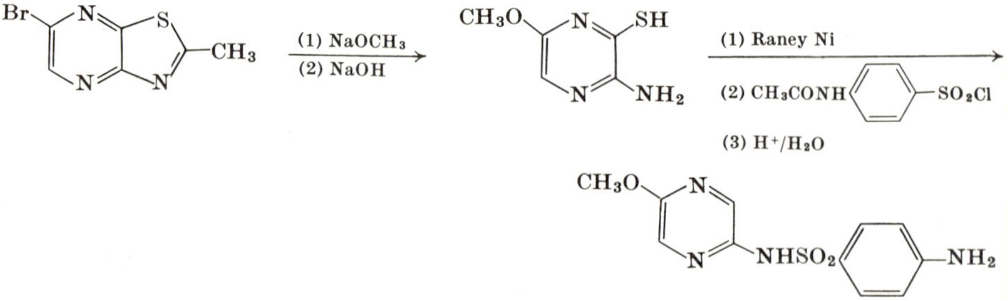

[354] K. Dickore, K. Sasse, and R. Wegler, Belgian Patent 610,601 (1962); *Chem. Abstr.* **57**, 13774 (1962).
[355] L. Holtzer, Austrian Patent 204,049 (1959); *Chem. Abstr.* **53**, 18070 (1959).

been prepared and converted into derivatives which are useful as pesticides. 5-Methoxy-2-sulfanilamidopyrazine (155), an antibacterial agent, is prepared from 2-amino-5-bromo-3-mercaptopyrazine (154) as shown.[356] A number of 5- and 6-substituted pyrazine 2-thio ethers have been prepared; 2-carboxamido-6-phenylthiopyrazine and its sulfone have been found to be especially good antituberculous drugs.[357] The thio ethers (157) obtained by reaction of the anion of 2-mercapto-3,6-dimethylpyrazine (156) and α-haloketones, undergo cyclization in polyphosphoric acid to give thiazolo[3,2-a]pyrazinium salts (158)[358,359]

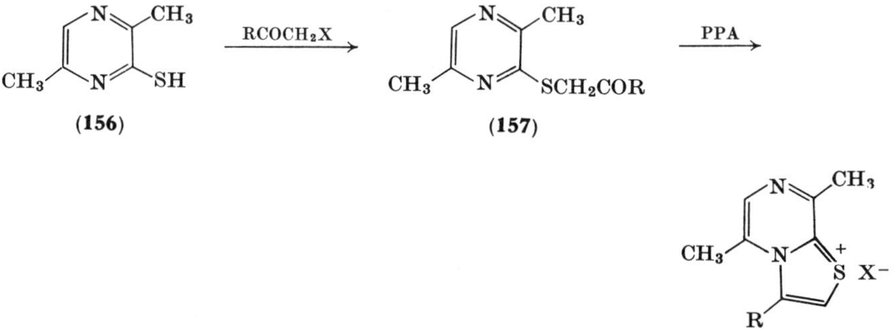

A series of O,O-dialkyl-O-2-pyrazinylphosphorothioates and related compounds have been prepared and tested as pesticides. Thionazin (15), the product of reaction of the sodium salt (159) and O,O-diethylphosphorochloridothioate (160) is used in agriculture.[360-364]

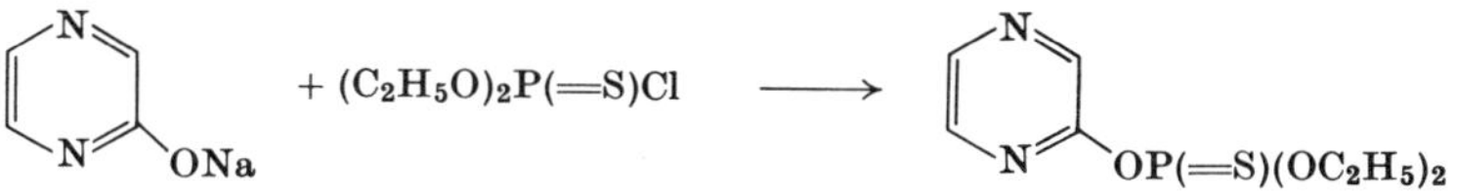

[356] British Patent 958,626 (1964); Chem. Abstr. 61, 5668 (1964).
[357] M. Asai, J. Pharm. Soc. Jap. 81, 1475 (1961); Chem. Abstr. 56, 8712 (1962).
[358] C. K. Bradsher and D. H. Lohr, J. Heterocycl. Chem. 4, 75 (1967).
[359] D. H. Lohr, Ph.D. Thesis, Duke University, 1965.
[360] F. M. Gordon, U.S. Patent 2,938,831 (1960); Chem. Abstr. 54, 17781 (1960).
[361] British Patent 984,522 (1964); Chem. Abstr. 60, 12027 (1964).
[362] G. N. Gagliardi, French Patent 1,476,886 (1967); Chem. Abstr. 68, 21952 (1968).
[363] U.S. Patent 3,172,888 (1965); Chem. Abstr. 64, 6669 (1966).
[364] R. J. Magee and J. B. Lovell, Ger. Offen. 1,944,923 (1970); Chem. Abstr. 72, 11513 (1970).

VI. Reduced Pyrazines

Derivatives of 1,2-, 1,4-, 2,3-, and 2,5-dihydropyrazine are known, but the structures of these compounds are not easy to establish because of the tendency of dihydropyrazines to isomerize, dimerize, and oxidize. Dihydro structures are stabilized by groups such as alkoxycarbonyl and phenyl; 1,4-dihydro-2,3,5,6-tetraethoxycarbonylpyrazine and 2,3-dihydro-5,6-diphenylpyrazine are relatively stable compounds.

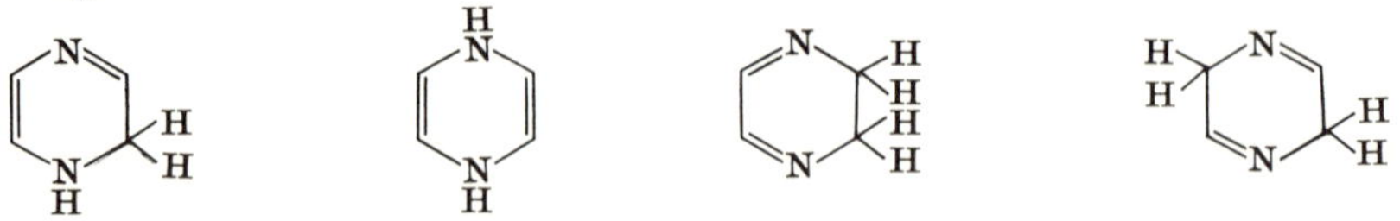

1,2-Dihydropyrazines

These compounds are formed by addition of lithium alkyls to alkylpyrazines, followed by hydrolysis (see Section V,A). They readily oxidize to pyrazines on exposure to air. Some early examples of 1,4-dihydropyrazines have recently been shown to have 1,2-dihydro structures.[364a]

1,4-Dihydropyrazines

The 1,4-dihydropyrazine formed by reduction of pyrazine with lithium and trimethylsilyl chloride, spontaneously decomposes on exposure to air.[364b]

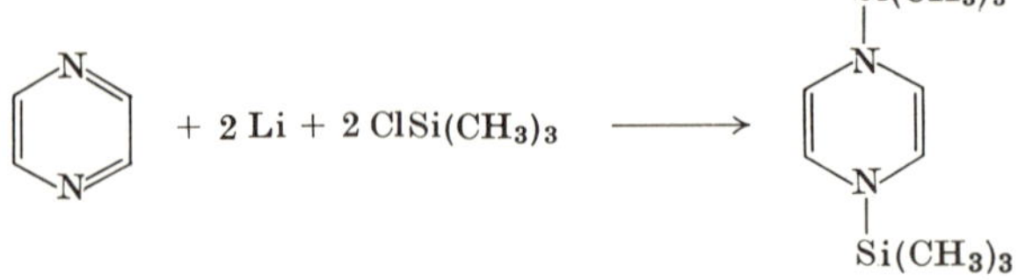

Mager and Berends found that tetraethoxycarbonylpyrazine (**161**) undergoes partial catalytic hydrogenation to give the tetrahydropyrazine (**162**) which spontaneously oxidizes in air to give the 1,4-dihydro derivative (**163**).[365] Compound **163** is more conveniently

[364a] S.-J. Chen and F. W. Fowler, *J. Org. Chem.* **35**, 3987 (1970).
[364b] R. A. Sulzbach and A. F. M. Iqbal, *Angew. Chem. Int. Ed. Engl.* **10**, 127 (1971).
[365] H. I. X. Mager and W. Berends, *Rec. Trav. Chim.* **78**, 109 (1959); **84**, 314 (1965).

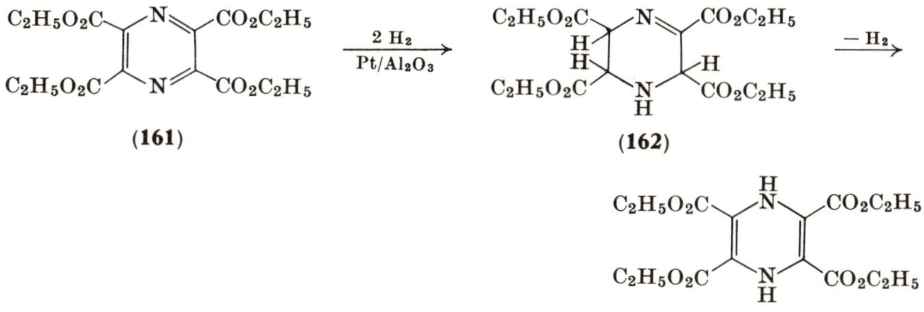

(161) → 2 H$_2$, Pt/Al$_2$O$_3$ → **(162)** → −H$_2$ → **(163)**

prepared by reduction of tetraethoxycarbonylpyrazine with alkaline sodium dithionite and yields of 85–90% are obtained.[366,367] Glover and Jones have prepared the bisiminium salt of 1,4-dihydropyrazine **(164)** by the condensation of pyridine-2-carboxaldehyde with aminoacetal followed by treatment of the resulting anil with hot concentrated hydrochloric acid. Reduction of compound **164** with sodium borohydride yields the piperazine **(165)**. The latter compound can also be prepared indirectly through the dilactam **(166)**.[368]

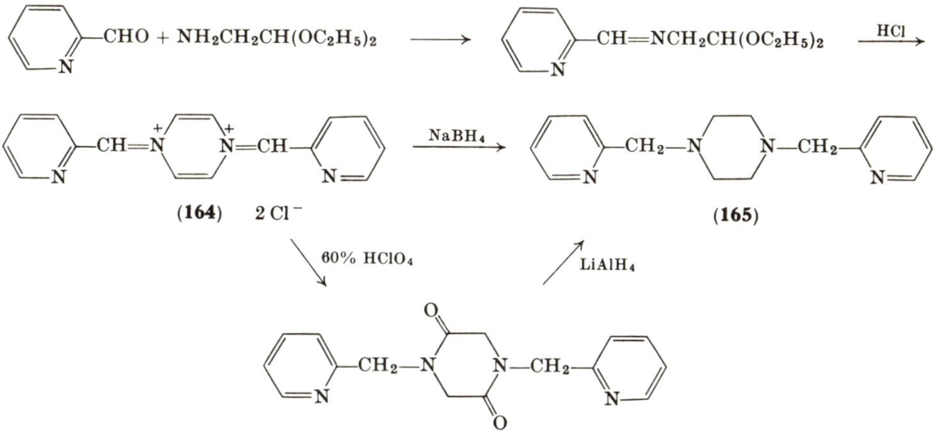

[366] H. I. X. Mager and W. Berends, *Rec. Trav. Chim.* **79**, 282 (1960).
[367] H. I. X. Mager and W. Berends, U.S. Patent 3,024,235 (1962); *Chem. Abstr.* **57**, 12507 (1962).
[368] E. E. Glover, G. Jones, and G. Trenholm, *J. Chem. Soc. C* 1209 (1966).

Methylated 1,4-dihydropyrazines (1,2,4-tri-, 1,2,4,5-tetra-, and 1,2,4,6-tetramethyl) as well as imidazolines have been isolated from the reaction of sucrose with methylamine at high temperatures (120°–260°), for prolonged reaction periods (18 hours) and in the presence of ammonium phosphate as catalyst. 1,4-Dihydropyrazines are also isolated from the reaction of sucrose and ethanolamine; sucrose–ammonia interaction yields a wide range of pyrazines and imidazoles.[369–372]

2,3-Dihydropyrazines

These compounds are synthesized (see Section III) by the condensation of α,β-dicarbonyl compounds with 1,2-diamines. Thus, 2,3-dihydro-2,3,5-trimethyl- and 2,3-dihydro-2,3,5,6-tetramethyl-pyrazines are isolated from the reaction of diacetyl with 1,2-diaminopropane and 2,3-diaminobutane, respectively.[373] Condensation of furil with ethylenediamine similarly gives 2,3-di(2-furyl)-5,6-dihydropyrazine (**167**).[374] The PMR spectrum of 2,3,5,5-tetramethyl-5,6-

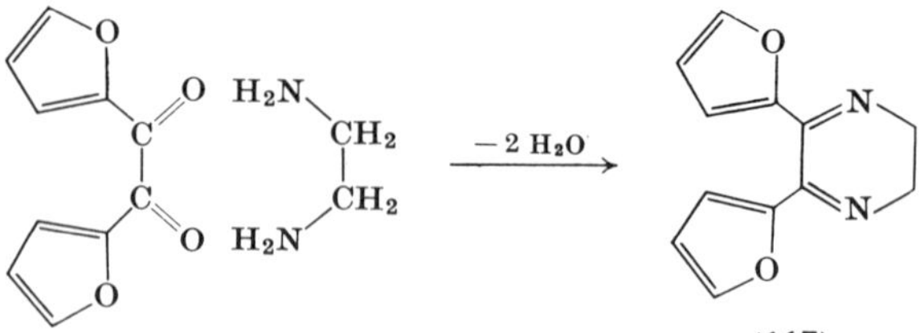

(**167**)

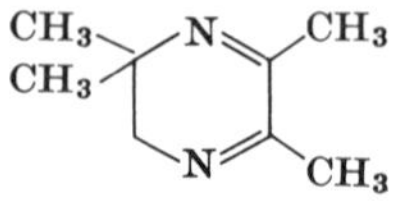

(**168**)

[369] I. Jezo and I. Luzak, *Chem. Zvesti* **17**, 865 (1963); *Chem. Abstr.* **61**, 2039 (1964).
[370] I. Jezo, *Chem. Zvesti* **17**, 126 (1963); *Chem. Abstr.* **60**, 4139 (1964).
[371] I. Jezo and I. Luzak, *Chem. Zvesti* **18**, 837 (1964); *Chem. Abstr.* **63**, 665 (1965).
[372] M. R. Grimmett, *Rev. Pure Appl. Chem.* **15**, 101 (1965).
[373] T. Ishiguro, M. Matsumura, and M. Awamura, *J. Pharm. Soc. Jap.* **78**, 751 (1958); *Chem. Abstr.* **52**, 18453 (1958).
[374] H. Saikachi and J. Matsuo, *J. Pharm. Soc. Jap.* **86**, 927 (1966); *Chem. Abstr.* **66**, 37882 (1967).

dihydropyrazine (**168**) shows a five-bond coupling of 1.85 Hz between the methylene protons at C-6 and the methyl protons at C-2.[375]

2,5-Dihydropyrazines

Wilen has shown by PMR measurements that 2,5-dihydro-3,6-dimethylpyrazine (**169**) is the major product from the reaction of α-aminoacetone hydrochloride with potassium hydroxide in the absence of air. PMR measurements also indicate the presence of minor amounts of either the isomeric 1,4-dihydro (**170**) or 1,2-dihydro derivative (**171**). Compound **169** is slowly converted on standing at room temperature into a dimer, $C_{12}H_{20}N_4$, the structure of which is still being investigated.[376] The self-condensation of two molecules of

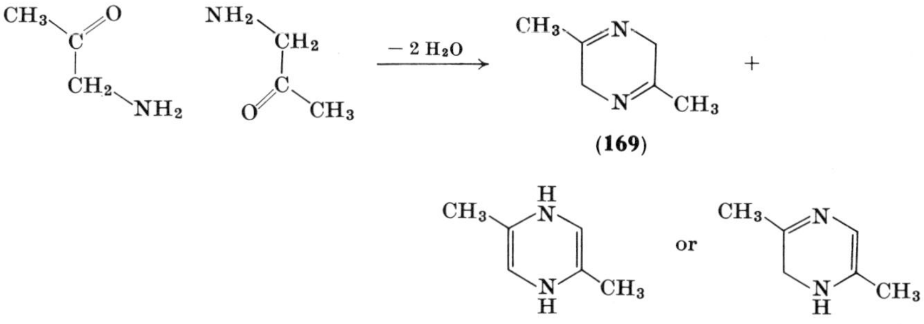

α-aminoethyl phenyl ketone (**172**) and α-amino-α-phenylacetone (**174**) gives the 2,5-dihydropyrazines (**173**) and (**175**), respectively.[377] This suggests that tautomerization is slow, and that the 1,4-dihydropyrazine (**176**) is, as expected, not an intermediate in their formation.

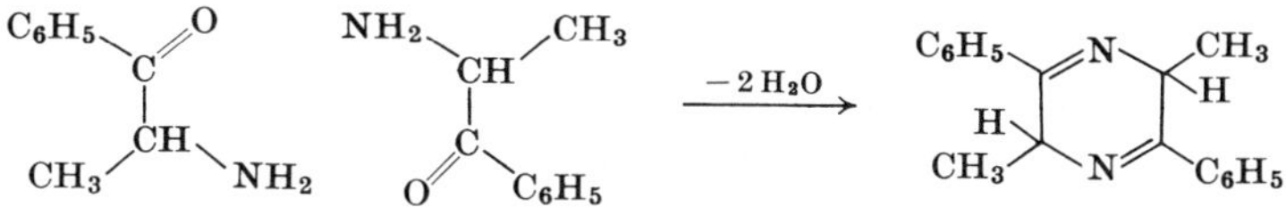

[375] J. M. Kliegman and R. K. Barnes, *Tetrahedron Lett.* 1953 (1969).
[376] S. H. Wilen, *J. Chem. Soc. D* 25 (1970).
[377] S. Gabriel, *Ber.* **41**, 1127 (1908); **44**, 57 (1911).

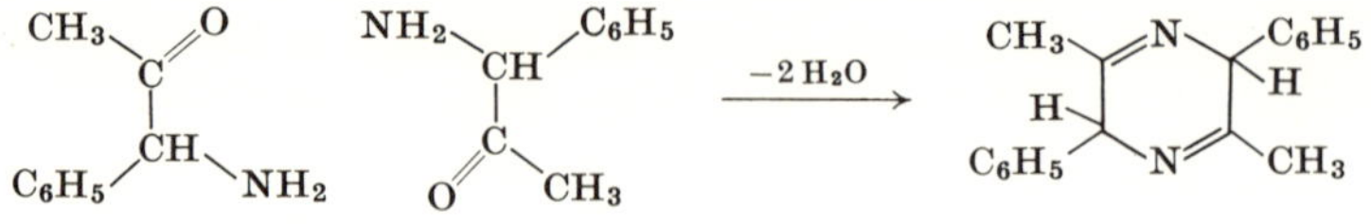

(174) — structures — (175)

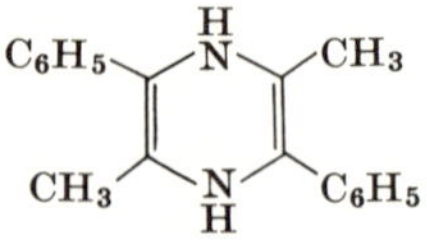

(176)

3-Phenylazirine (178), the photolysis product of α-azidostyrene (177), dimerizes on standing to 2,5-dihydro-3,6-diphenylpyrazine (179). The latter compound, unlike the isomeric 2,3-dihydro-5,6-diphenylpyrazine is readily aromatized by treatment with hydrogen peroxide in acetic acid.[378] 2,5-Dibenzyloxy-2,5-dihydropyrazine is prepared by

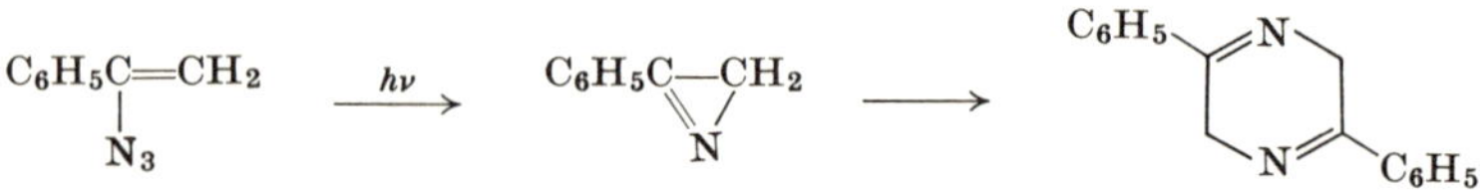

(177) → (178) → (179)

benzyl chloride alkylation of the silver salt of 2,5-diketopiperazine [Eq. (20)],[379, 380] and mention has already been made of the preparation of the corresponding diethoxydihydro compound by alkylation of the diketopiperazine with triethyloxonium fluoroborate (see Section III).

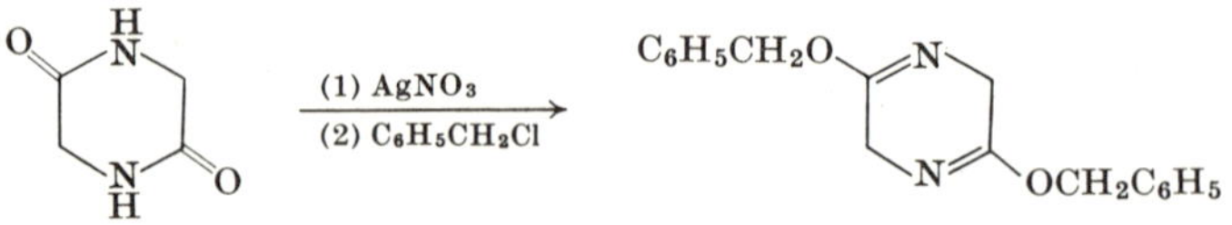

(20)

Kappe et al.[381] have prepared 3,6-bis(2-hydroxyphenyl)-2,5-dihydropyrazine (181) by alkaline treatment of 3-amino-4-hydroxycoumarin (180). Compound 181 is conveniently dehydrogenated over

[378] L. Horner, A. Christmann, and A. Gross, Chem. Ber. 96, 399 (1963).
[379] L. N. Akimove and Z. V. Kiryukhina, J. Gen. Chem. USSR 27, 1979 (1957).
[380] P. Karrer and C. Gränacher, Helv. Chim. Acta 6, 1108 (1923); 7, 763 (1924).
[381] T. Kappe, K. Burdeska, and E. Zeigler, Monatsh. Chem. 97, 77 (1966).

a palladium-on-carbon catalyst to the corresponding pyrazine and reduced with sodium and *n*-butanol to 2,5-bis(2-hydroxyphenyl)-piperazine. Reaction of compound **181** with acetic anhydride leads to 1,4-diacetyl-3,6-bis(2-acetoxyphenyl)-1,4-dihydropyrazine (**182**). Alkaline hydrolysis of the tetraacetyl compound (**182**) with exclusion of air regenerates the 2,5-dihydropyrazine (**181**). Preference in this

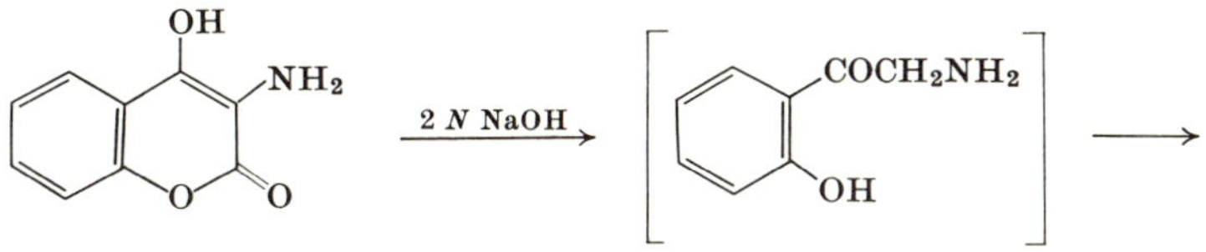

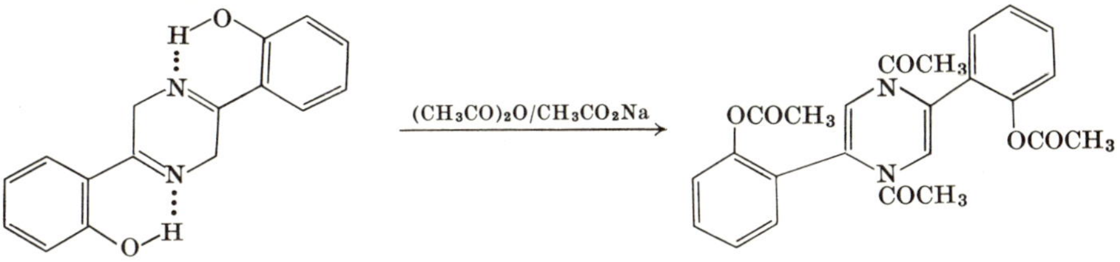

case for the 2,5- rather than the 1,4-dihydro form is attributed to stabilization of the 2,5-tautomer by intramolecular hydrogen bonding.[381]

The reaction of isonitriles and trihalogenoboranes gives hexahalogeno-2,5-diboradihydropyrazines of moderate stability (Scheme 42).[382]

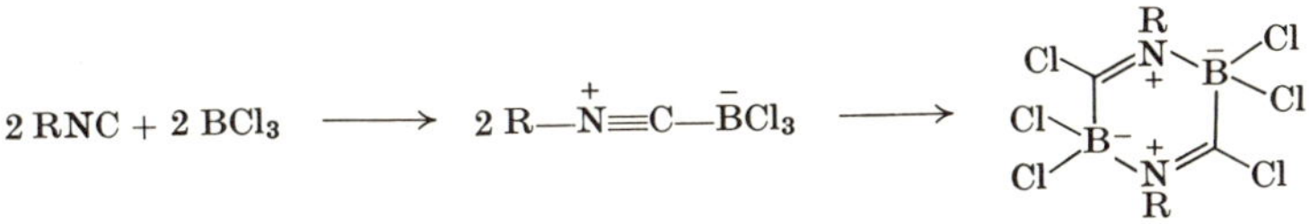

SCHEME 42

Tetrahydropyrazines

Stevens and his co-workers report the preparation of 2-phenyl-3,3-dimethyl-3,4,5,6-tetrahydropyrazine (**184**) in 91% yield from the reaction of the epoxy ether (**183**) and ethylenediamine. The structural assignment is consistent with the observation of strong N–H and

[382] A. Meller and H. Batka, *Monatsh. Chem.* **100**, 1823 (1969).

C═N absorptions in the infrared and the derivation of the corresponding piperazine on sodium borohydride reduction. Treatment of compound **184** with hydrochloric acid gives the aminoketone (**186**), and with acetic anhydride the 4-acetyl derivative (**185**) is obtained.[383]

$C_6H_5-C(O)-C(OCH_3)(CH_3)_2$ (**183**) $\xrightarrow{NH_2CH_2CH_2NH_2}$ (**184**) $\xrightarrow{(CH_3CO)_2O/C_5H_5N}$ (**185**)

(**184**) $\xrightarrow{6\ N\ HCl}$ $NH_2CH_2CH_2NHC(CH_3)_2COC_6H_5 \cdot 2\ HCl$ (**186**)

The *trans*-benzylidene derivative (**188**), obtained as the major product of condensation of 3-methylpiperazine 2,5-dione and benzaldehyde in the presence of acetic anhydride, undergoes photoisomerization to the cis isomer (**191**) on irradiation in methanol. Both isomers have been converted into tetrahydropyrazine imino ethers (**189**) and (**192**) by treatment with triethyloxonium fluoroborate. The trans compound reacts more slowly and gives a lower yield of imino ether and this is attributed to steric hindrance. Compounds **188** and **191** are deacetylated on treatment with methanolic 2 N potassium hydroxide. The trans and cis isomers (**187**) and (**190**), so produced are converted into 3-benzyl-2,5-dihydroxy-6-methylpyrazine (**144**) when heated at 100° with sodium hydroxide.[384] Treatment of the dichloroacetyl derivative of phenylalanine with methylamine gives 1-methyl-3-benzylidenepiperazine 2,5-dione with the stereochemistry shown.[384a]

$C_6H_5CH_2CH(NHCOCHCl_2)\ CO_2H \longrightarrow$

A number of biologically interesting compounds are piperazine 2,5-diones with a two-atom sulfur bridge between carbons 3 and 6.

[383] C. L. Stevens, K. G. Taylor, and M. E. Munk, *J. Org. Chem.* **29**, 3574 (1964).
[384] K. W. Blake and P. G. Sammes, *J. Chem. Soc. C* 980 (1970).
[384a] A. E. A. Porter and P. G. Sammes, *J. Chem. Soc. C* 2530 (1970).

Sec. VI.] PYRAZINE CHEMISTRY 189

It is therefore of interest that a simple derivative of this type has been synthesized.[384b]

Several references have already been made to the oxidation of 2,3-dihydropyrazines to pyrazines (see Section III). When the 2- and 3-positions are fully substituted as in 2,3-dihydrohexamethylpyrazine (**193**), aromatization is structurally precluded. Compound **193** is isolated as an unstable, yellow solid which rapidly dimerizes oxidatively in air and in the presence of hydrochloric acid to give a stable purple dihydrochloride (**194a**). Rearrangement of compound **194a** gives the conjugated tautomer (**194b**).[385] Mager and Berends have

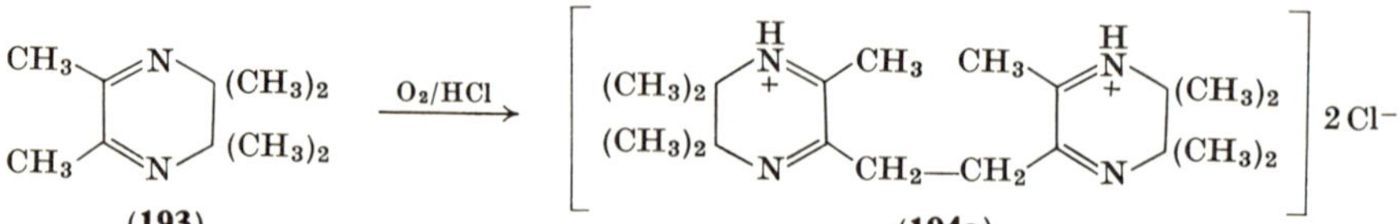

reexamined the effect of hydrogen peroxide on 5,6-diphenyl-2,3-dihydropyrazine. In boiling methanol, the main product is N,N-dibenzoylethylenediamine and a little 2,3-diphenylpyrazine is also formed. In aqueous acetic acid (Scheme 43), the main products are

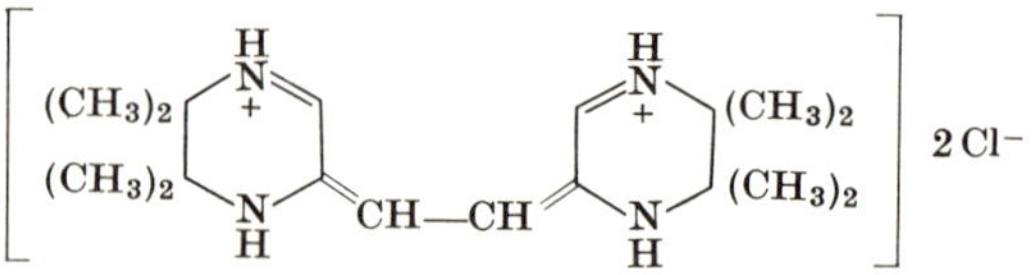

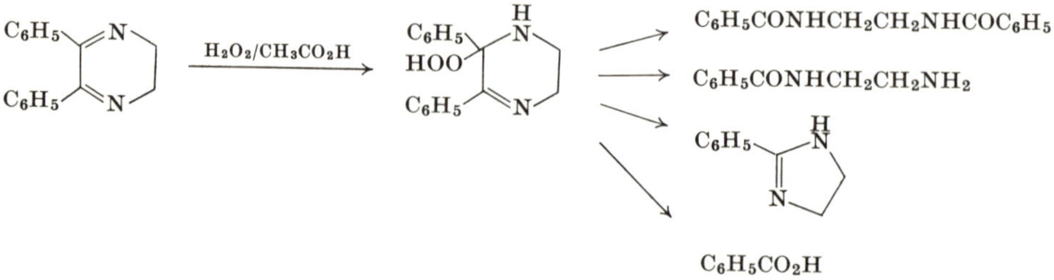

SCHEME 43

[384b] H. Poisel and U. Schmidt, *Angew. Chem. Int. Ed. Engl.* **10**, 130 (1971).
[385] M. Lamchen and T. Mittag, *Proc. Chem. Soc.* **10**, 302 (1963).

N-benzoyl- and N,N'-dibenzoylethylenediamine, 2-phenylimidazoline, and benzoic acid. The formation of these products is rationalized by mechanisms involving the initial addition of hydrogen peroxide to the dihydropyrazine to give the hydroperoxide. It appears therefore that the earlier report[386] that peracetic acid oxidation of 5,6-diphenyl-2,3-dihydropyrazine gives the 1,4-di-N-oxide is incorrect.[386a]

Hydrogen peroxide is produced during the autoxidation of some 2,5-dihydropyrazines and 1,2,3,6-tetrahydropyrazines, e.g., compound **195**. Hydrogen peroxide production is strongly catalyzed by manganese diphosphate complexes.[387,388]

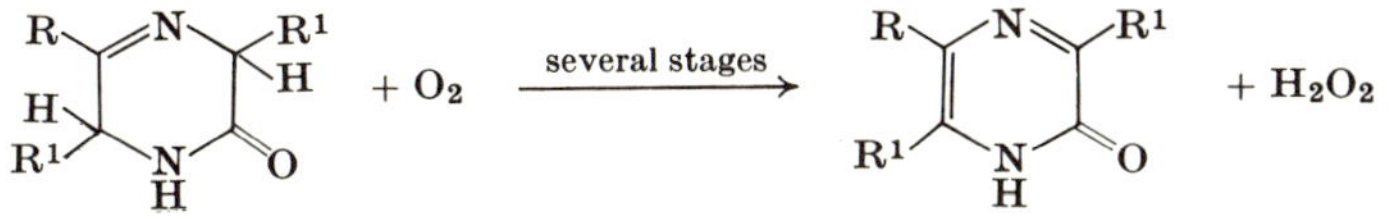

(R = C_2H_5, R^1 = $(CH_3)_2CHCH_2$)

The photolysis of 2,3-dihydro-5,6-diphenylpyrazine and related 2,3-dihydropyrazines gives imidazoles. Reaction proceeds through a diazahexatriene intermediate (**196**) as illustrated. The major product,

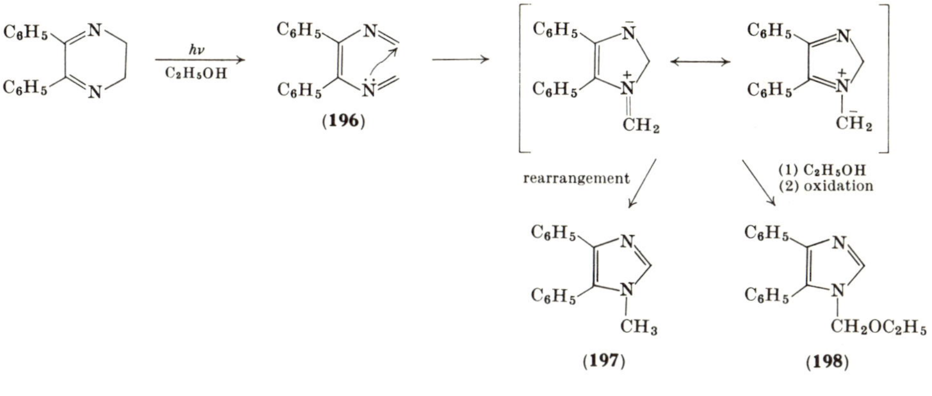

---

[386] J. K. Landquist, *J. Chem. Soc.* 1885 (1956).
[386a] H. I. X. Mager and W. Berends, *Rec. Trav. Chim.* **84**, 314 (1965).
[387] R. Zell and H. Erlenmeyer, *Helv. Chim. Acta* **49**, 1131 (1966).
[388] R. Zell, H. Brintzingler, B. Prijs, and H. Erlenmeyer, *Experientia* **20**, 117 (1964).

isolated in 75% yield, is 1-methyl-4,5-diphenylimidazole (**197**) and only a small amount of the ethoxymethylimidazole (**198**) is obtained.

The photolysis of 2,3-dihydro-2-isobutyl-5,6-diphenylpyrazine gives, in addition to 1-methyl-2-isobutyl-5,6-diphenylimidazole (**199**), 1,2-di(2-isobutyl-4,5-diphenylimidazoyl)ethane (**200**). The formation of compound **200** appears to involve oxidative dimerization of the intermediate triene.[389, 390] Photolysis of 2,3,5,5-tetramethylpyrazine also gives products which arise from initial ring cleavage to a diazahexatriene.[390a] A bridged-ring species has been reported to be formed by reaction of 2,3-diphenyl-5,6-dihydropyrazine with diethyl fumarate.[390b]

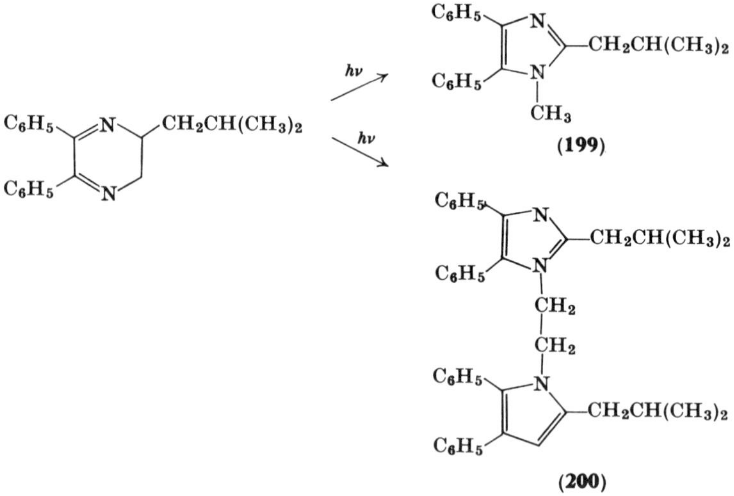

VII. Pyrazine N-Oxides

Pyrazine N-oxides are usually prepared by direct oxidation, and a common technique is to add hydrogen peroxide of varying strength (30–85% w/w) to a solution of the pyrazine in acetic acid. Alternatively oxidation is carried out in either formic acid or trifluoroacetic acid.

[389] P. Beak and J. L. Miesel, *J. Amer. Chem. Soc.* **89**, 2375 (1967).
[390] J. L. Miesel, *Diss. Abstr. B* **27**, 3860 (1967).
[390a] D. R. Arnold, V. Y. Abraitys, and D. McLeod, Jr., *Can. J. Chem.* **49**, 923 (1971).
[390b] M. Sakamoto and Y. Tomimatsu, *Yakugaku Zasshi* **90**, 544 (1970); *Chem. Abstr.* **73**, 35327 (1970).

For example, good yields of 2-chloropyrazine 4-oxide,[391,392] 2-carboxamidopyrazine 4-oxide,[392] and 2-methylpyrazine 1,4-dioxide,[391] have been obtained by oxidation of the appropriate pyrazine. In general, the preferred product from oxidation of monosubstituted pyrazines at lower temperatures is the monooxide formed by oxidation on the nitrogen remote from the substituent, whereas 1,4-dioxides are obtained by prolonged heating at higher temperatures.[155] Pertrifluoroacetic acid oxidation of 2,5-dichloro-3,6-dimethylpyrazine furnishes the di-N-oxide, whereas permaleic acid gives only the mono-N-oxide.[393]

Oxidation of methylpyrazine with one equivalent of hydrogen peroxide in acetic acid gives a mixture of the two mono-N-oxides from which the higher melting 2-methyl 1-oxide (**201**) can be separated by fractional crystallization. The 2-methyl 4-oxide (**202**) is isolated by treatment of the mono-N-oxide mixture with acetic anhydride. This selectively converts the 1-oxide into acetoxymethylpyrazine (**203**).[394] Crystallization of the resulting mixture of unreacted 4-oxide and acetoxymethylpyrazine then gives pure 2-methylpyrazine 4-oxide.[395] The N–O stretching frequencies of the 1- and 4-oxides are 1311 and

---

[391] B. Klein, N. E. Hetman, and M. E. O'Donnell, *J. Org. Chem.* **28**, 1682 (1963).
[392] L. Bernardi, G. Palamidessi, A. Leone, and G. Larini, *Gazz. Chim. Ital.* **91**, 1431 (1961).
[393] K. W. Blake and P. G. Sammes, *J. Chem. Soc. C* 1070 (1970).
[394] M. Asai, *J. Pharm. Soc. Jap.* **79**, 1273 (1959); *Chem. Abstr.* **54**, 4607 (1960).
[395] W. H. Gumprecht, T. E. Benkelman, and R. Paju, *J. Org. Chem.* **29**, 2477 (1964).

1330 cm^{-1}, respectively; it appears to be a general phenomenon that the oxide with the higher frequency absorption has the methyl group beta to the N-oxide function.[71] Although pyrazinecarboxylic acid and its sodium salt cannot be directly oxidized,[396] derivatives such as the ethyl ester,[71] amide,[392] and hydrazide[396] can be converted into 4-oxides by reaction with hydrogen peroxide in acetic acid. Oxidation of acetamidopyrazine with hydrogen peroxide in a mixture of acetic acid and acetic anhydride gives approximately equal amounts of the isomeric 1- and 4-oxides, together with a little of the 1,4-di-N-oxide. Acid hydrolysis of the acetamido-N-oxides yields the corresponding amino-N-oxides.[397] Two groups of workers have found that the mono-N-oxide of tetrachloropyrazine can be prepared by treatment of tetrachloropyrazine with hydrogen peroxide in a strongly acid medium (CF$_3$CO$_2$H or a mixture of H$_2$SO$_4$ and CF$_3$CO$_2$H).[397a, 397b]

Palamidessi and Bernardi have obtained 2-chloropyrazine 1-oxide by mild treatment of pyrazine 1,4-dioxide with phosphoryl chloride. The structure of the 1-oxide was confirmed by hydrolysis to 2-hydroxypyrazine 1-oxide, which was also prepared by direct synthesis from glyoxal and glycine hydroxamic acid.[398] This synthesis is illustrative of a general method for preparing 2-hydroxypyrazine 1-oxides by condensation of α,β-dicarbonyl compounds with α-aminohydroxamic acids. An analogous synthesis of 2-aminopyrazine 1-oxides has already

been mentioned (Section V, D). Condensation of 2-amino-2-deoxy-D-glucose oxime or 2-amino-2-deoxy-D-mannose oxime (204) with glyoxal gives 2-(D-*arabino*-tetrahydroxybutyl)pyrazine 4-oxide (205) and oxidation of the latter compound with potassium permanganate

[396] H. Foks and J. Sawlewicz, *Acta Pol. Pharm.* **21**, 429 (1964); *Chem. Abstr.* **62**, 7754 (1965).
[397] A. S. Elina, I. S. Musatova, and G. P. Syrova, *Khim. Geterotsikl. Soedin.* **4**, 725 (1968); *Chem. Abstr.* **70**, 37778 (1969).
[397a] H. Johnston, U.S. Patent 3,509,144 (1970); *Chem. Abstr.* **73**, 25517 (1970).
[397b] G. E. Chivers and H. Suschitzky, *J. Chem. Soc. D* **28** (1971).
[398] G. Palamidessi and L. Bernardi, *Gazz. Chim. Ital.* **93**, 339 (1963).

Sec. VII.] PYRAZINE CHEMISTRY 195

CHOCHO + CH(=NOH)CHNH₂(CHOH)₃CH₂OH ⟶ [pyrazine N-oxide with (CHOH)₃CH₂OH substituent]

(204) (205)

gives pyrazine-2-carboxylic acid 4-oxide.[399,400] Pyrazine 1,4-dioxides are formed by acid treatment of 1,2-hydroxylaminooximes, the latter compounds are obtained by isomerization of olefin–dinitrogen trioxide adducts, followed by partial reduction of the resulting 1,2-nitroximes with molecular hydrogen in the presence of a palladium-on-carbon catalyst.[401]

$$RCH=CHR \xrightarrow{N_2O_3} RCH(NO)CH(NO_2)R \rightleftharpoons RC(=NOH)CH(NO_2)R \xrightarrow{H_2/Pd-C}$$

$$RC(=NOH)CH(NHOH)R \xrightarrow{conc.\ H_2SO_4} \text{[pyrazine 1,4-dioxide with R groups]}$$

The pyrazine mono-N-oxides show close similarity in their reactions to the pyridine N-oxides.[402–404] In the latter compounds the electron-releasing ability of the N-oxide function is demonstrated by the activation of the α- and γ-ring carbon atoms to electrophilic attack. Pyrazine mono-N-oxides are predictably less activated to electrophilic substitution, and thus there have been no reports of the successful nitration of pyrazine N-oxides. The pyrazine ring is, however, activated by the N-oxide function to nucleophilic attack, especially if the positive charge of the N atom is enhanced by formation of an intermediate with an electron acceptor. Thus, phosphoryl chloride treatment of pyrazine 1-oxide yields chloropyrazine[391] and similar

---

[399] S. Fujii and H. Kobatake, *J. Org. Chem.* **34**, 3842 (1969).
[400] S. Fujii and H. Kushida, Japanese Patent 68 13469 (1968); *Chem. Abstr.* **70**, 68694 (1969).
[401] M. L. Scheinbaum, *J. Org. Chem.* **35**, 2790 (1970).
[402] A. R. Katritzky, *Quart. Rev.* **10**, 395 (1956).
[403] D. V. Joffe and L. S. Efros, *Russ. Chem. Rev.* **30**, 569 (1961).
[404] E. Ochiai, "Aromatic Amine Oxides." Elsevier, Amsterdam, 1967.

treatment of 2-pyrazinecarboxylic acid 4-oxide gives 6-chloropyrazine 2-carboxylic acid.[405]

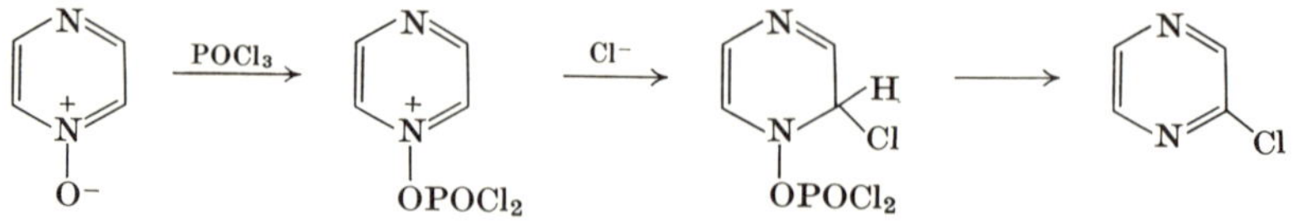

Pyrazine 1,4-dioxide reacts exothermically with phosphoryl chloride; if, after the initial reaction has subsided, reaction is completed by heating under reflux, 2,6-dichloropyrazine is obtained.[391,392] Elina and Musatova report that reaction of the dioxide with benzene sulfonyl chloride gives 2-chloropyrazine 1-oxide, m.p. 131°–132°, in low yield.[406] A melting point of 140°–146° has been reported by previous workers for this compound.[398]

Pyrazine 1-oxide, 2-methylpyrazine 4-oxide, and 2,6-dimethylpyrazine 4-oxide are unreactive toward acetic anhydride; but 2,5-dimethylpyrazine 1-oxide, 2,6-dimethylpyrazine 1-oxide, and 2,3,5,6-tetramethylpyrazine 1-oxide react to give 2-acetoxymethyl derivatives. The following mechanism (Scheme 44) is consistent with the observation that an acetoxymethyl derivative is only formed when a methyl group is adjacent to the N-oxide function.[407] However, in

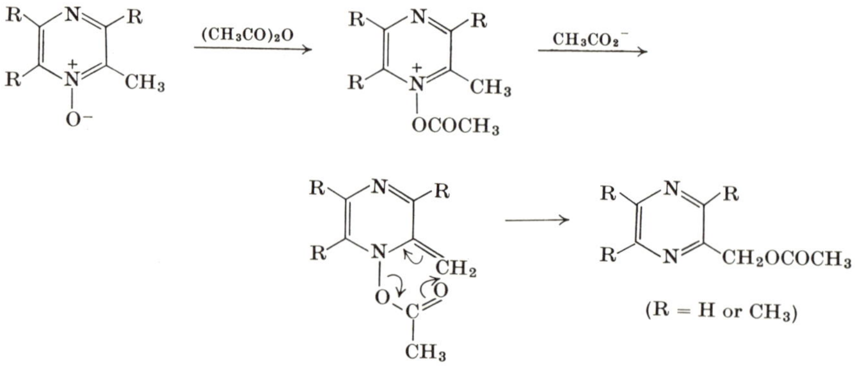

(R = H or CH₃)

SCHEME 44

[405] Y. Abe, Y. Shigeta, F. Uchimaru, S. Okada, and E. Ozasayma, Japanese Patent 69 12,898 (1969); *Chem. Abstr.* **71**, 112979 (1969).
[406] A. S. Elina and T. S. Musatova, *Chem. Heterocycl. Compounds* **3**, 127 (1967); *Chem. Abstr.* **67**, 64357 (1967).
[407] B. Klein, J. Berkowitz, and N. E. Hetman, *J. Org. Chem.* **26**, 127 (1961).

view of more recent work on the mechanisms of picoline N-oxide–acetic anhydride interaction, a diradical mechanism seems probable.

Thin-layer chromatography indicates that prolonged reaction of pyrazine 1,4-dioxide with acetic anhydride gives the mono-N-oxide, together with other unidentified products.[406] Pyrazine mono-N-oxide is also formed on reaction of pyrazine-2-carboxylic acid 4-oxide with acetic anhydride.[408]

2,5-Dimethylpyrazine 1-oxide is converted into 2,5-dimethyl-3-propylmercaptopyrazine on treatment with 1-propanethiol and acetic anhydride, and similarly the 1,4-dioxide yields the bisthio ether (**206**).[409]

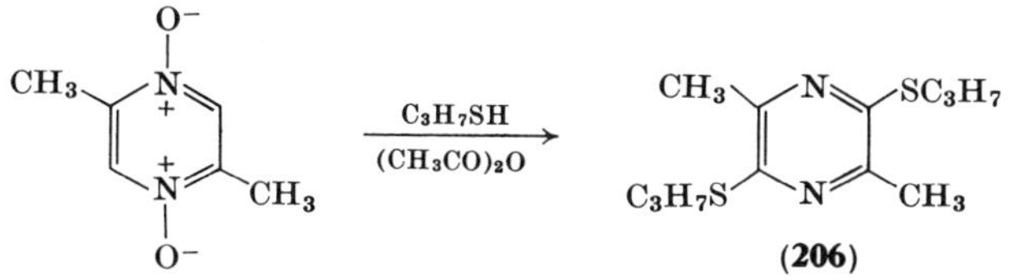

It is apparent from the preceding examples that electrophilic attack on the oxygen of the N-oxide function usually leads to deoxygenation. A further example of deoxygenative chlorination is the reaction of 2-morpholinocarbonylpyrazine 4-oxide with phosphoryl

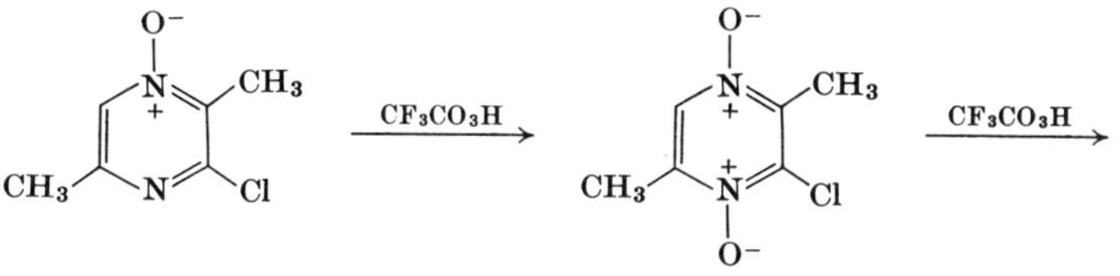

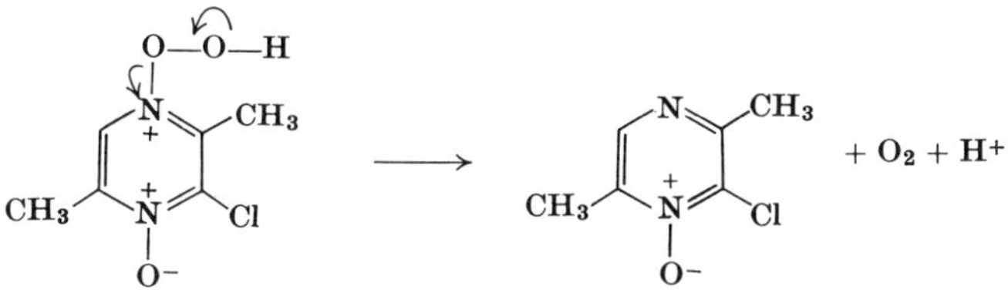

---

[408] H. Foks and J. Sawlewicz, *Acta Pol. Pharm.* **23**, 411 (1966); *Chem. Abstr.* **66**, 94996 (1967).
[409] L. Bauer and A. L. Hirsch, *J. Org. Chem.* **31**, 1210 (1966).

chloride to give 6-chloro-2-morpholinocarbonylpyrazine.[410] Blake and Sammes have reported a case of oxidative deoxygenation with pertrifluoroacetic acid. Thus, treatment of 2-chloro-3,6-dimethylpyrazine 4-oxide (**207**) with hydrogen peroxide in trifluoroacetic acid gives, in addition to the expected 1,4-dioxide (**208**), a little of the 1-oxide (**209**). The mechanism shown, involving oxygen formation is suggested.[393] The removal of the oxygen atom from pyrazine N-oxide has also been achieved by reaction with potassium in dimethoxyethane in high vacuum. Under these conditions pyrazine is reduced to the radical anion (Scheme 45).[411]

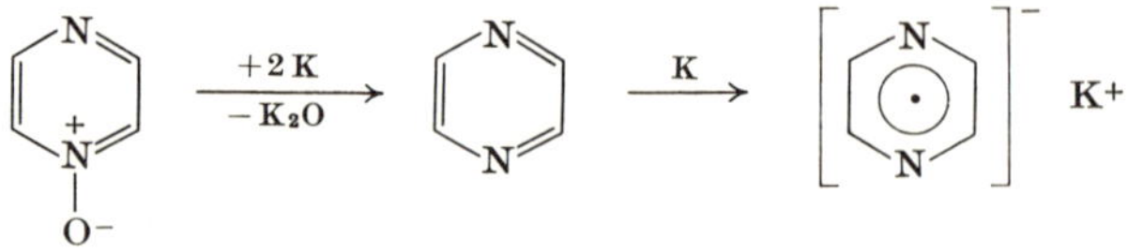

SCHEME 45

Ring substituents of pyrazine N-oxides show increased reactivity, and substituents in the α-position to the N-oxide function are more reactive than those in the β-position. Thus, 2-chloropyrazine 1-oxide is converted into the 2-hydroxy-1-oxide on mild alkali treatment,[398] but attempts to carry out a similar reaction with the 2-chloro-4-oxide were not successful.[412] Ammonolysis of the 2-chloro-4-oxide has been achieved, and nitrous acid treatment of the resulting 2-amino-4-oxide gives 2-hydroxypyrazine 4-oxide (Scheme 46). The chlorine atom of both isomeric 2-chloropyrazine N-oxides is readily displaced with sulfanilamide to give the corresponding sulfanilamidopyrazine N-oxides.[413,414]

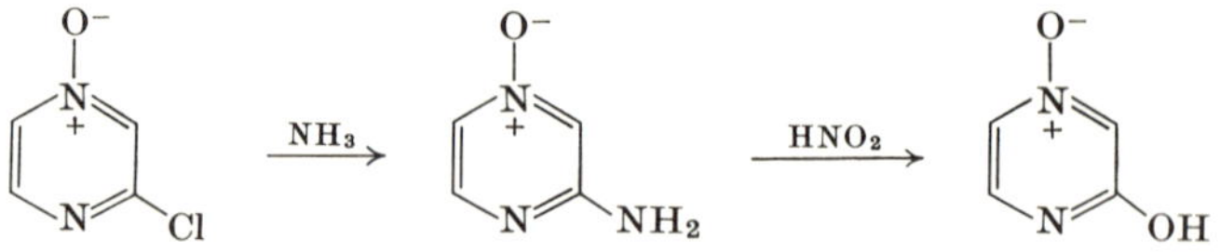

SCHEME 46

[410] H. Abe, Y. Shigeta, F. Uchimaru, S. Okada, and E. Kosasayama, Japanese Patent 69 20,345 (1969); *Chem. Abstr.* **71**, 124230 (1969).
[411] M. Itok and T. Okamoto, *Chem. Pharm. Bull.* **15**, 435 (1967).
[412] B. Klein, E. O'Donnell, and J. M. Gordon, *J. Org. Chem.* **29**, 2623 (1964).
[413] B. Camerino and G. Palamidessi, S. African Patent 67 03,272 (1968); *Chem. Abstr.* **71**, 49981 (1969).
[414] British Patent 1,140,655 (1969); *Chem. Abstr.* **70**, 106561 (1969).

2-Chloro-3-methylpyrazine 4-oxide (210) reacts with thiourea (preferably in 2 N sulfuric acid) to give a high yield of the corresponding thione (211), but the reaction of 2-chloro-3-methylpyrazine itself with thiourea gives only a moderate yield of the thione, thus further illustrating the activating effect of the N-oxide function.[415] The

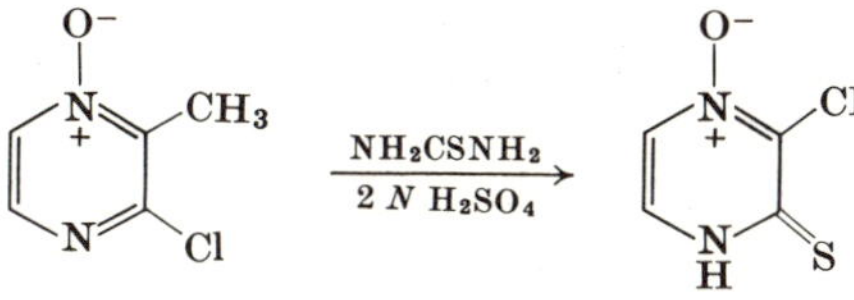

chlorine atom of the N-oxide (210) can also be readily displaced with piperidine and dimethylamine. Only a small quantity of 2-piperidino-3-methylpyrazine 1-oxide (213) is obtained by direct oxidation of 2-piperidino-3-methylpyrazine (212). The main product is formulated as 1-hydroxy-1-(3-methyl-2-pyrazinyl)piperidinium hydroxide but is probably the hydrate (214).[416]

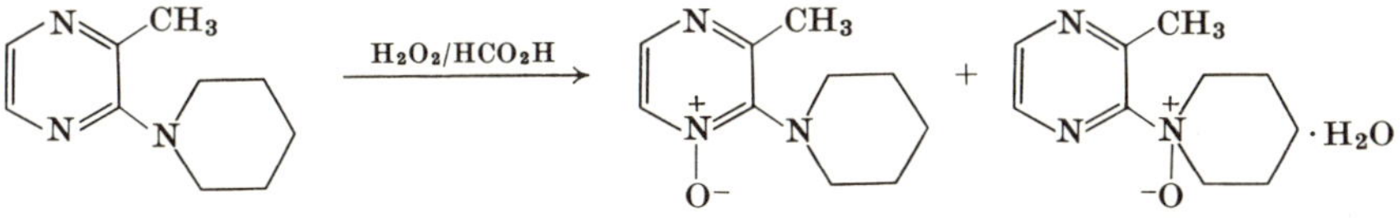

The marked activation of the N-oxide function on the chlorine atom of 2-chloropyrazine 4-oxide and 2-chloro-3,6-dimethylpyrazine 4-oxide is also demonstrated by the milder conditions under which these compounds react with ammonia and amines compared with chloropyrazine and 2-chloro-3,6-dimethyl pyrazine, respectively. Although 2-chloropyrazine 4-oxide undergoes the expected displacement reaction with ammonia on heating at 115°–120° for 2.5 hours, reaction at 140° for 16 hours gives 2,3-diaminopyrazine, possibly as a result of an addition–elimination reaction on the initially formed 2-aminopyrazine 4-oxide (Scheme 47).[417]

[415] J. Cullen and D. Harrison, *J. Chem. Soc. C* 495 (1966).
[416] W. B. Lutz, S. Lazarus, S. Klutchko, and R. I. Meltzer, *J. Org. Chem.* **29**, 1645 (1964).
[417] B. Klein, E. O'Donnell, and J. Auerbach, *J. Org. Chem.* **32**, 2412 (1967).

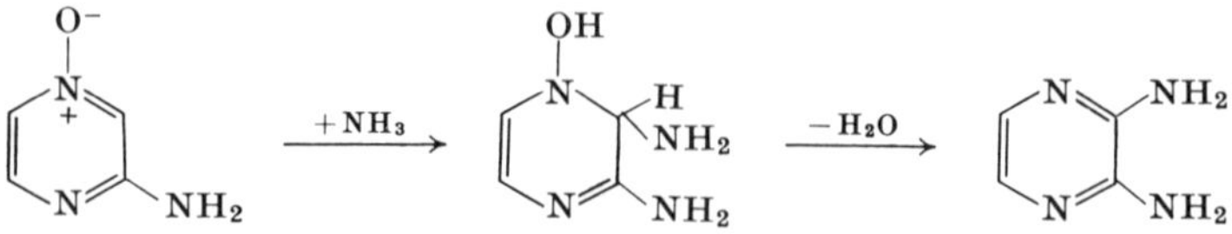

SCHEME 47

2-Methylpyrazine 1,4-dioxide and 2,3-dimethylpyrazine 1,4-dioxide condense with aromatic aldehydes under mild conditions to give styryl derivatives,[418] and treatment of the di-N-oxides with pyridine and iodine gives pyrazinyl methylpyridinium iodides (Scheme 48).[419]

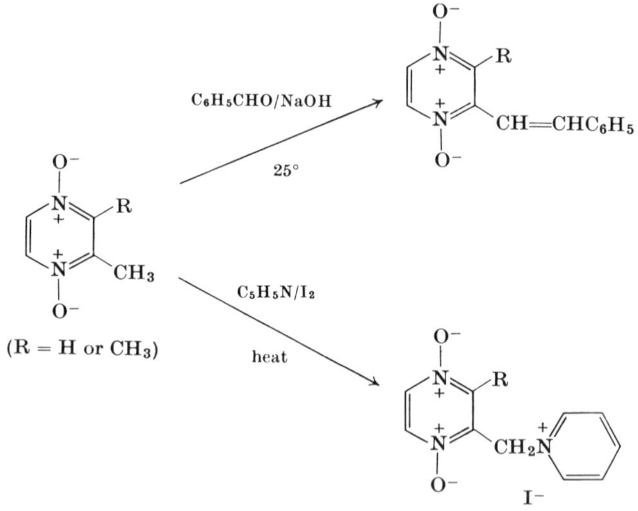

SCHEME 48

Alkaline hydrolysis of the di-N-oxide of 2,5-dichloro-3,6-dimethylpyrazine (215) readily yields the monohydroxamic acid (216), but the second chlorine atom is inert. Both chlorine atoms in compound 215 may be displaced by ethoxide or benzyloxide ion. Acid treatment of the dibenzyloxy-1,4-dioxide (217) yields compound 218 which with

[418] A. S. Elina and I. S. Musatova, *Khim. Geterotsikl. Soedin.* 419 (1967); *Chem. Abstr.* **70**, 87747 (1969).
[419] A. S. Elina, L. G. Tsyrulnikova, and I. S. Musatova, *Khim. Farm. Zh.* **1**, 10 (1967); *Chem. Abstr.* **68**, 78247 (1968).

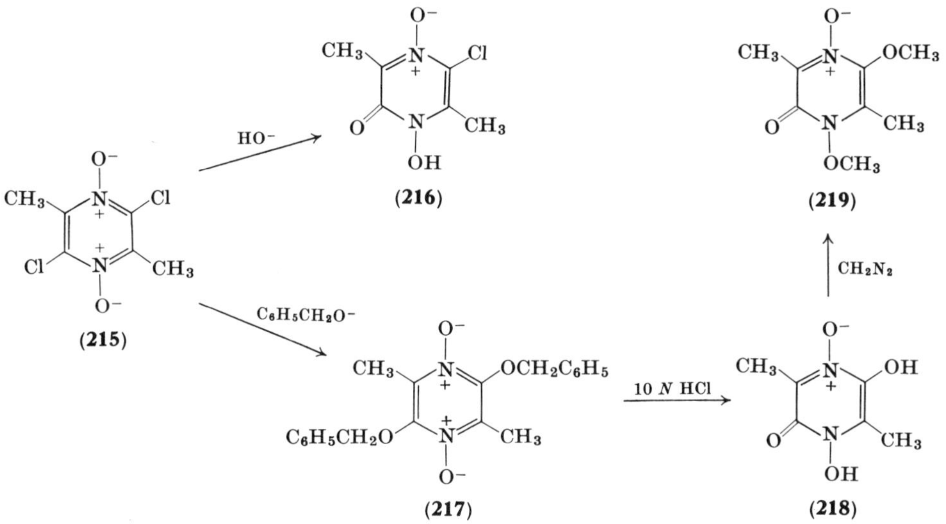

diazomethane gives the dimethyl derivative (**219**). Since compound **218** has a strong carbonyl absorption band at 1660 cm^{-1}, it is formulated as shown rather than as the tautomeric dihydroxypyrazine di-N-oxide.[393]

Recently, some interesting photochemical reactions of 2,5-dimethyl- and 2,5-diphenylpyrazine 1-oxide have been reported. Irradiation of a nitrogen-stirred benzene suspension of 2,5-dimethylpyrazine 1-oxide gives 5% 2-acetyl-4-methylimidazole (**220**) and 15% 2,4-dimethylimidazole (**221**). When the irradiation is carried out in aqueous solution, 10% 2-hydroxy-3,6-dimethylpyrazine (**126**) and 20% 1-acetamido-2-formamido-1-propene (**222**) is formed. The mechanisms in Scheme 49 are suggested for these transformations.

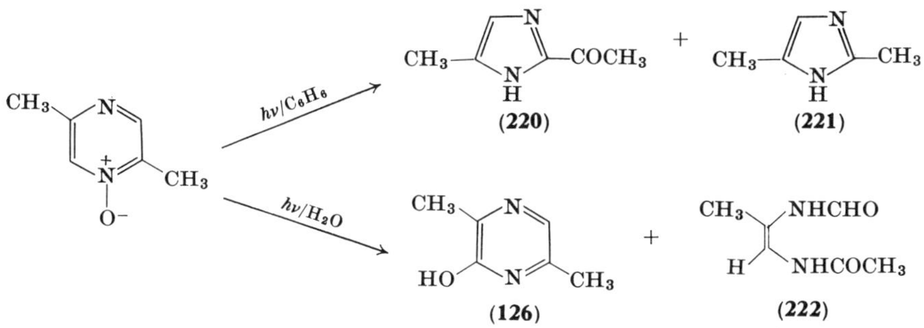

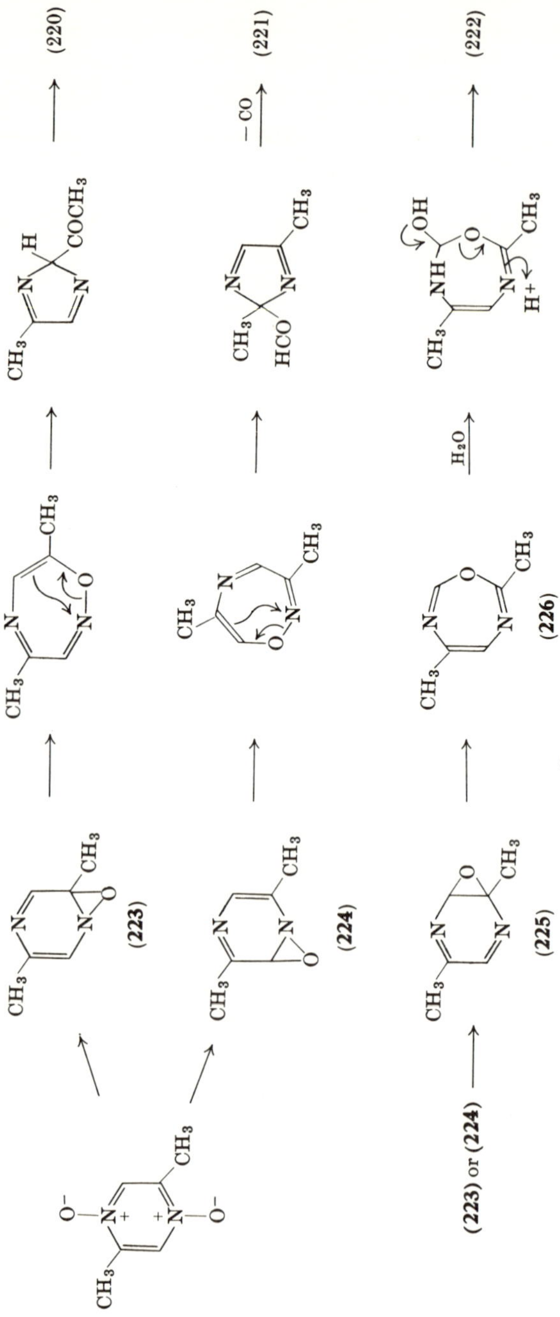

Scheme 49

Hydration of the diazaoxepine (**226**) takes place at the less hindered carbon–nitrogen double bond to give the substituted propene (**222**). The hydroxypyrazine (**126**) arises from isomerization of the oxaziranes (**224**) and (**225**). Ultraviolet irradiation of 2,5-diphenylpyrazine 1-oxide in benzene gives 3% 3,6-diphenyl-2-hydroxypyrazine, 5% 2-benzoyl-4-phenylimidazole, 29% 2,4-diphenylimidazole, and 50% 2,5-diphenylpyrazine.[420, 421]

Lamchen and Mittag have prepared hexamethyl-2,3-dihydropyrazine 1,4-dioxide (**228a**) by reaction of the dihydroxylamine (**227**) with butane-2,3-dione. Reduction of compound **228** with sodium borohydride gives the expected dihydroxypiperazine (**229**), but unlike the cyclic mononitrone (**230**), compound **228** fails to condense with aromatic aldehydes, to undergo cyanoethylation with acrylonitrile, or to undergo 1,3-cycloaddition, and it is oxidized only very slowly with selenium dioxide to a monoaldehyde. On the basis of these observations, the authors suggest that the canonical form (**228b**) is a major contributor to the resonance hybrid.[422] Later work did, however, show that compound **228** behaves as a typical nitrone on irradiation. Irradiation in dioxane with light of wavelength less than 300 nm gives a mixture of the *cis*- and *trans*-dioxaziridines (**231**) and (**232**); with light of wavelength greater than 300 nm the oxaziridine-nitrone (**233**) is obtained. Compound **233** undergoes dipolar 1,3-cycloaddition with acrylonitrile.[423]

As mentioned earlier, a number of naturally occurring pyrazines are cyclic hydroxamic acids.[9] Aspergillic acid (**3a**), the first compound of this group to be isolated, has antibiotic activity, but it is too toxic to be used for therapeutic purposes.[10] Tautomeric with aspergillic acid is the hydroxy-*N*-oxide (**3b**); the infrared spectrum of the aspergillic acid, however, shows carbonyl absorption at 1640 cm^{-1}, indicating that it exists mainly in the hydroxypyrazinone form (**3a**). Aspergillic acid is obtained from cultured filtrates of *Aspergillus flavus* and an improved method for its extraction is now available.[424] Hydroxyaspergillic acid (**5a, b**) is also obtained from *A. flavus* and both compounds **3** and **5** have been shown to be biosynthesized from a com-

---

[420] G. G. Spence, E. C. Taylor, and O. Buchardt, *Chem. Rev.* **70**, 231 (1970).
[421] N. Ikekawa, Y. Honma, and R. Kenkyuso, *Tetrahedron Lett.* 1197 (1967).
[422] M. Lamchen and T. W. Mittag, *J. Chem. Soc. C* 2300 (1966).
[423] M. Lamchen and T. W. Mittag, *J. Chem. Soc. C* 1917 (1968).
[424] S. Omata and T. Ueno, Japanese Patent 13,794 (1965); *Chem. Abstr.* **63**, 11589 (1965).

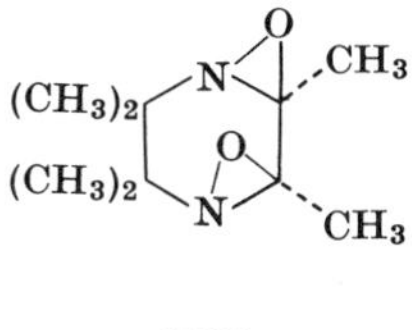

(231) (232) (233)

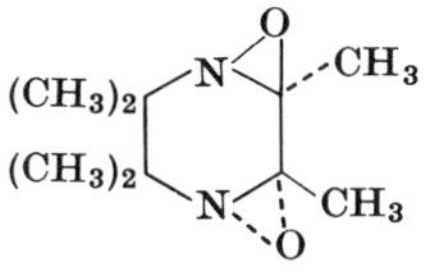

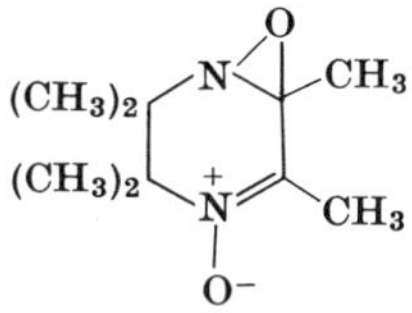

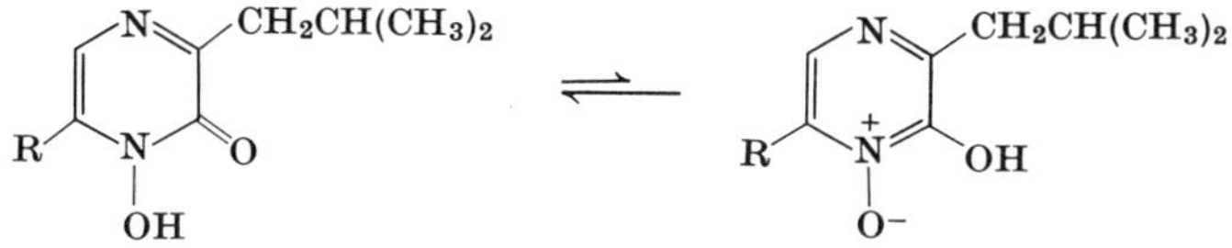

(3a) (R = CH₃ĊHCH₂CH₃) (3b) (R = CH₃ĊHCH₂CH₃)
(5a) (R = CH₃Ċ(OH)CH₂CH₃) (5b) (R = CH₃Ċ(OH)CH₂CH₃)

bination of leucine and isoleucine.[425,426] Neoaspergillic acid (8) and neohydroxyaspergillic acid (7) are synthesized by *A. sclerotiorum* entirely from leucine.[427] The proposed biosynthetic pathway is leucine → flavacol (4) → neoaspergillic acid (8) → neohydroxyaspergillic acid (7).

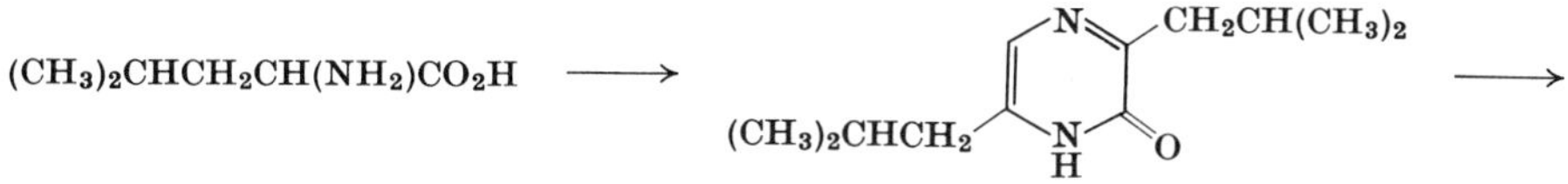

$(CH_3)_2CHCH_2CH(NH_2)CO_2H \longrightarrow$ (4) $\longrightarrow$

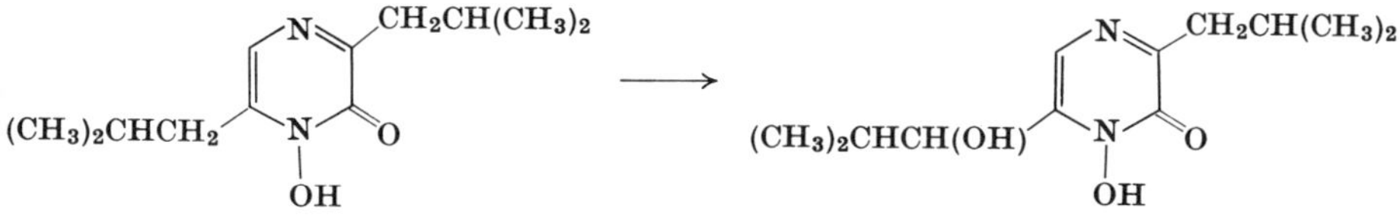

(8) (7)

A general synthesis of compounds with the aspergillic acid skeleton has been mentioned in Section III and this has been successfully applied to the synthesis of racemic aspergillic acid[124] and to neo-

[425] J. C. MacDonald, *J. Biol. Chem.* **236**, 512 (1961).
[426] J. C. MacDonald, *J. Biol. Chem.* **237**, 1977 (1962).
[427] R. G. Micetich and J. C. MacDonald, *J. Biol. Chem.* **240**, 1692 (1965).

aspergillic acid.[124] A further example is the following synthesis of 1-hydroxy-3-isobutyl-6(1-hydroxy-1-methylethyl)-2-pyrazinone (234). The synthetic material did not have an identical ultraviolet spectrum to natural mutaaspergillic acid for which structure 234 has been assigned. Further comparisons of the synthetic and naturally derived materials are being carried out to clarify this point.[125]

$$(CH_3)_2C(OH)-C(=NOH)-CH_2Cl + NH_2CHR \cdot CONH \cdot OCH_2C_6H_5 \longrightarrow$$

$$(CH_3)_2C(OH)-C(=NOH)-CH_2NHCHR \cdot CONH \cdot OCH_2C_6H_5 \longrightarrow$$

$$(CH_3)_2C(OH)COCH_2 \cdot NHCHR \cdot CONHOH \xrightarrow[(2)\text{ oxidation}]{(1)\ NH_3/CH_3OH}$$

[R = CH$_2$CH(CH$_3$)$_2$]

(234)

An alternative approach applicable both to the synthesis of racemic aspergillic acid[428] and neoaspergillic acid is illustrated by the synthesis (Scheme 50) of neoaspergillic acid (8).[429] The initial reaction of DL-leucine anhydride (235) with phosphoryl chloride produces, in addition to the required monochloro compound (236), some dichloro compound (237), and flavacol (4). Reaction of the latter compound with a mixture of phosphoryl chloride and phosphorus pentachloride yields further 2-chloro-3,6-diisobutylpyrazine (236). The remaining steps of the synthesis involve reactions discussed previously in this review, with the exception that the hydroxamic function is protected by diazomethane methylation and finally regenerated by ethanolic hydriodic acid treatment.[429]

2,5-Dichloro-3,6-diisobutylpyrazine (237) is the starting material for a very similar synthesis of pulcherriminic acid (10).[430] Pulcherriminic acid, as already mentioned in Section I, is derived from pulcherrimin (9), a metabolite of the yeast *Candida pulcherrima*. The key step in the synthesis of pulcherriminic acid is the initial step of

---

[428] A. Ohta and S. Fujii, *Chem. Pharm. Bull.* **17**, 851 (1969).
[429] A. Ohta, *Chem. Pharm. Bull.* **16**, 1160 (1968).
[430] A. Ohta, *Chem. Pharm. Bull.* **12**, 125 (1964).

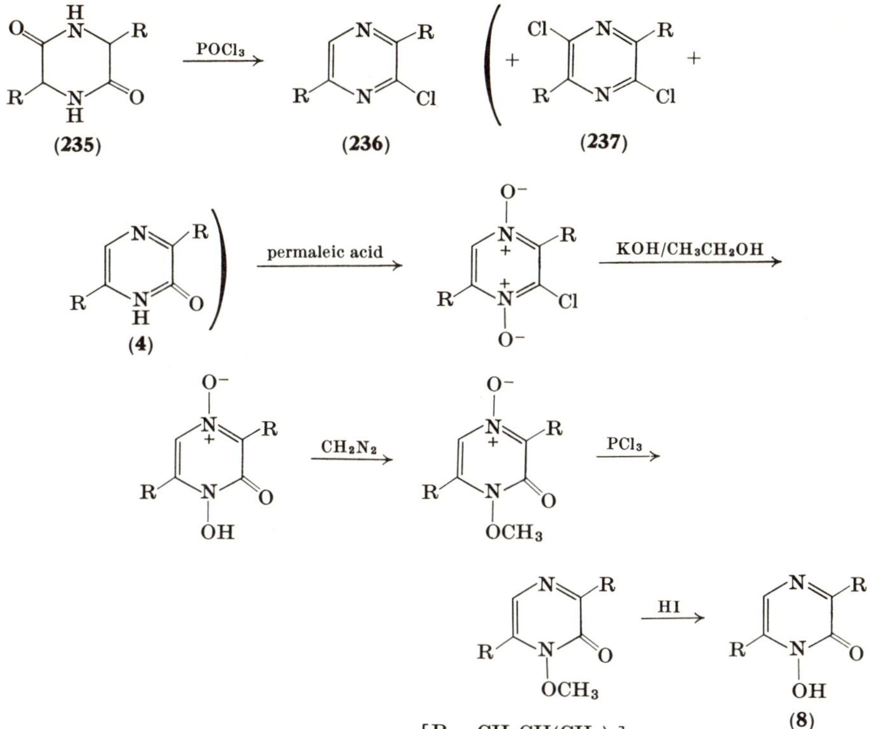

Sec. VII.] PYRAZINE CHEMISTRY

SCHEME 50

[R = CH$_2$CH(CH$_3$)$_2$]

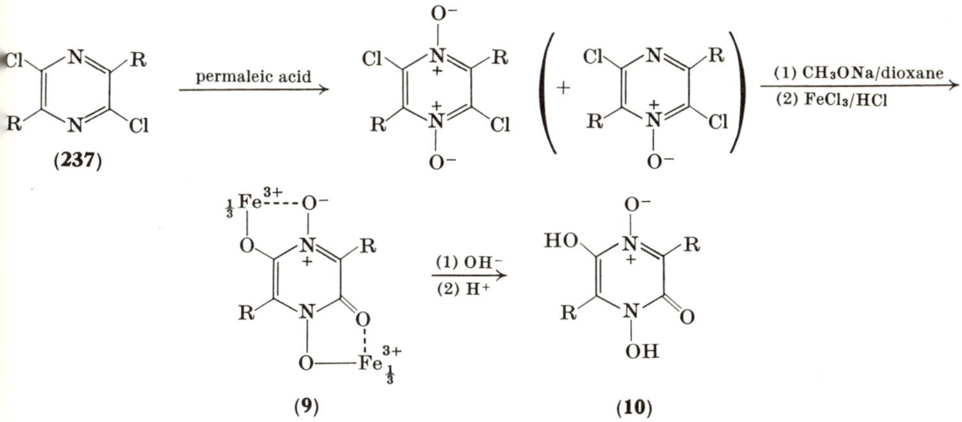

[R = CH$_2$CH(CH$_3$)$_2$]

oxidation which produces a mixture of mono- and di-*N*-oxides. Additional quantities of the required di-*N*-oxide are obtained by the further oxidation of the mono-*N*-oxide. Biosynthetic studies indicate that pulcherriminic acid is derived from L-leucine and that cyclo-L-leucyl-L-leucyl is an intermediate in the biosynthesis.[431]

VIII. Biological Activity

Reference has already been made in this article to pyrazines with antibacterial, antituberculous, antidepressant, diuretic, pesticidal, and herbicidal activity. A review further illustrates the wide range of biological activities of pyrazine derivatives.[432]

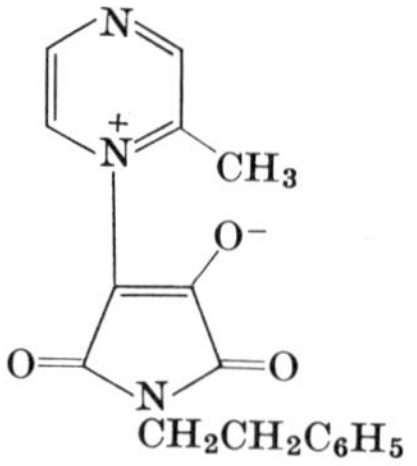

(238)

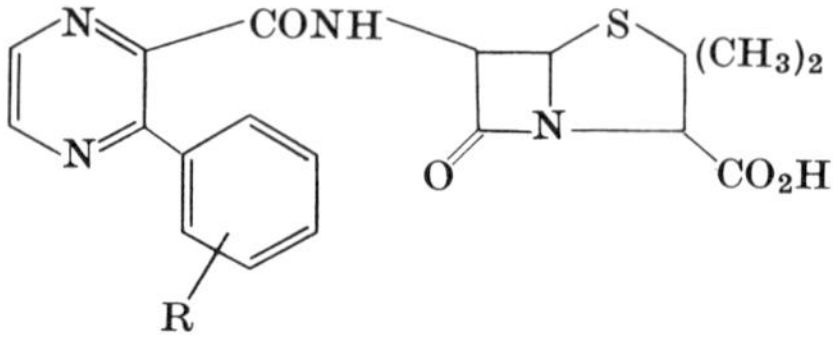

(239) (R = H, 2-Cl, 4-Cl, 4-OCH$_3$, 4-SCH$_3$, or 4-SO$_2$CH$_3$)

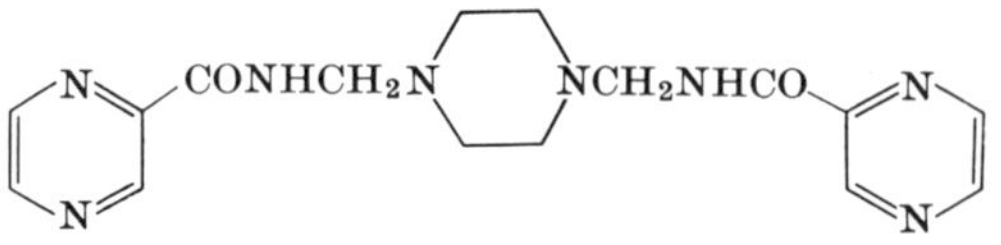

(240)

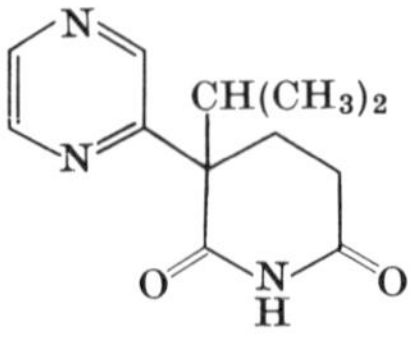

(241)

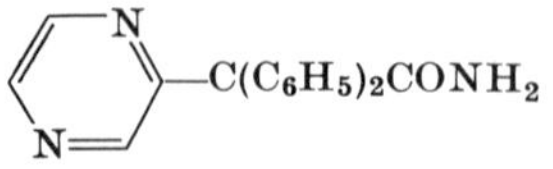

(242)

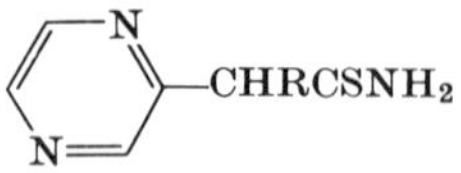

(243) (R = H, C$_2$H$_5$, or C$_6$H$_5$)

[431] J. C. MacDonald, *Biochem. J.* **96**, 533 (1965).
[432] L. Novacek, *Cesk. Farm.* **15**, 323 (1966).

Additional examples of pyrazines with antibacterial activity include the pyrazinium betaines of type (**238**)[433] and the penicillanic acids (**239**).[434,435] The pyrazinamide (**240**) has antituberculous activity[436] and pyrazinylpiperidines such as compound **241** are reported to be hypnotics.[437] The pyranzinylacetic acid derivative (**242**) has sedative and anticonvulsant activity[438] and pyrazinylthioamides of type (**243**) reportedly give protection against gastroduodenal ulcers.[439] A novel ring contraction of a pyrazine ring to an imidazole has been observed in the conversion of triazolo[4,3-*a*]pyrazines to imidazo[2,1-*c*]-*s*-triazoles under acidic conditions.[440] The triazolopyrazines were prepared for biological screening and various pyrido[3,4-*b*]pyrazines have been prepared in the search for anti-malarial agents.[441-443]

---

[433] S. L. Shapiro, L. Freedman, and M. J. Karten, U.S. Patent 3,129,225 (1964); *Chem. Abstr.* **61**, 7029 (1964).
[434] Netherlands Appl. 6,404,841 (1964); *Chem. Abstr.* **62**, 16270 (1965).
[435] T. Naito and S. Nagagawa, Japanese Patent 26,822 (1965); *Chem. Abstr.* **64**, 8208 (1964).
[436] G. Pato, L. Dede, and I. Grasz, Hungarian Patent 151,427 (1967); *Chem. Abstr.* **61**, 7028 (1964).
[437] A. M. Akkerman, G. C. van Leeuwen, and J. F. Michels, French Patent 1,404,515 (1965); *Chem. Abstr.* **63**, 18118 (1965).
[438] A. M. Akkerman, H. Kofman, and G. de Vries, Netherlands Patent 105,432 (1963); *Chem. Abstr.* **62**, 6495 (1965).
[439] C. Malen, B. Danfee, and X. Pascand, Ger. Offen. 1,934,392; *Chem. Abstr.* **72**, 90292 (1970).
[440] F. L. Rose, G. J. Stacey, P. J. Taylor, and T. W. Thompson, *J. Chem. Soc. D* 1524 (1970).
[441] C. Temple, Jr., A. G. Laseter, J. D. Rose, and J. A. Montgomery, *J. Heterocycl. Chem.* **7**, 1195 (1970).
[442] C. Temple, Jr., J. D. Rose, R. D. Elliott, and J. A. Montgomery, *J. Med. Chem.* **13**, 853 (1970).
[443] C. Temple, Jr., J. D. Rose, and J. A. Montgomery, *J. Med. Chem.* **13**, 1234 (1970).

Heterocycles by Ring Closure of Ortho-Substituted t-Anilines (The t-Amino Effect)

O. METH-COHN and H. SUSCHITZKY

*Department of Chemistry and Applied Chemistry,
University of Salford, Salford, England*

I. Introduction	. . . .	211
II. Interactions of the Ortho Substituent with the Nitrogen in t-Anilines.	. . . .	212
III. Interactions of the Ortho Substituent with the α-Methylene Group in t-Anilines	. . . .	225
A. o-Nitroso and o-Nitro Groups	. . . . .	226
B. o-Amino and o-Acylamino Groups	. . . . .	243
C. o-Azo Group	. . . .	252
D. o-Azomethine Group (—N═C)	. . . . .	256
E. o-Carbonyl (C═O) and o-Imine (C═N) Groups	. .	263

I. Introduction

The chemistry of N-dialkylanilines is, for the most part, predictable. However, the presence of an ortho substituent can give rise to unexpected and interesting reactions of synthetic value. Although such reactions have been reported sporadically during the last 75 years, their usefulness is little appreciated possibly because the relevant observations are buried in the literature. In some cases incorrect structures have been assigned to products arising from such reactions simply because the effect of a t-amino group was not understood.

The first example of a surprising reaction sequence due to a "t-amino effect" was observed by Pinnow as early as 1895 during his attempts to prepare the acetyl derivative (**3**) of *o*-aminodimethylaniline (**1**) by prolonged reflux in acetic anhydride. He obtained 1,2-dimethylbenzimidazole (**2**) instead, and on the basis of analogous observations in related reactions he suggested that the formation of a benzimidazole under "Pinnow conditions" could, in fact, be used to diagnose ortho nitration in dialkylanilines.[1-6]

It is the purpose of this review to collate all the available material relating to the "*t*-amino effect" and to attempt a rationalization of its operative influence in a mechanistic manner. The cyclizations of ortho-sustituted *t*-anilines are described in two main sections. The first discusses ring closures between the ortho substituent and the *t*-nitrogen (**4**), while the second chapter deals with those ring formations which involve the α-methylene groups attached to the nitrogen (**5**).

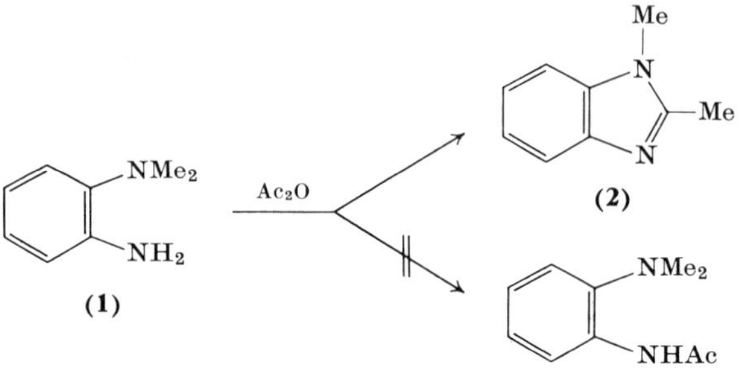

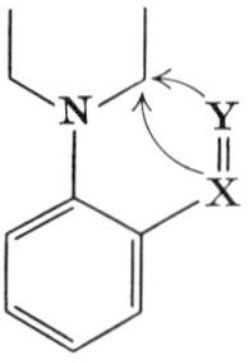

II. Interactions of the Ortho Substituent with the Nitrogen in *t*-Anilines

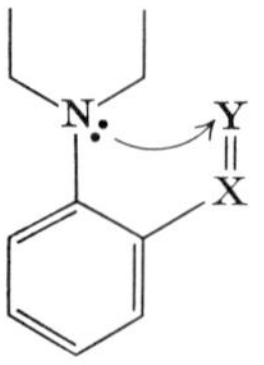

A variety of seemingly unrelated reactions have been reported in which a *t*-aniline with an ortho substituent of the type N⋯C⋯Y is converted into a benzimidazole with loss of a group from the *t*-nitrogen.

[1] J. Pinnow and G. Pistor, *Ber. Deut. Chem. Ges.* **27**, 602 (1894).
[2] J. Pinnow, *Ber. Deut. Chem. Ges.* **28**, 3039 (1895).
[3] J. Pinnow, *Ber. Deut. Chem. Ges.* **30**, 3119 (1897).
[4] J. Pinnow, *Ber. Deut. Chem. Ges.* **31**, 2982 (1898).
[5] J. Pinnow, *Ber. Deut. Chem. Ges.* **32**, 1666 (1899).
[6] A. Schuster and J. Pinnow, *Ber. Deut. Chem. Ges.* **29**, 1053 (1896).

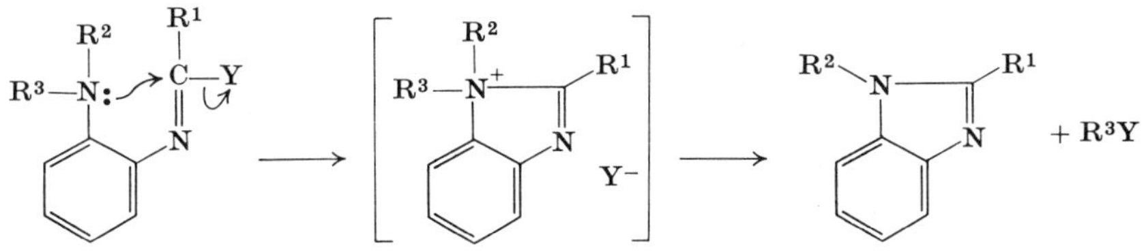

All these cyclizations can be rationalized by the intermediacy of an imine (6) bearing a good leaving group (Y). Ring closure yields an unstable quaternary benzimidazole (7), which by loss of one of the nitrogen substituents (R^3) attains aromaticity to give a 1,2-disubstituted benzimidazole (8).

The acetylation of o-aminodimethylaniline (1) already quoted[1-6] is a typical example. Its acetyl derivative (9) is capable of further acetylation, the O-acetyl derivative (10) being the required intermediate for cyclization. Elimination of methyl acetate, which was isolated by Pinnow, gives the required product (2).

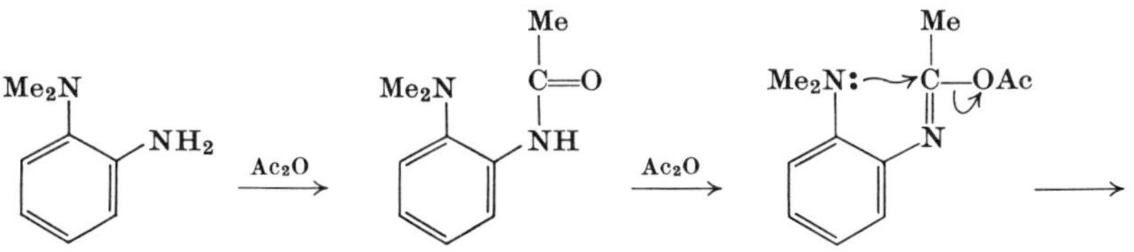

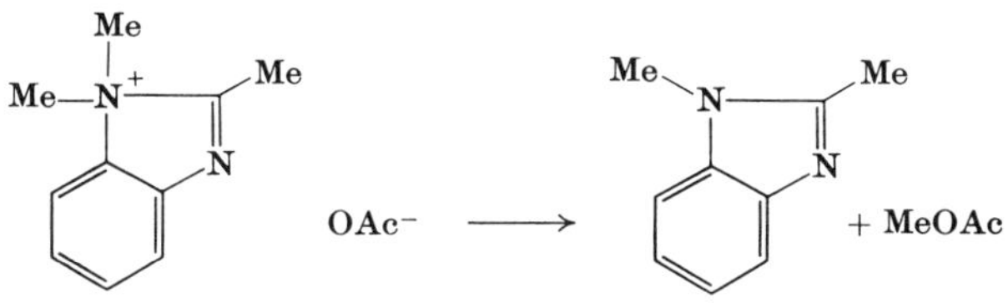

The above type of reaction is of particular interest when the N,N-dialkyl substituents are part of a ring system[7,8] (e.g., 11). It was found of advantage to replace the acetic anhydride by polyphosphoric

---

[7] O. Meth-Cohn and H. Suschitzky, *J. Chem. Soc.*, C 2609 (1964).
[8] R. Garner and H. Suschitzky, *J. Chem. Soc.*, C 1572 (1966).

acid (PPA). Although treatment of the corresponding pyrrole (12)[9,10] and the cyclohexyl derivative[11] (14) with this reagent gave the quinoxaline (13) and the tetrahydrophenanthridine (15), respectively, with the cyclic amines (11) the reaction follows a different course as outlined involving the iminopolyphosphate (16) and the benzimidazolium salt (17). Fission of the salt yields a carbonium ion (18) which readily forms the polyphosphate (19). Higher temperature or prolonged reaction time causes elimination of polyphosphoric acid from 19 to give an olefin (20) which under the prevailing Friedel–Crafts conditions cyclizes at the penultimate carbon to yield a tricyclic imidazole (21). The whole sequence (11)–(21) proceeds, often in good yield (see Table I) at temperatures of about 145°. No cyclization occurs with morpholine derivatives or with electron-withdrawing substituents at R (e.g., R = CF_3, $ClCH_2$) or bulky groups (R = CMe_3), and yields are best when R = Me, H, or Ph. The action of PPA on acetoacetyl and benzoylacetyl derivatives of o-aminodialkylanilines (22) failed to give benzimidazoles,[12] but produced 4-quinolones (23) and pyrones (24). Similarly, no interaction with the t-amine was observed in the case of the mandelanilides (25), the oxindoles (26) being the sole product.[13]

[9] G. W. H. Cheeseman and B. Tuck, *Chem. Ind. (London)* 1382 (1965).
[10] G. W. H. Cheeseman and B. Tuck, *J. Chem. Soc., C* 852 (1966).
[11] D. A. Denton, R. K. Smalley, and H. Suschitzky, *J. Chem. Soc.* 2421 (1964).
[12] R. Garner and H. Suschitzky, *J. Chem. Soc., C* 186 (1966).
[13] E. B. Mullock, R. Searby, and H. Suschitzky, *J. Chem. Soc., C* 829 (1970).

The ring size of the *t*-amino function is significant as it influences the rate of disappearance of the starting material at 145°. Thus, with $n = 4$–6 (cf. **11**) no starting material remains after 45 minutes. With $n = 7$, 90 minutes are required for completion of the reaction and with

TABLE I

BENZIMIDAZOLES (**21**) DERIVED[a] FROM *o*-ACYLAMINODIALKYLANILINES (**11**)

(**11**)   (**21a**)

Aniline (**11**)		Heating time (hours)	Product (**21a**)					
							M.p. or b.p.	
n	R^1		n	R^2	R^3	Yield (%)	[(°C)/mm]	Ref.
4	H	0.5	2	Me	H	27	(125/0.5)	7
4	Me	6	2	Me	H	57	140	7
4	Ph	0.5	2	Me	H	10	—	7
5	H	1	3	Me	H	58	(157/2)	7
5	Me	1	3	Me	H	90	117	7
5	Ph	1	3	Me	H	55	130	7
6	H	1	4	Me	H	10	—	7
6	Me	1	4	Me	H	35	120	7
6	Ph	1	4	Me	H	14	102	7
7	Me	1.5	3	Et	H	35	96	8
8	Me	4.5	4	Me	Et	10	70–76	8
			3	Me	*n*-Pr			

[a] By the action of polyphosphoric acid at 140°–145°.

$n = 8$, the reaction is incomplete even after 2 hours indicating a progressive steric inhibition in the cyclization step (16)–(17). The steps (17)–(19) proceed readily in all cases, whereas the sequence (19)–(21) occurs much more slowly in the pyrrolidine [(**11**) $n = 4$] than in the higher homologs. Only 20% of the cyclized product [(**21**) $n = 4$] was isolated, although all the starting compound was consumed after

45 minutes. The remaining water-soluble product was the polyphosphate (**19**), which on hydrolysis with dilute acid gave the corresponding 1-(4-hydroxybutyl)benzimidazole. The yield of cyclized product (**21**) increased with heating time because the 4-hydroxybutyl compound cyclized under these conditions. With the higher homologs [(**11**) $n = 7$ or 8], cyclization does not occur at the penultimate carbon of the chain (which would lead to medium-size rings), but takes place only after the chain has rearranged to contain a tertiary carbonium ion. Thus, the heptamethylenimine derivative [(**11**) $n = 7$] gave the product **27**, whereas the octamethylenimine [(**11**) $n = 8$)] gave a mixture of **28** and **29**.

TABLE II

NAPHTHIMIDAZOLES (**31**) DERIVED[a] FROM 1-ACYLAMINO-2-DIALKYLAMINONAPHTHALENES (**30**)[b]

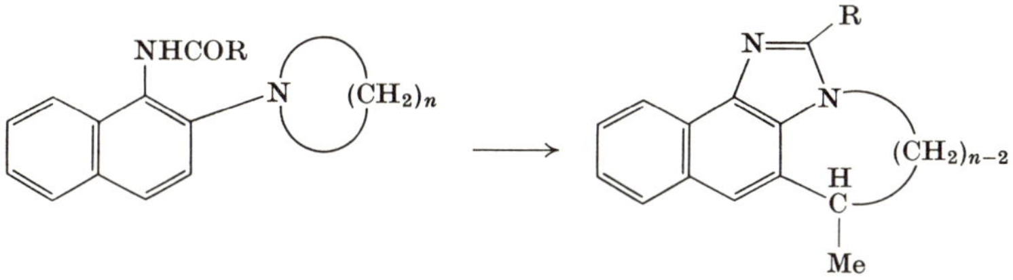

Naphthylamine (30)		Heating time (hours)	Product (31)	
n	R		Yield (%)	M.p. (°C)
4	H	0.75	2	106
4	H	17	6	106
4	Me	0.75	2	132
4	Me	17	10	132
5	H	0.75	50	112
5	Me	0.75	50	119
5	Ph	0.75	20	143
6	H	0.75	20	172
6	Me	0.75	30	161
6	Ph	0.75	10	187

[a] Garner and Suschitzky.[8]
[b] By the action of polyphosphoric acid at 145°.

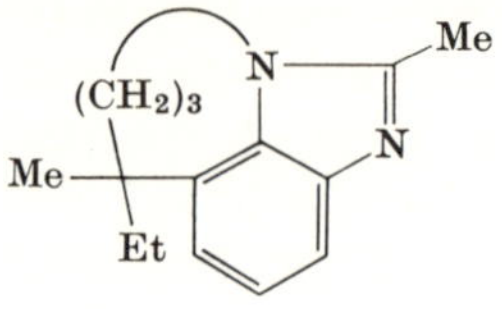

(27) 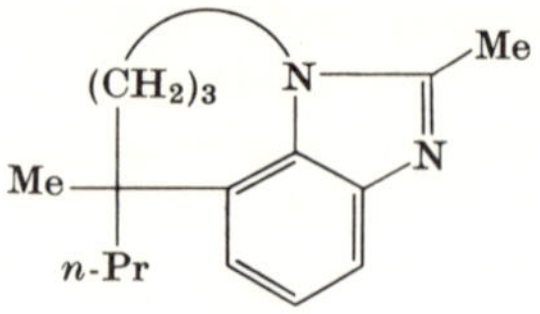 (28) 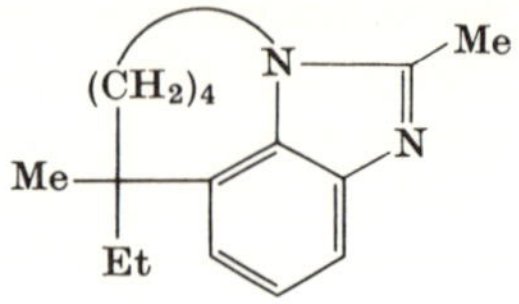 (29)

This type of cyclization was successful with the corresponding naphthalene derivatives [(30) $n = 4$–6], although yields were lower than in the benzene series[8] (see Tables II and III). The isomeric series (32) was of considerable interest since pericyclization occurred in the final step (33) or (34). With the pyrrolidine and piperidine derivatives

TABLE III

NAPHTHIMIDAZOLES (33) OR (34) DERIVED[a] FROM 2-ACYLAMINODIALKYLAMINONAPHTHALENES (32)[b]

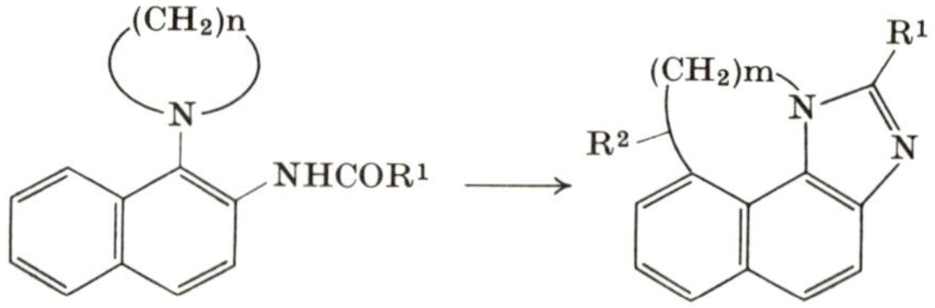

Naphthylamine (32)		Product (33) or (34)			
n	R^1	m	R^2	Yield (%)	M.p. (°C)
4	Me	2	Me	5	124
4	Ph	—	—	0	—
5	H	3	Me	30	108
5	Me	3	Me	50	162
5	Ph	3	Me	15	179
6	H	4	Et	Trace	—
6	Me	4	Et	50	157
6	Ph	—	—	0	—
7	Me	—	—	0	—

[a] Garner and Suschitzky.[8]
[b] By the action of polyphosphoric acid at 145° for 0.75 hour.

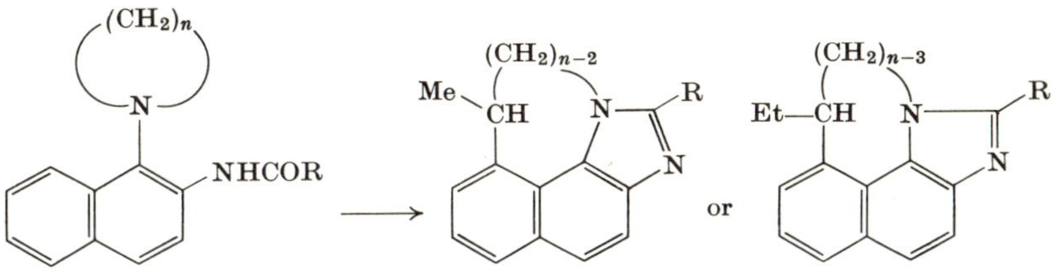

[(**32**) $n = 4$ or 5, respectively] the expected product from pericyclization at the penultimate carbon of the liberated chain occurred [(**33**) $n = 4$ or 5, respectively]. However, with the analogous hexamethylenimine [(**32**) $n = 6$] a further rearrangement was involved to yield the system **34**. Extension of this reaction to the quinoline series[13a]

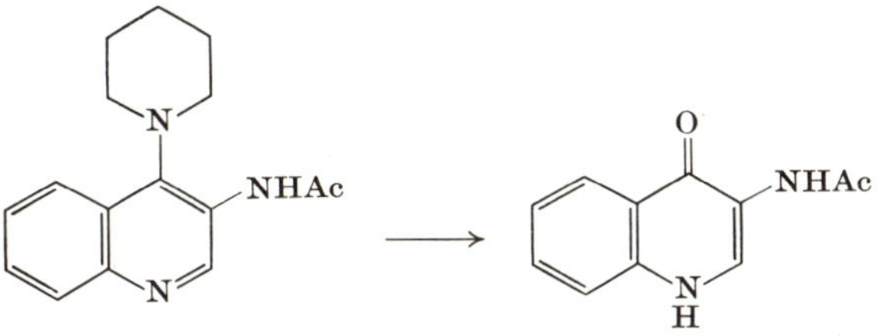

(e.g., **35**) failed when on treatment with PPA the amine substituent was hydrolyzed to give the 4-quinolone (**36**). This limitation is not unexpected in view of the nucleophilicity of the 4-position in quinoline. In fact, even boiling water removed the piperidino group to give **36**.

A recently reported reaction[14] operating by a similar mechanism involves the treatment of the benzoylaminoethylindoline (**37**) with PPA to give the imidazoline (**39**) by analogous steps (**38**)–(**39**).

It has long been known[15] that primary aromatic amines react with chloral hydrate and hydroxylamine hydrochloride (or preformed chloral oxime) in aqueous solution to give isonitrosoacetanilides (e.g., **40**). The reaction is improved by addition of certain inorganic ions (e.g., sulfate[16]) and is of value as a route to isatins obtained by

---

[13a] G. V. Garner and H. Suschitzky, unpublished results.
[14] J. B. Hester, *J. Org. Chem.* **35**, 547 (1970).
[15] T. Sandmeyer, *Helv. Chim. Acta* **2**, 234 (1919).
[16] M. Akahoshi, *J. Chem. Soc. Jap.* **72**, 805 (1951).

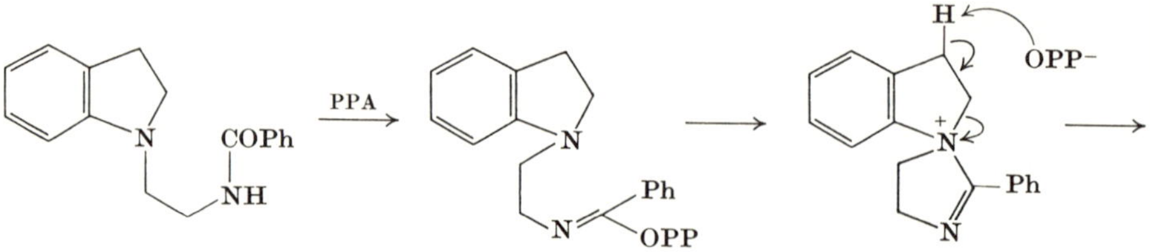

the further action of concentrated sulfuric acid on **40**. However, the interaction of chloral oxime with o-aminodialkylanilines (**41**) was

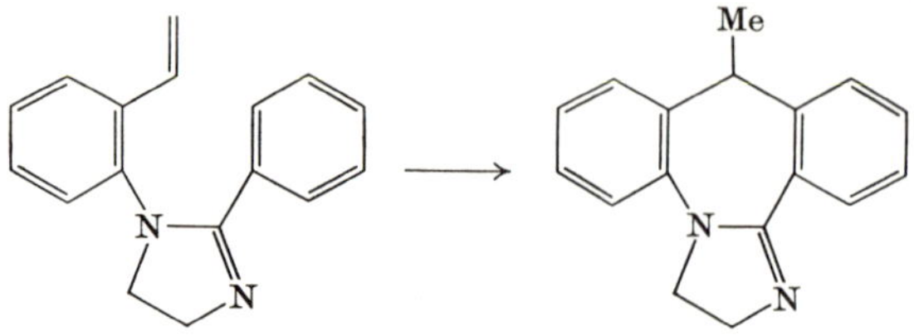

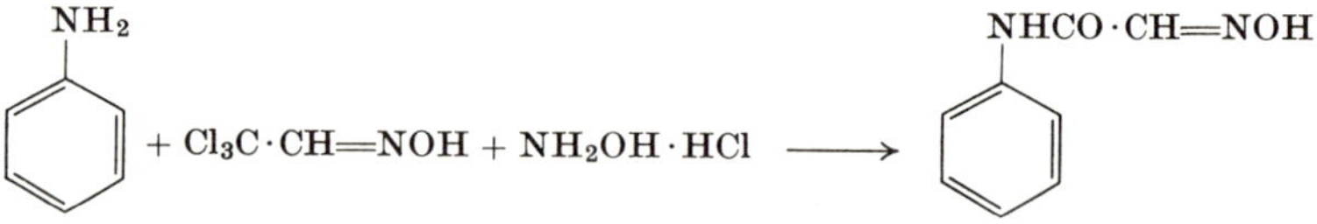

shown by Somin and Petrov[17] to give 1-alkyl-2-oximinoformyl-benzimidazoles (**42**) in good yields (50–73%). They correctly envisaged that the reaction involved the chlorimino derivative (**43**) and the

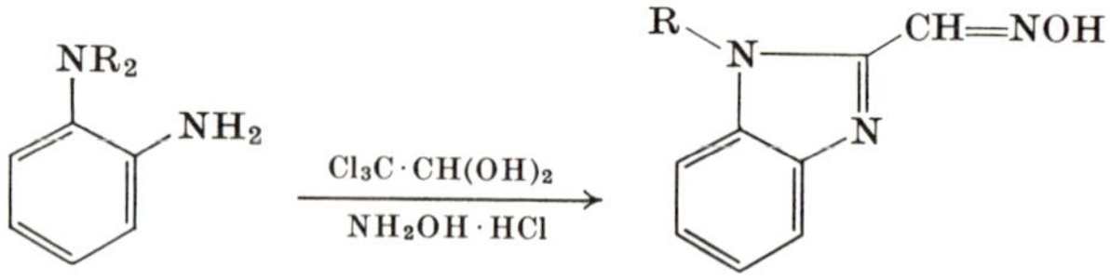

benzimidazolium salt (**44**) which aromatized with loss of alkyl chloride to give **42**. This mechanism accords with the general scheme already

---

[17] I. N. Somin and A. S. Petrov, *Zh. Obshch. Khim.* **34**, 3131 (1964); *J. Gen. Chem. USSR* **34**, 3177 (1964).

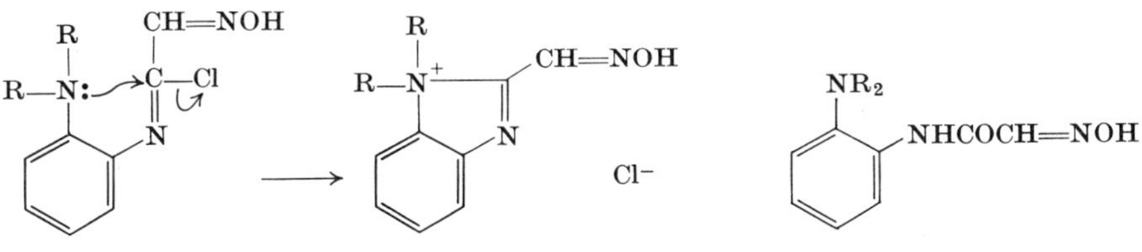

outlined (cf. **6–8**). The same workers[18] showed that by limiting the acidity of the solution, the isonitrosoacetanilide (**45**) was formed rather than the benzimidazole. The value of this reaction was underlined by the conversion, in essentially quantitative yield, of the oximes (**42**) into the unknown 2-cyanobenzimidazoles (**46**) with thionyl chloride[19] or acetic anhydride[20] and into the amides (**47**).

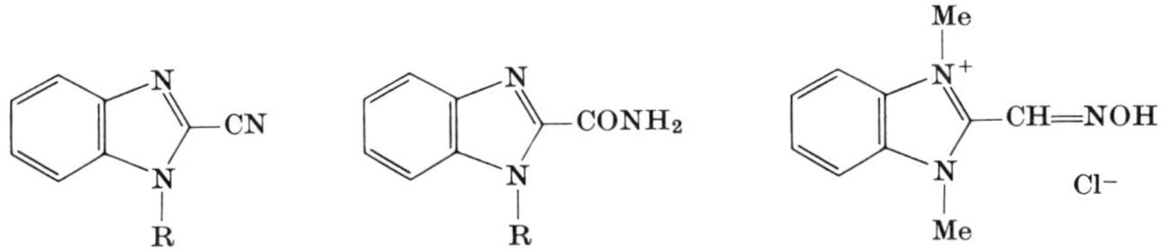

Chloral oxime in ethanol was particularly effective for the formation of the oximinoformylbenzimidazoles and gave the parent system [(**42**) R = H] with o-phenylenediamine.[21] Both N,N'-dimethyl- and N,N,N'-trimethyl-o-phenylenediamine gave 1,3-dimethyl-2-oximinoformylbenzimidazolium chloride (**48**) under these conditions.

The application of this reaction to the polymethylene analogs (**49**) was conducted independently by two groups of workers.[20, 22] Both observed a similar sequence of reactions resulting in good yields of the 1-(ω-haloalkyl)benzimidazoles (**50**). The ready substitution of the halogen by nucleophiles (e.g., NMe_2, NEt_2, pyrrolidino, Br, I, ONO_2,

[18] A. S. Petrov, I. N. Somin, and S. G. Kuznetsov, *Zh. Org. Khim.* **1**, 1434 (1965); *J. Org. Chem. USSR* **1**, 1454 (1965).
[19] A. S. Petrov and I. N. Somin, *Khim. Geterotsikl. Soedin* 472 (1966).
[20] R. Garner and H. Suschitzky, *Chem. Commun.* 129 (1967); *J. Chem. Soc., C* 2536 (1967).
[21] S. G. Kuznetsov, A. S. Petrov, and I. N. Somin, *Khim. Geterotsikl. Soedin.* 146 (1967).
[22] A. S. Petrov, I. N. Somin, and G. S. Kuznetsov, *Khim. Geterotsikl. Soedin.* 152 (1967).

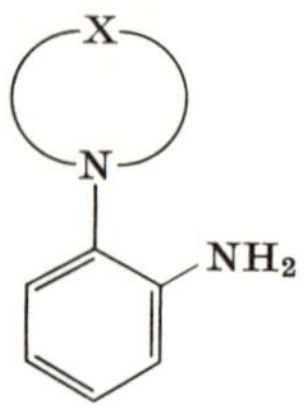

$X = (CH_2)_{4-6}$ or $(CH_2)_2O(CH_2)_2$

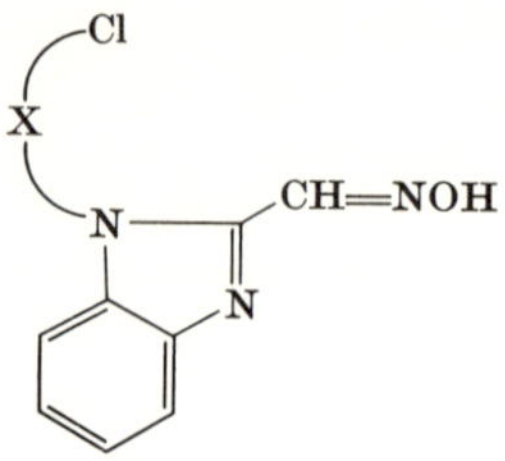

TABLE IV

BENZIMIDAZOLES (**51**) BY THE ACTION OF CHLORAL
DERIVATIVES ON o-AMINO-t-ANILINES (**49**)

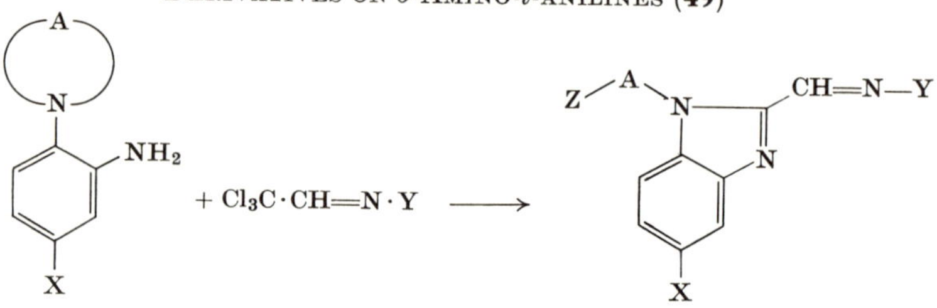

Substituent				Product (**51**)		
A	X	Y	Z	Yield (%)	M.p. (°C)	Ref.
Me₂	H	OH	Me[a]	61	218	17
Et₂	H	OH	Et[b]	52	209	17
(CH₂)₄	H	OH	Cl	32	178	22
(CH₂)₅	H	OH	Cl	52	165.5–166	20, 22
(CH₂)₅	H	OH	Br	42	164	20
(CH₂)₅	H	OH	I	—	162 (dec.)	20
(CH₂)₅	Cl	OH	Cl	35	195	20
(CH₂)₅	NO₂	OH	Cl	38	191	20
(CH₂)₅	H	NHCONH₂	Cl	32	194 (dec.)	20
(CH₂)₂O(CH₂)₂	H	OH	Cl	16	155	20
(CH₂)₂O(CH₂)₂	H	OH	Cl	49, 60	167–167.5	22
(CH₂)₆	H	OH	Cl	48	159.5–160	22

[a] A–Z = Me.
[b] A–Z = Et.

or NO_2) and the efficient conversion of the oxime function into a nitrile, amidine, amide, or acid rendered these compounds particularly attractive intermediates. Furthermore, chloral may be replaced by bromal, giving the corresponding ω-bromoalkyl derivative (but not by trifluoroacetaldehyde which remains inert) and the hydroxylamine by other "carbonyl reagents." Thus, semicarbazide and phenylhydrazine hydrochloride gave the corresponding semicarbazone (**51a**) and phenylhydrazone (**51b**) (for results see Table IV). It is

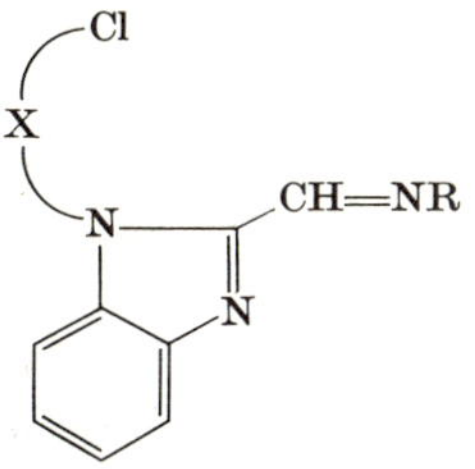

(**51a**) R = $NHCONH_2$
(**51b**) R = NHPh

noteworthy that in the presence of a high concentration of a "foreign" nucleophile, the reaction with chloral does not give the chloroalkyl derivatives (**50**). Thus, with added sodium bromide the bromoalkyl compound (**54**) results,[20] indicating that the ring-opening step (**52**)–(**54**) involves $S_N 1$ attack (**53**). However, when the foreign ion is sulfate,

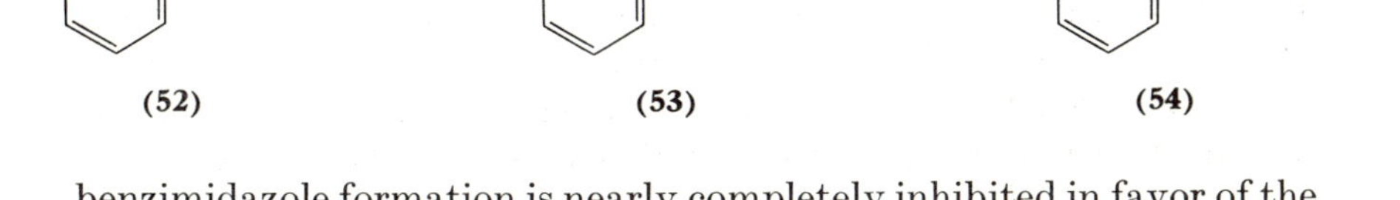

(**52**)   (**53**)   (**54**)

benzimidazole formation is nearly completely inhibited in favor of the isonitrosoacetanilide. This probably arises from attack of the chlorimine (**43**) by the sulfate ion and subsequent hydrolysis to the anilide (**45**).

Another reaction under this heading[21] involves the action of benzotrichloride on tetramethyl-o-phenylenediamine (55) which occurs with the loss of two molecules of methyl chloride to give the

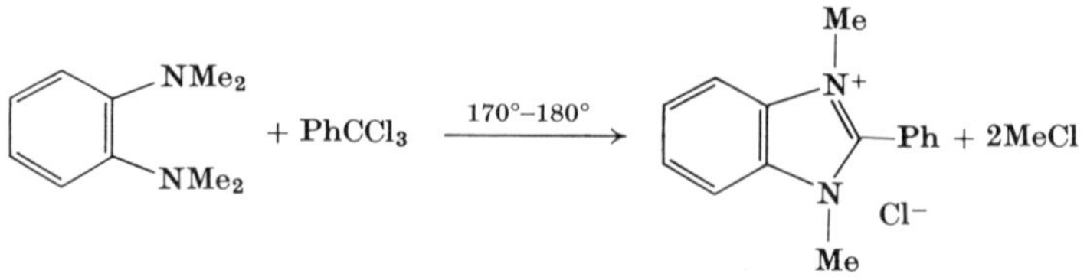

benzimidazolium salt (56). It probably involves a chloroimine intermediate, but its mechanism has not yet been fully explored.

A related reaction was recently described by the Russian school.[23] It is well known[24] that amides react with thionyl chloride to give chloroimines. As a logical extension of previous work, a series of acetyl

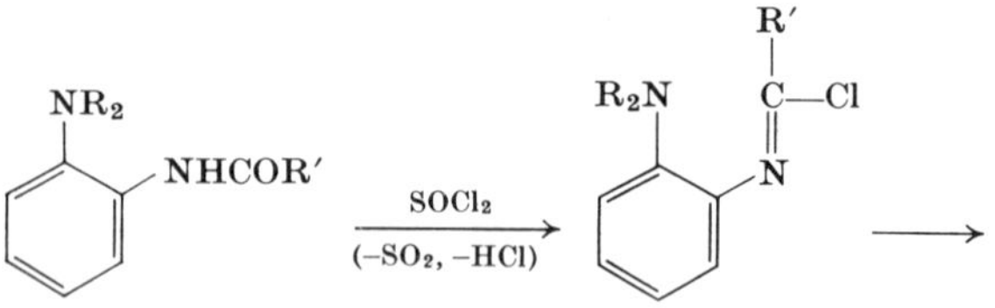

and benzoyl derivatives [(57) R' = Me or Ph] was treated with boiling thionyl chloride and gave the analogous dealkylated benzimidazole (60) by way of the chloroimine (58) and the salt (59). The piperidine derivative [(57) $R_2 = (CH_2)_5$] in an analogous manner gave a good yield of the 1-(5-chloropentyl)benzimidazoles [(60) $R = Cl(CH_2)_5$].

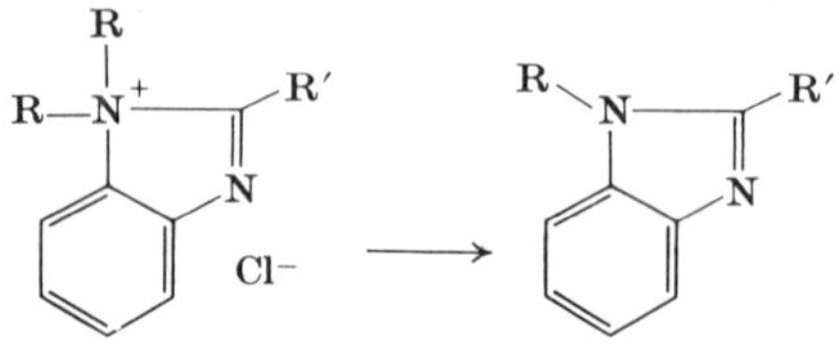

[23] A. S. Petrov, I. N. Somin, and G. S. Kuznetsov, *Khim. Geterotsikl. Soedin.* 328 (1969).
[24] Houben-Weyl "Methoden der organische Chemie," Vol. 8 (3), p. 673. Thieme, Stuttgart 1952.

III. Interactions of the Ortho Substituent with the α-Methylene Group in *t*-Anilines

The reactions which fall into this category conform to a basic mechanism as outlined below for an unsaturated ortho substituent (A=B) capable of mesomerism (**61**) ↔ (**62**). The polarized mesomer

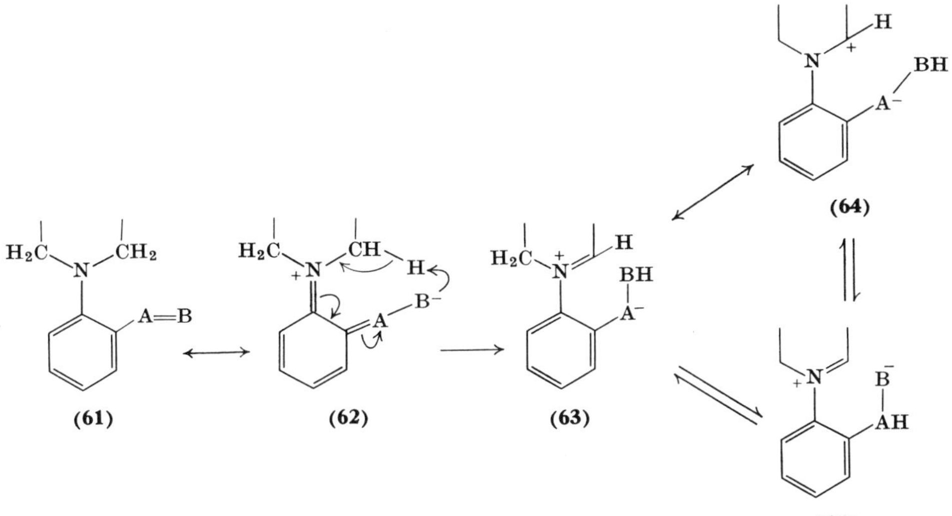

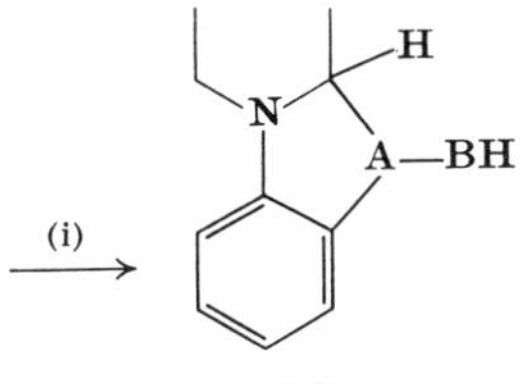

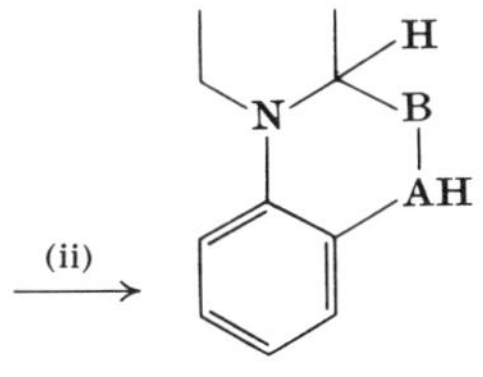

(62) is obviously set up for further reaction since the hydrogens of the α-methylene group are rendered labile by the charged nitrogen and are consequently abstracted by the nucleophilic center (B⁻). The resulting immonium ion (63) mesomeric with a carbonium ion (64) may finally cyclize in modes (i) and (ii) yielding a 5- (66) or 6-membered (67) hetero-ring system. The reactions are frequently acid-catalyzed since the charge on A may be neutralized by the acid. Similar schemes involving radicals or "excited states" may be envisaged to account for some uncatalyzed thermal and photochemical ring closures in this series.

Most of the known examples involve the following groups (—A=B): —N=O (including —NO$_2$), —N=C, —N=N, —C=N, and —C=O. Many observations were unexpected and unwanted results arising from simple laboratory procedures such as reduction or acylation. The reactions, therefore, were occasionally misinterpreted and as a consequence incorrect structures were assigned to the products.

The following discussion will be divided into sections according to the functional group situated at the ortho position.

A. o-Nitroso and o-Nitro Groups

Pinnow[2-5, 25-27] was the first to observe the interaction of the nitro group with a tertiary amino function. In a series of papers from 1895 onward he noticed that during the reduction of o-nitrodimethylanilines (68) with tin and hydrochloric acid, significant quantities of a 1-methylbenzimidazole (69) were formed together with the amine (70) and a chlorinated product (71). He proposed that the benzimidazole arose by reduction of the nitro to a nitroso group which then

[25] J. Pinnow and C. Saemann, *Ber. Deut. Chem. Ges.* **32**, 2181 (1899).
[26] J. Pinnow, *J. Prakt. Chem.* **63**, 352 (1901).
[27] J. Pinnow, *J. Prakt. Chem.* **65**, 579 (1902).

Sec. III. A.] THE *t*-AMINO EFFECT 227

condensed with the *N*-methyl function. Similar observations were made by Spiegel and Kaufmann[28] with the dinitro compound (**72**)

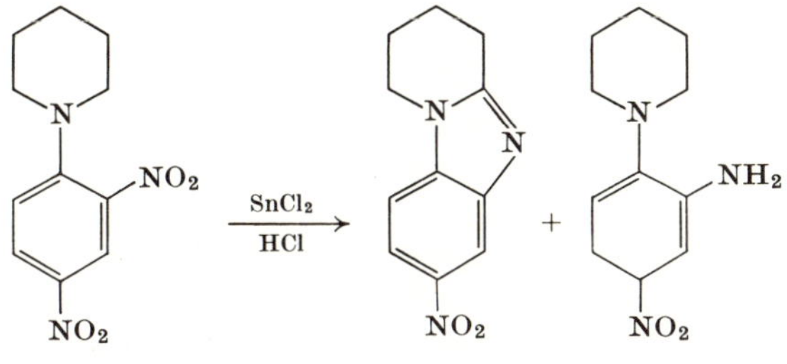

and also by Nair and Adams[29] with several analogous *o*-nitro-*t*-anilines. However, in all cases the yields were low.

This reaction was recently exploited[30] employing titanous chloride in acid solution as the reducing agent. By the slow addition of 2 moles of the reducing agent (equivalent to the reduction of a nitro to a nitroso group) very high yields of benz- (**75**), naphth- (**76**) and (**77**),

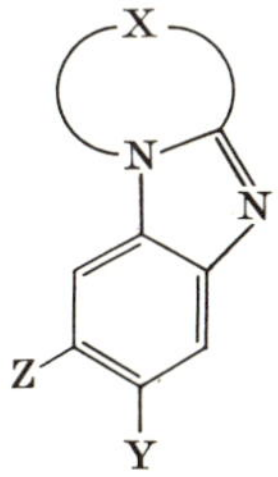

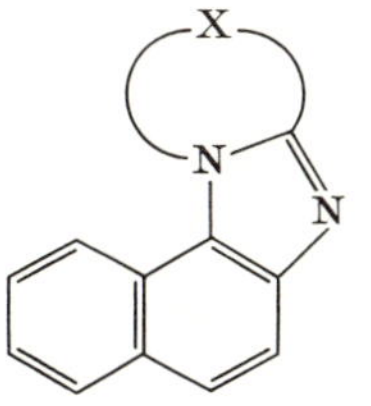

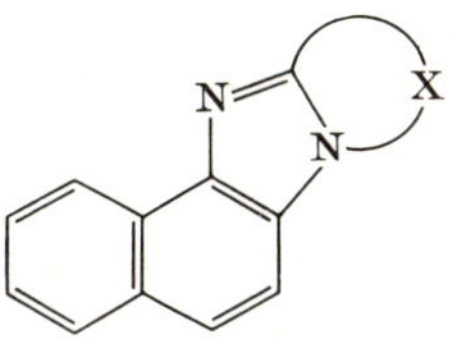

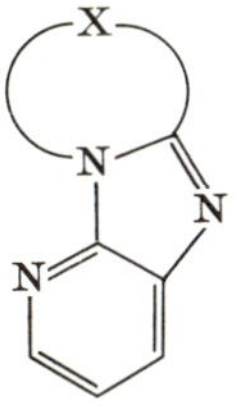

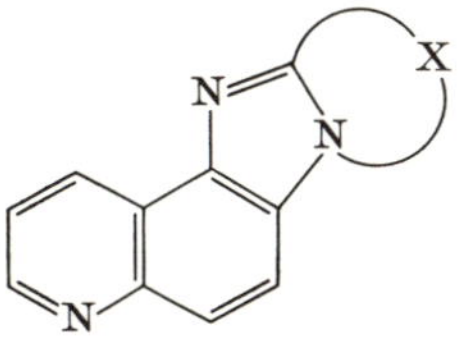

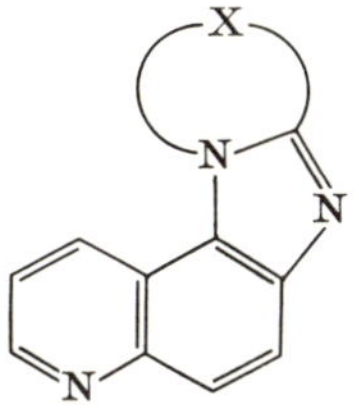

[28] L. Spiegel and H. Kaufmann, *Ber. Deut. Chem. Ges.* **41**, 679 (1908).
[29] M. D. Nair and R. Adams, *J. Amer. Chem. Soc.* **82**, 3518 (1961).
[30] H. Suschitzky and M. E. Sutton, *Tetrahedron* **24**, 4581 (1968).

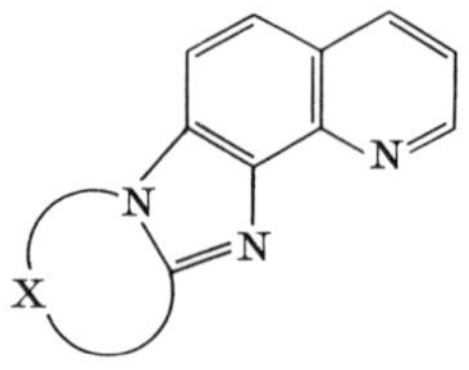

(81)

pyrido- (**78**), and quinolinoimidazoles (**79**)–(**81**) were isolated with very little amine being formed. The reaction failed with morpholine or piperazine derivatives and also with 8-nitro- and 4-dialkylamino-3-nitroquinolines which yielded only the corresponding amine.

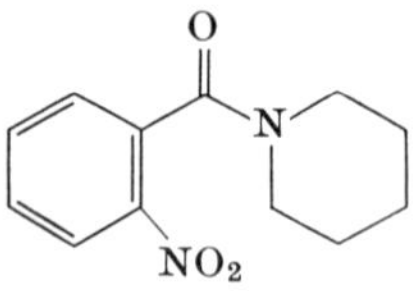

(**82**) 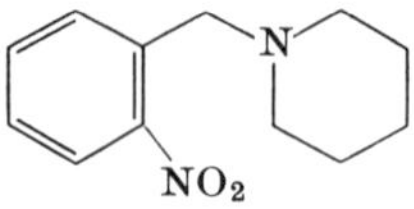   (**83**) 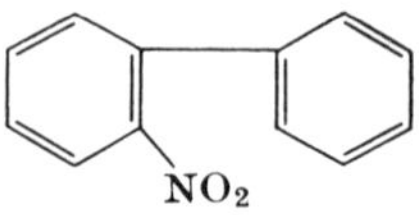   (**84**)

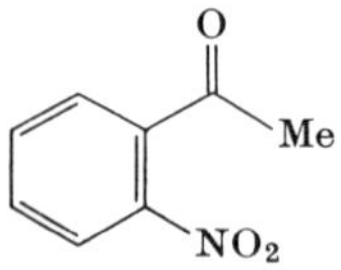

(**85**)

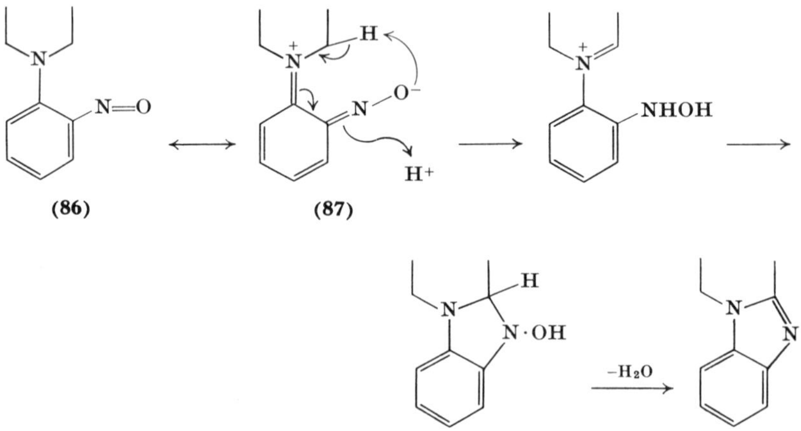

(**86**)   (**87**)

(**88**)

Unfavorable complexing of the reagent is probably the cause for the failure. This method is, in our experience, by far the best for preparing the rather inaccessible pyridoimidazoles. With both the pyridine and the quinoline systems an extra mole of reducing agent was necessary, presumably to complex the basic nitrogen first. The essential role of the presence as well as the position of the t-nitrogen in these reactions is clearly indicated by the unsuccessful attempts to cyclize the nitro compounds (82)–(85) with titanous chloride. There is difficulty in assigning with certainty a mechanism to this reaction. It is tempting to suggest that the reduction proceeds to the nitroso compound (86) which then cyclizes under acid catalysis (87)–(88) in accord with the general scheme already outlined. Alternatively, cyclization assisted by the transition metal may precede reduction. Examples of this latter type of reaction of o-nitro-t-anilines will be dealt with later in this section. It is significant that no o-nitroso-t-anilines have been reported despite numerous attempts to prepare them either by reduction of the analogous nitro compounds or oxidation of the appropriate amine or by electrophilic substitution.[2] Claims[31] for their isolation appear to be unfounded[32] and the alleged synthesis[33] of an o-nitroso-t-aniline by nitrosation of p-benzyldimethylaniline (89) on reinvestigation gave only the nitro compound (90).[32]

The bisulfite reduction of the nitro compound (91) has been reported to give the benzimidazole (93),[34] purportedly via the nitroso compound (92). Lithium aluminum hydride reduction of aromatic nitro compounds generally gives azo compounds, believed to be formed

---

[31] I. L. Finar, "Organic Chemistry," 5th ed., Vol. 1, p. 621. Longmans, Green, London, 1967.
[32] O. Meth-Cohn, unpublished observations.
[33] J. Limpricht, Ann. Chem. 307, 305 (1899).
[34] W. M. Lauer, M. M. Sprung, and C. M. Langkammerer, J. Amer. Chem. Soc. 58, 225 (1936).

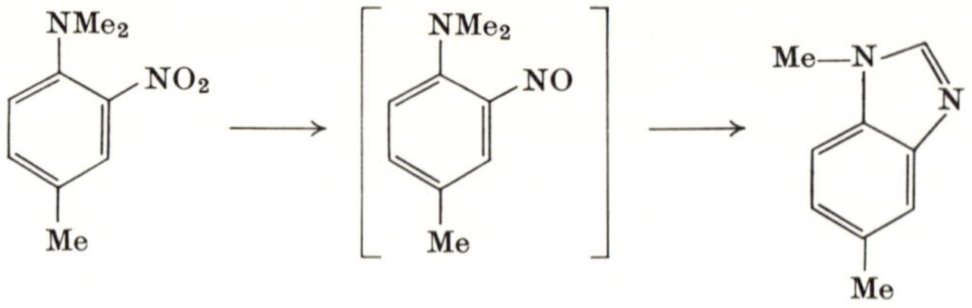

from the intermediate nitrosobenzene.[35] Thus, reduction of o-nitrophenylpiperidine with lithium aluminum hydride in ether gave primarily the azo compound (94) together with the benzimidazole (95), again probably involving the corresponding nitroso compound.[32] The benzimidazole (95) and its homologs are also isolated in low yields from the hydrogenation of the same nitro compounds in the presence of Raney nickel.[32] Huisgen and Rist observed that the benzimidazole (95) was formed together with other products from the action of lithium piperidide in ether on nitrobenzene and proposed the nitroso compound as an intermediate.[36] Finally, the electrolytic reduction of o-nitrodimethylaniline [(68) X = H] in 2N hydrochloric acid gives a mixture of the expected amine [(70) X = H] and 1-

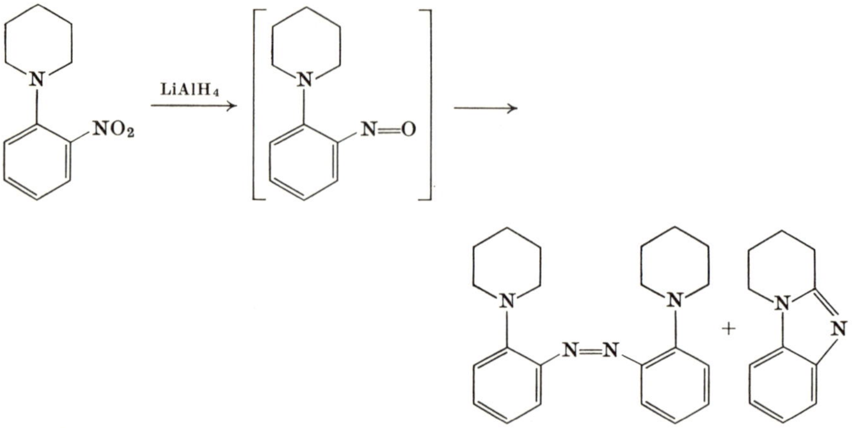

[35] N. V. Sidgwick, I. T. Millar, and H. D. Springall, "The Organic Chemistry of Nitrogen," 3rd ed., pp. 387–391. Oxford Univ. Press (Clarendon), London and New York, 1966.
[36] R. Huisgen and H. Rist, *Ann.* **594**, 159 (1955).

methylbenzimidazole (**69**), believed to be derived from the intermediate nitroso compound.[36a]

The cyclization of the above nitro compounds to benzimidazoles is also observed thermally.[37] Indeed, the mass spectra of *o*-nitro-*t*-anilines reveal the thermal cyclization to benzimidazoles in the probe.[32] Analogous interactions of *o*-nitro groups in the mass spectrometer are well known.[38]

For preparative purposes the cyclization is best conducted by heating the nitro compound mixed with sand or, even better, dispersed on glass beads to a temperature of 220°–240° and excellent yields are often obtained. The presence of electron-withdrawing groups in the benzene ring (NO_2, CO_2H, CF_3, or Cl) causes the reaction to be vigorous and complete in 30 minutes, whereas electron-donating functions (NHAc or Me) need longer reaction times (1–4 hours) and give lower yields. The use of a solvent (e.g., diphenyl ether or PPA) instead of sand gave similar results. The mechanism (**96**)–(**88**) can again be rationalized by analogy to the general scheme (**61**)–(**66**),

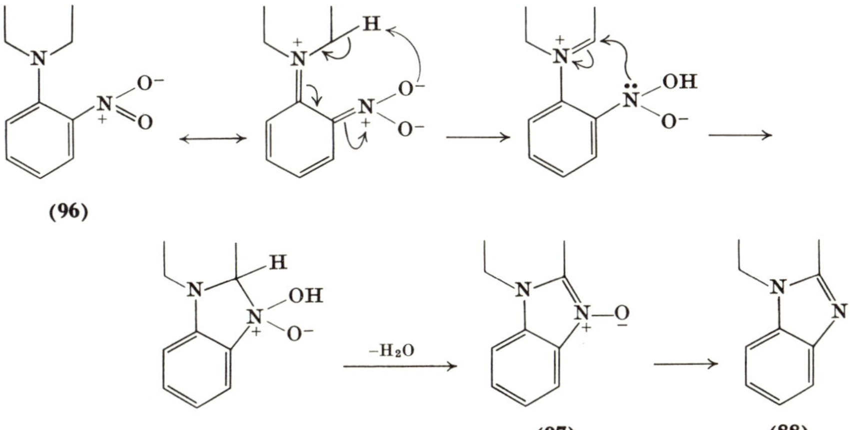

in which the benzimidazole *N*-oxide (**97**) is an intermediate. (For results see Table V.)

[36a] M. L. Guyader and D. Peltier, *Bull. Soc. Chim. Fr.* 2695 (1966).
[37] H. Suschitzky and M. E. Sutton, *Tetrahedron Lett.* 3933 (1967).
[38] J. Seibl and J. Voellmin, *Org. Mass. Spectrom* **1**, 713 (1968) and references cited therein.

TABLE V

Imidazoles (**75**), (**76**), (**77**) AND (**81**) Obtained[a] by
Heating the Appropriate Nitro Compounds[b]

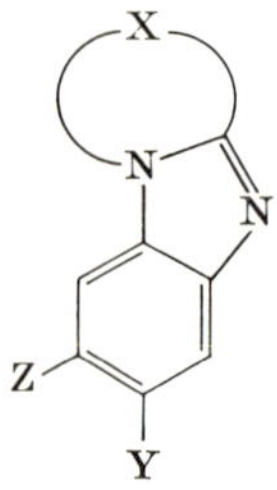

(**75**) 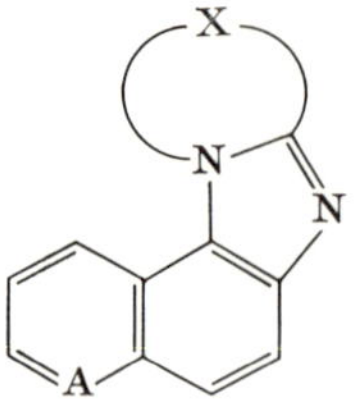   (**76**) 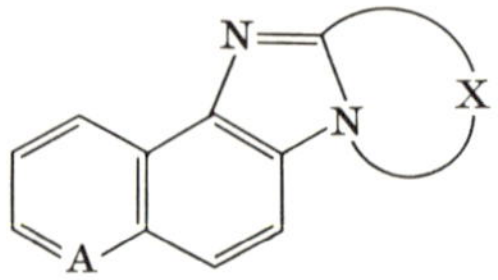   (**77**) 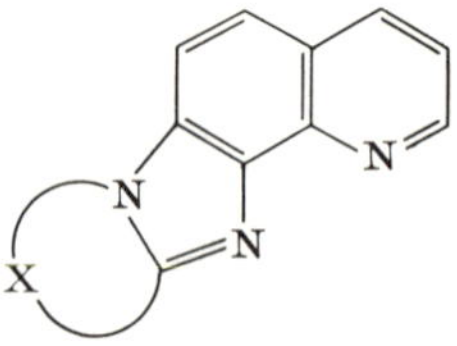   (**81**)

Imi-dazole	Substituent				Reaction time (hours)	Yield (%)	M.p. (°C)
	A	X	Y	Z			
75	—	$(CH_2)_4$	CO_2H	H	0.5	88	298
75	—	$(CH_2)_4$	NO_2	H	0.5	82	209
75	—	$(CH_2)_4$	Cl	H	0.75	70	152
75	—	$(CH_2)_3$	CF_3	H	0.5	65	148
75	—	$(CH_2)_4$	NHAc	H	1	40	222
75	—	$(CH_2)_4$	Me	H	3	55	126
75	—	$(CH_2)_4$	H	H	1.5	10	100
75	—	$(CH_2)_4$	H	H	4	62	100
75	—	$(CH_2)_4$	H	NHAc	1	60	238
76	CH	$(CH_2)_3$	—	—	0.25	77	151
76	N	$(CH_2)_3$	—	—	0.5	80	213
77	CH	$(CH_2)_3$	—	—	0.25	70	160
77	CH	$CH_2 \cdot O \cdot CH_2 \cdot CH_2$	—	—	1	8	147
77	CH	$CH_2 \cdot O \cdot CH_2 \cdot CH_2$	—	—	3	32	147
77	N	$(CH_2)_3$	—	—	0.5	88	184
81	—	$(CH_2)_4$	—	—	0.5	72	232

[a] Suschitzky and Sutton.[37]
[b] In sand at about 240°.

The thermal cyclization of o-nitro-t-anilines is considerably facilitated by the presence of mineral acid, when the cyclization proceeds at temperatures sufficiently low to allow the isolation of the intermediate N-oxide.[39,40] Indeed, the benzimidazole N-oxides (**98**) and (**99**) are obtained in good yield simply by refluxing the relevant

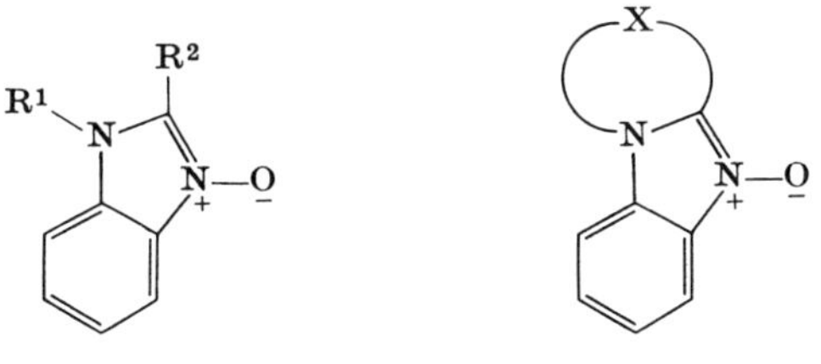

(**98**)   (**99**)

nitro compound in constant-boiling hydrochloric acid.[39] Since benzimidazole N-oxides cannot be prepared by direct oxidation,[41,42] this method offers a convenient route to these somewhat inaccessible compounds (cf. Table VI).

An interesting side reaction accompanies this synthesis leading to denitration or rearrangement of the nitro group.[40] This behavior is again a feature of the t-amino group and we have proposed the intermediacy of the hydrofuroxan (**101**) to account for it (cf. Schönberg

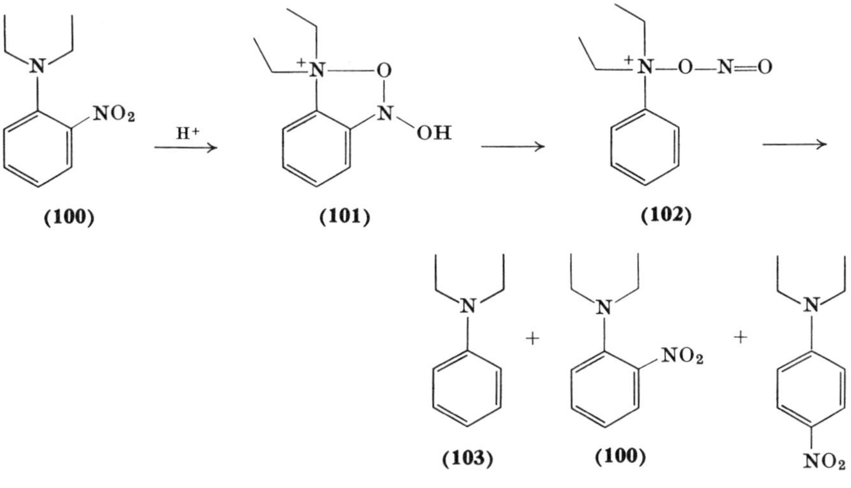

(**100**)   (**101**)   (**102**)

(**103**)   (**100**)   (**104**)

[39] R. Fielden, O. Meth-Cohn, D. Price, and H. Suschitzky, *Chem. Commun.* 772 (1969).
[40] R. Fielden, O. Meth-Cohn, and H. Suschitzky, *Tetrahedron Lett.* 1229 (1970).
[41] O. Meth-Cohn and H. Suschitzky, *J. Chem. Soc.* 4666 (1963).
[42] S. Takahashi and H. Kano, *Chem. Pharm. Bull.* **11**, 1375 (1963), **14**, 1219 (1966) and references cited therein.

TABLE VI

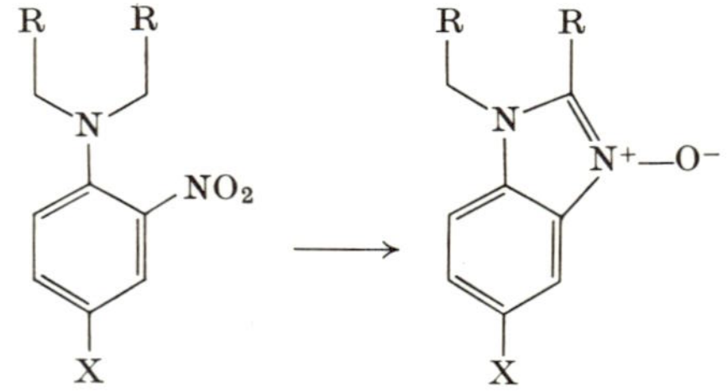

BENZIMIDAZOLE N-OXIDES (117) BY THE ACTION OF HYDROCHLORIC ACID ON THE o-NITRO-t-ANILINES (96)[a]

Substituent		Reaction temp. (°C)	Reaction time (hours)	Product (117)			Unreacted (96) (%)	Remarks[b]
R	X			Yield (%)	M.p. (°C)			
(CH$_2$)$_2$	H	110	20	51	224		18.5	HCl
(CH$_2$)$_2$	Cl	110	20	70	182		0	—
(CH$_2$)$_2$	NO$_2$	110	20	32	212		47	HCl

(CH$_2$)$_3$	H	110	20	Trace	—	89	—
(CH$_2$)$_3$	H	160	7	61	202–204	10	HCl
(CH$_2$)$_3$	Cl	110	72	27	131	0	—
(CH$_2$)$_3$	NO$_2$	110	20	20	228	80	HCl
CH$_2$OCH$_2$	H	110	20	26	201	68	HCl
CH$_2$OCH$_2$	H	150	12	30	201	66	HCl
CH$_2$OCH$_2$	NO$_2$	110	20	13	215–218	87	HCl
(CH$_2$)$_4$	H	110	1	12	212	—	HCl
(CH$_2$)$_4$	H	110	2	16	212	—	HCl
(CH$_2$)$_4$	H	110	4	25	212	—	HCl
(CH$_2$)$_4$	H	110	8	38	212	14.5	HCl
(CH$_2$)$_4$	H	110	20	61	212	0	HCl
(CH$_2$)$_4$	H	110	40	74	212	0	HCl
(CH$_2$)$_4$	Cl	110	48	62	129	80	HCl
(CH$_2$)$_4$	NO$_2$	110	20	16	206	19	HCl
(CH$_2$)$_4$	NO$_2$	150	12	76	206	75	HCl
Me	H	110	20	8	240	10	HCl
Me	H	150	12	47	240	0	HCl
Et	Cl	110	48	56	101	0	—
(CH$_2$)$_6$	Cl	110	72	61	125	0	—
(CH$_2$)$_7$	Cl	110	72	52	110	35	—

[a] Fielden et al.[39]
[b] HCl indicates isolation of N-oxides as hydrochlorides.

et al.,[43] Shabarov *et al.*[44]). "Crossing" experiments indicate the intramolecularity of the rearrangement (**100**) → (**104**), which is also observed when the *N*-oxide (**105**) is treated with nitrosyl chloride.

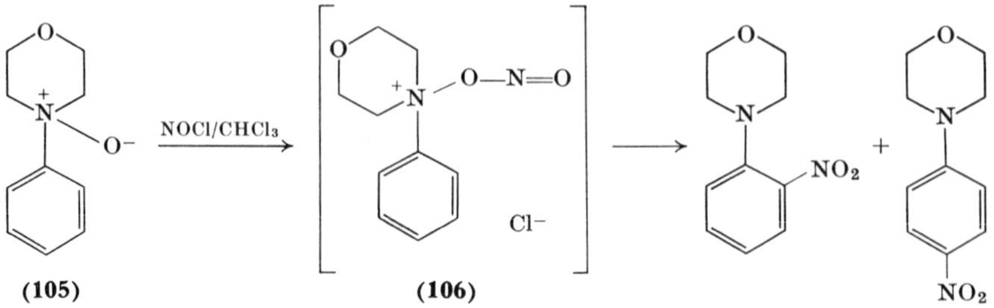

An observation reminiscent of this denitration is the corresponding ready decarboxylation or desulfonation of the acids (**107**) by action of hot mineral acid.[45] This reaction evidently requires the *o-t*-amino group since the 2-nitro-4-carboxylic acid (**107a**), the corresponding sulfonic acid (**107a**), as well as *p*-nitrophenylanilines (**104**) are un-

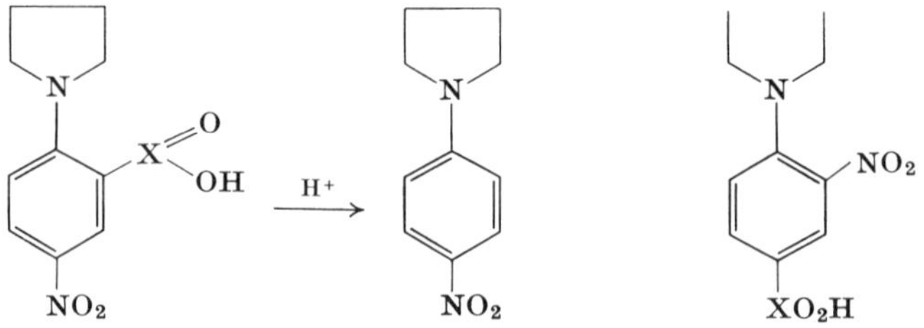

affected by prolonged boiling with mineral acid. Intermediates analogous to **101** in which the *t*-nitrogen participates may be visualized to account for this surprising result. Elimination of the ortho substituent occurs in the order $COOH > SO_3H > NO_2$ with regard to ease.

[43] A. Schönberg, G. O. Schenk, and O-A. Neumüller, "Preparative Organic Photochemistry," p. 266 and references cited therein. Springer, Berlin, 1968.
[44] Y. S. Shabarov, S. S. Mochalov, and I. P. Stepanov, *Dokl. Akad. Nauk. SSSR* **189**, 1028 (1969).
[45] J. Martin, O. Meth-Cohn, and H. Suschitzky, unpublished observations.

Sec. III. A.] THE *t*-AMINO EFFECT 237

The above cyclization by hot mineral acid has a parallel in the action of an alkyl halide on this system.[32] Thus, methyl iodide and

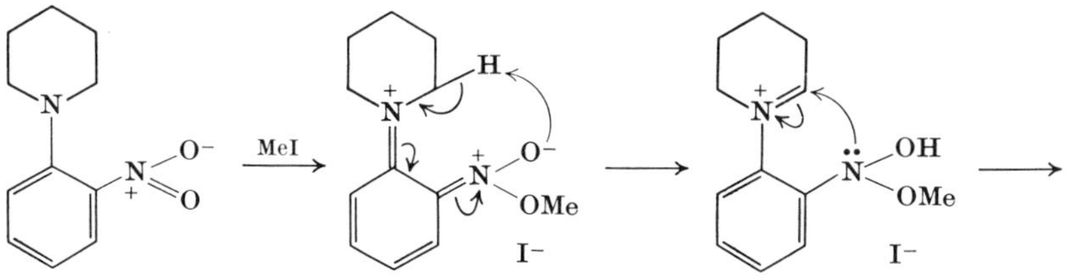

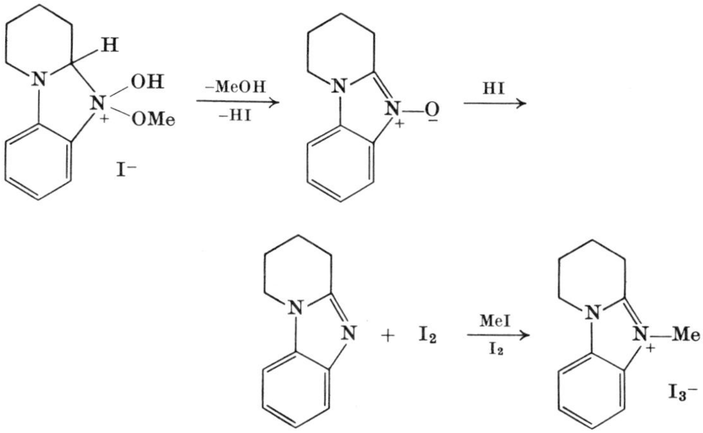

o-nitrophenylpiperidine (**108**) interact to yield the brown methylbenzimidazolium periodide (**109**) which is identical to that obtained by addition of iodine in ethanol to the corresponding methiodide. It is probable that the intermediate *N*-oxide oxidizes hydriodic acid to liberate iodine.

The acid-catalyzed cyclization to benzimidazole *N*-oxides may also be conducted photolytically with good results.[40] Denitration is not observed here and very clean reaction products are obtained. In a few cases, however, the corresponding benzimidazole and not its *N*-oxide is the sole product. Since these *N*-oxides are photostable under the reaction conditions, it appears that a different mechanism operates for the photocyclization. A plausible alternative (**101**)–(**88**) may be visualized by analogy to the photolytic rearrangement of *o*-nitro to *o*-nitroso compounds.[43] The *o*-nitroso-*N*-oxide (**110**) may

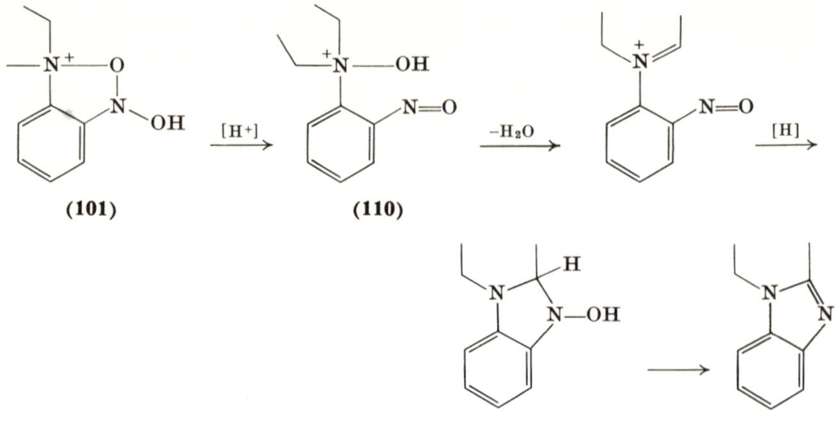

238   O. METH-COHN AND H. SUSCHITZKY   [Sec. III. A.

(101)   (110)

(88)

then cyclize to the benzimidazole in the manner shown. The photochemical formation of benzimidazole N-oxides is very slow without acid and the presence of a second nitro group leads to secondary photoproducts, particularly in polar solution (cf. Wubbels et al.,[46] Pacifici et al.,[47] Döpp,[48] Jones et al.,[49] Roth and Adomeit[50]). The thermal cyclization is, therefore, the preferred route when photolabile substituents such as the nitro group are present.

The effect of a Lewis acid on the o-nitro-t-aniline system is well illustrated by the unexpected result from the attempted Friedel–Crafts acetylation of these compounds. In 1926, van Romburgh and co-workers[51, 52] observed that the action of acetic anhydride and zinc chloride on the o-nitrodimethylaniline (111) did not give the acetylated product (112), but instead the benzimidazolone (113). Even more surprising was their alleged observation that the same reagents converted the diethyl compound (114) into the quinoxaline

---

[46] G. W. Wubbels, R. R. Hautala, and R. L. Letsinger, *Tetrahedron Lett.* 1689 (1970).

[47] J. G. Pacifici, G. Irick, and C. G. Anderson, *J. Amer. Chem. Soc.* **91**, 5654 (1969).

[48] D. Döpp, *Chem. Commun.* 1284 (1968).

[49] L. B. Jones, J. C. Kudrna, and J. P. Foster, *Tetrahedron Lett.* 3263 (1969).

[50] J. J. Roth and M. Adomeit, *Tetrahedron Lett.* 3201 (1969).

[51] P. van Romburgh and H. W. Huyser, *Versl. Akad. Wetensch. Amsterdam* **30**, 845 (1926); *Chem. Abstr.* **21**, 382 (1927).

[52] P. van Romburgh and H. W. Huyser, *Rec. Trav. Chim.* **49**, 165 (1930).

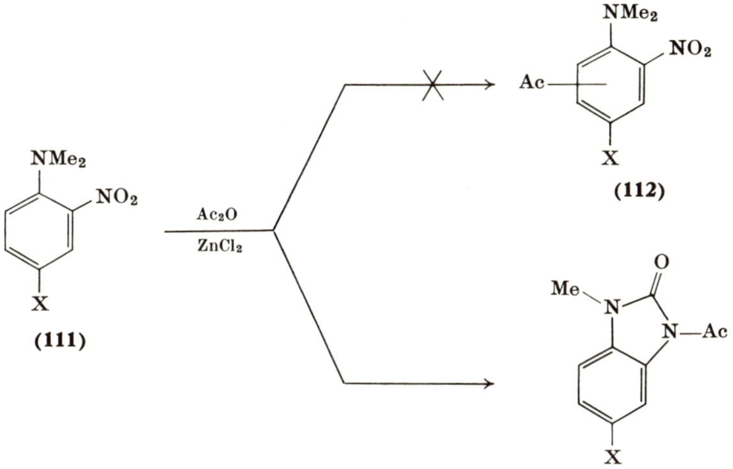

(115), which was supported by an assumedly unambiguous synthesis.[53] In view of the deviation of this reaction from our general scheme, a reinvestigation was carried out[54] which confirmed the course of the

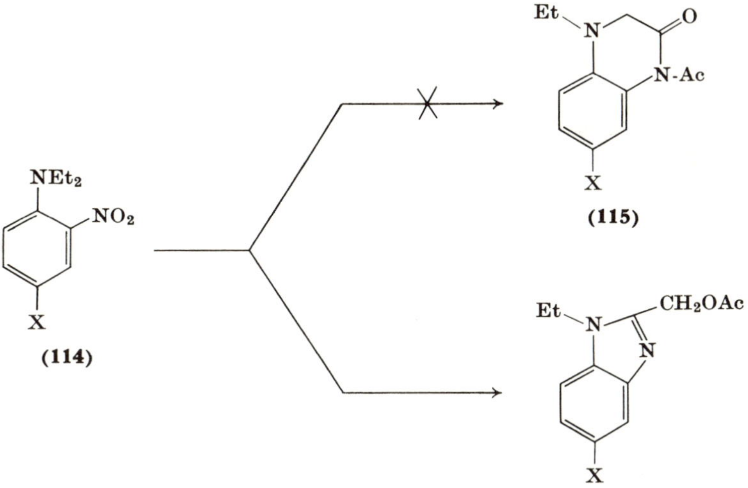

[53] P. van Romburgh and W. B. Deys, *Proc. Acad. Sci. Amsterdam* **34**, 1004 (1931); *Chem. Abstr.* **26**, 989 (1932).
[54] R. K. Grantham and O. Meth-Cohn, *J. Chem. Soc., C* 70 (1969).

first reaction (**111**)–(**113**), but showed the product from the diethyl compound to be the 2-acetoxymethylbenzimidazole (**116**). The formation of these products can again be rationalized by the general mechanism (**96**)–(**88**) entailing the *N*-oxide intermediate (**117**) which is acetylated by the acetic anhydride and then rearranges. This reaction course is in agreement with the conversion of 2-unsubstituted benzimidazole *N*-oxides into benzimidazolones and 2-alkylbenzimidazole *N*-oxides into 2(α-acetoxy)alkylbenzimidazoles with acetic anhydride.[55, 56] It is evident that the reaction is Lewis acid-catalyzed and that intermediates such as **118** may be visualized. The reaction is equally applicable to the cyclic analog of **114** derived from pyrrolidine, piperidine, etc., when the products analogous to **116** may be isolated, often in good yields.[54]

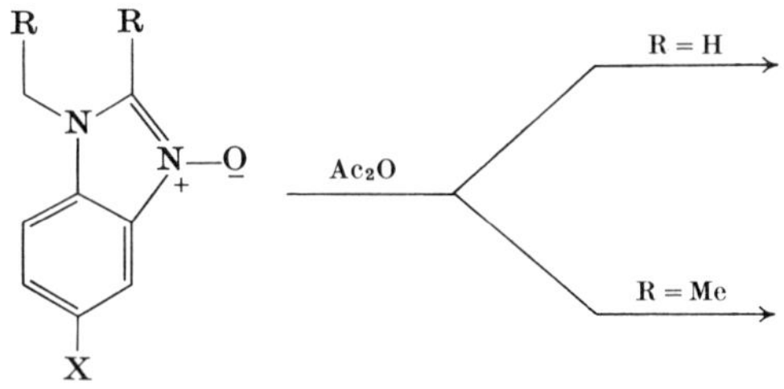

(**117**)

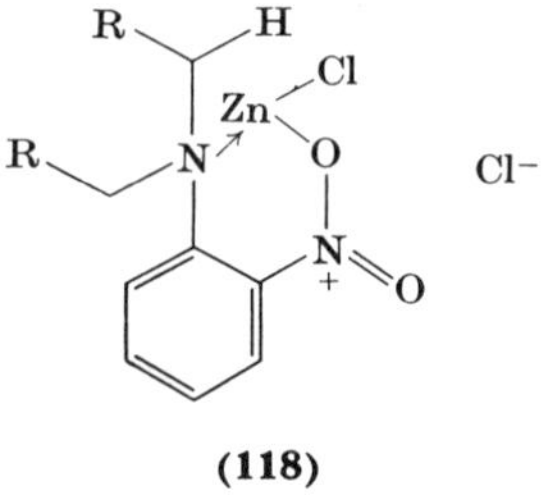

(**118**)

A reaction worth noting in the context of nitro-group interactions with *t*-amino functions involves the rapid photolytic breakdown of 2,4-dinitrophenyl derivatives of α-amino acids. The lability of such derivatives in sunlight has long been known to protein chemists who

[55] S. Takahashi and H. Kano, *Chem. Pharm. Bull.* **12**, 783 (1964).
[56] S. Takahashi and H. Kano, *Chem. Pharm. Bull.* **14**, 1219 (1966).

have utilized 2,4-dinitrohalobenzenes in protein and peptide degradations.[57,58] Akabori et al. observed that decarboxylation occurred at the same rate for a variety of 2,4-dinitrophenyl-α-amino acids.[59] Russell[60-65] and Neadle and Pollitt[66a-67] have identified the products from this reaction and have shown that at low pH a benzimidazole N-oxide (121) is formed, whereas at higher pH the 2-nitroso-4-nitroaniline (120) is isolated. The nitroso compound (120) reacted with aldehydes under acid catalysis to give the corresponding benzimidazole N-oxide (121).[64] The reaction is relevant to the present review with N-alkylamino acid derivatives [(119) R' = alkyl] and is best explained by analogy to the foregoing photolytic reactions.[68]

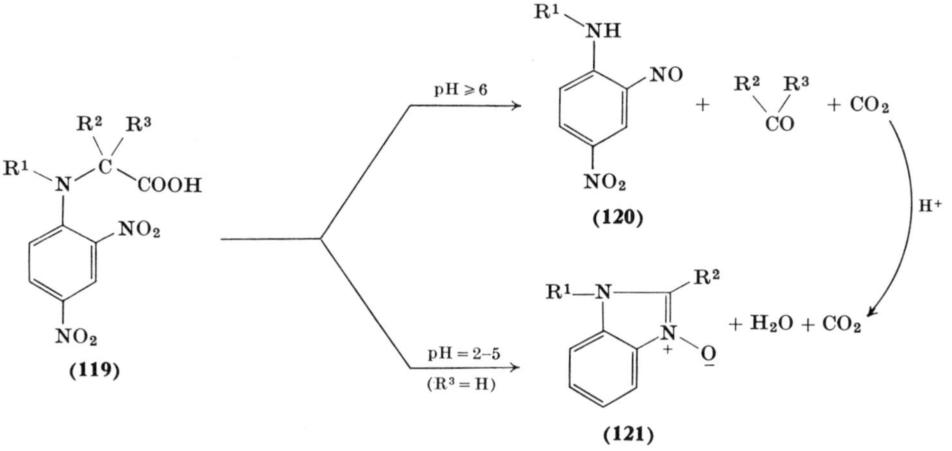

[57] F. Sanger, *Biochem. J.* **45**, 563 (1949).
[58] S. Blackburn, *Biochem. J.* **45**, 579 (1949).
[59] S. Akabori, T. Ikena, Y. Okada, and K. Kohno, *Proc. Imp. Acad. (Tokyo)* **29**, 509 (1953).
[60] D. W. Russell, *Biochem. J.* **83**, 8 (1962).
[61] D. W. Russell, *Biochem. J.* **87**, 1 (1963).
[62] D. W. Russell, *J. Chem. Soc.* 894 (1963).
[63] D. W. Russell, *J. Chem. Soc.* 2829 (1964).
[64] D. W. Russell, *Chem. Commun.* 498 (1965).
[65] D. W. Russell, *J. Med. Chem.* **10**, 984 (1967).
[66a] R. J. Pollitt, *Chem. Commun.* 262 (1965).
[66b] D. J. Neadle and R. J. Pollitt, *J. Chem. Soc., C* 1764 (1967).
[67] D. J. Neadle and R. J. Pollitt, *J. Chem. Soc., C* 2127 (1969).
[68] O. Meth-Cohn, *Tetrahedron Lett.* 1235 (1970).

The key intermediate is the hydrofuroxan (**122**) which facilitates decarboxylative fragmentation in the same manner as in a β-bromo acid anion.

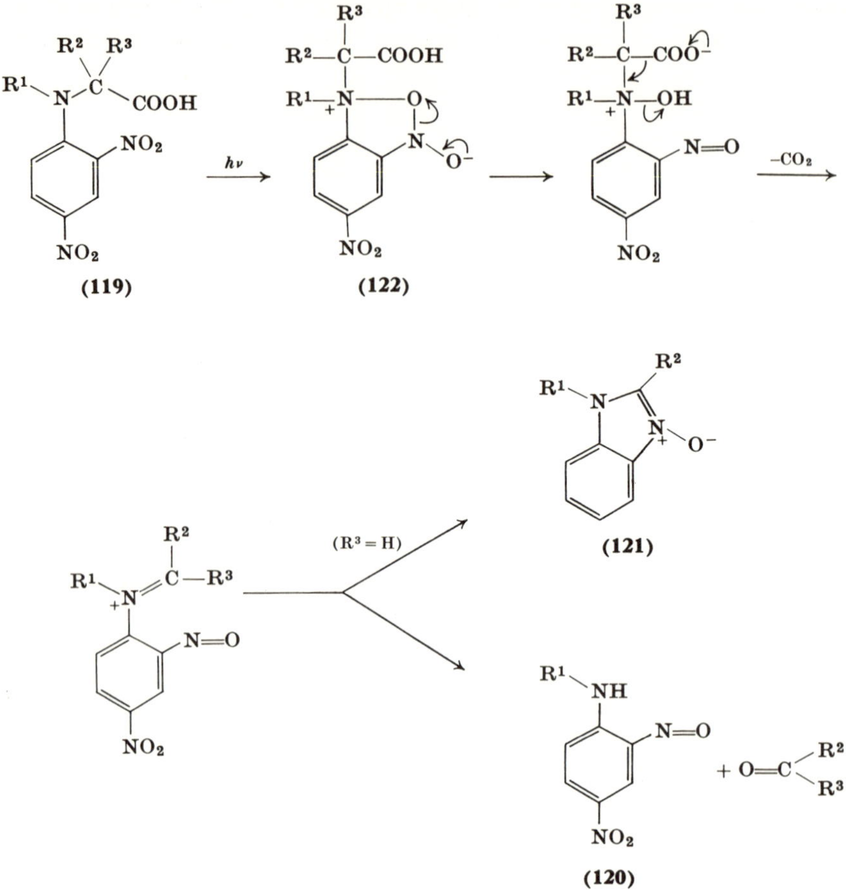

An interesting photoreaction of picrylaziridines (**123**) has recently been reported which is relevant to this discussion.[69] The authors suggest the following mechanism (**123**)–(**125**) in which an intermediate *o*-nitrosoanil (**124**) cyclizes to an *N*-hydroxybenzimidazole (**125**).

[69] H. W. Heine, G. J. Blosick, and G. B. Lowrie, *Tetrahedron Lett.* 4801 (1968).

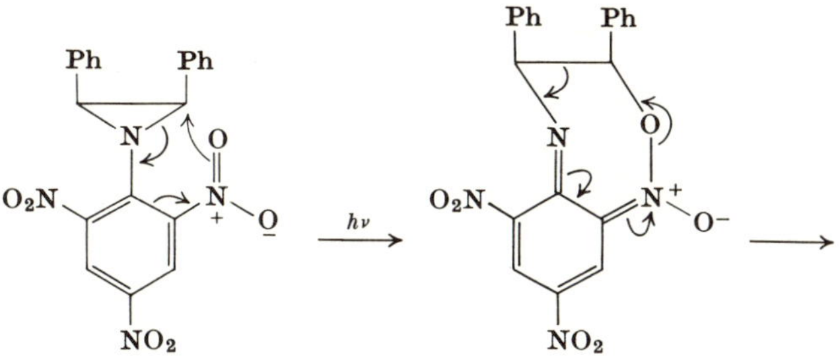

[Scheme showing compound (123) → intermediate → and (124) + PhCHO → (125) 95%]

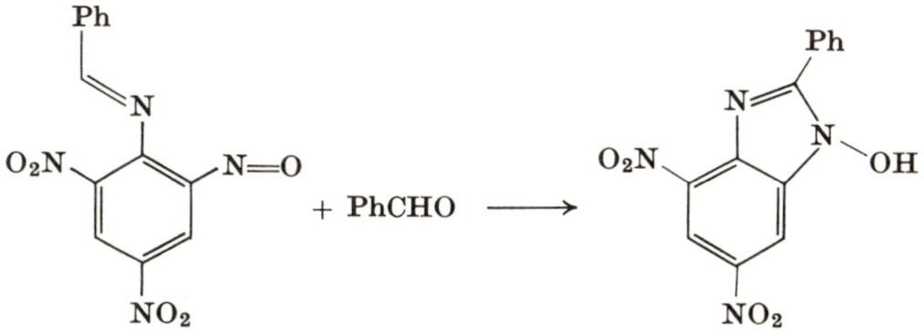

Similarly, the 2,4-dinitrophenylaziridine (126) gave the same type of product (127) and benzoic acid.[69]

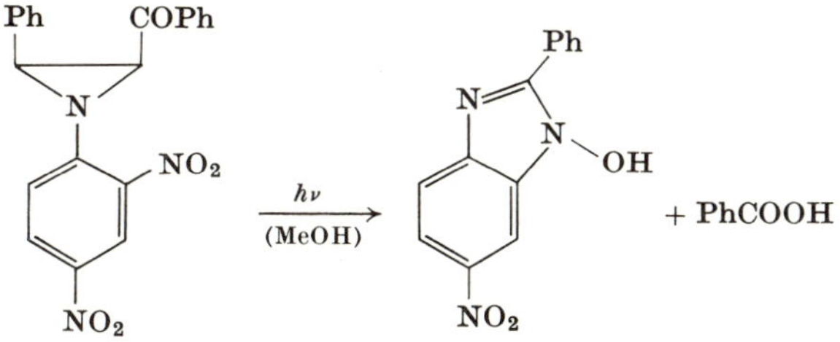

[Scheme: (126) →hν (MeOH)→ (127) 96% + PhCOOH]

B. o-AMINO AND o-ACYLAMINO GROUPS

In 1908, Spiegel and Kaufmann[28] tried to prove that the course of the unexpected cyclization of 2,4-dinitrophenylpiperidine (72) with stannous chloride (see p. 227) involved a nitroso compound, by

TABLE VII
BENZIMIDAZOLES (130) BY THE ACTION OF A PERACID ON o-AMINO-t-ANILINES (129)

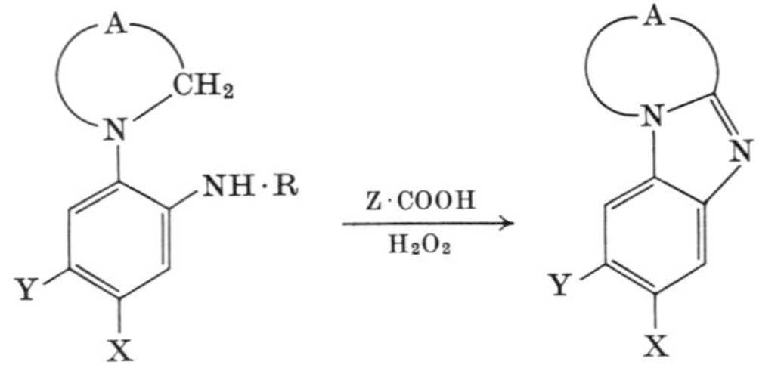

A	Substituent					Product (130)			
	X	Y	R	Z		Yield (%)	M.p. (°C)	Ref.	
$(CH_2)_3$	H	H	H	CF_3		81	115	29	
$(CH_2)_3$	H	H	H	H		85–95	115	41	
$(CH_2)_3$	H	H	Ac	CH_3		92	115	32	
$(CH_2)_3$	H	H	H	H		74	115	41	
$(CH_2)_3$	H	H	COPh	H		85	115	41	
$(CH_2)_3$	Cl	H	H	CF_3		75	133–134	29	
$(CH_2)_3$	CH_3	H	H	CF_3		86	170–171	29	
$(CH_2)_3$	NO_2	H	H	CF_3		72	209–210	29	

$(CH_2)_3$	COOEt	H	CHO	H	62	139	69a
$(CH_2)_3$	NHAc	H	Ac	H	43	253	69b
$(CH_2)_3$	H	NHAc	Ac	H	91	236	69b
$(CH_2)_4$	H	H	H	CF_3	58	99–100	29
$(CH_2)_4$	Cl	H	H	CF_3	66	152	29
$(CH_2)_4$	CH_3	H	H	CF_3	60	126	29
$(CH_2)_4$	NO_2	H	H	CF_3	95	219–220	29
$(CH_2)_4$	COOEt	H	CHO	H	75	134	69c
$(CH_2)_4$	NHAc	H	Ac	H	60	222	69b
$(CH_2)_4$	H	NHAc	Ac	H	84	238	69b
$(CH_2)_4$	Cyclohexyl	H	H	H	70	141–142	69d
$CH_2CH_2OCH_2$	H	H	H	CF_3	73	129–130	29
$CH_2CH_2OCH_2$	Cl	H	H	CF_3	62	196	29
$CH_2CH_2OCH_2$	CH_3	H	H	CF_3	61	170–171	29
$CH_2CH_2OCH_2$	NO_2	H	H	CF_3	76	214–215	29
$CH_2CH_2OCH_2$	COOEt	H	Ac	H	52	138	69c
$CH_2CH_2OCH_2$	NHAc	H	Ac	H	17	210	69b
$CH_2CH_2OCH_2$	H	NHAc	Ac	H	67	258	69b
$(CH_2)_5$	H	H	H	CF_3	91	124–125	29
$(CH_2)_5$	COOEt	H	H	H	58	108	69c
$(CH_2)_5$	NHAc	H	Ac	H	58	252	69b
$(CH_2)_5$	H	NHAc	Ac	H	66	222	69b

69a C. Perera, M.Sc. Thesis, Universtiy of Salford, 1970.
69b R. Garner and H. Suschitzky, *J. Chem. Soc.*, *C* 74 (1967).
69c R. C. Perera and R. K. Smalley, *J. Chem. Soc.*, *C* in press (1971).
69d R. K. Smalley, Ph.D. Thesis, University of Salford, 1964.

oxidizing the corresponding amine (74) with Caro's acid, which was known to convert aniline into nitrosobenzene. The benzimidazole (73) was, in fact, obtained in reasonable yield (although they reported

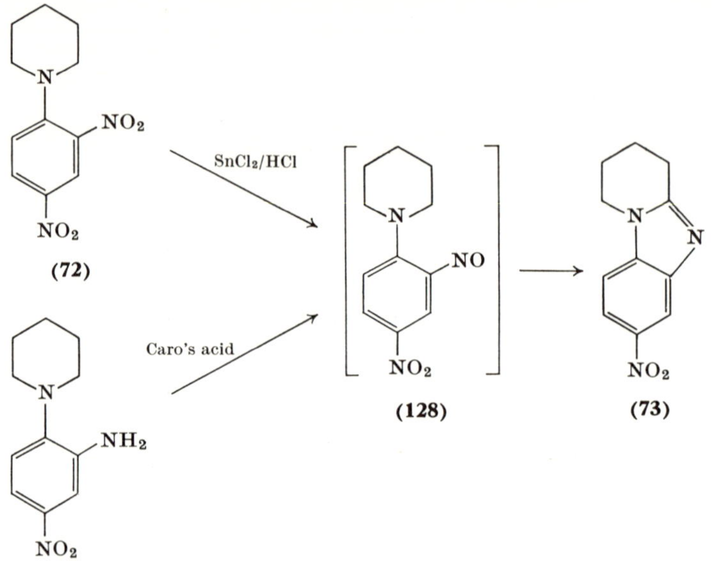

failure in the absence of the *p*-nitro group) which, they concluded, supported the proposed pathway. This reaction "lay fallow" for more than 50 years until it was reexamined by Nair and Adams[29] using the then new oxidant, peroxytrifluoroacetic acid, prepared *in situ* from hydrogen peroxide and trifluoroacetic acid. They obtained good yields of the benzimidazoles (130) from the corresponding amines (129),

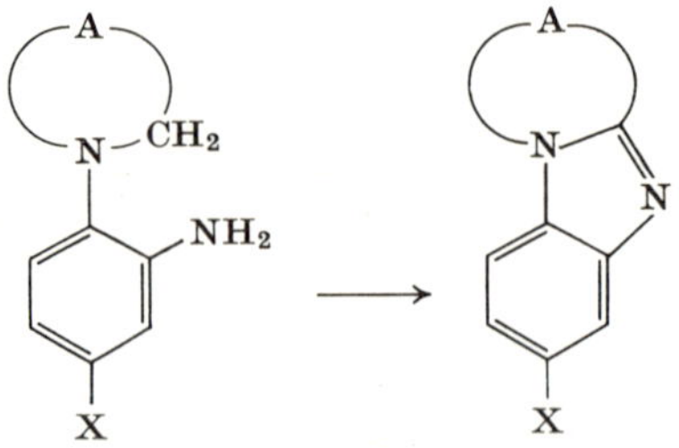

A = $(CH_2)_3$, $(CH_2)_4$, $(CH_2)_5$, or $(CH_2)_2OCH_2$
X = H, Cl, CH_3, or NO_2

Sec. III. B.]   THE *t*-AMINO EFFECT   247

particularly with electron-withdrawing substituents at X. These authors favored a nitroso intermediate (e.g., **128**), although they conceded the possibility of the involvement of a *t*-amine oxide.

Meth-Cohn and Suschitzky[41] examined this reaction further and observed that the acylamines (**131**) reacted as easily and more cleanly than the parent amines, especially with performic acid (see

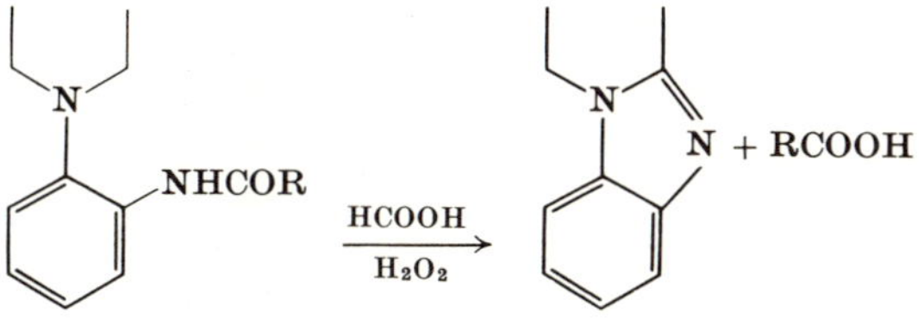

(**131**) $\xrightarrow{\text{HCOOH}}$ + RCOOH
H_2O_2

R = H, Me, or Ph

Table VII). The reaction was equally applicable to the preparation of

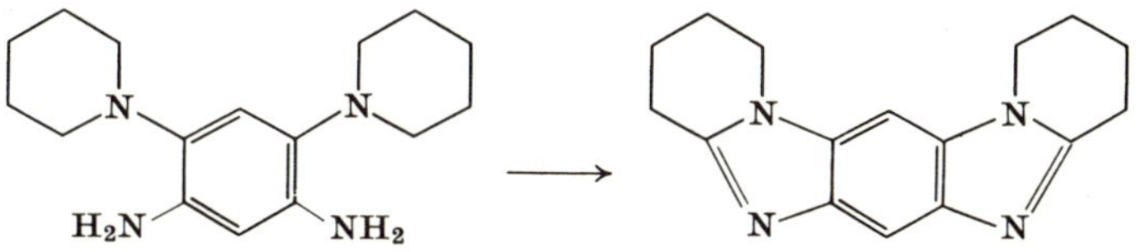

(**132**)

the bisimidazole system (**132**) and was extended[70] to the synthesis of

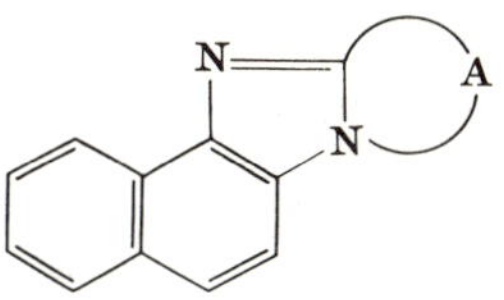

(**133**) 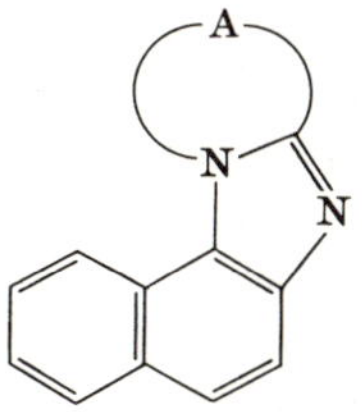   (**134**)

the naphthimidazoles (**133**) and (**134**), albeit with lower yields (see Table VIII), probably because the starting material was oxidized to

[70] H. Suschitzky and M. E. Sutton, *J. Chem. Soc., C* 3058 (1968).

TABLE VIII

FUSED IMIDAZOLES BY THE ACTION OF A PERACID ON
o-AMINO- AND o-ACYLAMINO-t-ARYLAMINES

Imidazole	A	Yield (%)	M.p. (°C)	Ref.
132	—	74	285–286	41
133	$(CH_2)_3$	72 (82)[a]	100	30
133	$(CH_2)_4$	85 (92)[a]	162	30
133	$(CH_2)_5$	84	115	30
133	$CH_2O(CH_2)_2$	0	147	30
134	$(CH_2)_3$	5[a]	147	30
134	$(CH_2)_4$	5[a]	128	30
134	$(CH_2)_5$	8[a]	155	30
135	$(CH_2)_3$	0	125	69,[a] 71
135	$(CH_2)_4$	43 (15)	100	71
135	$(CH_2)_5$	45	93	71
135	$(CH_2)_2OCH_2$	44	128	71
136	$(CH_2)_3$	18	184	30
136	$(CH_2)_4$	20	125	30
136	$(CH_2)_5$	25	152	30
137	—	5	213	30
138	—	15	210	30
139	—	15	238	30
140	—	88	152–153	30

[a] Prepared from benzoyl derivative.

quinones. Thus, these compounds are better prepared from the thermal or reductive treatment of the appropriate nitro compounds (Section III,A). Formation of the imidazopyridines (135) proceeded in low yield accompanied by breakdown since ammonia was formed.[71]

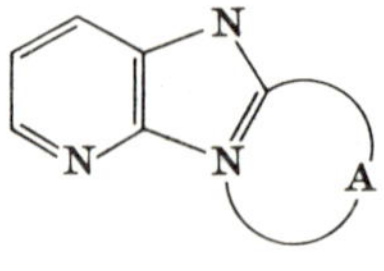

(135) 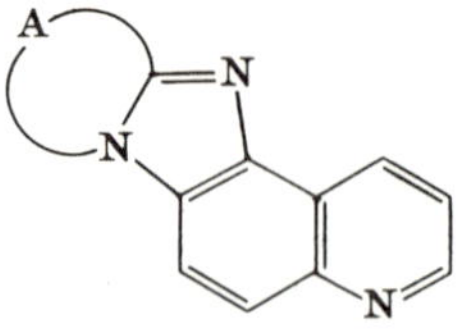 (136)

[71] R. K. Smalley, J. Chem. Soc., C 80 (1966).

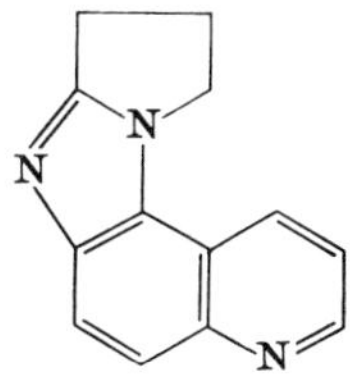

(137)

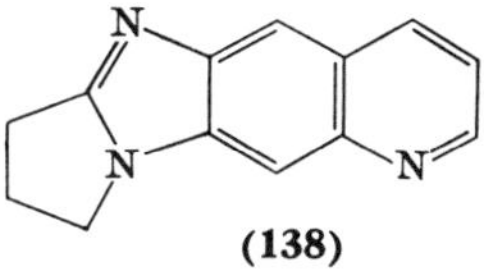

(138)

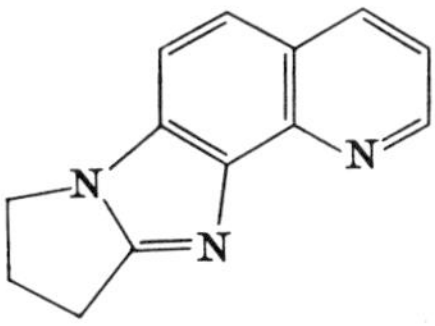

(139)

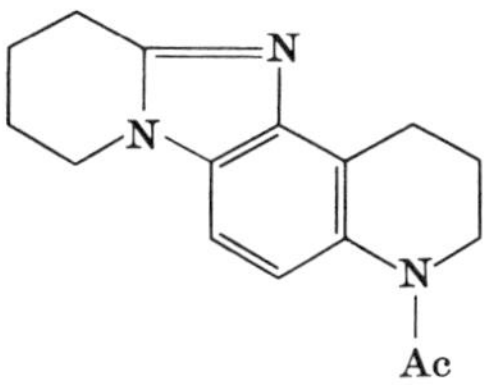

(140)

Similarly, the imidazoquinolines (136)–(139) were isolated in low yield, although the tetrahydroquinoline (140) was obtained in high yield (see Table VIII).[70]

Again intermediacy of an o-nitroso group is possible in the case of the amine oxidation. It is, however, unlikely with the acylamines since formyl-, acetyl-, and benzoylaniline remained unchanged under the reaction conditions. On the other hand, phenylpiperidine and o-nitrophenylpiperidine were rapidly converted into their N-oxides by treatment with performic acid.[41] It would, therefore, appear that an o-substituted N-oxide is instrumental in the cyclization which we suggest[41] to be an intramolecular Polonovski reaction.[72] In this an N-oxide is treated with an acylating agent or other electrophile to give a carbinolamine which suffers dealkylation and thereby converts a t-amine into its secondary analog. In the present case the required electrophile is evidently the built-in ortho substituent (NHCOR) and a plausible mechanism may be envisaged to proceed as shown (cf. 141 → 142 → 130).

Recently, the postulated N-oxide intermediate [(142) R = Ph] has been isolated by peracid oxidation of the benzoyl derivative [(14) R = Ph] in pyridine solution.[73] Brief heating of this N-oxide with dilute

[72] M. Polonovski and M. Polonovski, Bull. Soc. Chim. Fr. 41, 1190 (1927).
[73] O. Meth-Cohn, J. Chem. Soc., C 1356 (1971).

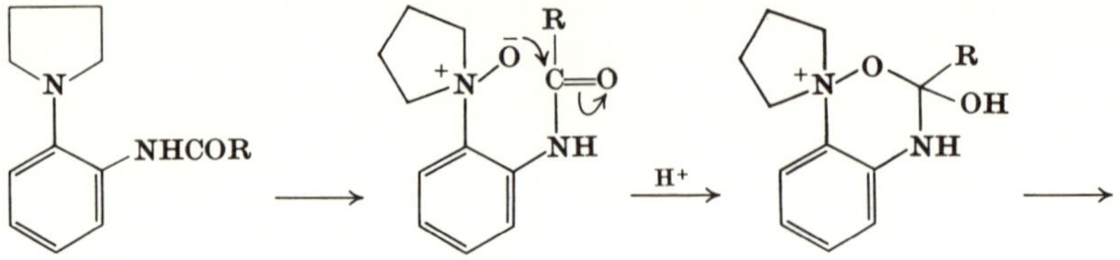

mineral acid gave the appropriate benzimidazole (**130**) in high yield, thus substantiating the proposed mechanism.

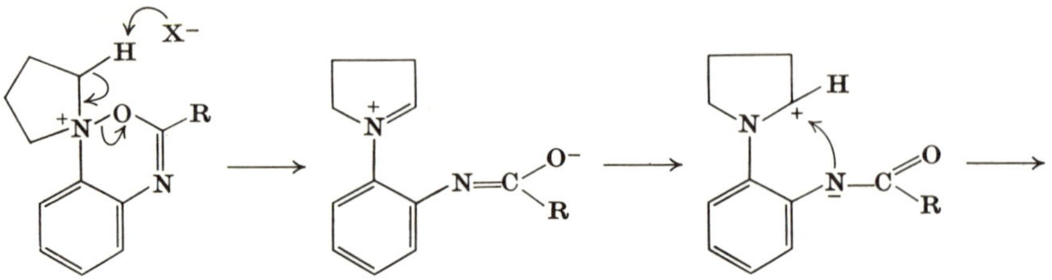

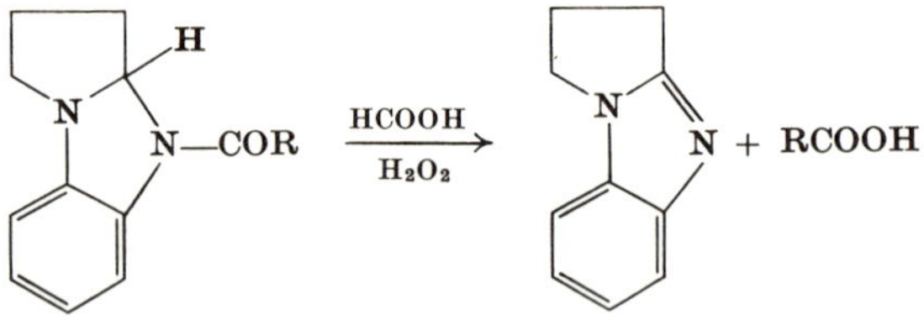

Ring closure of the *o*-dimethylaminoacetanilide (**144**) can also be brought about with ozone, presumably by a related mechanism.[74]

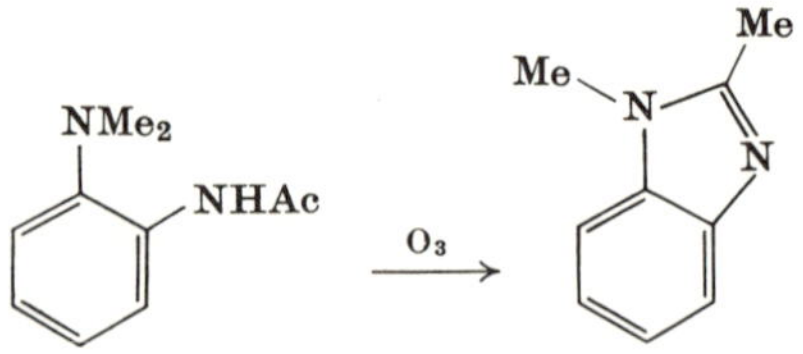

[74] G. H. Kerr and O. Meth-Cohn, *J. Chem. Soc.*, C 1369 (1971).

The intermediate acyldihydrobenzimidazoles (**143**) are known and are readily transformed into the benzimidazoles by peracid.[75] The detailed mechanism of the Polonovski reaction is still in dispute, but is unlikely to be very different from that proposed.[76, 77] The role of the ortho substituent as an internal acylating agent is underlined by the observation that the *o*-carboxylic acid (**146**) is quantitatively

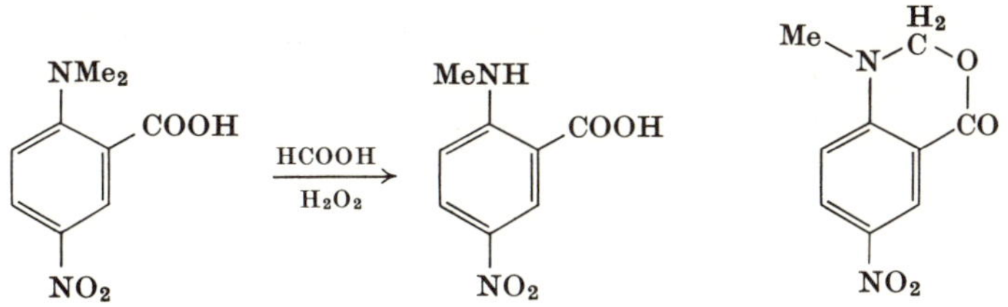

demethylated by performic acid presumably via the benzoxazone (**147**). By analogy the cyclization of the amines with peracid can be

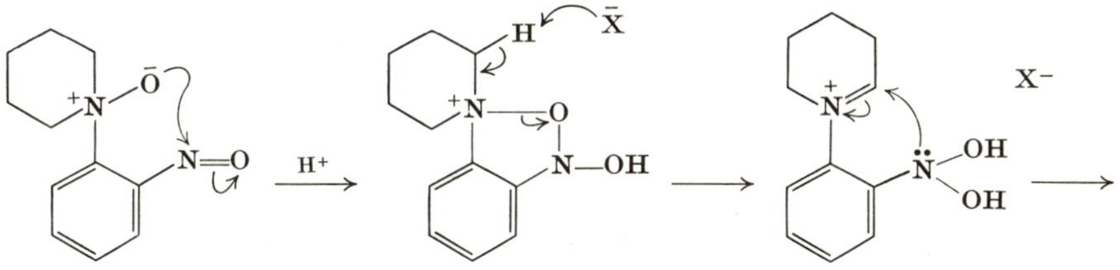

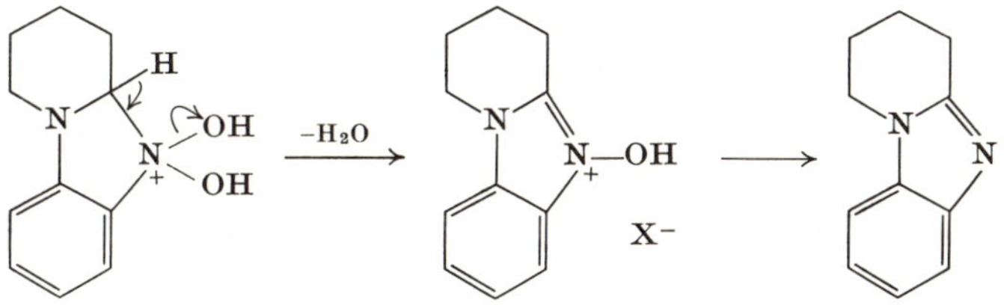

explained by the intermediacy of the *o*-nitroso-*N*-oxide (**148**) in which the nitroso group is the required electrophile. Subsequent cyclization

---

[75] O. Meth-Cohn, R. K. Smalley, and H. Suschitzky, *J. Chem. Soc.* 1666 (1963).
[76] R. Huisgen, F. Bayerlein, and W. Heydkamp, *Chem. Ber.* **94**, 2462 (1961).
[77] S. Oae, T. Kitao, and Y. Kitaoka, *J. Amer. Chem. Soc.* **84**, 3366 (1962).

(148) → (149) and dehydration gives the N-oxide salt (150) which is known to be deoxygenated under the action of a peracid.[41]

C. o-Azo Group

As in many other o-substituted t-amino systems, the interaction of the o-azo group was also discovered accidentally. Price[78, 79] observed that the ortho substituted azobenzenes (X and Y = OH, NH_2, NHMe, etc.) form stable 2:1 cobalt chloride complexes with anhydrous cobalt chloride in ethanol, which were of interest as dyestuffs. However, the corresponding complexes (152) from the bis(dimethylamino)-azobenzenes [(151) X = Y = NMe_2] were slowly transformed by heating in dry ethanol into the cobalt complex of 1-methylbenzimidazole (153) and o-aminodimethylaniline (154). Similar results were observed with the diethylamino analog [(151) X = Y = NEt_2] and their corresponding hydrochlorides. The azo compounds (151) reacted differently with hydrated cobalt chloride to give the benzimidazole (157), o-aminodimethylaniline, and cobalt hydroxide. The cobalt hydroxide arose presumably from the hydrated cobalt chloride

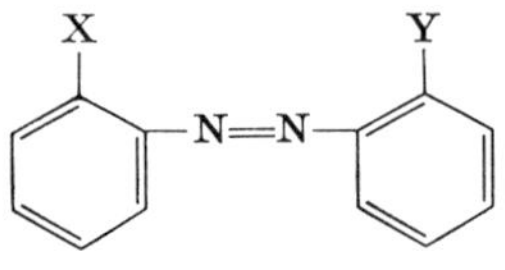

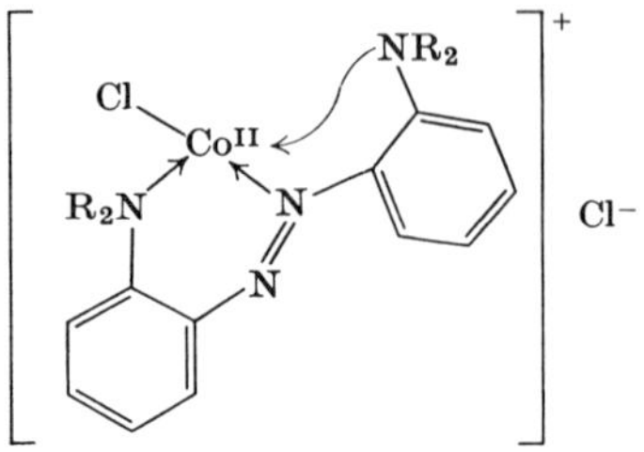

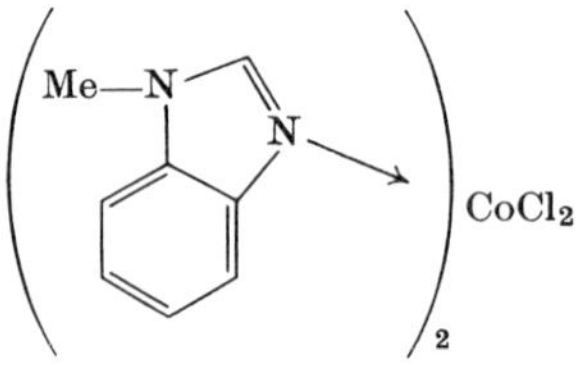

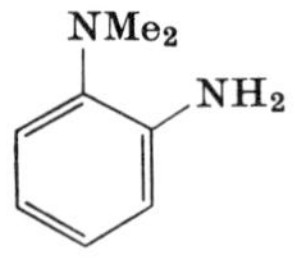

[78] R. Price, *J. Chem. Soc.*, A 521 (1967).
[79] R. Price, *J. Chem. Soc.*, A 2048 (1967).

Sec. III. C.]   THE *t*-AMINO EFFECT   253

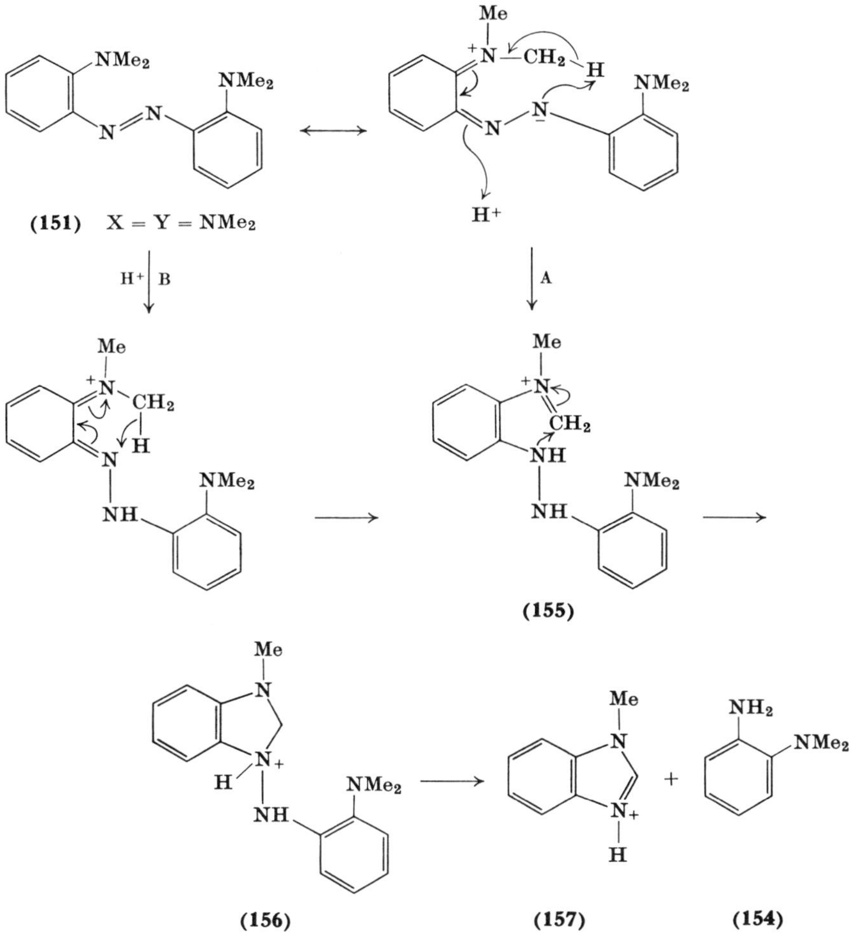

by loss of hydrochloric acid which, in turn, catalyzed the transformation of the azo compound to the benzimidazole.[78] This was further substantiated when it was found that ethanolic hydrochloric acid alone brought about cyclization. We classify this reaction mechanistically as analogous to the acid- and transition metal-catalyzed reactions of *o*-nitro-*t*-anilines (Section III,A). Similar cyclizations have been noted for the cyclic analogs [(**151**) $X = Y =$ pyrrolidino or piperidino] and in the case of 2,2′-bis(pyrrolidino)-azobenzene the reaction proceeds rapidly in aqueous hydrochloric acid in the cold.[32] Either of the two nitrogens may abstract the α-methylene proton (pathway A or B) to give the immonium ion (**155**) which can then cyclize to the dihydrobenzimidazole (**156**). The known hydride lability of this system (see later) accounts for the cleavage into benzimidazole (**157**) and amine (**154**). The role of the transition metal in bringing about the same result from the complex (**152**) can then be rationalized on the basis of a similar polarization of the *o*-dimethylaminoazo system to that effected by the acid.

Unexpected cyclizations were also observed during the attempted preparation of *o*-piperidinophenylhydrazine by Fischer's method in

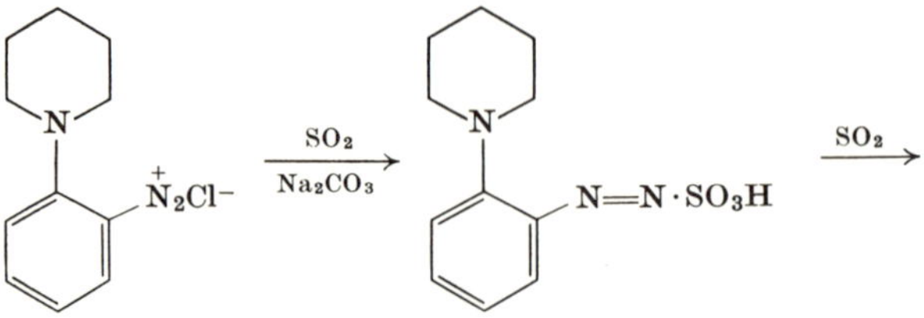

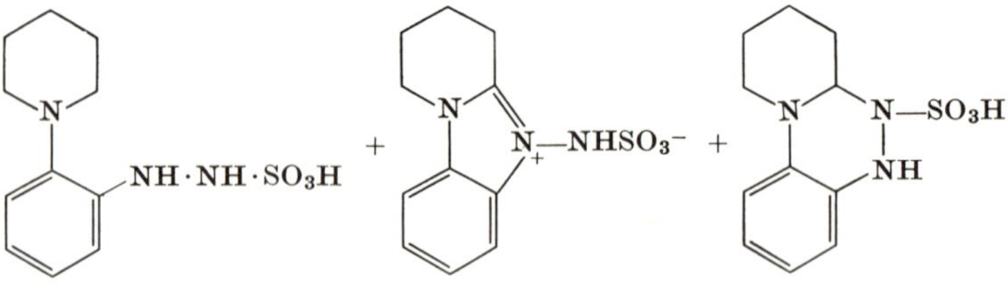

which the diazonium salt (**158**) was reduced with alkaline sulfur dioxide. Ainsworth, Meth-Cohn, and Suschitzky[80] observed various products in the course of this reaction. Brief treatment of the diazo compound (**158**) with sulfur dioxide gave the expected stable diazosulfonate (**159**). This compound, by the further action of sulfur dioxide (during which the solution became acid), gave a mixture of three compounds (**160–162**). Although the first was the expected product, the second and third clearly arose from a "*t*-amino effect." The benzimidazolium sulfamate (**161**) was also formed by the action of acid on the diazosulfonate (**159**). The presence of the cyclic products

[80] D. P. Ainsworth, O. Meth-Cohn, and H. Suschitzky, *J. Chem. Soc., C* 923 (1968).

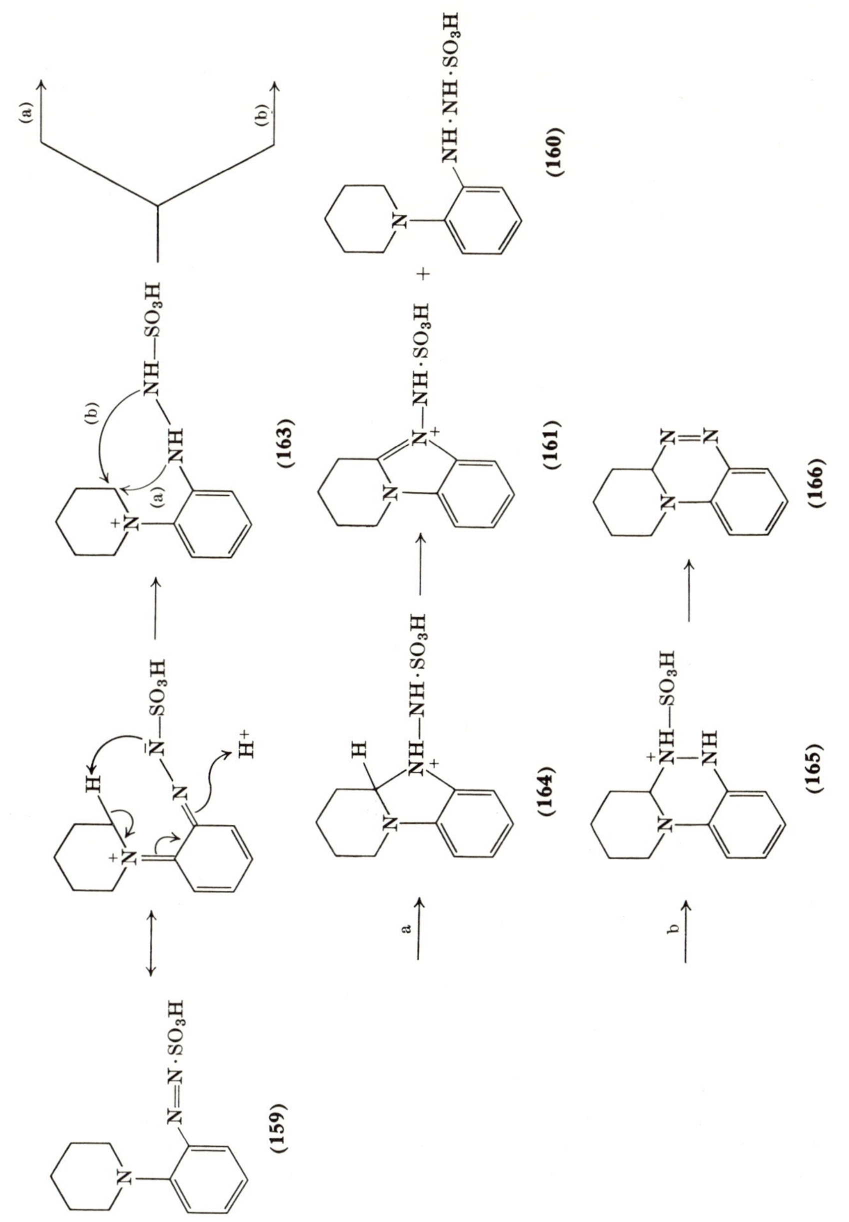

Sec. III. C.] THE *t*-AMINO EFFECT 255

is again rationalized by a mechanism employing the immonium ion (**163**), in which there are two nucleophilic centers, yielding either the 6-membered triazine (**165**) or the 5-membered imidazole (**164**). The subsequent disproportionation of the dihydrobenzimidazole (**164**) to the benzimidazole (**161**) and the hydrazinesulfonic acid (**160**) by the action of acid is in accord with the known properties of dihydrobenzimidazoles (see later).[81] The triazine sulfonic acid readily eliminated the elements of sulfurous acid on treatment with alkali to give the parent dihydrotriazine (**166**) which was characterized by spectroscopic methods.

D. o-AZOMETHINE GROUP (—N=C)

The acid-catalyzed cyclization of the o-azomethine function to a t-aniline was first observed in 1938 by Rudy and Cramer,[82] although this reaction was not recognized until recently and has been a subject of considerable research and of erroneous conclusions. The simple anils, exemplified by compound **167**, were shown more recently to

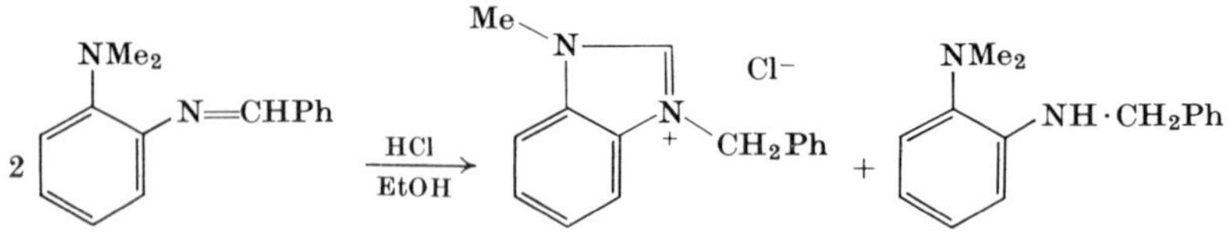

react rapidly and nearly quantitatively in the cold with acid to yield the benzimidazolium salt (**168**) and the amine (**169**) in equal amounts.[81, 83–85] A closer investigation of this reaction[81] revealed

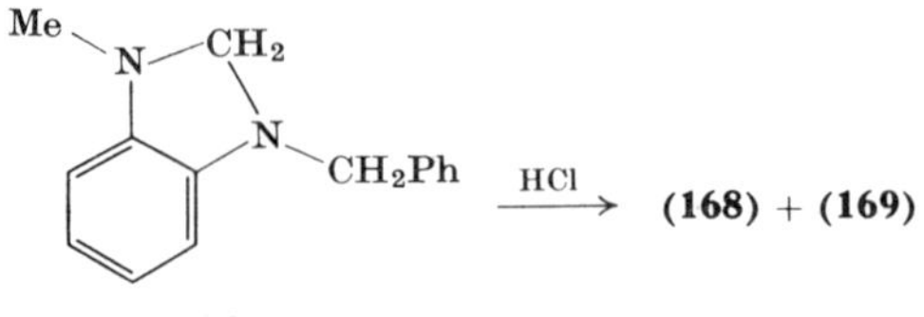

[81] R. K. Grantham and O. Meth-Cohn, *J. Chem. Soc.*, C 1444 (1969).
[82] H. Rudy and K.-E. Cramer, *Ber. Deut. Chem. Ges.* **71**, 1234 (1938).
[83] O. Meth-Cohn and M. A. Naqui, *Chem. Commun.* 1157 (1967).
[84] R. K. Grantham and O. Meth-Cohn, *Chem. Commun.* 500 (1968).
[85] R. K. Grantham, O. Meth-Cohn, and M. A. Naqui, *J. Chem. Soc.*, C 1438 (1969).

that the reaction proceeded first to a dihydrobenzimidazole (**170**), which rapidly disproportionated with acid in accord with the known hydride lability of these compounds.[86] This second reaction normally proceeds more rapidly than the first, rendering the intermediate (**170**) nonisolable. However, with the pyrrolidines (**171**), cyclization was rapid, but the disproportionation of the dihydrobenzimidazole (**172**)

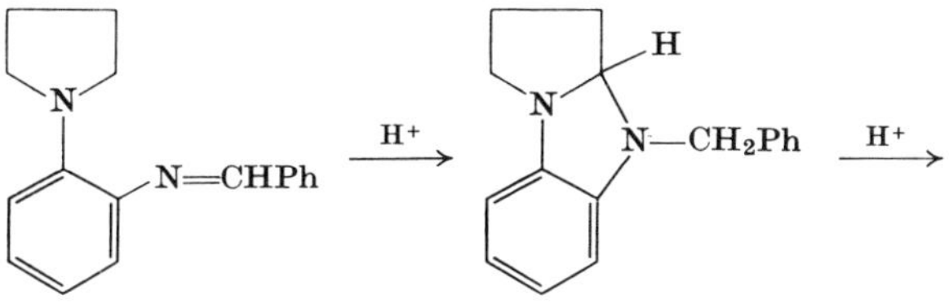

was slow (presumably owing to the greater difficulty in aromatizing the imidazoline ring because of the fusion of two 5-membered rings), thus allowing isolation of the dihydrobenzimidazoles in high yield. The intramolecular nature of the cyclization was demonstrated both by the nonincorporation of deuterium (except at the NH group) when cyclization was conducted in deuteriated solvents and by the formation of the deuteriated dihydrobenzimidazole (**176**) from the corresponding anil (**175**) on treatment with acid in ethanol solution.[85] The transfer of a deuteron from the α-pyrrolidine position to the benzylic methylene had occurred. This was confirmed by kinetic studies since the deuteriated anil (**175**) cyclized nearly six times slower than the protium analog.[87] Furthermore, the reaction proved the bimolecular course of the disproportionation step, as had been proposed,[86] because the amine (**178**) was fully deuteriated at the α-positions of the pyrrolidine ring.

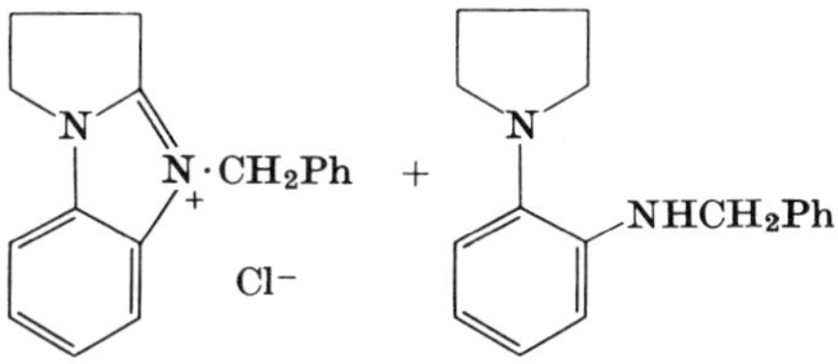

[86] A. V. El'tsov, *J. Org. Chem. USSR* **3**, 191 (1967).
[87] D. J. Barraclough, O. Meth-Cohn, and E. Warburton, unpublished results.

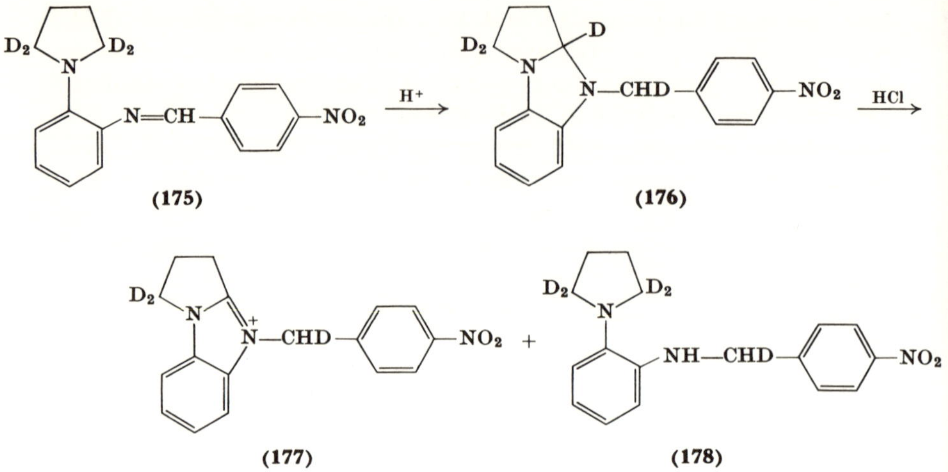

Thus, the anil may be seen to react by way of its mesomeric form (**179**) involving proton abstraction by the aldehyde CH group.

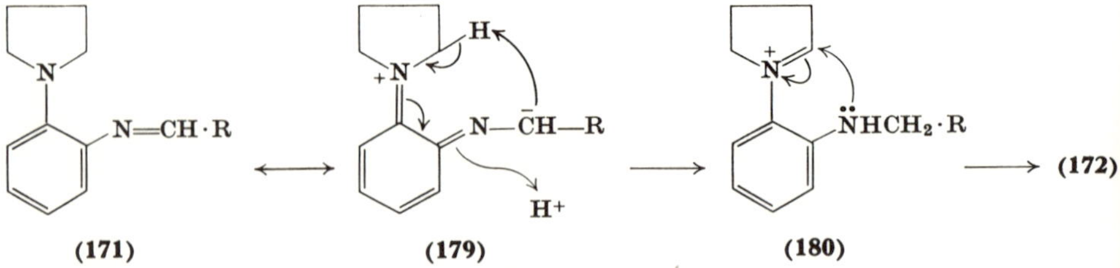

Subsequent cyclization of the immonium ion (**180**) gives the dihydrobenzimidazole (**172**).

The reaction of Rudy and Cramer referred to earlier involved the preparation of the alloxan anil (**183**), analogous to the known blue-black para isomer (**184**).[88] However, unlike the latter compound, the major product was pale yellow and did not revert to the starting materials (**181**) and (**182**) with acid. Rudy and Cramer decided that the compound was the anil (**183**) and that its unusual properties were due to the ortho substituent.[82, 89–91] They observed[91] a second product

---

[88] O. Piloty and K. Finckh, *Ann. Chem.* **333**, 37 (1904).
[89] H. Rudy and K.-E. Cramer, *Ber. Deut. Chem. Ges.* **72**, 227 (1939).
[90] H. Rudy and K.-E. Cramer, *Oesterr. Chem. Ztg.* **42**, 329 (1939).
[91] H. Rudy and K.-E. Cramer, *Ber. Deut. Chem. Ges.* **72**, 728 (1939).

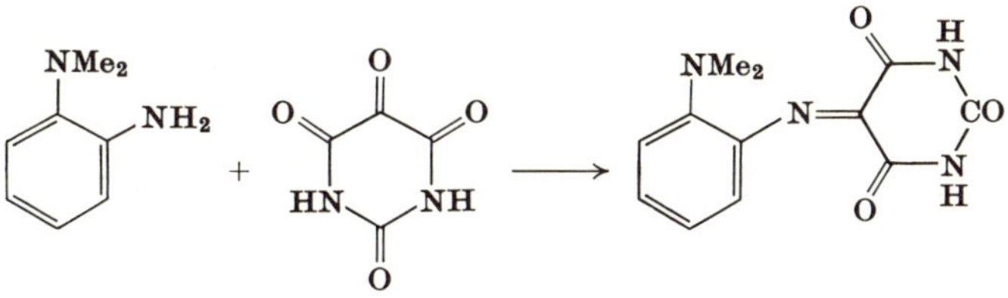

THE t-AMINO EFFECT

(181) **(182)** **(183)** 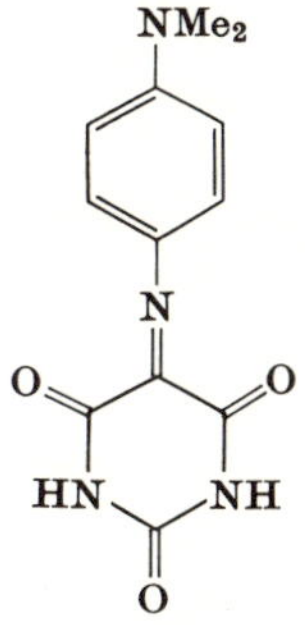 **(184)**

which was also produced by aerial oxidation of the supposed anil (183) to which they assigned the highly problematical structure (185), and

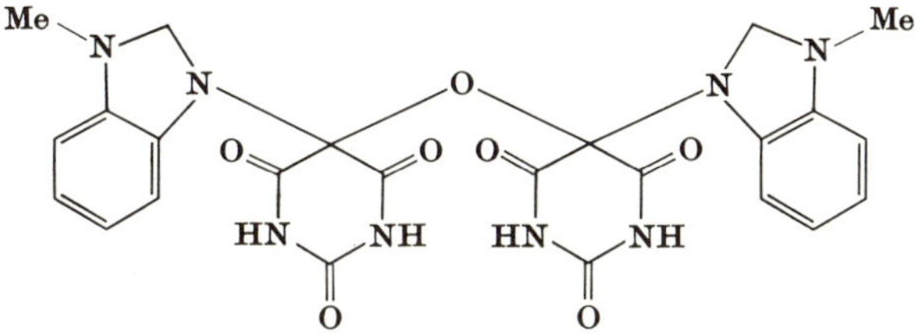

(185) 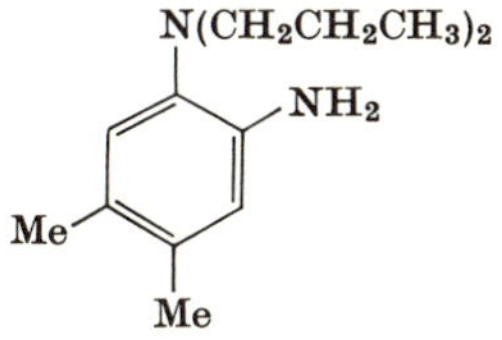 **(186)**

noted that other *o*-amino-*t*-anilines (e.g., **186**) formed this type of product exclusively. Other workers[92,93] prepared analogous compounds from *o*-amino-*t*-anilines and proposed different structures,[94,95] which were only resolved more recently by spectral methods. Clark-Lewis and co-workers[94–100] showed that the supposed anil was, in

[92] R. B. Barlow, *J. Chem. Soc.* 2225 (1951).
[93] R. B. Barlow, H. R. Ing, and I. M. Lewis, *J. Chem. Soc.* 3242 (1951).
[94] F. E. King and J. W. Clark-Lewis, *J. Chem. Soc.* 172 (1953).
[95] J. W. Clark-Lewis, J. A. Edgar, J. S. Shannon, and M. J. Thompson, *Austr. J. Chem.* **17**, 877 (1964).
[96] F. E. King and J. W. Clark-Lewis, *J. Chem. Soc.* 3080 (1951).
[97] J. W. Clark-Lewis, J. A. Edgar, J. S. Shannon, and M. J. Thompson, *Austr. J. Chem.* **18**, 907 (1965).
[98] J. W. Clark-Lewis, *Chem. Ind. (London)* 241 (1966).
[99] J. W. Clark-Lewis, K. Moody, and M. J. Thompson, *Austr. J. Chem.* **23**, 1249 (1970).
[100] J. W. Clark-Lewis and K. Moody, *Austr. J. Chem.* **23**, 1229 (1970).

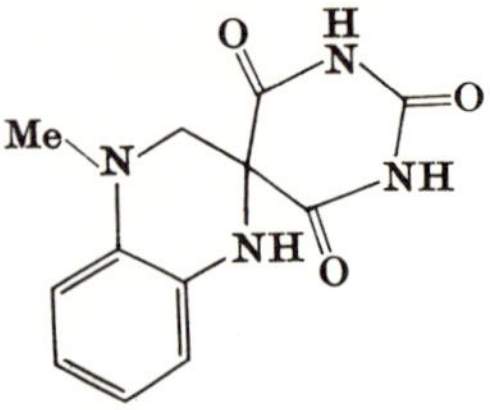

(187) 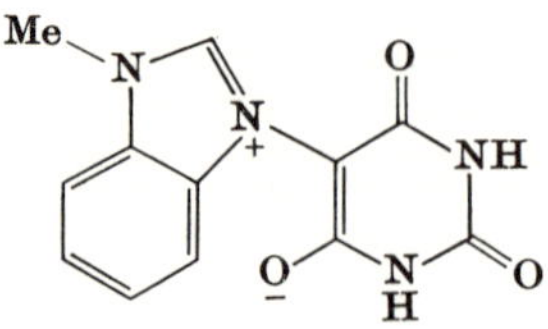 **(188)** 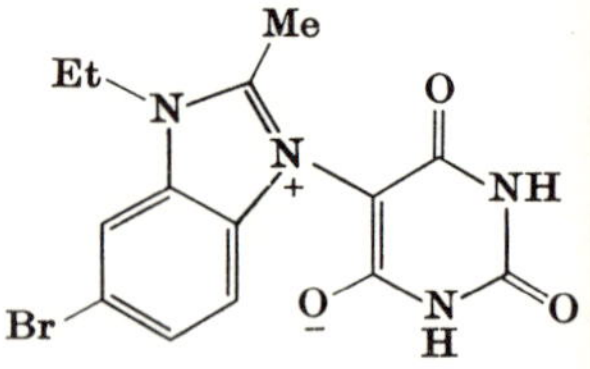 **(189)**

fact, a spiroquinoxaline (**187**), and they and others[101] assigned to the second product the betaine structure (**188**). The latter system was

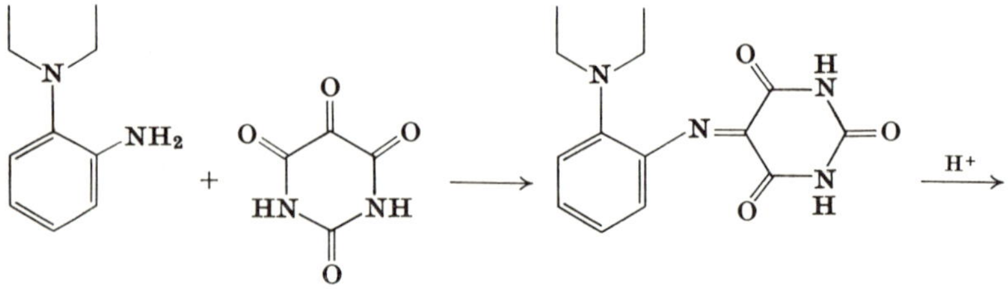

(190) **(182)** **(191)**

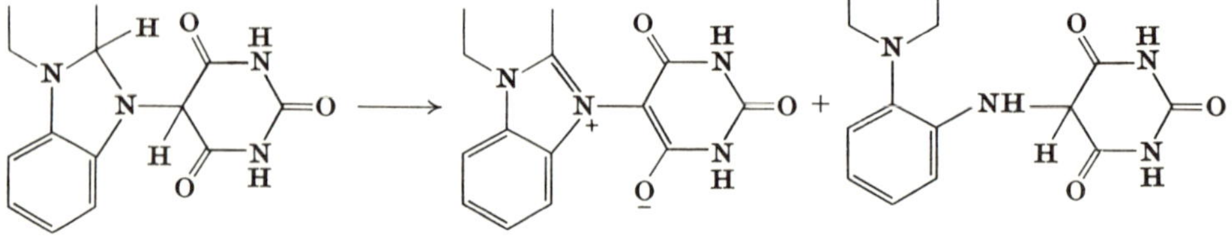

(192) **(193)** **(194)**

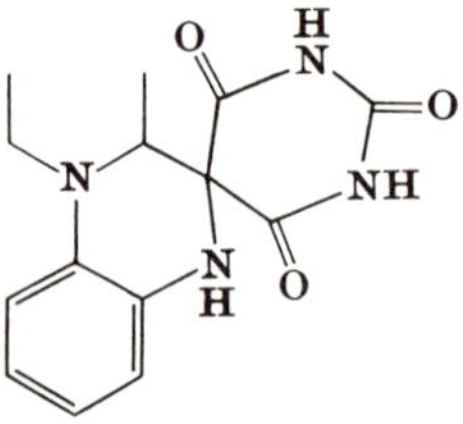

(195)

[101] R. K. Grantham, Ph.D. Thesis, University of Salford, 1969.

Sec. III. D.] THE t-AMINO EFFECT 261

conclusively verified by X-ray diffraction of compound **189** derived from 2-amino-5-bromo-N,N-diethylaniline and alloxan.[102]

In the light of the acid-catalyzed cyclizations of anils described above, the benzimidazolium betaine formation may be rationalized (with alloxan acting as the required acid catalyst). It has been shown[99,100] that oxidation of the amine (**194**) with alloxan (which is a good oxidizing agent) gives the cyclized products (**193**) and (**195**), presumably via the anil (**191**) and the dihydrobenzimidazole (**192**). However, the quinoxaline formation (**195**) must require a different mechanism, and would appear to be a feature of a ketone flanked by carbonyl groups. To throw light on this reaction,[85] the anil (**197**), derived from N-methylisatin (**196**), was treated with acid. The

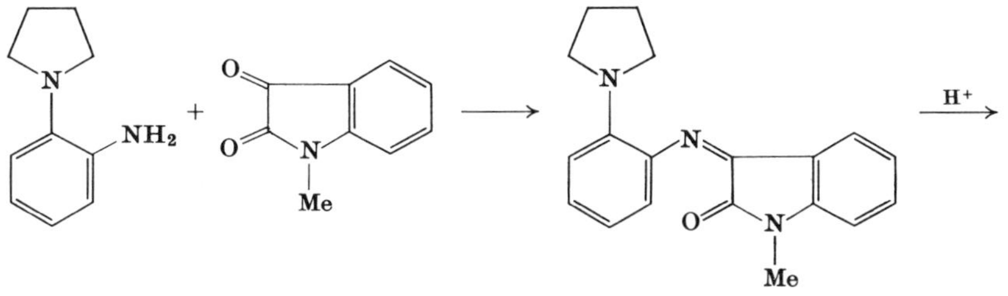

(**196**)   (**197**)

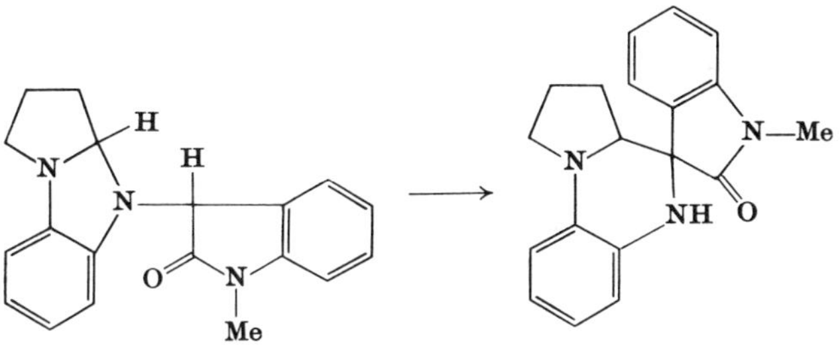

(**198**)   (**199**)

dihydrobenzimidazole (**198**) readily formed, and on further acid treatment gave the spiroquinoxaline (**199**), in agreement with the results obtained with alloxan. The o-dimethylaminoanil (**200**) underwent a similar sequence of reactions[101] to give the benzimidazolium salt (**202**) and the amine (**203**), but no quinoxaline.

[102] B. W. Mathews, *Acta Crystallog.* **18**, 151 (1965).

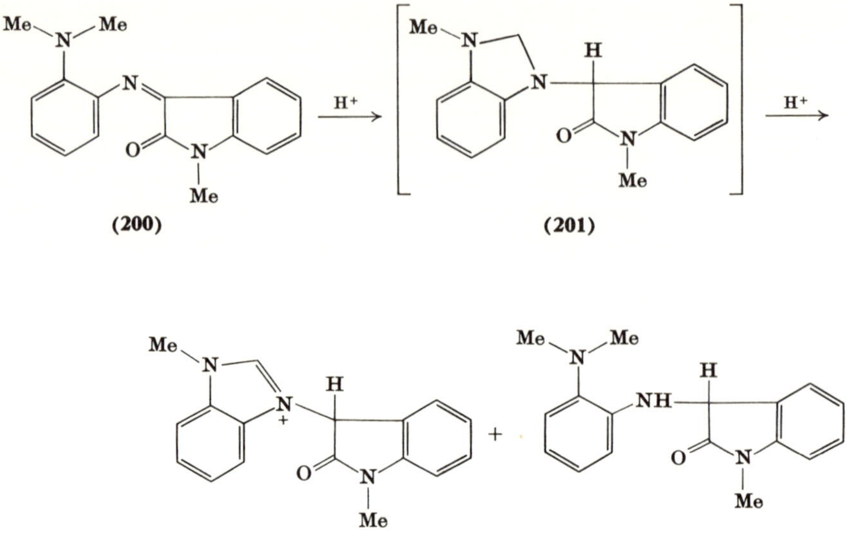

(200) → [(201)] →ᴴ⁺

(202) + (203)

The tendency to form a quinoxaline is ascribable to a reduced hydride lability in the intermediate dihydrobenzimidazole (e.g., **198**). 2-Alkyl groups should enhance this lability inductively, whereas annelated 5-membered rings should reduce it. Clark-Lewis[99] has recently shown that 2-amino-N-alkyl-N-mesitylanilines (**204**) (in which hydride lability is reduced by the presence of bulky aryl

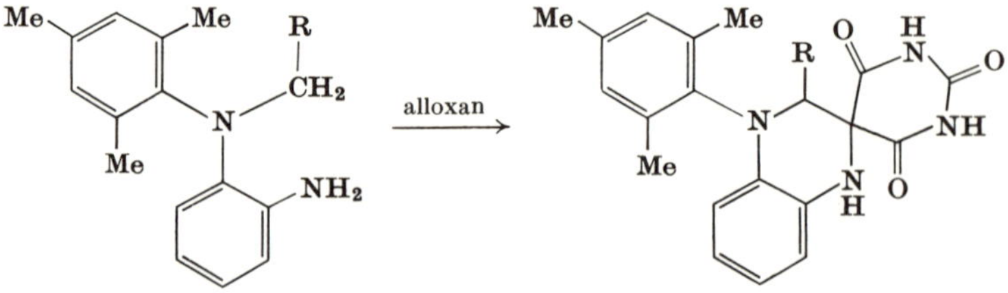

(204) →alloxan (205)

substituents) yield exclusively spiroquinoxalines (**205**) in further support of this proposition. Thus, we have the overall scheme (**183**) → (**187**) and (**188**) by way of the mesomer (**206**).

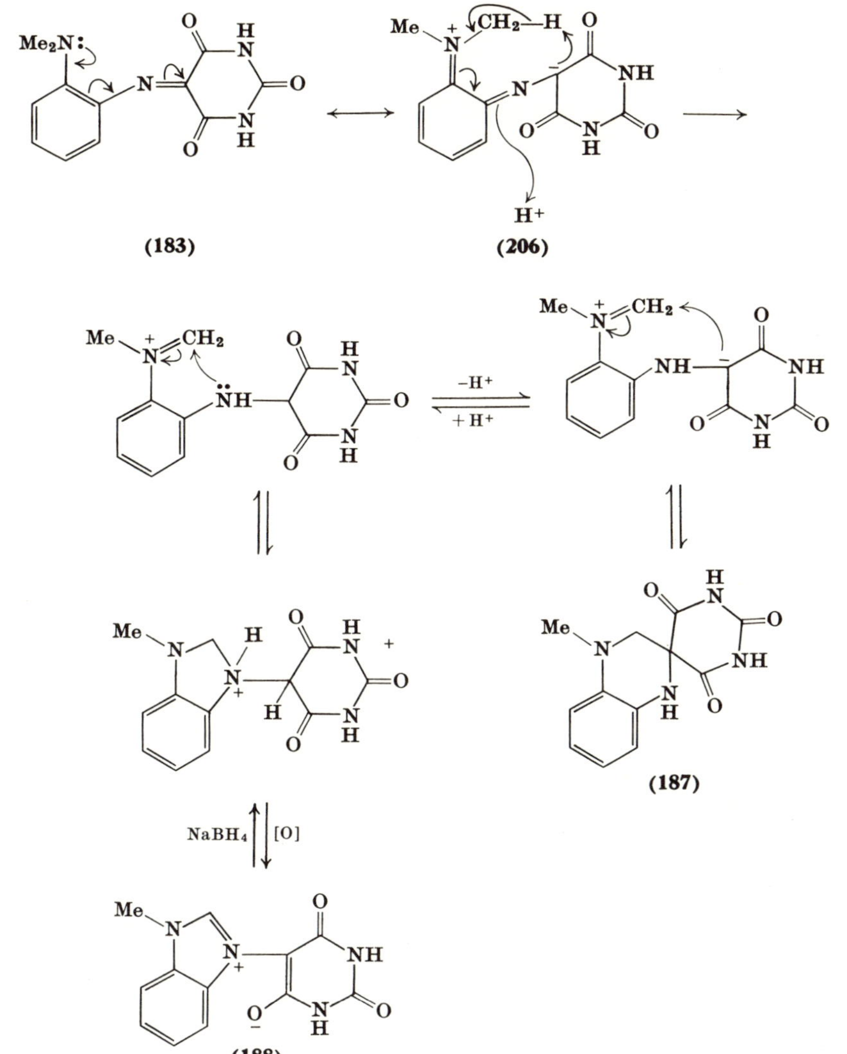

Sec. III. E.] THE *t*-AMINO EFFECT 263

(183) (206)

(187)

(188)

E. *o*-Carbonyl (C=O) and *o*-Imine (C=N) Groups

Although no examples of aldehyde or ketone interaction with a *t*-amino group are known in the benzene series, 1-*t*-aminoanthraquinones are reported to undergo unexpected reactions. Thus,

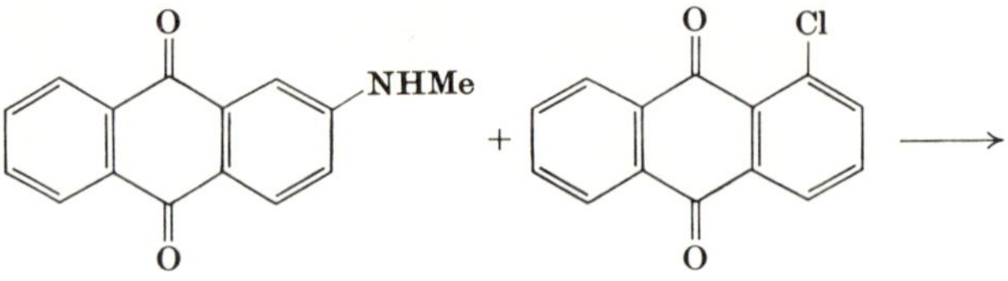

Bradley and Leete[103] in 1951, observed that attempts to prepare 1,2′-dianthraquinolyl methylamine (**209**) by the condensation of 1-chloroanthraquinone (**208**) with 2-methylaminoanthraquinone (**207**) resulted in formation of the secondary amine (**210**) and formaldehyde. Condensation of a series of secondary amines (**211**) with 1-chloroanthraquinone (**208**) at temperatures of 160°–190° gave analogous results.[104] By contrast dimethylamine and piperidine condensed normally with 1-chloroanthraquinone at water-bath temperatures.

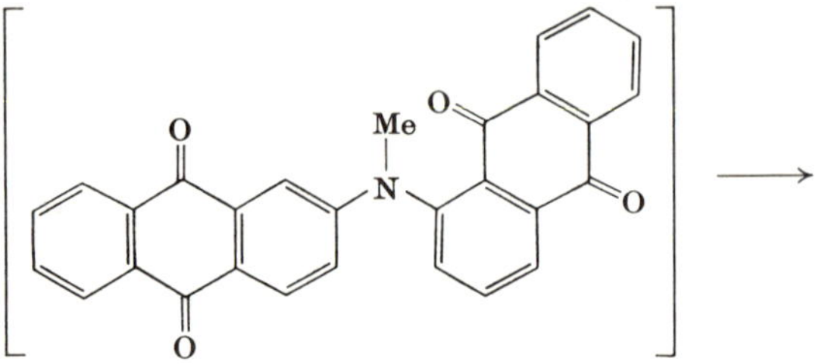

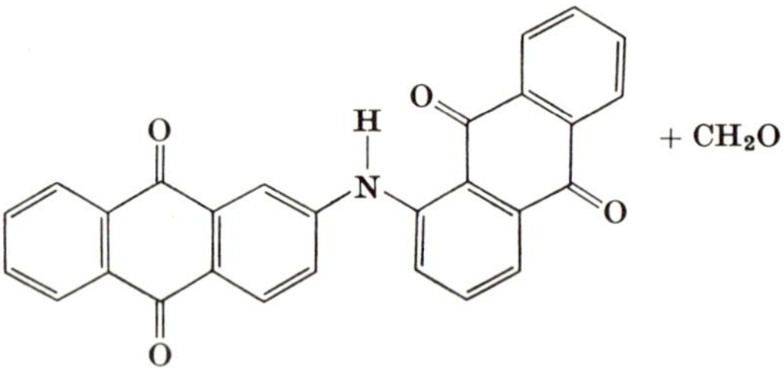

Fokin and his co-workers[105, 106] demonstrated that the cleavage

---

[103] W. Bradley and E. Leete, *J. Chem Soc.* 2147 (1951).
[104] W. Bradley and R. F. Maisey, *J. Chem. Soc.* 247 (1954).
[105] E. P. Fokin, O. I. Andrievskaya, and V. V. Russkikh, *Izv. Sib. Otd. Akad. Nauk. SSSR, Ser. Khim. Nauk* 103 (1962).
[106] V. V. Russkikh and E. P. Fokin, *Izv. Sib. Otd. Akad. Nauk. SSSR, Ser. Khim. Nauk.* 91 (1966).

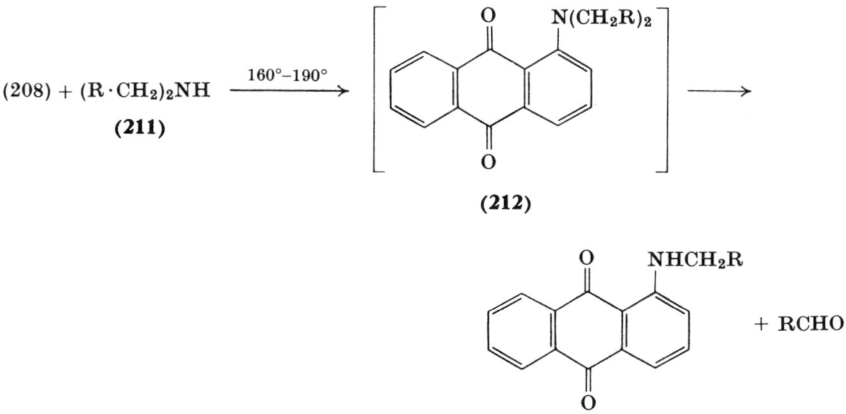

(208) + (R·CH₂)₂NH →[160°–190°] [structure (212): 1-N(CH₂R)₂-anthraquinone] →

(211)

→ [structure (213): 1-NHCH₂R-anthraquinone] + RCHO

R = Me, n-Pr, or Ph

was temperature-dependent and that heating of a pyridine solution of a 1-dialkylaminoanthraquinone, followed by treatment with air, gave a 1-monoalkylaminoanthraquinone and an aliphatic aldehyde.[107] Similar treatment of 1-polymethyleniminoanthraquinones (214)

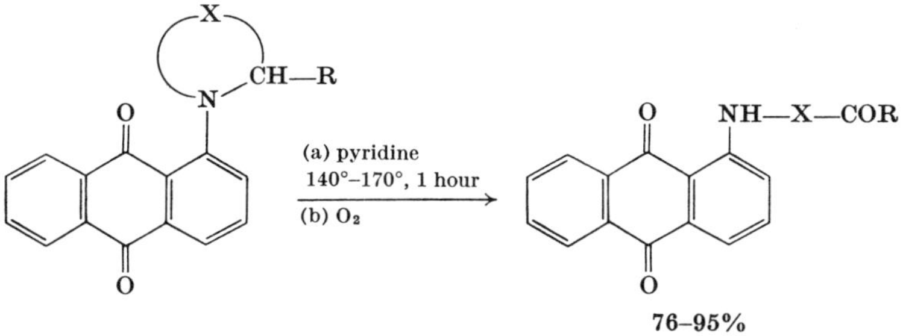

(214) [1-(cyclic amine)anthraquinone with CH–R] →[(a) pyridine 140°–170°, 1 hour; (b) O₂]→ (215) [1-NH–X–COR anthraquinone] 76–95%

X = (CH₂)₃, (CH₂)₄, (CH₂)₅, (CH₂)₃, (CH₂)₄, CH₂CH₂OCH₂
R = H,   H,   H,   Et,   Me,   H

resulted in the formation of the ring-opened aldehyde or ketone (215) in high yield.[108] The reactions were conducted in a sealed tube and it was observed that the deep-red starting material became yellow

[107] E. P. Fokin and V. V. Russkikh, *Izv. Sib. Otd. Akad. Nauk. SSSR, Ser. Khim. Nauk* 126 (1965).
[108] E. P. Fokin and V. V. Russkikh, *Zh. Org. Khim.* **2**, 907 (1966); *J. Org. Chem. USSR* **2**, 902 (1966).

on completion of heating and rapidly became red again on opening the tube. The unstable yellow intermediates could be trapped as acetyl derivatives by treatment of the pyridine solution with acetic anhydride.[109] Thus, 1-piperidinoanthraquinone gave the oxazine (217) which with acid and air gave the aldehyde [(215) R = H]. Reductive acetylation of the aldehyde gave the same oxazine.

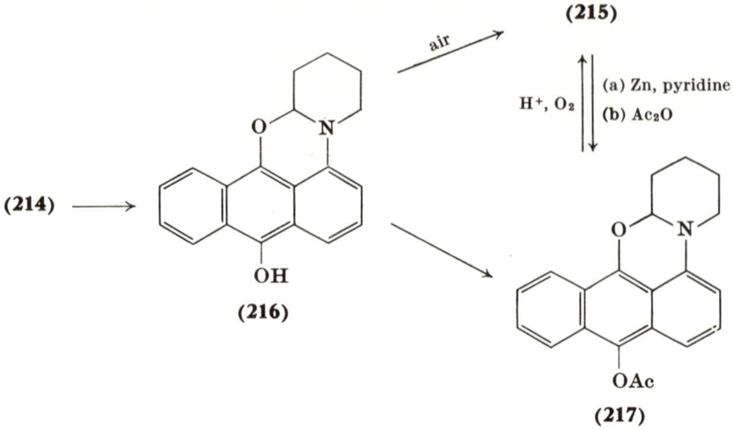

The dealkylation or ring opening of 1-*t*-aminoanthraquinones may thus be viewed mechanistically as another example of the "*t*-amino effect" and can be rationalized as shown below (214)–(206).[110]

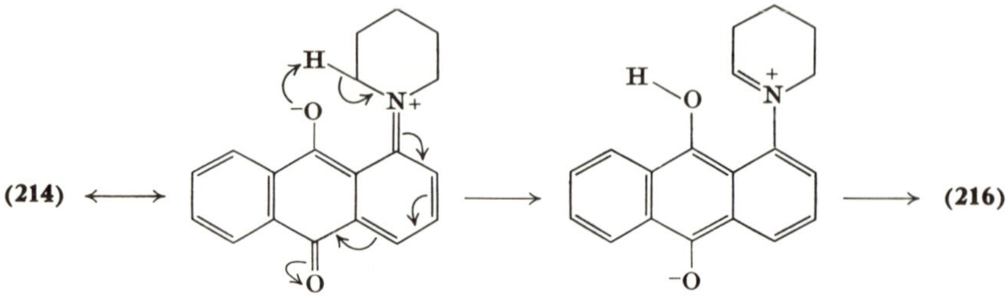

The 1-*t*-aminoanthraquinones are also photolabile giving numerous products.[111] In fact, Fokin[112] was able to demonstrate that plastics

[109] E. P. Fokin and V. V. Russkikh, *Zh. Org. Khim.* **2**, 912 (1966); *J. Org. Chem. USSR* **2**, 907 (1966).
[110] J. Lynch and O. Meth-Cohn, *Tetrahedron Lett.* 161 (1970).
[111] J. Lynch and O. Meth-Cohn, unpublished results.
[112] I. Y. Kalontarov, E. P. Fokin, and N. N. Kiseleva, *Kh. Prikl. Khim.* **41**, 2809 (1968).

dyed with 1-*t*-aminoanthraquinone dyes underwent color changes in light ascribable to a similar sequence of reactions. The claim that such changes in solution or in a polymer matrix are the result of photoreduction must, therefore, be viewed with caution, particularly when the results are based on ultraviolet absorption data, without identification of products.[113,114]

Fokin[115] extended his work to 2-substituted 1-*t*-aminoanthraquinones and concluded erroneously that 2-acetylamino- and 2-benzoylamino-1-piperidinoanthraquinone [(**218**) R = Me or Ph] gave

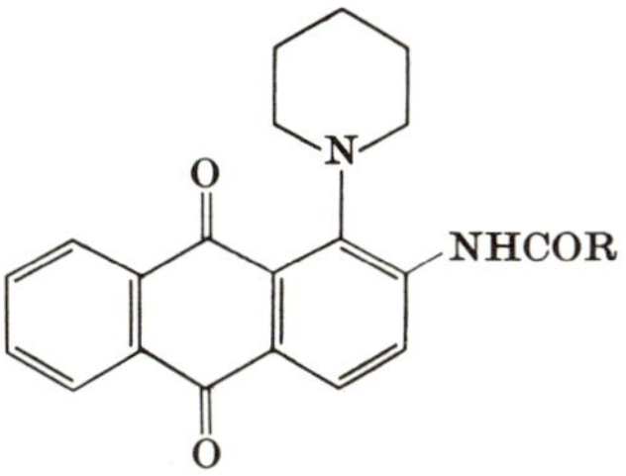

(**218**)

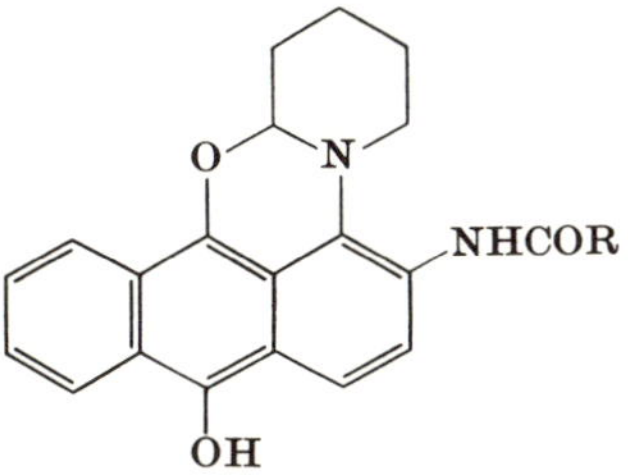

(**219**)

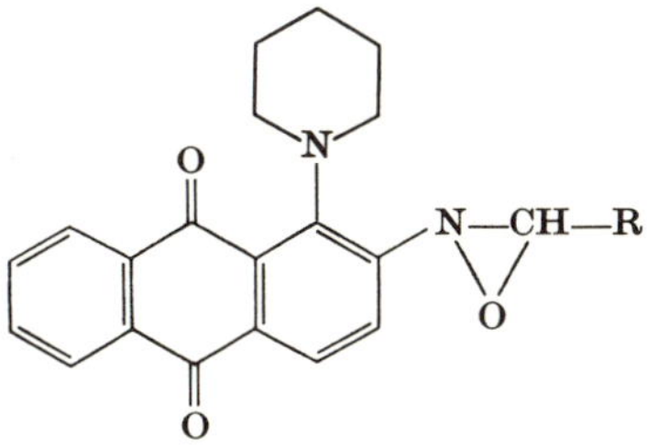

(**220**)

R = Me or Ph

the 2-oxaziridino products (**220**) and not an oxazine (**219**) or its ring-opened derivative. However, on reinvestigation[110] the product was found to be a dihydroimidazoanthraquinone (**221**) and its formation was rationalized as follows.

[113] G. S. Egerton, N. E. N. Assaad, and N. D. Uffindell, *J. Soc. Dyers. Colour.* **83**, 409 (1967).
[114] G. S. Egerton and N. E. N. Assaad, *J. Soc. Dyers Colour.* **86**, 203 (1970).
[115] E. P. Fokin and V. Ya. Denisov, *Zh. Org. Khim.* **4**, 1486 (1968); *J. Org. Chem. USSR* **4**, 1428 (1968).

(218) ⟶

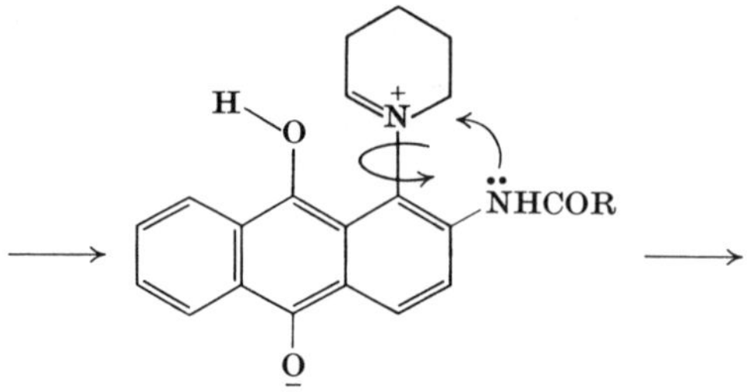

(222)

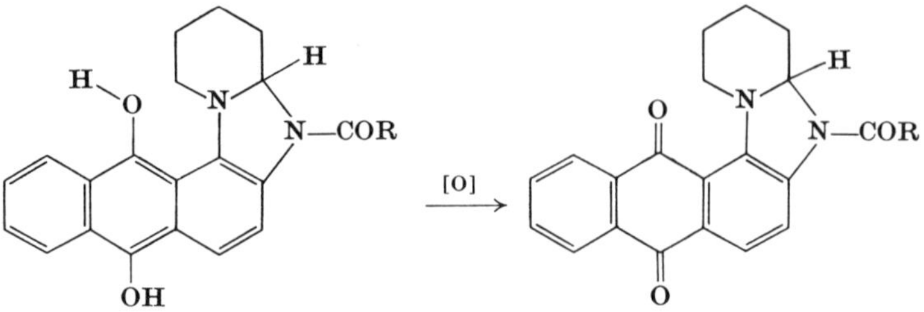

(223) [O]⟶ (221)

Proton abstraction by the *o*-carbonyl group, as before, yields the immonium ion (222), which then interacts preferably with the acylamino group rather than the OH group to give the reduced product (223). It is significant that the product in the sealed tube is bright red and rapidly becomes purple on exposure to the atmosphere, presumably because oxidation to the anthraquinone occurs only in air.

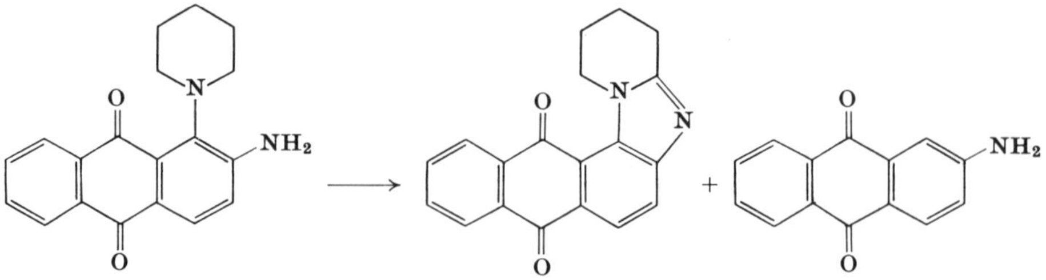

(224) (225) (226)

In a similar way 1-piperidino-2-aminoanthraquinone (224) gives the benzimidazole (225) and, surprisingly, 2-aminoanthraquinone (226).[110] This also applies to other dialkylamino derivatives.[111] In the "unambiguous synthesis" of the supposed oxaziridine, the anil (**227**) was treated with peroxyacetic acid, conditions which with the benzene analog (**174**) gave the benzimidazole (**130**) (cf. p. 247). In fact, we were able to isolate only traces of the imidazole (**225**) under these conditions, together with several other products, none of which were the true oxaziridine.[110] Recently Fokin and Denisov[116] has shown that thermal or hydrolytic decomposition of the products (**223**) lead to the benzimidazole (**225**).

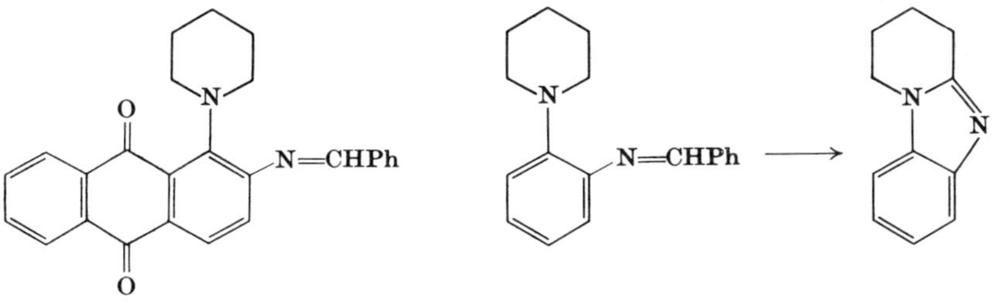

(**227**)   (**174**)   (**130**)

A further development of this reaction was the recent observation that the dimethylaminoanthraquinones (**228**) react differently from the cyclic analogs in that both the oxazole [(**229**) R = Ph] and the

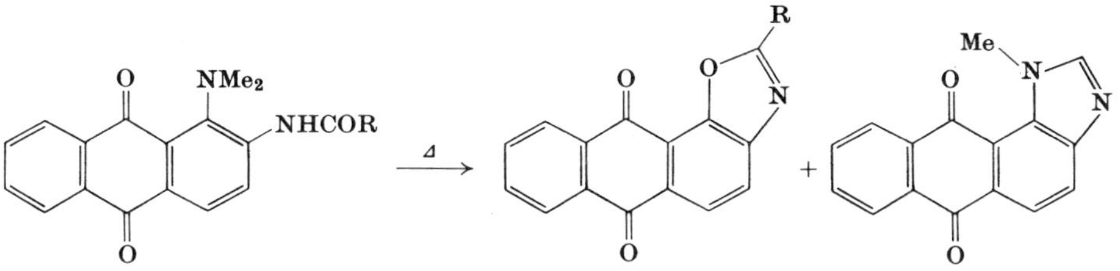

(**228**)   (**229**)   (**230**)

R = Me or Ph

imidazole (**230**) are formed from the acetyl derivative [(**228**) R = Me], while the benzoyl derivative gives only the oxazole [(**229**)) R = Ph].[117] The oxazole formation, involving displacement of the dimethylamino

[116] E. P. Fokin and V. Ya. Denisov, *Khim. Geterotsikl. Soedin.* **321** (1970).
[117] E. P. Fokin, Y. Denisov, and L. N. Anishina, *Izv. Sib. Otd. Akad. Nauk. SSSR, Ser. Khim. Nauk.* **83** (1970).

group, is analogous to the well-documented oxazole formation from 1-halo-2-acylaminoanthraquinones and related systems.[118-121]

[NOTE ADDED IN PROOF: We find[111] that while the benzoyl derivative (**228**) R = Ph gave the corresponding oxazole (**229**), the acetyl derivative gave the oxazole (**229**) R = Me together with the imidazole (**230a**) and no imidazole (**230**). The formyl derivative

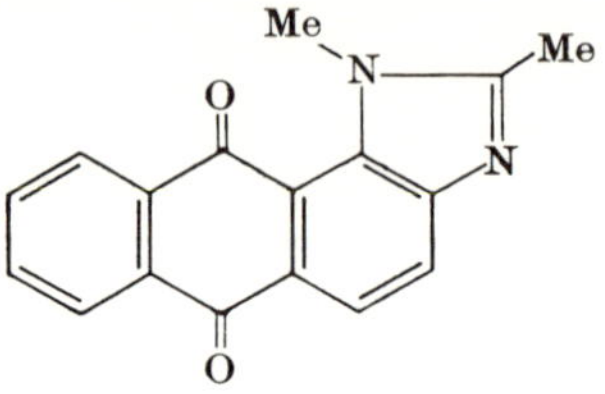

(**230a**) 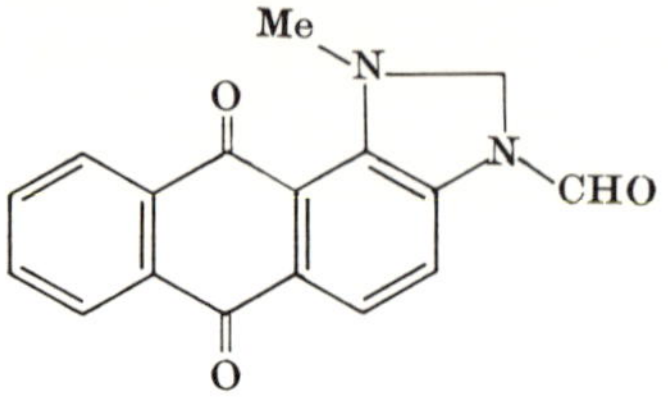 (**230b**)

(**228**) R = H gave no oxazole [(**229**) R = H] but the formyldihydroimidazole (**230b**) together with the imidazole (**230**). It is evident that the course of these reactions is critically dependent upon steric factors.]

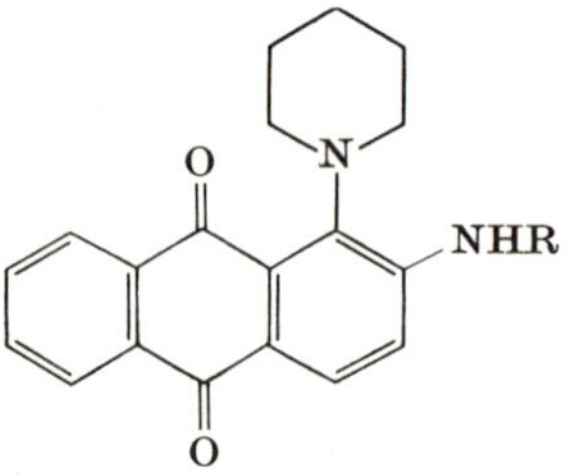

(**231**) 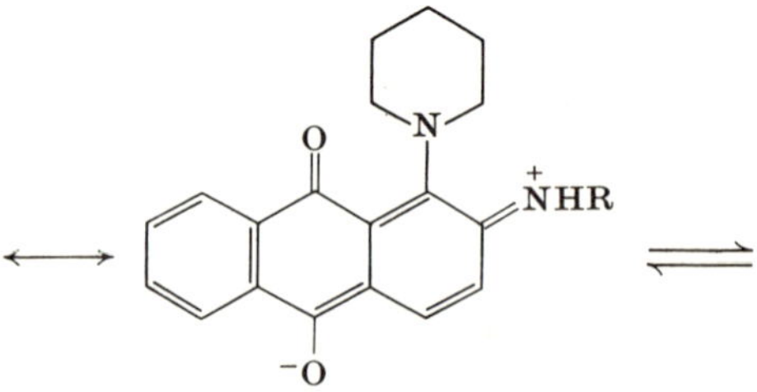 (**232**)

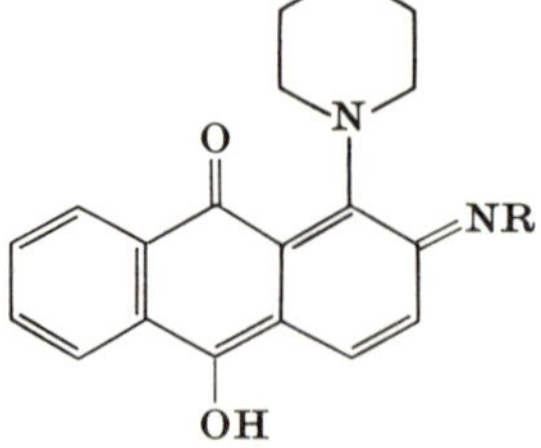

(**233**)

[118] F. Ullmann and W. Junghans, *Ann. Chem.* **399**, 335 (1913).
[119] M. Fries and P. Ochwat, *Ber. Deut. Chem. Ges.* **56**, 1291 (1923).
[120] German Patents 286,093 and 286,094 (1915); *Chem. Zentr.* **3**, 567 (1915).
[121] F. I. Carrol and J. T. Blackwell, *Chem. Commun.* 923 (1969).

Sec. III. E.] THE *t*-AMINO EFFECT 271

The 2-amino- and 2-acylamino-1-*t*-aminoanthraquinones (**231**) are also photolabile, giving imidazoles of type **225** or acyldihydroimidazoles (**221**).[111] This may suggest that the photoreaction follows a different mechanism from that of the thermal process, and is more related to the reactions of benzoquinones described below. Thus, mesomeric or tautomeric forms such as **232** and **233** may be invoked to explain the photoreactions, by analogy with the corresponding benzoquinone imines described below.

The photolability of 1-*t*-aminoanthraquinones is paralleled in 2-substituted benzo- (**234**) and naphthoquinones (**235**). Although these systems are not strictly isosteric with the foregoing *o*-*t*-amino compounds, they undergo mechanistically related reactions. Indeed, it has been experimentally demonstrated that the benzoquinones (**234**) exhibit considerable quadrupolar character due to the contribution of canonical forms such as **236**.[122-124]

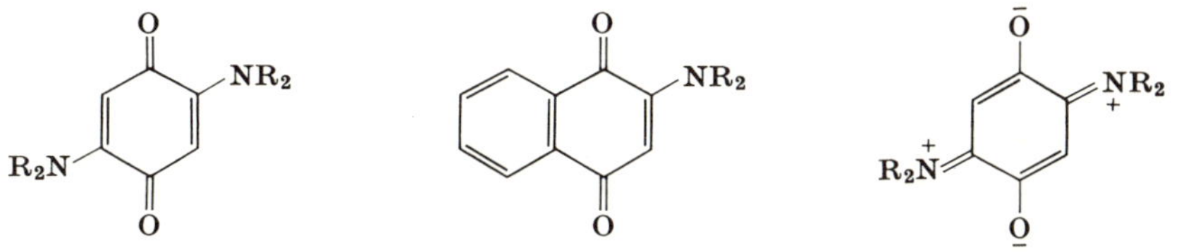

The dealkylation of 2-(*N*-alkylanilino)naphthoquinones (**237**)–(**238**) in sunlight in an inert atmosphere was first observed by

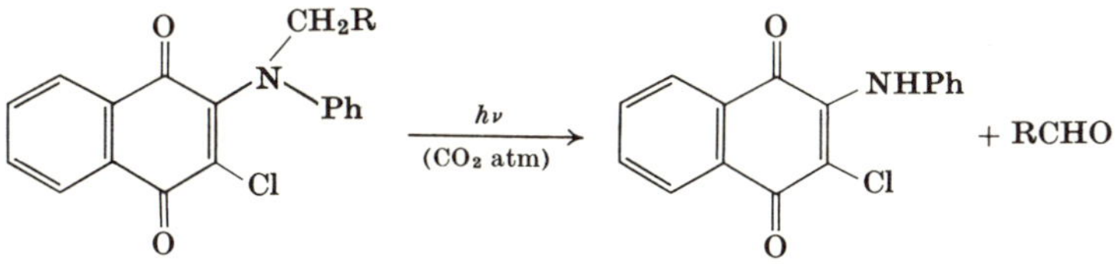

---

[122] S. Kulpe, *Tetrahedron* **26**, 2899 (1970).
[123] D. W. Cameron, R. G. F. Giles, and M. H. Pay, *Tetrahedron Lett.* 2047 (1970) and references cited therein.
[124] D. W. Cameron, R. G. F. Giles, and M. H. Pay, *Tetrahedron Lett.* 2049 (1970).

Vladimirtsev[125] and co-workers and further examined by Fokin and Prudchenko.[126]

Irradiation of the piperidino and the morpholino derivatives (239) at 70°–100° produced the enamines (240), which on acid treatment gave the corresponding aminoaldehyde (241).[127]

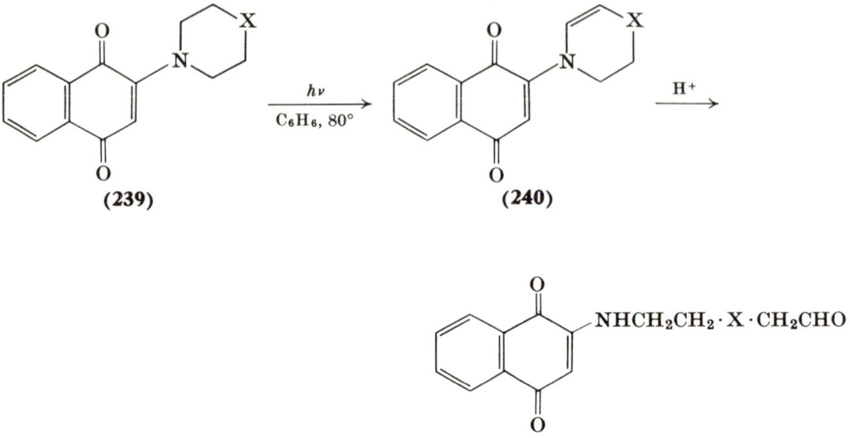

X = CH$_2$ or O

The corresponding reactions in the benzoquinone field were discovered by Cameron and Giles[128] in 1965, who reported rapid decoloration of the violet-black quinones (242) on exposure to sunlight. The product was the benzoxazoline (243) which with acid treatment followed by oxidation, gave the quinone (245).[129] A second irradiation process yielded the bis(monomethylamino)benzoquinone (246). Analogous benzoxazolines (248) were isolated from the pyrrolidino-, the piperidino-, and the perhydroazepinobenzoquinones (247).

[125] I. F. Vladimirtsev, I. Ya. Postovskii, and L. F. Trefilova, *Zh. Obsch. Khim.* **24**, 181 (1954); *J. Gen. Chem. USSR* **24**, 183 (1954).
[126] E. P. Fokin and E. P. Prudchenko, *Izv. Sib. Otd. Akad. Nauk. SSSR, Ser. Khim. Nauk.* 98 (1966); *Chem. Abstr.* **66**, 37, 809 (1967).
[127] E. P. Fokin and A. M. Detsina, *Izv. Sib. Otd. Akad. Nauk. SSSR, Ser. Khim. Nauk.* 95 (1969).
[128] D. W. Cameron and R. G. F. Giles, *Chem. Commun.* 573 (1965).
[129] D. W. Cameron and R. G. F. Giles, *J. Chem. Soc., C* 1461 (1968).

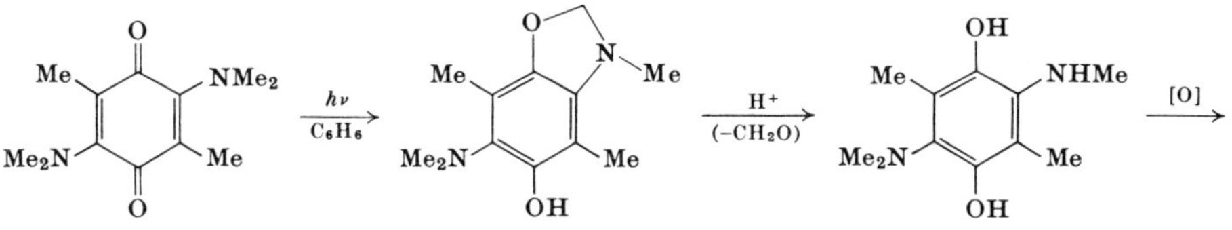

Sec. III. E.] THE t-AMINO EFFECT 273

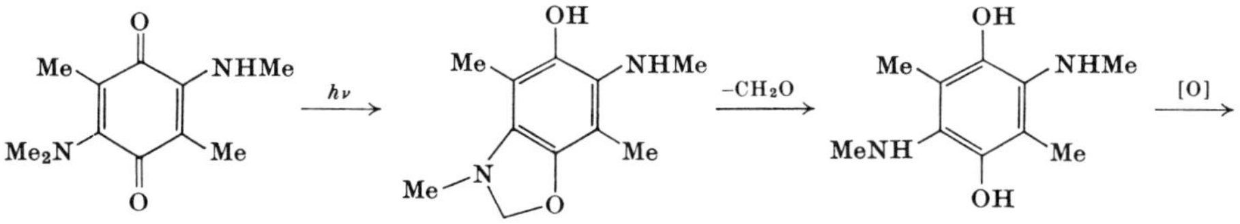

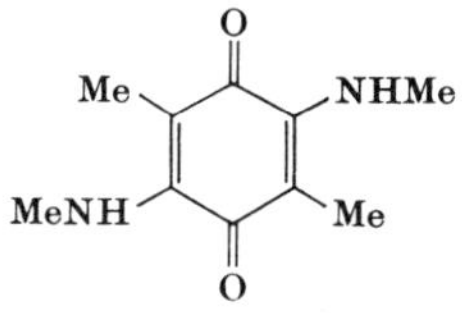

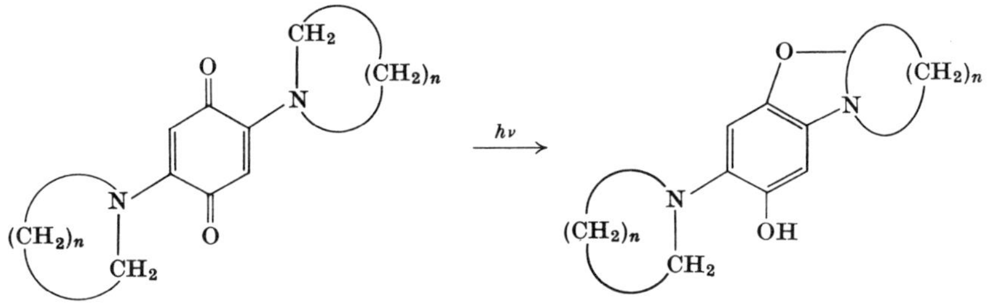

Baxter and Cameron[130, 131] extended this work to the quinonediimides (e.g., **249**) from which benzimidazoles (e.g., **250**) were readily

[130] I. Baxter and D. W. Cameron, *Chem. Ind. (London)* 1403 (1967).
[131] I. Baxter and D. W. Cameron, *J. Chem. Soc., C* 1747 (1968).

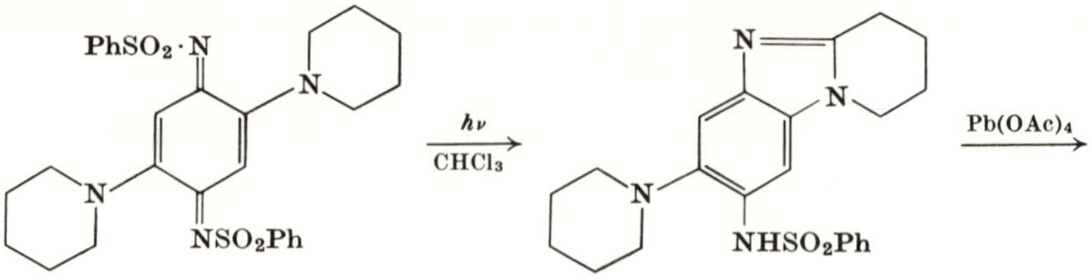

obtained. Oxidation of this product and a second photolysis gave the pentacyclic bisimidazole (**251**). In each photolysis, the leuco compound (e.g., **252**) was also isolated and the cyclized product was obtained as its benzenesulfonate salt. Analogous products were obtained from the dimethylamino compound (**253**). The piperazine derivative (**254**) underwent a double cyclization to give **255**, together with the leuco compound. No isomer (**256**) was found.

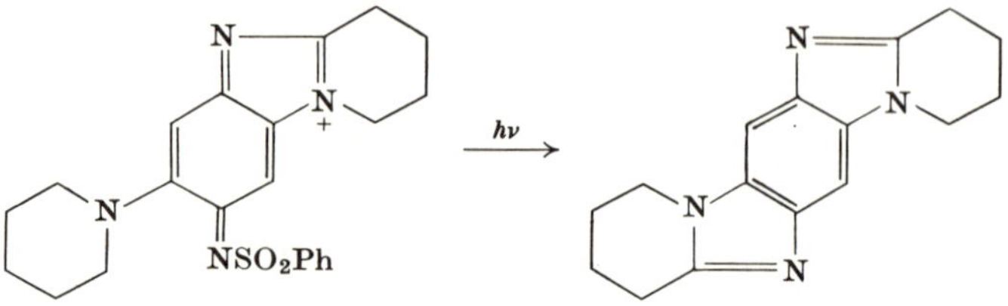

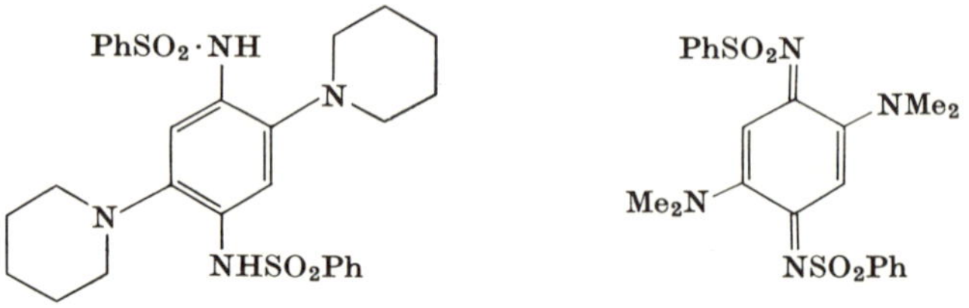

Significantly, neither the methoxy (**257**) nor the thio (**258**) analogs of the quinones below or quinoneimines underwent photochemical cyclizations. We have found that generally in the reactions reported

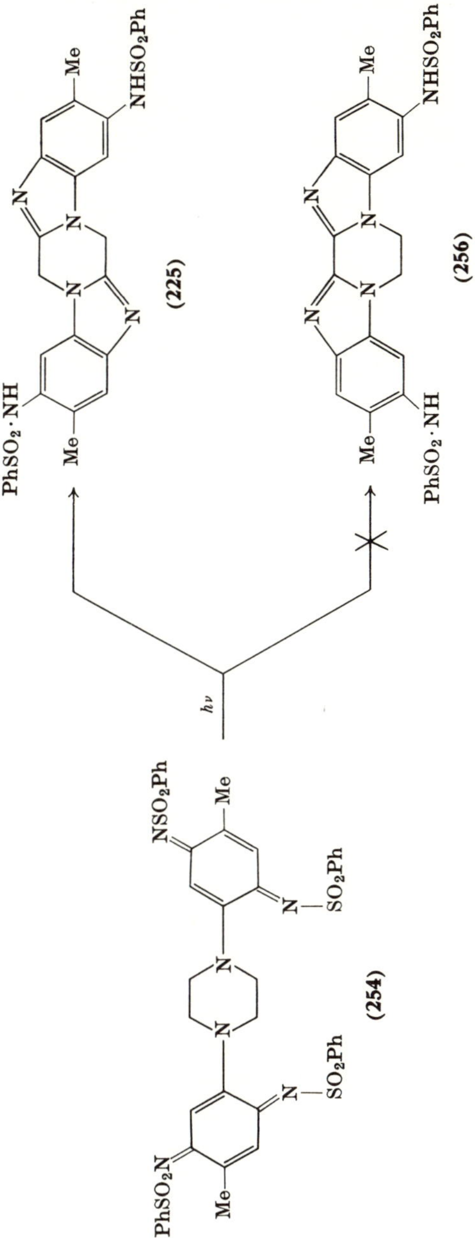

Sec. III. E.]   THE *t*-AMINO EFFECT   275

in this review methoxy or alkylsulfide substituents cannot successfully replace the *t*-amino group.

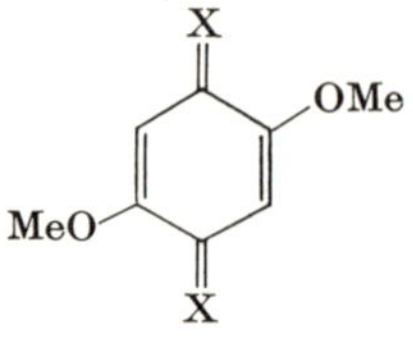

(257) X = O or NSO₂Ph

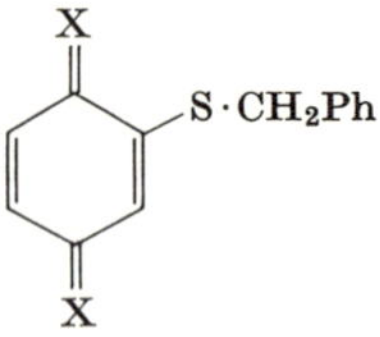

(258)

The above reactions involving e.g., **247** and **249** may be rationalized as shown below.

(247) ⟷ (236) —hv→ ⟶

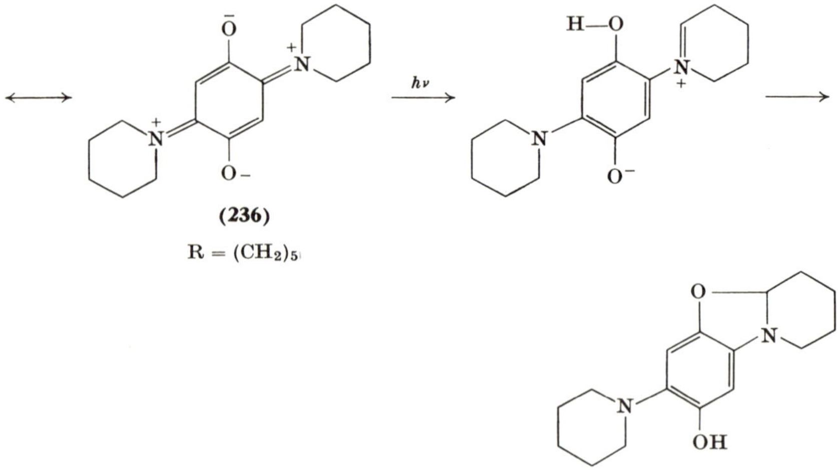

R = (CH₂)₅

(248)

The initial cyclization of the quinones or quinoneimines may arise either by proton (**249a**) or hydride (**249**) abstraction. The latter course is supported by the isolation of **255**, exclusively, from the piperazine (**254**), since the second cyclization involves abstraction of the less acidic of the protons at the two available sites. The hydride lability of dihydroimidazoles and hydride reducibility of quinones are the probable reasons for the isolation of the leuco compound (**261**) and benzimidazole salt (**262**). Another possibility is an acid-catalyzed disproportionation of the dihydro compound (**259**) analogous to that described earlier.[81] The sulfonamide group would supply the necessary acidity. The benzimidazole sulfonate (**262**) is probably derived hydrolytically during work-up from a salt such as **260** which is readily hydrolyzed. Baxter and Cameron accounted for the leuco product by reduction of benzenesulfinic acid (presumably formed by elimination) or photoreduction. However, N-acyldihydrobenzimidazoles are stable in the absence of oxidants and do not normally aromatize by elimination.[75]

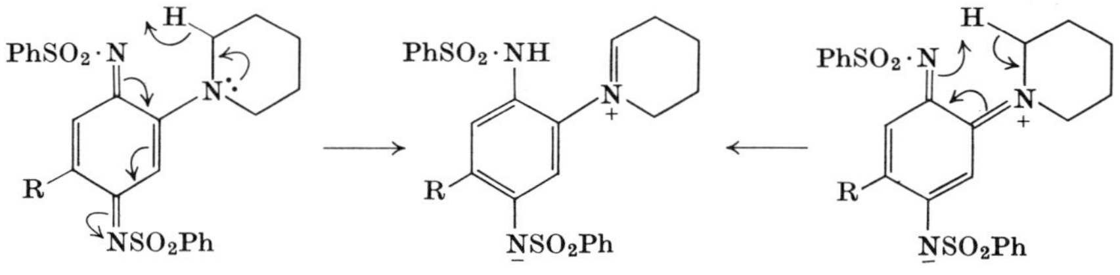

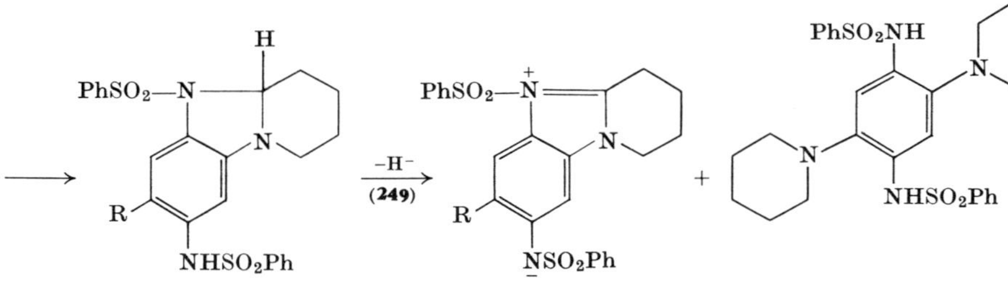

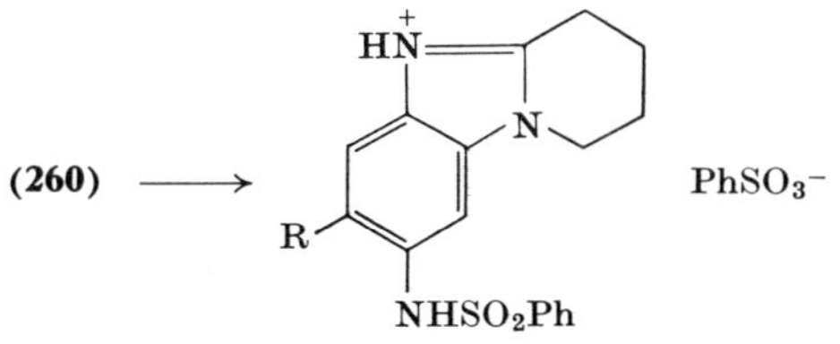

As mentioned earlier, the quadrupolar character of 2,5-bisaminated quinones has been demonstrated experimentally. Cameron, Giles, and Pay[123] have quantified this polarity by observation of coalescence temperature data of the NMR absorption for the N-methylene groups. Also oxidation potentials[124] and shielding effects[132] on the quinonoid protons have been similarly applied. From these measurements the significance of the polarized mesomer decreases in the order azetidino > pyrrolidino > dimethylamino > piperidino > aziridino. This order agrees with our qualitative observations on the relative ease of several cyclizations mentioned in Section III, B in which the effectiveness of lone-pair donation by the nitrogen is probably the deciding factor in determining the rate of cyclization. However, further quantitative data are required to clarify this point.

Finally, the interaction of an o-t-amino group with a carboxylic acid function should be considered. This type of reaction has already been referred to (p. 251) in the context of peracid oxidative cyclizations. A further example is found in the ozonization of the o-dimethylaminobenzoic acid (263) from which the benzoxazinone (264), earlier

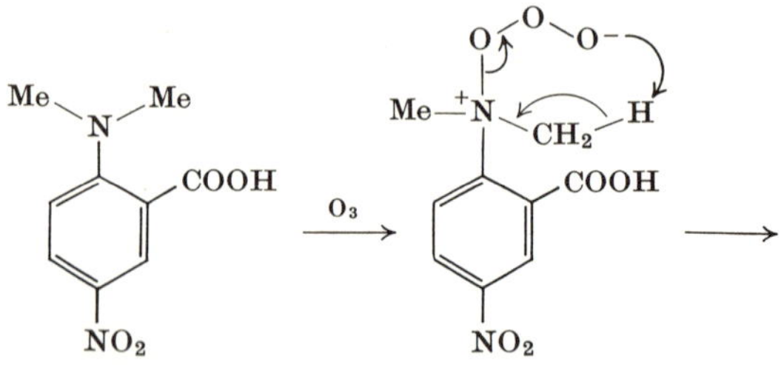

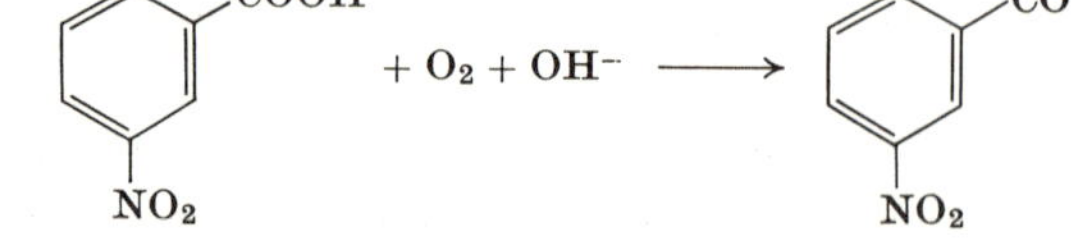

proposed as an intermediate in the related peracid reaction, was isolated as the major product.[133]

---

[132] S. Dähne, J. Ranft, and H. Paul, *Tetrahedron Lett.* 3335 (1964).
[133] P. Kolsaker and O. Meth-Cohn, *Chem. Commun.* 423 (1965).

1,2-Dihydroisoquinolines

S. F. DYKE

School of Chemistry and Chemical Engineering
Bath University of Technology, Bath, England

I. Introduction 279
II. Formation 280
    A. Reduction of Isoquinolinium Salts 280
    B. Other Nucleophilic Additions to C-1 of Isoquinolinium Salts 283
    C. Cyclization of Benzylaminoacetaldehyde Dialkyl Acetals . 285
    D. Miscellaneous 288
III. Stability 289
IV. Detection and Estimation 294
V. Reactions 294
    A. Oxidation 294
    B. Reduction 295
    C. Hydroboration 295
    D. Enamine Reactions 296
    E. Rearrangements 319
    F. 1,2-Dihydroisoquinolines in Biosynthesis . . . 326

I. Introduction

The α,β-unsaturated amine system (**1**) was termed[1] "enamine" in 1927 to emphasize the structural similarity between it and an enol (**2**). Although some enamines have been known for many years,[2] the synthetic potential of such compounds was not fully appreciated until quite recently.[3-5] The true structure of an enamine function may be regarded as a resonance hybrid of the canonical forms (**1**) and (**3**), so that electrophilic reagents might be expected to attack at

[1] G. Wittig and H. Blumenthal, *Ber.* **60**, 1085 (1927).
[2] R. Robinson, *J. Chem. Soc.* **109**, 1038 (1916).
[3] J. Szmuskovicz, *Advan. Org. Chem.* **4**, 1 (1963).
[4] K. Blaha and O. Cervinka, *Advan. Heterocycl. Chem.* **4**, 147 (1966).
[5] A. G. Cook, ed., "Enamines." Dekker, New York, 1969.

$$\underset{(1)}{\overset{}{\underset{\beta\ \alpha}{>C=C-N<}}} \qquad \underset{(2)}{>C=C-O-H}$$

$$\updownarrow$$

$$\underset{(3)}{>\overset{-}{C}-C=\overset{+}{N}<} \qquad \underset{(4)}{>\overset{H}{\underset{|}{C}}-C=\overset{+}{N}<}$$

either the nitrogen atom or at C-β. In the derived imminium ion (4), nucleophilic reagents should attack at C-α.

Various heterocyclic compounds contain an α,β-unsaturated amine moiety in their structure, for example, indoles, 1,2,3,4-tetrahydropyridines, 1,4-dihydroquinolines, and 1,2-dihydroisoquinolines. It is the purpose of this review to summarize our knowledge of the chemistry of 1,2-dihydroisoquinolines and to evaluate those reactions that may be interpreted in terms of enamine character.

II. Formation

A. Reduction of Isoquinolinium Salts

Isoquinolinium salts are reduced to 2-alkyl-1,2-dihydroisoquinoline derivatives by sodium dithionite,[6] lithium aluminum hydride[7,8] (LAH), or dialkyl aluminum hydrides.[9] LAH is preferred[7] to dithionite since it leads to purer products, and will reduce those salts that are inert to dithionite (e.g., papaverine methiodide). Isoquinoline itself[10,11] and 3-methyl-6,7-methylenedioxyisoquinoline[12] (5) can be reduced to the 1,2-dihydroisoquinoline with LAH, as can isoquinoline

[6] P. Karrer, *Helv. Chim. Acta* **21**, 2233 (1938); P. Karrer, F. W. Kahnt, R. Epstein, W. Jaffe, and T. Ishii, *Helv. Chim. Acta* **21**, 223 (1938).
[7] H. Schmid and P. Karrer, *Helv. Chim. Acta* **32**, 960 (1949).
[8] M. Sainsbury, S. F. Dyke, and A. R. Marshall, *Tetrahedron* **22**, 2445 (1966).
[9] W. P. Neumann, *Angew. Chem.* **70**, 401 (1958); *Ann.* **618**, 90 (1958).
[10] W. Traber, M. Hubmann, and P. Karrer, *Helv. Chim. Acta* **43**, 265 (1960).
[11] D. I. Packham and L. M. Jackman, *Chem. Ind.* 360 (1955).
[12] T. Kametani, K. Fukumoto, and T. Katagi, *Chem. Pharm. Bull.* **7**, 567 (1959); *Yakugaku Zasshi* **80**, 1288 (1960) [*Chem. Abstr.* **55**, 3589 (1961)].

N-oxide.[10] 1,2-Dihydroisoquinolines themselves are slowly reduced by LAH to form 1,2,3,4-tetrahydroisoquinolines (8).[13]

When isoquinolinium salts are reduced with sodium borohydride (NaBH$_4$) under the usual aqueous–ethanolic solvent conditions, tetrahydroisoquinolines (8) are produced. The 1,2-dihydroisoquinoline (6) is formed first, and this can[14] be protonated by the solvent to 7

and reduced further to 8. This is in agreement with the behavior of simple ketone enamines which are[15] reduced by NaBH$_4$ only after acetic acid has been added to the solution. If a nonprotonic solvent such as pyridine[16] or dimethylformamide[17] is used, then 1,2-dihydroisoquinolines can be isolated. In some cases[18] the use of anhydrous methanol enables selective reduction of isoquinolinium salts to be achieved. The salt (9) is reduced only to the 1,2-dihydroisoquinoline even with aqueous alcoholic NaBH$_4$ solution.[18] The isomeric ester (10)[19] and the nitro compound[20] (11) are also reduced to the 1,2-dihydroisoquinoline stage by NaBH$_4$. In view of the measured[21]

---

[13] S. F. Dyke and M. Sainsbury, *Tetrahedron* **21**, 1907 (1965).
[14] R. Mirza, *J. Chem. Soc.* 4400 (1957).
[15] J. A. Marshall and W. S. Johnson, *J. Org. Chem.* **28**, 595 (1963).
[16] D. H. R. Barton, R. H. Hesse, and G. W. Kirby, *J. Chem. Soc.* 6379 (1965).
[17] P. Bichaut, G. Thuillier, and P. Rumpf, *C. Rend. Acad. Sci.* 1550 (1969).
[18] D. W. Brown, M. Sainsbury, S. F. Dyke, and W. G. D. Lugton, *Tetrahedron*, in press.
[19] W. Wiegrabe and W. D. Sasso, *Arch. Pharm. (Weinheim)* **303**, 145 (1970).
[20] J. L. Neumeyer, M. McCarthy, K. W. Weinhardt, and P. L. Levins, *J. Org. Chem.* **33**, 2890 (1968).
[21] E. J. Stamhuis, W. Maas, and H. Wynberg, *J. Org. Chem.* **30**, 2160 (1965).

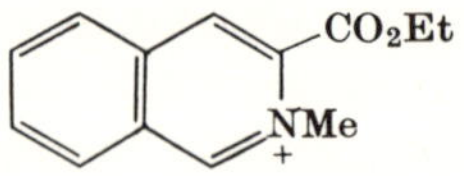

(9) 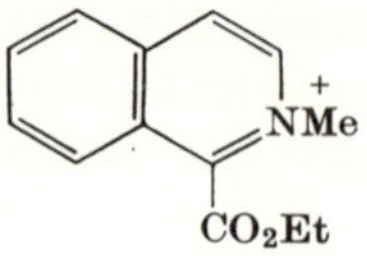 (10) 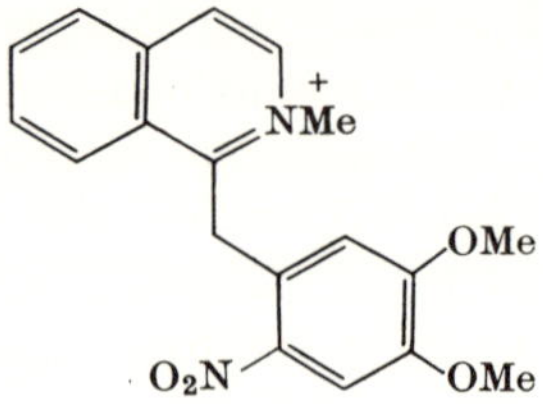 (11)

basicities of some β-substituted enamines, the ease of C-4 protonation, and consequently of reduction, of C-4-substituted 1,2-dihydroisoquinolines by $NaBH_4$ will depend very largely on the nature of the C-4 substituent. Thus, compounds of the type (12a–f) can be prepared by reducing the corresponding isoquinolinium salts with $NaBH_4$ in ethanol.[18, 22]

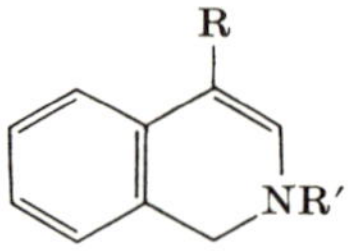

(12a) R = $CONH_2$
(12b) R = CO_2Et
(12c) R = CN
(12d) R = [benzisoxazolyl]
(12e) R = —CH_2—C_6H_5
(12f) R = CH_3

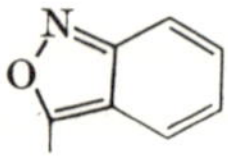

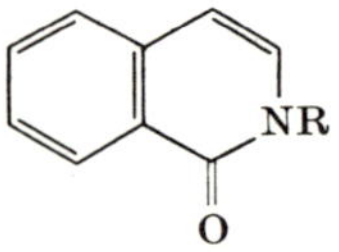

(13)

Isocarbostyrils (13) can also be reduced by LAH to 1,2-dihydroisoquinolines,[23] and since the former are accessible from isocoumarins,[24] the method is of practical importance.

[22] G. Thuillier, B. Marcot, J. Cruanes, and P. Rumpf, *Bull. Soc. Chim. Fr.* 4770 (1967).
[23] G. Hardy, Ph.D. Thesis, Bath University, 1969.
[24] R. D. Barry, *Chem. Rev.* **64**, 229 (1964).

B. Other Nucleophilic Additions to C-1 of Isoquinolinium Salts

The reduction of isoquinolinium salts by complex metal hydrides may be regarded as examples of nucleophilic addition to the $>\!\!\overset{+}{C^1}\!\!=\!\!N\!\!<$ grouping, and several other methods for generating 1,2-dihydroisoquinolines can be conveniently considered together.

Treatment of a quaternary salt (**14**) with concentrated alkali causes disproportionation to occur to yield an isocarbostyril (**13**) and a 1,2-dihydroisoquinoline (**6**).[25] The reaction, which probably proceeds via the pseudobase (**15**), has been used[26] in a modified berbine ring

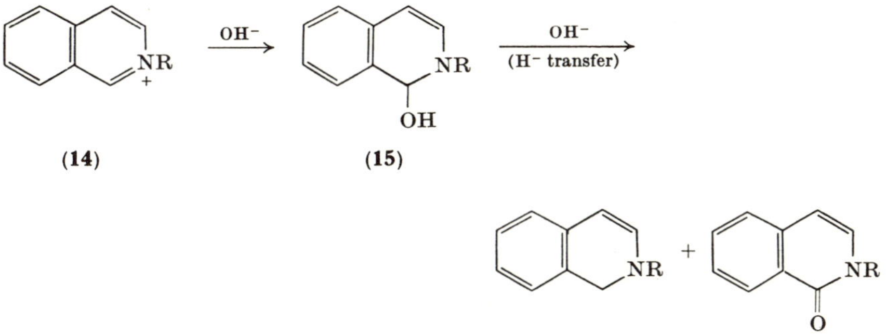

synthesis (Section V, D, 2, c). It is possible for the pseudobase (**15**) itself to exhibit some enamine character[18, 27] (Section V, D, 1, d). Isoquinoline methiodide condenses with o-nitrotoluenes in alkaline solution to yield 1,2-dihydroisoquinolines, which have been used[28] in the synthesis of the aporphine ring system.[29]

[25] W. J. Gensler, in "Heterocyclic Compounds" (R. C. Elderfield, ed.), Vol. 4, p. 344. Wiley, New York, 1952.
[26] D. W. Brown and S. F. Dyke, *Tetrahedron* **22**, 2429 (1966).
[27] M. Sainsbury, S. F. Dyke, D. W. Brown, and W. G. D. Lugton, *Tetrahedron* **24**, 427 (1968).
[28] J. Gadamer, M. Oberlin, and A. Schoeler, *Arch. Pharm. (Weinheim)* **263**, 81 (1925).
[29] M. Shamma and W. A. Slusarchyk, *Chem. Rev.* **64**, 59 (1964); M. Shamma, in "The Alkaloids" (R. H. F. Manske, ed.), Vol. 9, p. 2. Academic Press, New York, 1967.

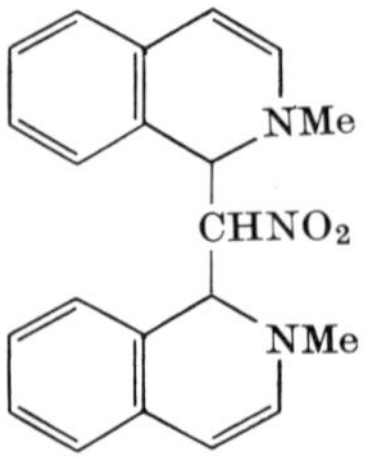

(16)

1-Substituted 1,2-dihydroisoquinolines are formed when isoquinolinium salts are treated with Grignard reagents[30] or organo-

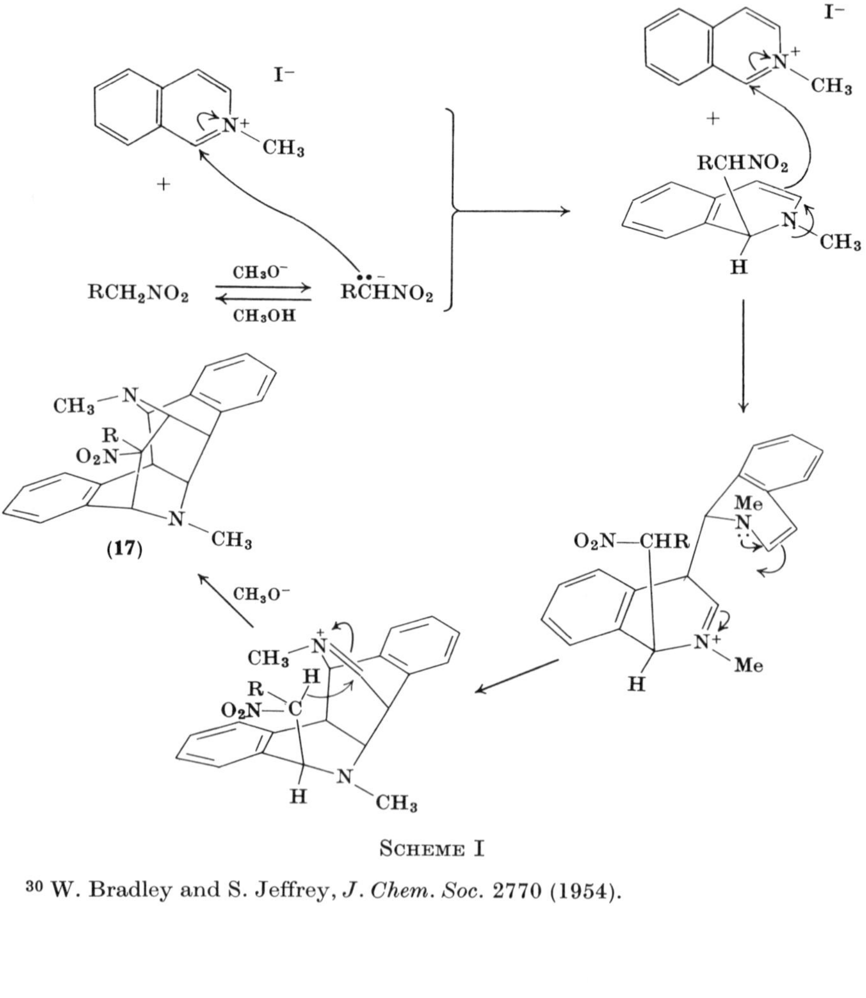

Scheme I

[30] W. Bradley and S. Jeffrey, J. Chem. Soc. 2770 (1954).

lithium compounds.[31-33] The addition of active methylene compounds (e.g., acetone, nitromethane) to isoquinolinium salts has also been reported[25] to yield 1-substituted 1,2-dihydroisoquinolines, but complications can arise. Thus, when nitromethane is added to isoquinoline methiodide, one of the several products is 2-nitronaphthalene[34] and a second compound, formulated originally[34] as the dimer (**16**), has been shown[35] to be **17** (R = H), formed as indicated in Scheme I. These addition products of active methylene compounds have been utilized in syntheses of the berberine group.[36]

C. Cyclization of Benzylaminoacetaldehyde Dialkyl Acetals

One of the most useful methods for generating 1,2-dihydroisoquinolines, especially those unsubstituted at nitrogen, involves treating a benzylaminoacetaldehyde dialkyl acetal (**18**) with 6 N hydrochloric acid.[37-39] At room temperature or below, 4-hydroxy-1,2,3,4-tetrahydroisoquinolines [(**19**) R' = H] (see chapter by J. M. Bobbitt[39a]) can be isolated,[40] but at higher temperatures (80°–100°), 1,2-dihydroisoquinolines (**20**) are formed.[41] Although these intermediates have not been isolated under these conditions, their existence is inferred from subsequent reactions (Section V, D, 1, d). Those derivatives that have been generated in this way are summarized in Table I. The 4-hydroxytetrahydroisoquinolines [(**19**) R' = H] may also, in a separate step, be dehydrated to **20**.[41] An oxygen function (hydroxyl or alkoxyl) at the position meta to the side chain in **18** is

[31] E. Bergmann, O. Blum-Bergmann, and A. F. Christiani, *Ann.* **483**, 80 (1930).
[32] K. Ziegler and K. Zeiser, *Ann.* **485**, 174 (1931).
[33] W. Slotta and G. Haberland, *Angew. Chem.* **46**, 766 (1933).
[34] N. J. Leonard and G. W. Leubner, *J. Amer. Chem. Soc.* **71**, 3405 (1949).
[35] W. R. Schleigh, *Tetrahedron Lett.* 1405 (1969).
[36] R. Gaze, *Arch. Pharm. (Weinheim)* **228**, 607 (1890).
[37] J. M. Bobbitt, K. L. Khanna, and J. M. Kiely, *Chem. Ind.* 1950 (1964).
[38] J. M. Bobbitt, J. M. Kiely, K. L. Khanna, and R. Ebermann, *J. Org. Chem.* **30**, 2247 (1965).
[39] J. M. Bobbitt, D. N. Roy, A. Marchand, and C. W. Allen, *J. Org. Chem.* **32**, 2225 (1967).
[39a] J. M. Bobbitt, *Advan. Heterocycl. Chem.* **15**, to be published.
[40] J. M. Bobbitt and J. C. Sih, *J. Org. Chem.* **33**, 856 (1968).
[41] D. W. Brown, S. F. Dyke, and M. Sainsbury, *Tetrahedron* **25**, 101 (1969).

TABLE I

1,2-Dihydroisoquinolines Generated by Bobbitt's Procedure

\multicolumn{6}{c}{Substituents}	
1	2	3	6	7	8	Refs.[a]
H	H	H	OH	OMe	H	1
H	H	H	OMe	OH	H	1
H	H	H	H	OMe	OH	1–3
H	H	H	OMe	OMe	H	3–5
H	H	H	H	OMe	OMe	2, 3, 5
H	H	H	O–CH$_2$–O		H	2, 4
H	H	H	H	O–CH$_2$–O		2, 6
H	H	H	H	OMe	H	2

[a] Key to references:
1. J. M. Bobbitt, D. P. Winter, and J. M. Kiely, *J. Org. Chem.* **30**, 2459 (1965).
2. D. W. Brown, S. F. Dyke, and M. Sainsbury, *Tetrahedron* **25**, 101 (1969).
3. S. F. Dyke, M. Sainsbury, D. W. Brown, and M. N. Palfreyman, *Tetrahedron* **25**, 5365 (1969).
4. S. F. Dyke, M. Sainsbury, and B. J. Moon, *Tetrahedron* **24**, 1467 (1968).
5. S. F. Dyke and M. Sainsbury, *Tetrahedron* **23**, 3161 (1967).
6. M. Sainsbury, S. F. Dyke, and B. J. Moon, *J. Chem. Soc. C* 1797 (1970).

necessary for cyclization. Catalytic hydrogenation of the acid solution of **18** or **19** (R′ = H) yields 1,2,3,4-tetrahydroisoquinolines,[37–39] and

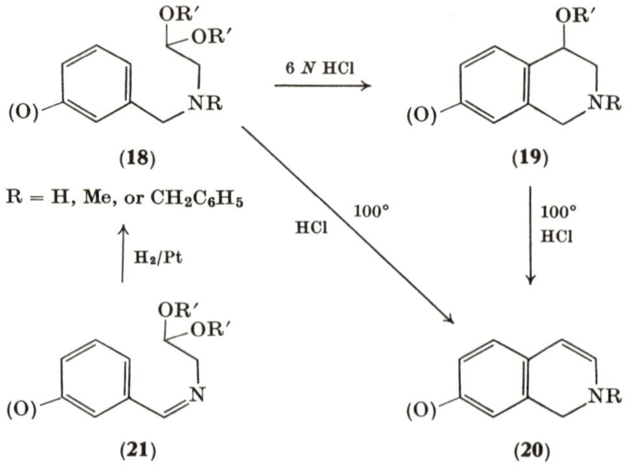

R = H, Me, or CH$_2$C$_6$H$_5$

this procedure constitutes a valuable modification of the classical Pomeranz–Fritsch synthesis [42] of isoquinolines in which a benzalaminoacetaldehyde dialkyl acetal (**21**) reacts with concentrated sulfuric acid. An early attempt to prepare the 1,2-dihydroisoquinoline (**23**) by reaction of the benzylaminoacetal (**22**) with phosphorus

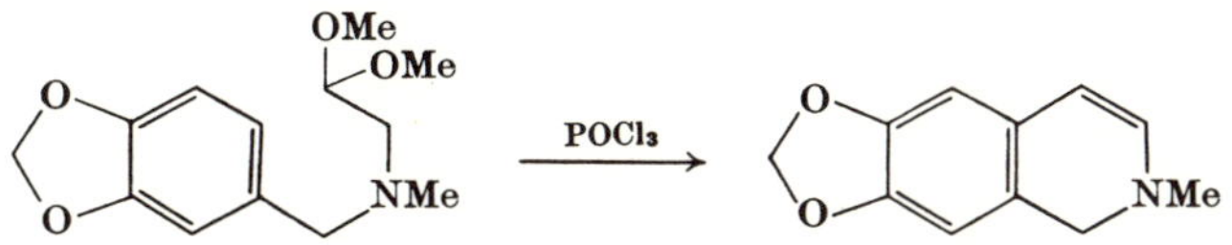

oxychloride was inconclusive.[43] Modifications to the method of formation of the benzylaminoacetals (**18**) have been described by Bobbitt et al.[44] One very useful modification involves [39] the reaction of an N-alkylbenzylamine (**24**) with glycidol, followed by periodate cleavage of the resulting diol (**25**), and acid-promoted ring closure of the aldehyde (**26**) to the 1,2-dihydroisoquinoline (**27**). This sequence

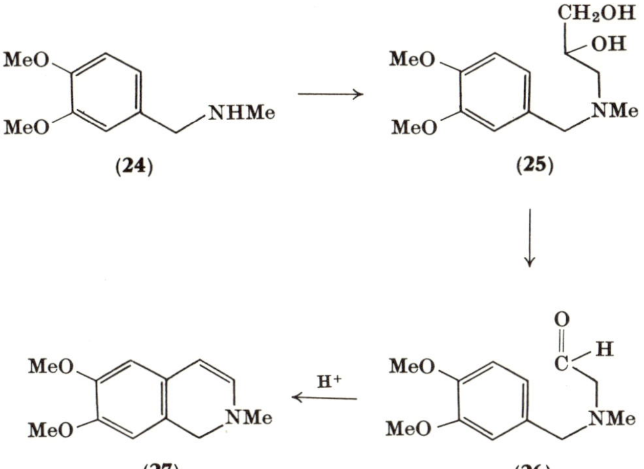

---

[42] W. J. Gensler, *Org. Reactions* **6**, 191 (1951).
[43] P. C. Young and R. Robinson, *J. Chem. Soc.* 275 (1933).
[44] (a) J. M. Bobbitt and C. P. Dutta, *J. Org. Chem.* **34**, 2001 (1969); (b) J. M. Bobbitt, A. S. Steinfeld, K. H. Weisgraber, and S. Dutta, *J. Org. Chem.* **34**. 2478 (1969).

has been utilized in some recent berbine syntheses,[45] as well as in some isoquinoline preparations.[46]

4-Alkoxy-1,2,3,4-tetrahydroisoquinolines [(**19**) R′ = alkyl], which can be prepared by application of the Pictet–Spengler reaction[47, 48]

(**28**)   (**29**)   (**30**)

using phenylethylamines of the type (**28**), or by treating[49] benzylaminoacetals (**18**) with boron trifluoride, can also be converted into 1,2-dihydroisoquinolines by reaction with mineral acid or $POCl_3$.[12]

D. Miscellaneous

1,2-Dihydroisoquinolines are reported to be formed when (*a*) the 1,2,3,4-tetrahydro derivatives are oxidized with oxygen in pyridine solution[50] or with palladium chloride,[51] (*b*) when **29** is treated with dry sodium hydroxide,[52] or (*c*) when the benzylaminopropyne [(**30**) R = H or Me] is heated with polyphosphoric acid.[53] When isoquinoline is reduced with sodium in liquid ammonia, one of the products was

[45] D. W. Brown, S. F. Dyke, G. Hardy, and M. Sainsbury, *Tetrahedron Lett.* 5177 (1968).
[46] M. Sainsbury, S. F. Dyke, D. W. Brown, and R. G. Kinsman, *Tetrahedron* **26**, 5265 (1970).
[47] W. M. Whaley and T. R. Govindachari, *Org. Reactions* **6**, 151 (1951).
[48] T. Kametani, K. Fukumoto, H. Agui, H. Yagi, K. Kigasawa, H. Sugahara, M. Hiiragi, T. Hayasaka, and H. Ishimaru, *J. Chem. Soc. C* 112 (1968); T. Kametani, H. Agui, and K. Fukumoto, *Chem. Pharm. Bull.* **16**, 1285 (1968); T. Kametani, K. Kigasawa, M. Hiiragi, and H. Ishimaru, *ibid.* **17**, 2353 (1969).
[49] R. Quelet and N. Vinot, *C. R. Acad. Sci.* **244**, 919 (1957); N. Vinot and R. Quelot, *ibid.* **246**, 1712 (1958); *Bull. Soc. Chim. Fr.* 1164 (1959); N. Vinot, *Ann. Chim. (Paris)* **3**, 461 (1958); *Bull. Soc. Chim. Fr.* 617 (1960).
[50] W. Bartok and H. Pobiner, *J. Org. Chem.* **30**, 274 (1965).
[51] G. W. Cooke and J. M. Gulland, *J. Chem. Soc.* 872 (1939).
[52] J. Hagimiwa, I. Murakoshi, and Y. Obe, *Yakugaku Zasshi* **79**, 1578 (1959).
[53] Japanese Patent 4489 (1964) [*Chem. Abstr.* **61**, 5618h (1964)]; S. F. Dyke and D. N. Harcourt, unpublished; J. R. Brooks and D. N. Harcourt, *J. Chem. Soc. C* 625 (1969).

formulated[54] as the trimer (**31**), formed from 1,2-dihydroisoquinoline via the imine tautomer. 1,2-Dihydroisoquinolines are formed by catalytic hydrogenation of isoquinoline or isoquinoline methiodide,[55, 56] but these have been trapped by reaction with aldehydes also present in the reaction mixture. Some aspects of this work have

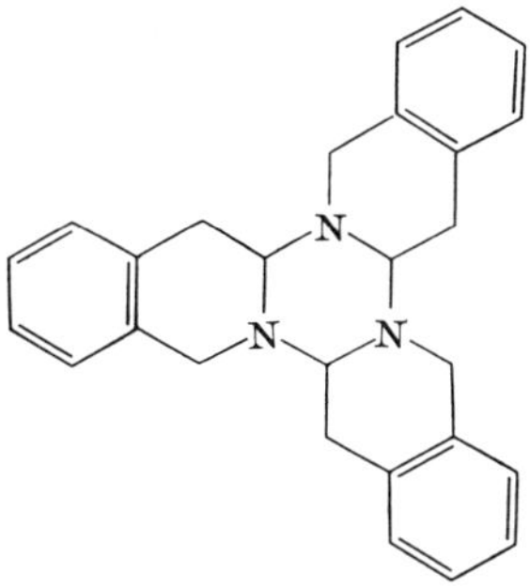

(**31**)

been corrected.[57, 58] Similar results have been obtained using $NaBH_4$ instead of by catalytic reduction.[17, 59] The original[60] description of 1,2-dihydropapaverine is erroneous.[43]

III. Stability

Although 3,4-dihydroisoquinolines are well known, and are reasonably stable, easily characterizable compounds, the isomeric 1,2-dihydroisoquinolines have often been considered[12, 17, 22, 61, 62] to be unstable and difficult to isolate in a pure state. This is certainly true in some cases, but stability does depend on (*a*) whether the nitrogen is secondary or tertiary, (*b*) the position and nature of

[54] W. Huckel and G. Graner, *Chem. Ber.* **90**, 2017 (1957).
[55] R. Grewe, W. Kruger, and E. Vangermain, *Chem. Ber.* **97**, 119 (1964).
[56] G. Thuillier, A. Vilar, and P. Rumpf, *C. R. Acad. Sci.* 1131 (1967).
[57] D. W. Brown and S. F. Dyke, *Tetrahedron* **22**, 2437 (1966).
[58] P. Garside and A. C. Ritchie, *J. Chem. Soc.* 2140 (1966).
[59] P. Bichaut, G. Thullier, and P. Rumpf, *C. R. Acad. Sci.* 993 (1969).
[60] J. S. Buck, *J. Amer. Chem. Soc.* **52**, 3610 (1930).
[61] M. Freund and G. Bode, *Ber.* **42**, 1746 (1909).
[62] M. Freund and K. Lederer, *Ber.* **44**, 2353 (1911).

substituents—both in the homocyclic and heterocyclic rings, (c) the pH of the solution in which the enamine is dissolved, and (d) whether oxygen is present.

Exposure of some 1,2-dihydroisoquinolines to air results in rapid polymerization or in oxidation to isocarbostyrils. In the presence of acids, 1,2-dihydroisoquinolines may undergo disproportionation, dimerization, or polymerization (see Section V, D, 1, a).

There is a discrepancy in the literature concerning 1,2-dihydroisoquinoline itself. Thus, Huckel and Graner[54] report its trimerization to **31** (m.p. 138°, the same melting point that Packham and Jackman[11] ascribe to the monomer). It is thought[63] that in a nonpolar solvent, 1,2-dihydroisoquinoline is relatively stable, but in methanol, protonation and trimerization occur; certainly the mass spectrum of the compound (m.p. 138°) described by Packham and Jackman indicates that it is trimeric.[63] The series of 1,2-dialkyl-1,2-dihydroisoquinolines described by Bradley and Jeffry[30] was purified by distillation under reduced pressure. The stability of these compounds is quite remarkable in view of the known tendency for 1,2-dihydroisoquinolines to undergo disproportionation. Some other 1,2-dialkyl-1,2-dihydroisoquinolines have been described,[7] as well as 1-aryl derivatives.[7] The derivative (**32**) when heated with triethyl phosphite is transformed by an unknown mechanism, in 37% yield, into **33**.[64]

(**32**)   (EtO)₃P, 160°, 3 hours →   (**33**)

The relative stabilities of some 1,2-dihydroisoquinolines may be related to their ease of protonation at C-4. Thus, introduction of a C-3 methyl group,[12, 41] a C-4 methyl group,[53] a C-4 benzyl group,[18, 65]

---

[63] S. F. Dyke and R. G. Kinsman, unpublished data.
[64] T. Kametani, T. Yamanaka, and K. Ogasawara, *Chem. Commun.* 7862 (1968); *J. Org. Chem.* **33**, 4446 (1968).
[65] S. F. Dyke, M. Sainsbury, D. W. Brown, and M. N. Palfreyman, *Tetrahedron* **25**, 5365 (1969).

Sec. III.] 1,2-DIHYDROISOQUINOLINES 291

or a C-3 aryl group[18, 66] enhances stability considerably. Compounds such as dihydroberberine[67] (**34**) and anhydrocryptopine (**35**)[68–70] are quite stable, easily characterizable compounds.

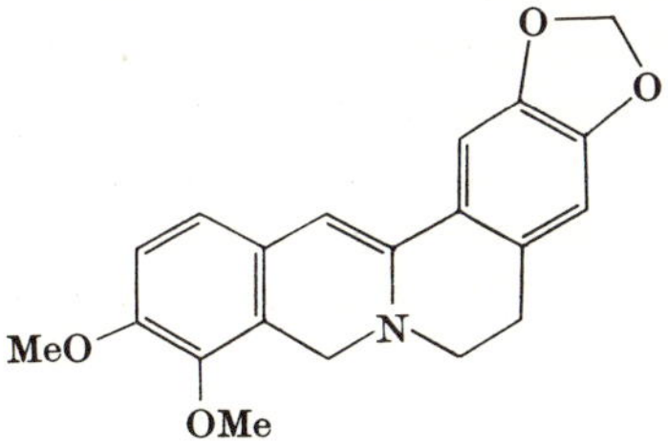

(**34**) 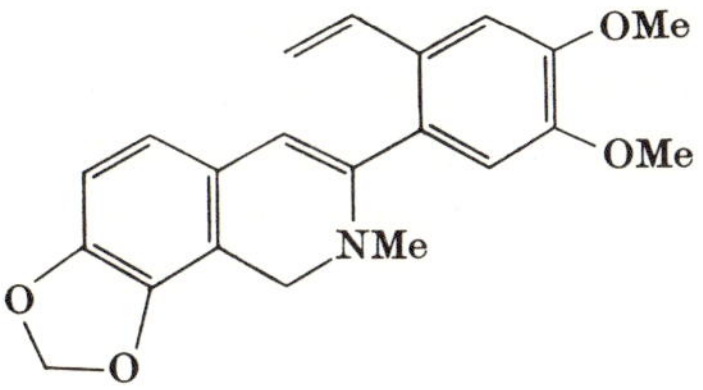   (**35**)

2-Cyano-1,2-dihydroisoquinolines[71] [(**36**) R = CN; Z = CN, OH, etc.] and Reissert compounds[72] [(**36**) R = COC$_6$H$_5$; Z = CN] are formally 1,2-dihydroisoquinolines; they are stable compounds. The enamine (**37**) is unstable (i.e., elemental analysis could not be ob-

(**36**)   (**37**)

tained), although its NMR spectrum has been recorded.[73] 6-Methoxy-2-methyl-1,2-dihydroisoquinoline is reported[74] to cause a marked and sustained fall in blood pressure in hypertensive dogs; presumably the compound is a stable one.

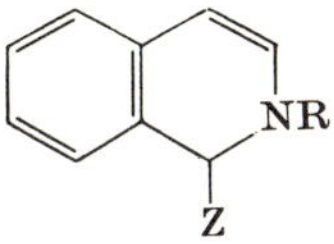

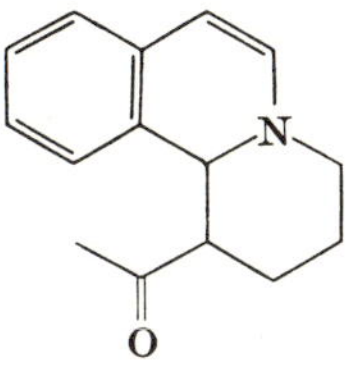

[66] S. F. Dyke, M. Sainsbury, D. W. Brown, M. N. Palfreyman, and E. P. Tiley, *Tetrahedron* **24**, 6703 (1968).
[67] R. H. F. Manske, *in* "The Alkaloids" (R. H. F. Manske and H. L. Holmes, eds.), Vol. 4, p. 78. Academic Press, New York, 1954.
[68] W. H. Perkin, *J. Chem. Soc.* **109**, 815 (1916).
[69] S. F. Dyke and D. W. Brown, *Tetrahedron* **24**, 1455 (1968).
[70] S. F. Dyke and D. W. Brown, *Tetrahedron* **25**, 5375 (1969).
[71] M. D. Johnson, *J. Chem. Soc.* 200 (1964).
[72] W. E. McEwen and R. L. Cobb, *Chem. Rev.* **55**, 511 (1955).
[73] R. M. Wilson and F. DiNenno, *Tetrahedron Lett.* 289 (1970).
[74] R. A. McLean, *in* "Medicinal Chemistry" (A. Burger, ed.), 2nd ed., p. 600. Wiley (Interscience), New York, 1960.

The 1,2-dihydroisoquinolines that are sufficiently stable for characterization to have been described are listed in Table II; Reissert compounds have been omitted.

Numerous modifications to the Pomeranz–Fritsch synthesis of isoquinolines have been studied over the years without much success.[42] Some of these cyclizations, had they been successful, would have generated 1,2-dihydroisoquinolines; it is possible, at least in some cases, that the overall sequence failed because the conditions of cyclization were too harsh for the 1,2-dihydroisoquinolines formed to survive.

TABLE II

STABLE 1,2-DIHYDROISOQUINOLINES

			Substituent[a]				
1	2	3	4	6	7	8	Refs.[b]
Me	Me	H	H	H	H	H	1
Et	Me	H	H	H	H	H	1
n-Pr	Me	H	H	H	H	H	1
Me	Et	H	H	H	H	H	1
Et	Et	H	H	H	H	H	1
n-Pr	Et	H	H	H	H	H	1
Me	n-Pr	H	H	H	H	H	1
Et	n-Pr	H	H	H	H	H	1
Me	isoPr	H	H	H	H	H	1
Et	isoPr	H	H	H	H	H	1
H	Me	H	C_6H_5	H	H	H	2
H	H	H	H	H	H	H	3
H	Me	H	Me	H	H	H	4
Allyl	Me	Me	H	H	H	H	5
Allyl	H	H	H	H	H	H	5, 6
C_6H_5	Me	H	H	H	H	H	7, 8
A	Me	H	H	OMe	OMe	H	9
H	H	Me	H	O–CH_2–O		H	10
B	Me	H	H	H	H	H	11, 12
H	Me	H	C	H	H	H	13
H	Me	H	$CONH_2$	H	H	H	14
H	Me	H	CO_2Et	H	H	H	14
H	Me	H	CN	H	H	H	14
H	H	Me	H	H	H	H	15
H	H	C_6H_5	H	H	H	H	15

TABLE II—continued

Substituent[a]							
1	2	3	4	6	7	8	Refs.[b]
H	Me	H	D	H	H	H	16
H	Me	H	E	H	OMe	OMe	17
H	Me	H	E	OMe	OMe	H	18
Allyl + A	Me	H	H	OMe	OMe	H	19
H	Me	CO_2Et	H	H	H	H	20
H	Me	H	H	OMe	H	H	21

[a] A, 3,4-$(OMe)_2C_6H_3CH_2-$; B, 3,4-$(OMe)_2$-6-$NO_2C_6H_2CH_2-$; C, 2,6-$Cl_2C_6H_3CH_2-$; D,
 ; and E, 3,4-$CH_2O_2C_6H_3CH_2-$.

[b] KEY TO REFERENCES:
1. W. Bradley and S. Jeffrey, *J. Chem. Soc.* 2770 (1954).
2. I. G. Hinton and F. G. Mann, *J. Chem. Soc.* 599 (1959).
3. D. I. Packham and L. M. Jackman, *Chem. Ind.* 360 (1955).
4. Japanese Patent 4489 (1964) [*Chem. Abstr.* **61**, 5618h (1964)]; S. F. Dyke and D. N. Harcourt, unpublished; J. R. Brooks and D. N. Harcourt, *J. Chem. Soc. C* 625 (1969).
5. S. F. Dyke and R. G. Kinsman, unpublished data.
6. J. Knabe and H.-D. Holtje, *Arch. Pharm. (Weinheim)* **303**, 404 (1970).
7. H. Schmid and P. Karrer, *Helv. Chim. Acta* **32**, 960 (1949).
8. P. R. Brook and P. Karrer, *Helv. Chim. Acta* **40**, 260 (1957).
9. C. Schopf and K. Thierfelder, *Ann.* **497**, 22 (1932).
10. T. Kametani, K. Fukumoto, and T. Katagi, *Chem. Pharm. Bull.* **7**, 567 (1959); *Yakugaku Zasshi* **80**, 1288 (1960) [*Chem. Abstr.* **55**, 3589 (1961)].
11. J. L. Neumeyer, M. McCarthy, K. W. Weinhardt, and P. L. Levins, *J. Org. Chem.* **33**, 2890 (1968).
12. T. Kametani, T. Yamanaka, and K. Ogasawara, *Chem. Commun.* 7862 (1968); *J. Org. Chem.* **33**, 4446 (1968).
13. P. Bichaut, G. Thuillier, and P. Rumpf, *C. R. Acad. Sci.* 1550 (1969).
14. G. Thuillier, B. Marcot, J. Cruanes, and P. Rumpf, *Bull. Soc. Chim. Fr.* 4770 (1967).
15. J. Hagimiwa, I. Murakoshi, and Y. Obe, *Yakugaku Zasshi* **79**, 1578 (1959).
16. S. F. Dyke, M. Sainsbury, D. W. Brown, and M. N. Palfreyman, *Tetrahedron* **25**, 5365 (1969).
17. W. J. Gensler, K. T. Shamasundar, and S. Marburg, *J. Org. Chem.* **33**, 2861 (1968).
18. S. F. Dyke, M. Sainsbury, D. W. Brown, M. N. Palfreyman, and D. W. Wiggins, *Tetrahedron* **27**, 281 (1971).
19. S. F. Dyke, R. G. Kinsman, J. Knabe, and H. D. Holtje, *Tetrahedron* in press.
20. D. W. Brown, M. Sainsbury, S. F. Dyke, and W. G. D. Lugton, *Tetrahedron* in press.
21. R. A. McLean, *in* "Medicinal Chemistry" (A. Burger, ed.), 2nd ed., p. 600. Wiley (Interscience), New York, 1960.

IV. Detection and Estimation

1,2-Dihydroisoquinolines can often be detected by the red coloration that develops when they are exposed to air or to mineral acids, and also by the ease with which they reduce silver nitrate solution.[7] The most convenient method, however, is by UV spectral measurement. Due to the instability of simple 1,2-dihydroisoquinolines, reliable quantitative spectral data have been difficult to obtain, although some have been reported.[20, 64, 87] Nevertheless, the qualitative UV absorption curve characteristically contains a medium-intensity broad band at 330–340 nm (in ethanol)[75, 144, 145], and this is extremely useful for the detection of 1,2-dihydroisoquinoline structures.

The Δ^3-double bond exhibits a sharp, medium-intensity band at 1610–1640 cm^{-1} in the infrared, and this is shifted to 1650–1675 cm^{-1} in the imminium ion, a characteristic property of enamines.[3] Insufficient NMR spectral data are available for generalizations to be made, and the mass spectral fragmentation patterns are virtually unknown.

2-Methyl-1,2-dihydroisoquinoline has been estimated[76] both iodometrically and by UV spectroscopy, using λ_{max}: 235 nm (ϵ_{max}: 7.9×10^3) as standard.

V. Reactions

A. Oxidation

When exposed to air, or better to manganese dioxide, 1,2-dihydroisoquinolines are readily oxidized to isocarbostyrils,[8, 105] whereas with iodine,[77] mercuric acetate,[78] or N-bromosuccinimide,[106] the products are the fully aromatic isoquinolinium salts.

---

[75] A. R. Battersby, D. J. Le Count, S. Garratt, and R. I. Thrift, *Tetrahedron* **14**, 46 (1961).
[76] K. T. Shamasundar, Ph.D. Thesis, Boston University, 1967.
[77] W. H. Perkin, *J. Chem. Soc.* **113**, 492 (1918).
[78] J. Knabe, J. Kubitz, and H. Roloff, *Arch. Pharm.* **298**, 401 (1965).

B. REDUCTION

1,2-Dihydroisoquinolines are easily reduced to 1,2,3,4-tetrahydroisoquinoline derivatives catalytically,[55] or by means of $NaBH_4$ in the presence of a proton source. Gensler[25] ascribes to Robinson and Shinoda[79] a report that the reduction of **37a** with stannous chloride gives **37b**, but the reduction actually described[79] is of **38** and is entirely unexceptional.

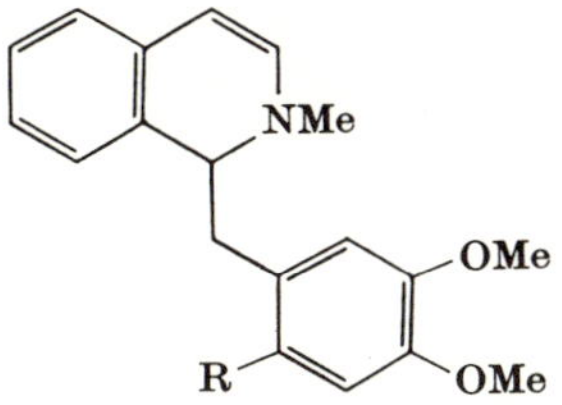

(37a) R = NO_2
(37b) R = NH_2

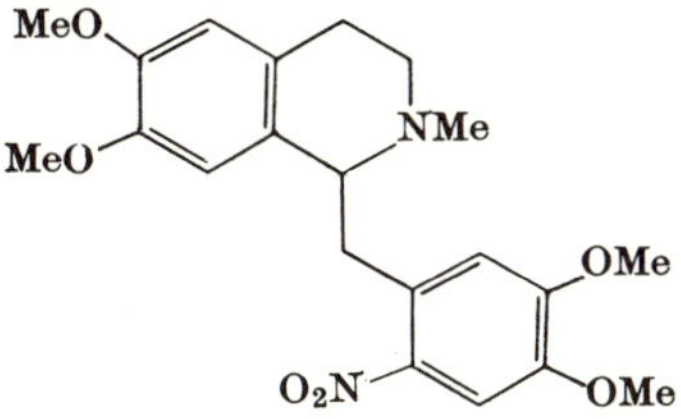

(38)

C. HYDROBORATION

It has been reported[80] that cyclohexanone enamines (**39**) can be converted into the β-hydroxyamines (**40**) by successive treatment with diborane and hydrogen peroxide. The method has been used[81] to convert berberine [(**41**) R + R = CH_2O_2] into the ophiocarpines (**42**)

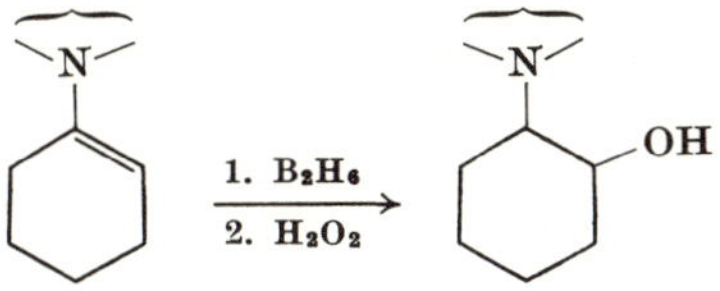

(39) 1. B_2H_6 2. H_2O_2 (40)

---

[79] R. Robinson and J. Shinoda, *J. Chem. Soc.* 1987 (1926).
[80] J. W. Lewis and A. A. Pearce, *Tetrahedron Lett.* 2039 (1964); I. J. Borowitz and G. J. Williams, *J. Org. Chem.* **32**, 4157 (1967).
[81] I. W. Elliot, *J. Heterocycl. Chem.* **4**, 639 (1967).

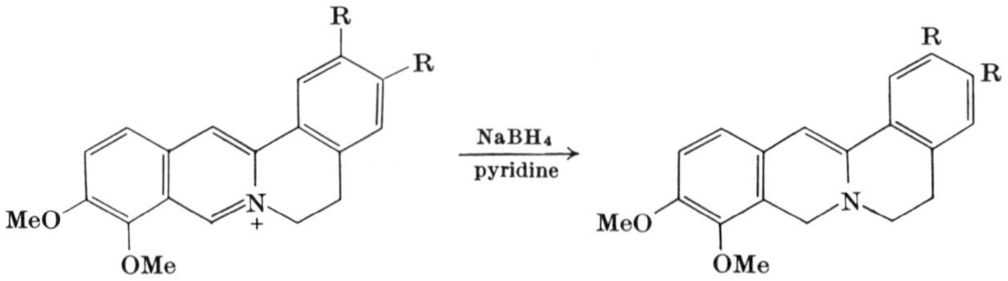

via dihydroberberine (**34**), and to prepare[82] **44** from 2-methyl-1,2-dihydropapaverine [(**43**) R = Me].

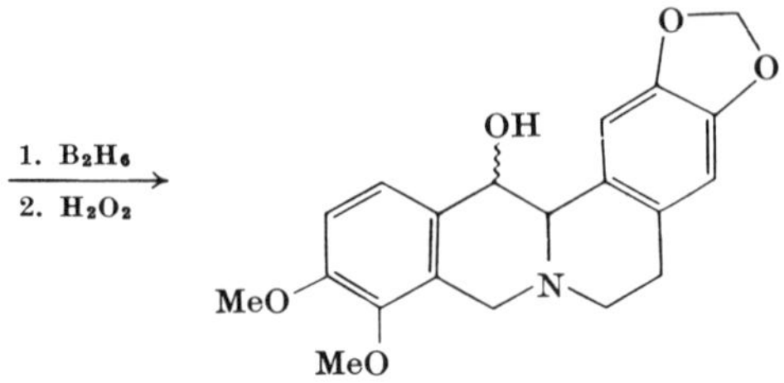

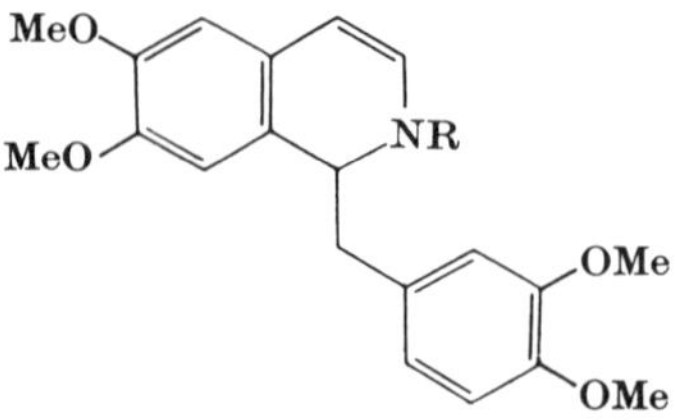

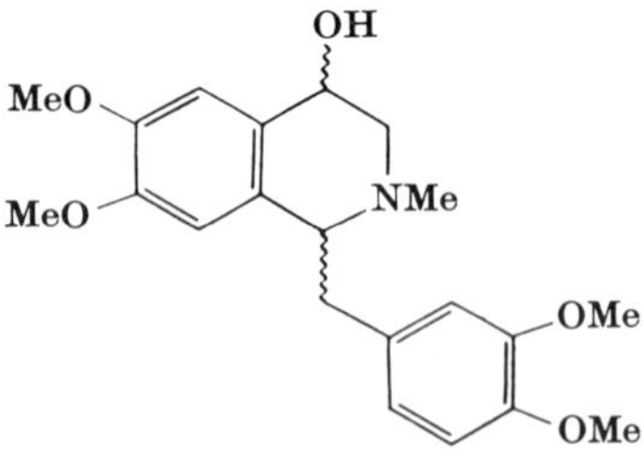

D. Enamine Reactions

It was pointed out explicitly by Battersby et al.[83,84] that 1,2-dihydroisoquinolines should exhibit enamine character, and should be susceptible to electrophilic attack at C-4, (**6**) → (**45**) → (**46**), and,

[82] S. F. Dyke and A. C. Ellis, *Tetrahedron* **27**, 3803 (1971).
[83] A. R. Battersby, R. Binks, and P. S. Uzzell, *Chem. Ind.* 1039 (1955).
[84] A. R. Battersby and R. Binks, *J. Chem. Soc.* 2888 (1955).

particularly when E^+ is a proton, to nucleophilic attack at C-3, **(45)** → **(47)**.

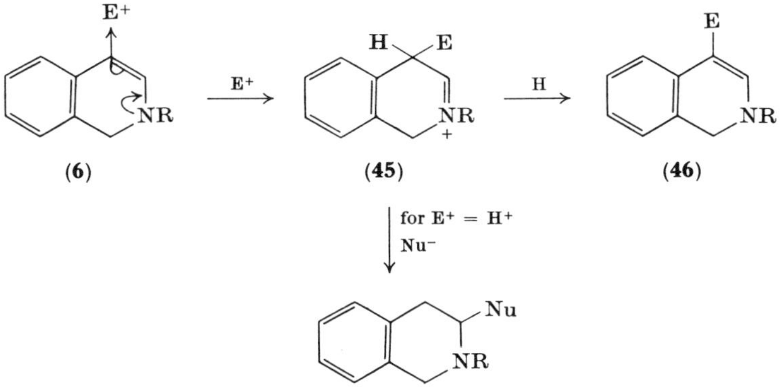

(6)   **(45)**   **(46)**

(47)

1. Electrophilic Attack at C-4

a. *Protonation.* In general, protonation of an enamine may occur at the nitrogen atom or at the C-β position, and unless steric or other factors are involved, the latter seems to be favored. A mixture of N- and C-β-protonation has been observed, for example, with **48** to

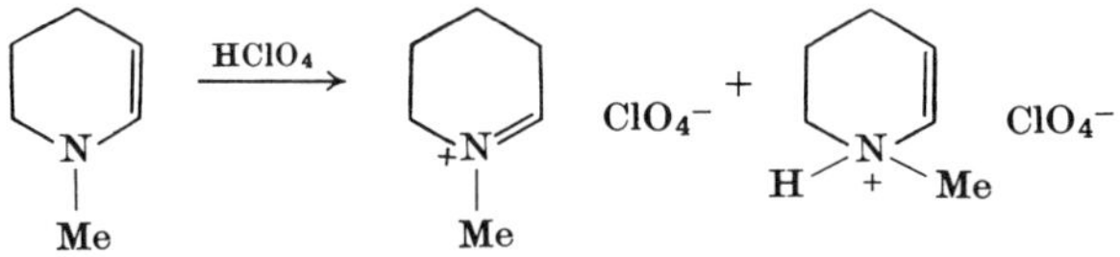

(48)   **(49)**   **(50)**

give **49** and **50**.[85] With appropriately substituted 1,2-dihydroisoquinolines, the formation of pavines (Section V, D, 2, b) and berbines (Section V, D, 2, c) upon acid treatment indicates that protonation at C-4 is occurring in these cases. The migration of a benzyl group (Section V, E, 1) or an allyl group (Section V, E, 2) from C-1 to C-3 also requires protonation at C-4 of the 1,2-dihydroisoquinoline.

Disproportionation to an equimolecular mixture of the 1,2,3,4-tetrahydroisoquinoline and the fully aromatic derivative is a common consequence of the interaction of 1,2-dihydroisoquinolines with acids. It is possible that protonation at C-4 occurs first (Scheme II), but with

[85] N. J. Leonard and F. D. Hauck, *J. Amer. Chem. Soc.* **79**, 5279 (1957).

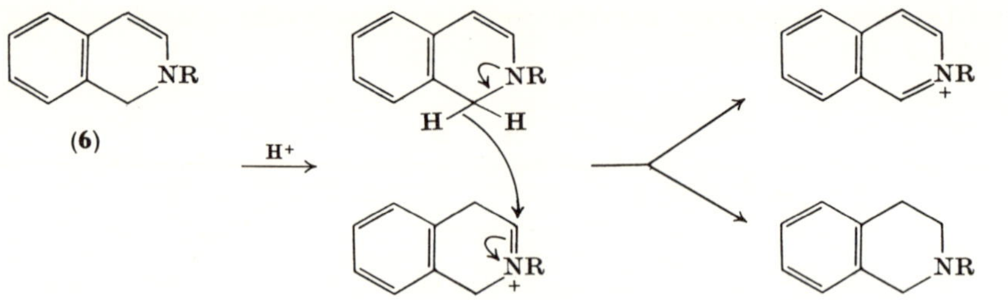

SCHEME II

compounds such as **51**[86] and **52**,[87] where cyclization, and therefore presumably protonation at C-4, does not occur, disproportionation is

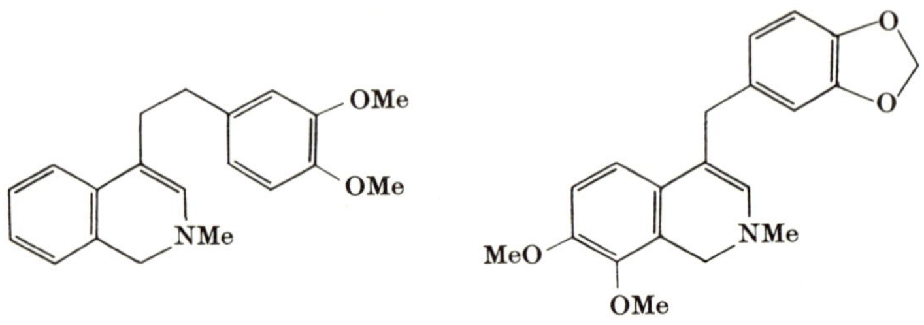

(51)

(52)

nevertheless still observed. A different mechanism (Scheme III) may be operating in these cases.

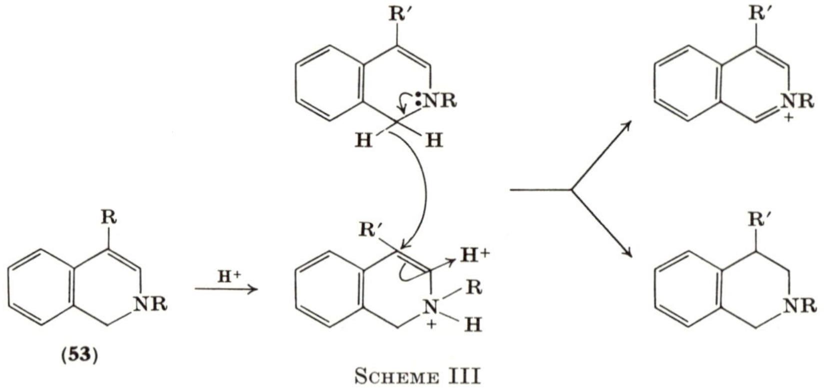

SCHEME III

[86] S. F. Dyke, M. Sainsbury, and B. J. Moon, *Tetrahedron* **24**, 1467 (1968).
[87] S. F. Dyke, M. Sainsbury, D. W. Brown, M. N. Palfreyman, and D. W. Wiggins, *Tetrahedron* **27**, 281 (1971).

Sec. V. D.1] 1,2-DIHYDROISOQUINOLINES 299

A further pathway for the iminium ion involves dimerization, and this has been observed quite frequently[38,57] (Scheme IV). Trimerization of 1,2-dihydroisoquinolines has also been reported.[38,54]

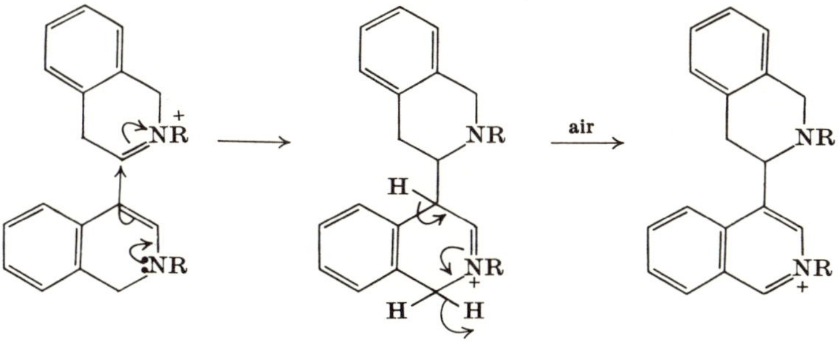

SCHEME IV

b. *Alkylation.* 2-Methyl-1,2-dihydroisoquinoline [(6) R = Me] reacts[27,88] with a variety of activated halides to yield, after oxidation of the intermediate 4-alkyl-1,2-dihydroisoquinoline with iodine, isoquinolinium salts (54). The yields of products, although modest (usually 10–20%), compare favorably with those resulting from the

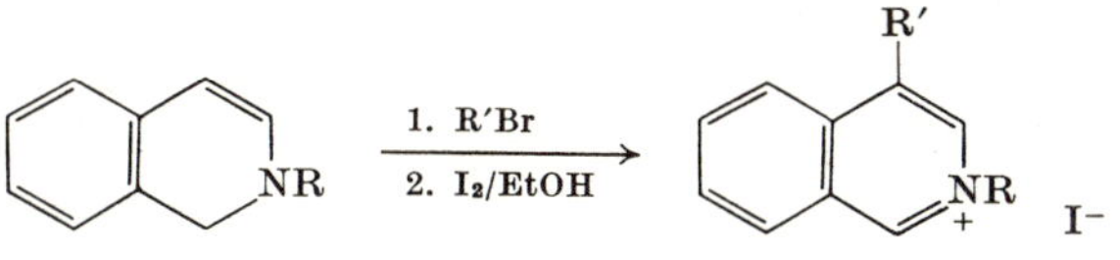

more traditional isoquinoline syntheses. Ethyl chloroformate, allyl bromide, propargyl bromide and ethyl bromide failed[88] to react with this enamine.

A closely related reaction with berberines has been known[89] for some time. Palmatine [(41) R = OMe] condenses with acetone to yield the 1,2-dihydroisoquinoline derivative [(55) R = OMe] which, with methyl iodide, followed by acid-catalyzed removal of the acetone residue, yields 56. Reduction then gives corydaline [(57) R = OMe]. Dihydroberberine (34) is alkylated by methanol in a similar reaction

[88] M. Sainsbury, D. W. Brown, S. F. Dyke, R. D. J. Clipperton, and W. R. Tonkyn, *Tetrahedron* **26**, 2239 (1970).
[89] F. Bruckhausen, *Arch. Pharm.* **261**, 28 (1922).

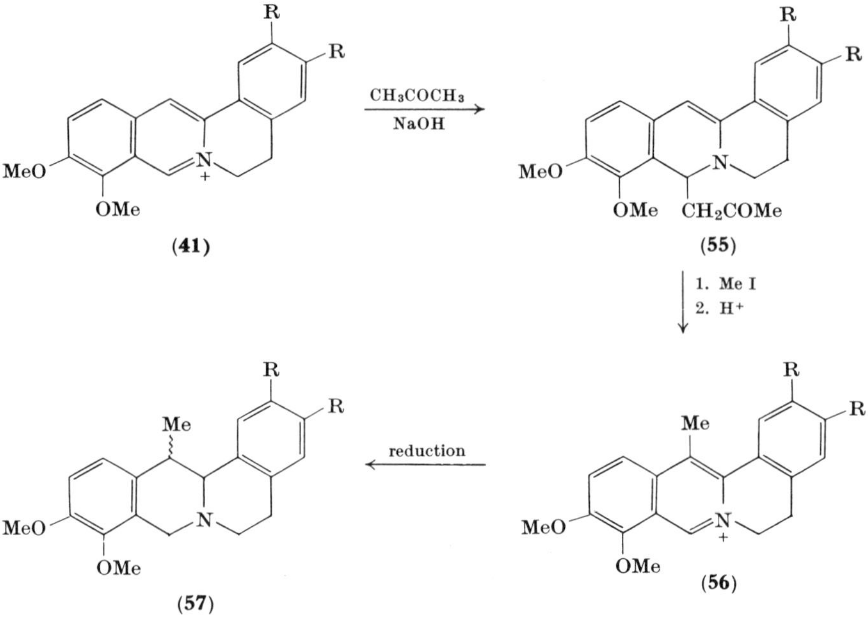

sequence to yield **57** (R + R = CH$_2$O$_2$).[90] Other examples of alkylation of dihydroberberine derivatives have been recorded.[91–93] The conversion of anhydrocryptopine (**35**) into epicryptopirubin chloride[69] (**58**) may be regarded as an example of an intramolecular alkylation reaction.

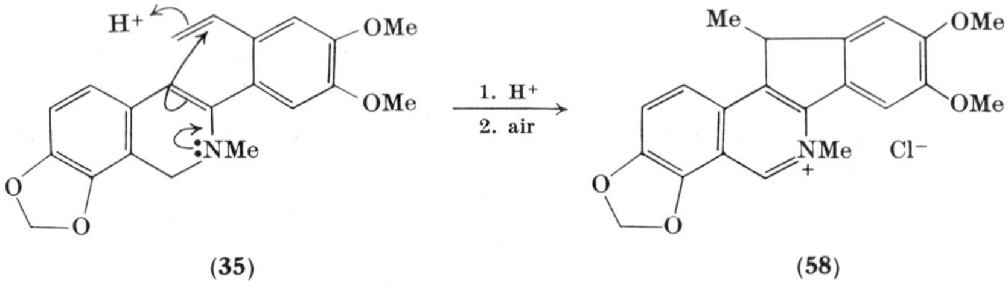

c. *Acylation.* Simple enamines, e.g., **39**, react readily with acid chlorides to yield β-diketones (**60**); the intermediate vinylogous

[90] T. Takemoto and Y. Kondo, *Yakagu Zasshi* **82**, 1408 (1962).
[91] M. Freund and K. Fleischer, *Ann.* **409**, 188 (1915).
[92] G. M. Robinson and R. Robinson, *J. Chem. Soc.* **111**, 958 (1917).
[93] H. W. Bersch, *Arch. Pharm.* **283**, 192 (1950).

amides (**59**) are not usually isolated. Normal benzoylation products have been described,[94] together with the dibenzoyl compound (**61**).[95] With cyclohexanone enamines (**39**) and o-nitrophenylacetyl chlorides,

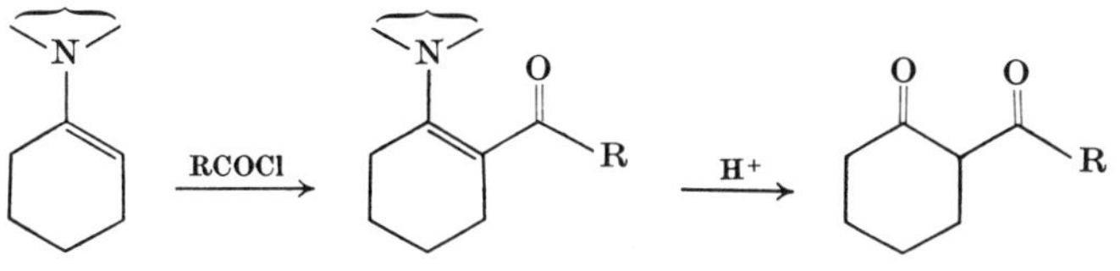

(**39**)   (**59**)   (**60**)

the expected product (**62**) is obtained,[96] but with α,β-unsaturated acid chlorides, the abnormal product (**63**) has been reported.[97]

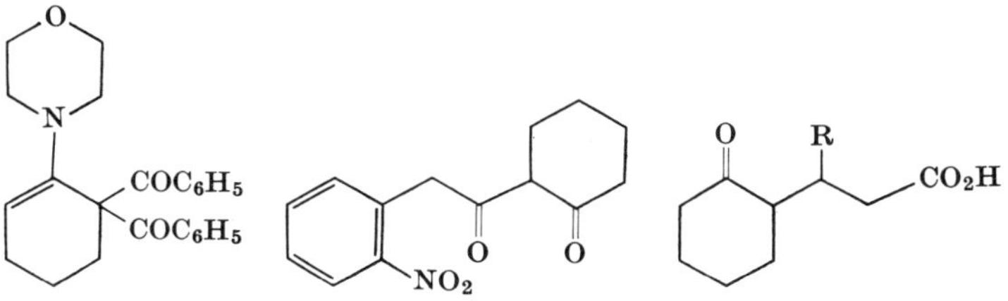

(**61**)   (**62**)   (**63**)

2-Methyl-1,2-dihydroisoquinoline has been reacted[8, 66] with a variety of acid chlorides (Table III); the expected vinylogous amides (**64**) were isolated in most cases. 2-Benzyl-1,2-dihydroisoquinoline behaves similarly. The acylation reaction fails with simple aliphatic acid chlorides. Sometimes, the reaction of the enamine with the acid in the presence of dicyclohexylcarbodiimide succeeds. The 1,2-dihydroisoquinoline (**65**) also reacts with ethoxalyl chloride to yield **66**.[66]

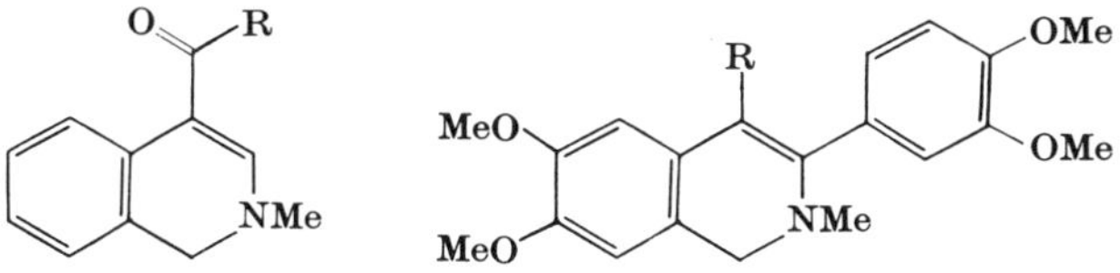

(**64**)

(**65**) R = H
(**66**) R = COCO$_2$Et

[94] R. Helmers, *Tetrahedron Lett.* 1905 (1966).
[95] R. D. Campbell and J. A. Jung, *J. Org. Chem.* **30**, 3711 (1965).
[96] P. Rosenmund and W. H. Haase, *Chem. Ber.* **99**, 2504 (1966).
[97] P. W. Hickmott, *Proc. Chem. Soc.* 287 (1964); P. W. Hickmott and B. J. Hopkins, *J. Chem. Soc. C* 2918 (1968) and references therein.

TABLE III

ACYLATION OF 2-METHYL-1,2-DIHYDROISOQUINOLINE

R	Yield (%)[a] I	II	R	Yield (%)[a] I	II
C_6H_5CO	33[b]	—	$C_6H_5CH_2CO$	30	34
o-$NO_2C_6H_4CO$	65	6	o-$NO_2C_6H_4CH_2CO$	20	10
m-$NO_2C_6H_4CO$	85	—	3,4-$(MeO)_2C_6H_3CH_2CO$	14	26
p-$NO_2C_6H_4CO$	73	—	2-Furoyl	33	—
2-NO_2-3-$MeOC_6H_3CO$	5	—	$ClCH_2CO$	20	27
3,4-$(OMe)_2C_6H_3CO$	33	—	EtO_2C–CO	21	—
3,4-$Cl_2C_6H_3CO$	11	—	$3NO_2C_6H_4SO_2$	28	—
3,4-$Me_2C_6H_3CO$	14	—	Indolyl-3-acetyl	—	24
H·CO	33[c]	—	CH_3CO[c]	36	—

[a] Based on isoquinoline methiodide, I from the acid chloride, II from the acid plus carbodiimide.
[b] Isocarbostyril.
[c] Vilsmeier reaction.

The enamine (**37**) has also been benzoylated.[73] The acylated enamines (**64**), which are easily oxidized by air or manganese dioxide to isocarbostyrils (**67**), are protonated at oxygen to give ions of the type (**68**),[8] in agreement with other experience.[98]

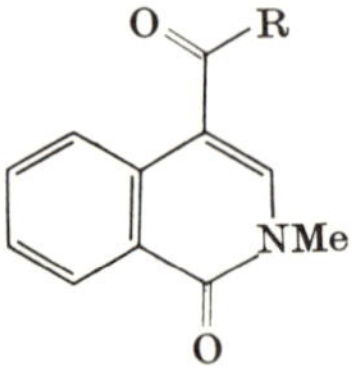

(**67**) 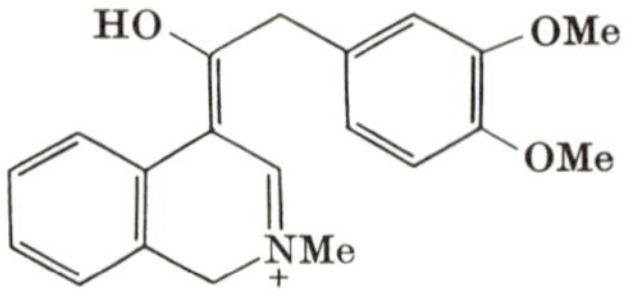   (**68**) 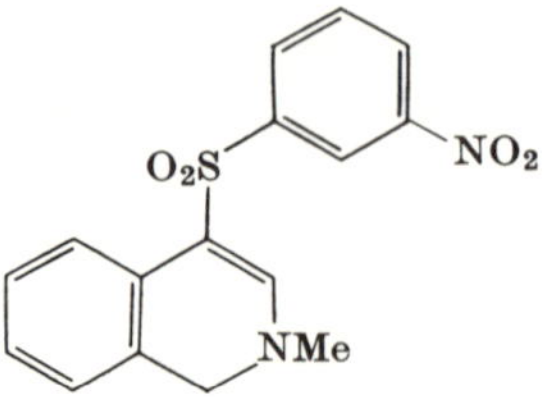   (**69**)

m-Nitrobenzene sulfonyl chloride reacts with 2-methyl-1,2-dihydroisoquinoline[66] to give the vinylogous sulfonamide (**69**).

[98] G. H. Alt and A. J. Speciale, *J. Org. Chem.* **30**, 1407 (1965); A. I. Meyers, A. H. Reeve, and R. Gault, *ibid.* **34**, 698 (1969).

Sec. V. D.1] 1,2-DIHYDROISOQUINOLINES 303

The synthetic potential of 4-acylisoquinoline derivatives is illustrated by the conversion of **70** into **71**[66] and of **72** into **73**.[87]

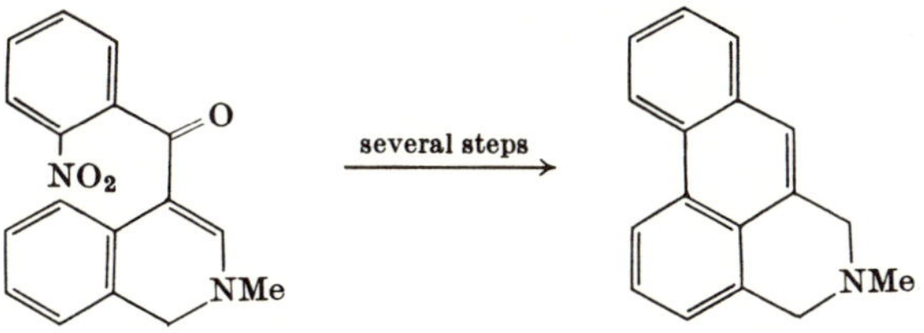

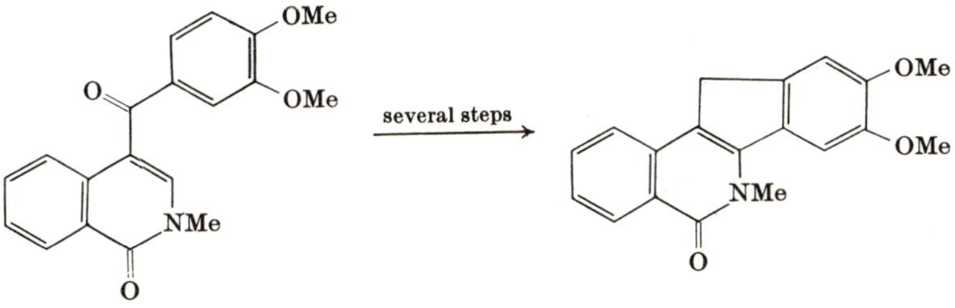

Other heterocyclic enamines, such as **74**[99] and **75**,[100] have been acylated, as well as some indoles[101] and also 1-ethyl-2-methylene-1,2-dihydroquinoline.[102]

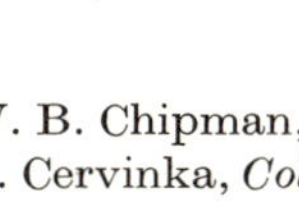

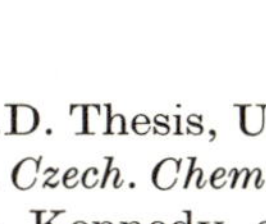

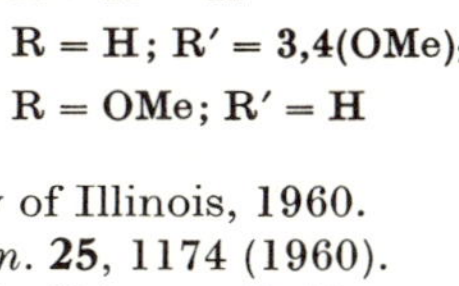

(76) R = R' = H
(77) R = H; R' = 3,4(OMe)$_2$C$_6$H$_3$
(78) R = OMe; R' = H

[99] W. B. Chipman, Ph.D. Thesis, University of Illinois, 1960.
[100] O. Cervinka, *Collect. Czech. Chem. Commun.* **25**, 1174 (1960).
[101] J. I. De Graw, J. G. Kennedy, and W. A. Skinner, *J. Heterocycl. Chem.* **3**, 9 (1966).
[102] G. H. Alt, *J. Org. Chem.* **31**, 2384 (1966).

The Vilsmeier reaction,[103] which has been successfully applied[104] to the formation of 3-acylindoles, has yielded the 4-formyl-1,2-dihydroisoquinolines (**76**)–(**78**).[105–107]

d. *Reactions with Aldehydes.* Cyclopentanone enamine gives the

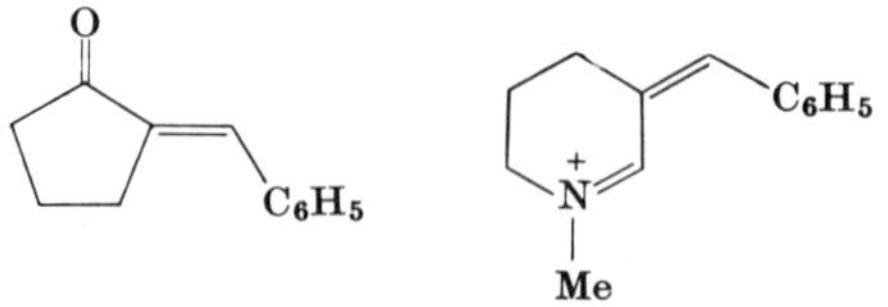

(**79**)   (**80**)

expected product (**79**) with benzaldehyde,[108] and **80** has been obtained from N-methyl-Δ^2-piperidine.[109]

When 1,2,3,4-tetrahydroisoquinoline is heated in acetic acid with benzaldehyde, 4-benzylisoquinoline can be isolated in 34% yield. It was suggested[110] that, by analogy with the products obtained from the condensation of benzaldehyde with piperidine, the reaction proceeds through a 1,2-dihydroisoquinoline intermediate (Scheme V).

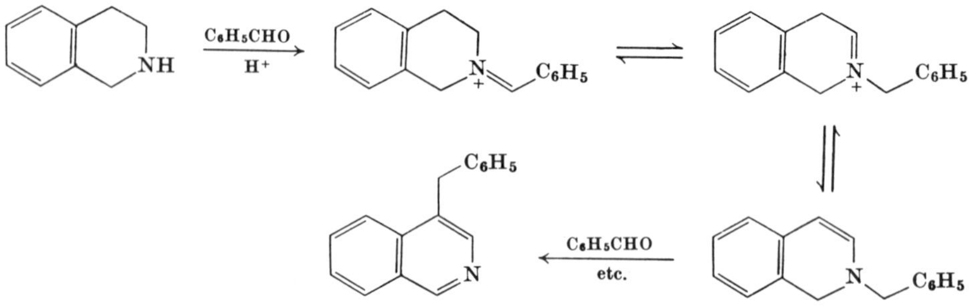

SCHEME V

[103] V. I. Minkin and G. N. Dorofeenko, *Russ. Chem. Rev.* **29**, 599 (1960).
[104] W. C. Anthony, *J. Org. Chem.* **25**, 2049 (1960); J. M. Muchowski and D. G. Horning, *Can. J. Chem.* **48**, 193 (1970).
[105] S. F. Dyke, M. Sainsbury, D. W. Brown, and R. D. J. Clipperton, *Tetrahedron* **26**, 5969 (1970).
[106] B. J. Moon, Ph.D. Thesis, Bath University, 1969.
[107] S. F. Dyke and P. A. Bather, to be published.
[108] L. Birhofer, S. M. Kim, and H. D. Engels, *Chem. Ber.* **95**, 1495 (1962); J. W. Lewis, P. L. Myers, and M. J. Readhead, *J. Chem. Soc. C* 771 (1970).
[109] C. Schopf, R. Klug, and R. Rausch, *Ann.* **616**, 151 (1958).
[110] W. D. Burrows and E. P. Burrows, *J. Org. Chem.* **28**, 1180 (1963).

Each step of Scheme V has been separately tested.[27] Although the details of the steps involved in Scheme V were not explicitly stated,[110] it is probably the first condensation reaction between a 1,2-dihydroisoquinoline and an aromatic aldehyde to be described. 4-Benzylisoquinoline has also been obtained[111, 112] by heating N-benzoyl-1,2,3,4-tetrahydroisoquinoline with benzaldehyde under pressure.

In 1964 Grewe et al.[55] reported that when a mixture of isoquinoline methiodide and benzaldehyde is hydrogenated in the presence of alkali, a 64% yield of 4-benzyl-2-methyl-1,2,3,4-tetrahydroisoquinoline can be isolated. It was postulated that reduction of isoquinoline methiodide occurs to form the 1,2-dihydroisoquinoline, which then condenses with the benzaldehyde to form **81** and this is immediately further reduced. With isoquinoline itself and a variety

(81)

of carbonyl compounds, a similar reaction was claimed,[55] but it has been shown[27, 58] that reductive alkylation occurs predominantly at nitrogen, rather than at C-4, with aromatic aldehydes, and entirely at nitrogen with aliphatic carbonyl compounds. 4-Alkyl-1,2,3,4-tetrahydroisoquinolines have also been obtained in good yield by reducing mixtures of isoquinoline methiodide and aldehydes with $NaBH_4$.[17, 22, 56]

Bobbitt et al.[113] reported that when benzylaminoacetaldehyde dialkyl acetals (**18**) are heated under reflux with concentrated hydrochloric acid and ethanol, in the presence of benzaldehyde, good yields of the 4-benzylisoquinoline derivatives (**85**), via the 1,2-dihydroisoquinoline (**82**), can be isolated. A mechanism for this reaction[114] is summarized in Scheme VI. With cyclohexanone enamine and aldehydes, the intermediate has been formulated[115] as **86**. It was originally reported[114] that an intermediate of the

[111] L. Rugheimer, Ann. **326**, 261 (1903).
[112] L. Rugheimer and E. Albreckt, Ann. **326**, 297 (1903).
[113] J. M. Bobbitt, D. P. Winter, and J. M. Kiely, J. Org. Chem. **30**, 2459 (1965).
[114] S. F. Dyke and M. Sainsbury, Tetrahedron **23**, 3161 (1967).
[115] L. Birkhofer, S. M. Kim, and H. D. Engels, Chem. Ber. **95**, 1495 (1962).

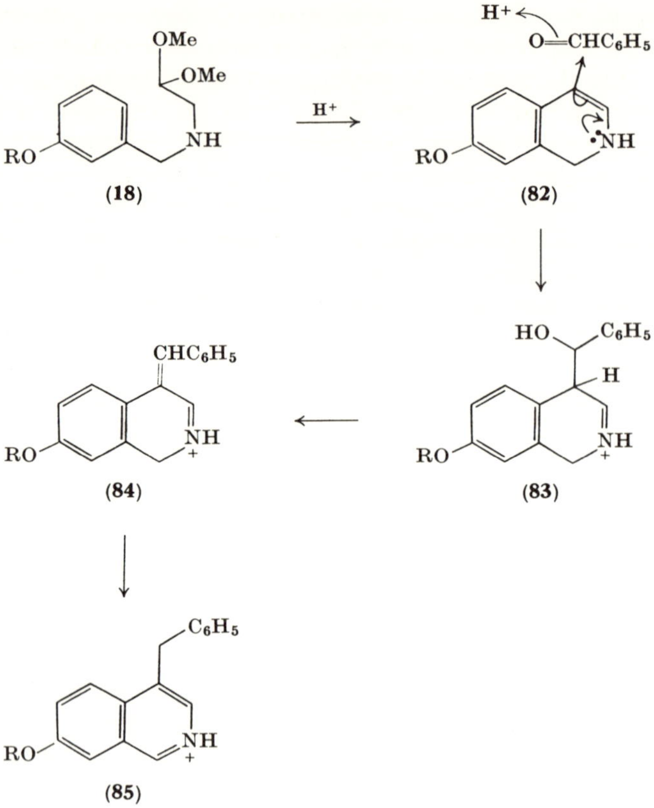

SCHEME VI

type (**83**) has been isolated, in one case, but it is now known[18] that the correct structures are of the 4-benzylidene-1,4-dihydroisoquinoline type (**84**). These latter compounds are easily isomerized to 4-benzylisoquinolines by acids or bases. The compounds of both types (**84**) and (**85**) reported are collected into Tables IV and V. Included in Table V are the compounds obtained when isoquinolinium salts are reacted with aldehydes in the presence of base. 2-Methyl-1,2-dihydroisoquinoline, 2,3-dimethyl-1,2-dihydroisoquinoline, and the ester (**87**) have also been reacted with various aldehydes to give 2-methyl-4-substituted isoquinolinium salts.[41, 116]

[116] S. F. Dyke and J. R. Evans, to be published.

TABLE IV
4-SUBSTITUTED ISOQUINOLINES FROM BENZYLAMINOACETALS

			Substituent[a]				Yield	
1	N	3	4	6	7	8	(%)	Refs.[b]
H	H	H	$C_6H_5CH_2$	H	OMe	OMe	30	1
H	H	H	$C_6H_5CH_2$	OMe	OMe	H	35	1
H	H	H	$pMeOC_6H_4CH_2$	OMe	OMe	H	47	1
H	H	H	$3,4\text{-}(OMe)_2C_6H_3COCH_2$	OMe	OMe	H	82	1
H	H	H	$3,4\text{-}(OMe)_2C_6H_3COCH_2$	H	OMe	OMe		1
H	H	H	CH_2CO_2H	$O\text{-}CH_2\text{-}O$		H	68	2
H	H	H	CH_2CO_2H	OMe	OMe	H	67	2
H	H	H	$3,4(OMe)_2C_6H_3COCH_2$	H	OMe	OMe	20	3
H	H	H	CH_2CO_2H	H	$O\text{-}CH_2\text{-}O$		46	3, 4
H	H	H	$3\text{-}OH\text{-}4\text{-}OMeC_6H_3CH_2$	H	OMe	OH	60	3
H	H	H	$3\text{-}OH\text{-}4\text{-}OMeC_6H_3CH_2$	H	OMe	H	60	3
H	H	H	A	OMe	OMe	H	10	5
H	H	H	B	OMe	OMe	H	18	5
H	H	H	A	H	OMe	OMe	13	5
H	H	H	A	H	OMe	OH	26	5
H	H	H	C	OMe	OMe	H	9	5
H	H	H	$C_6H_5CH_2$	OH	OMe	H	63	6
H	H	H	$C_6H_5CH_2$	H	OMe	OH	61	6
H	H	H	$C_6H_5CH_2$	MeO	OH	H	54	6

[a] A, [benzisoxazole with two OMe groups]; B, [benzisoxazole with methylenedioxy]; and C, [benzisoxazole].

[b] KEY TO REFERENCES:
1. S. F. Dyke and M. Sainsbury, *Tetrahedron* **23**, 3161 (1967).
2. S. F. Dyke, M. Sainsbury, and B. J. Moon, *Tetrahedron* **24**, 1467 (1968).
3. D. W. Brown, S. F. Dyke, and M. Sainsbury, *Tetrahedron* **25**, 101 (1969).
4. M. Sainsbury, S. F. Dyke, and B. J. Moon, *J. Chem. Soc. C* 1797 (1970).
5. S. F. Dyke, M. Sainsbury, D. W. Brown, and M. N. Palfreyman, *Tetrahedron* **25**, 5365 (1969).
6. J. M. Bobbitt, D. P. Winter, and J. M. Kiely, *J. Org. Chem.* **30**, 2459 (1965).

TABLE V

4,BENZYLIDINE-1,4-DIHYDROISOQUINOLINES [a]

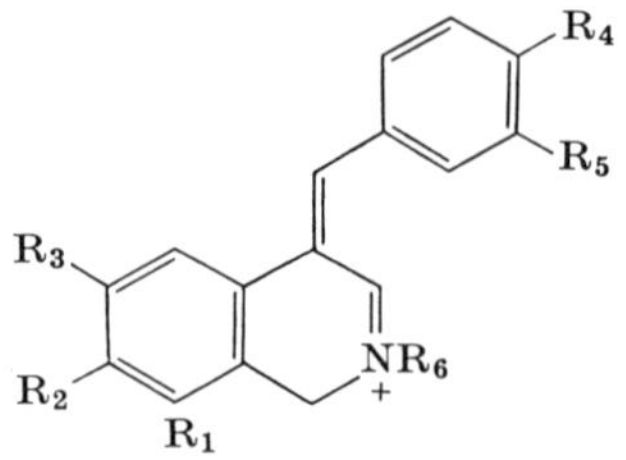

Conditions	Substituent						Yield (%)
	R_1	R_2	R_3	R_4	R_5	R_6	
Acid	H	OMe	OMe	H	H	H	69
Acid	OMe	OMe	H	H	H	H	60
Acid	OH	OMe	H	H	H	H	30
Acid	H	OMe	OMe	OMe	OMe	H	75
Acid	OMe	OMe	H	O–CH_2–O		H	60
Acid	H	OMe	OMe	O–CH_2–O		H	53
Acid	H	OMe	OMe	O–CH_2–O		Me	20
Acid	H	OMe	OMe	H	H	Me	70
Acid	H	OMe	OMe	O–CH_2–O		Me	70
Acid	OMe	OMe	H	O–CH_2–O		Me	65
NaOEt	H	H	H	H	H	Me	47
NaOEt	H	H	H	Me	H	Me	34
NaOEt	H	H	H	OMe	H	Me	20
NaOEt	H	H	H	H	OMe	Me	26
NaOEt	H	H	H	Cl	H	Me	6

[a] From Brown et al.[18]

Various types of by-products have been identified in some of these reactions between 1,2-dihydroisoquinolines and aldehydes. Thus,

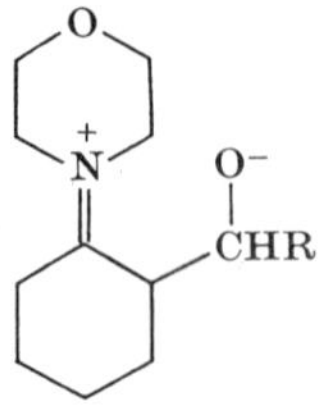

(86) 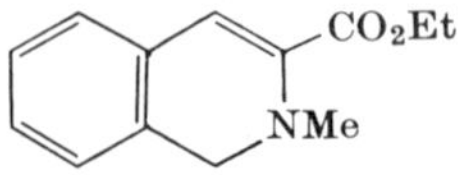   (87)

with arylglyoxals, the N-alkylated product (**88**) can predominate unless the conditions, especially the ratio of concentrated hydrochloric acid to ethanol, are carefully controlled.[114] This type of side reaction can be eliminated entirely by using the N-alkylaminoacetal instead

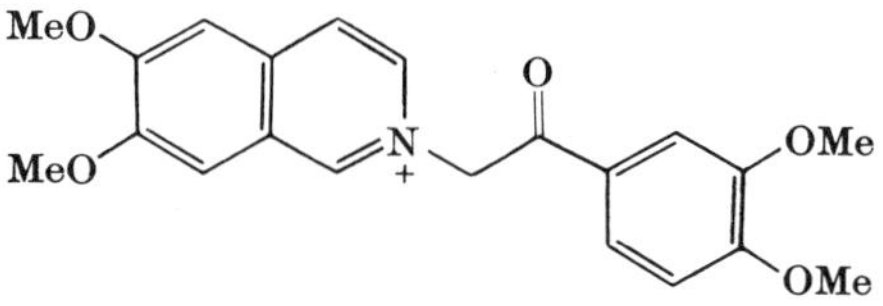

(**88**) 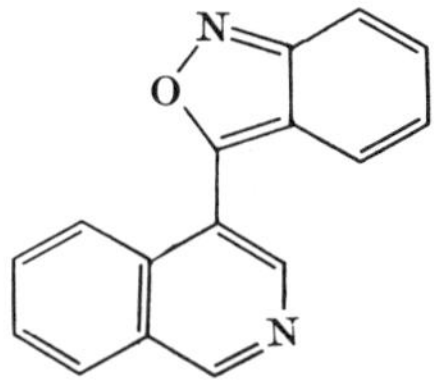　(**89**)

of the secondary amine. With o-nitrobenzaldehydes, anthranils (**89**) are formed exclusively,[65] and with veratraldehyde and piperonal, small amounts of indeno[1,2-c]isoquinolines (**90**) (Section V,D,2,e) can be isolated.[87, 117]

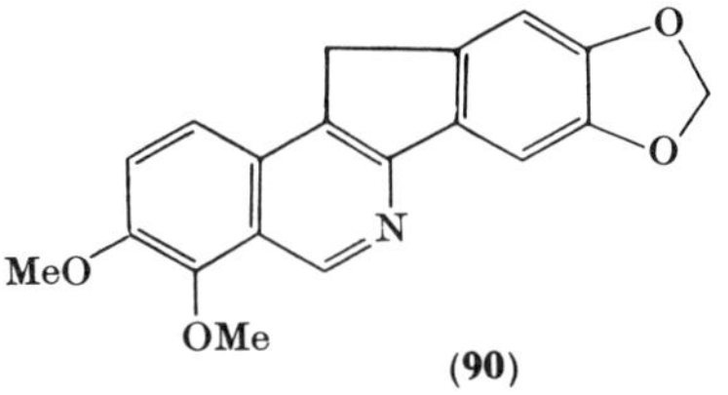

(**90**) 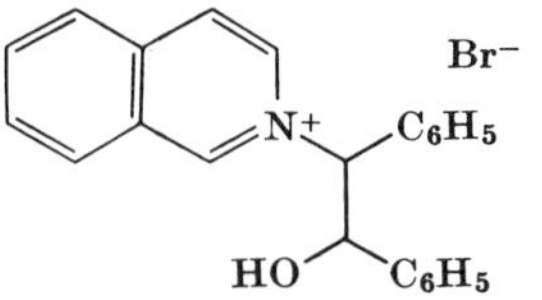　(**91**)

In 1935 Krohnke[118] reported that benzaldehyde condenses, under alkaline conditions, with 2-benzylisoquinolinium bromide to yield **91**, but it was shown much later that a mixture of products is formed, one of which is **91**. Another substance present was found to be **94** (R = CH$_2$Ph), which may be regarded as arising from the pseudobase (**92**) (Scheme VII) acting as an enamine.[27] Elimination of water from the intermediate (**93**) then yields the observed product. With isoquinoline methiodide in place of 2-benzylisoquinolinium bromide, this C-4-alkylation reaction is of preparative significance.[18] The products are **94** (R = Me) or the 4-benzylidene-1,4-dihydroisoquinolines (**84**) (Table V), depending on the amount of alkali used. The latter type of compound can best be accounted for by assuming that the initially formed pseudobase undergoes disproportionation,

[117] W. J. Gensler, K. T. Shamasundar, and S. Marburg, *J. Org. Chem.* **33**, 2861 (1968).
[118] F. Krohnke, *Ber.* **68**, 1351 (1935).

and that it is the 1,2-dihydroisoquinoline thus formed that then reacts with the aldehyde.

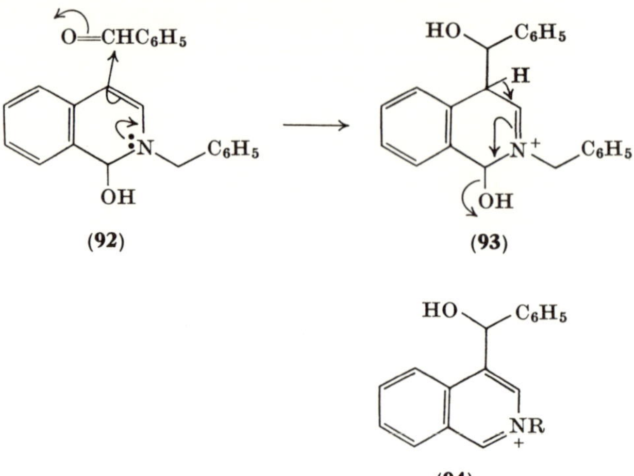

SCHEME VII

It has been reported[119] that when 2-methyl-1,2,3,4-tetrahydroisoquinoline-3-carboxylate is heated to about 200° in the presence of benzaldehyde, 4-benzylisoquinoline is formed and can be isolated in 73% yield. It is probable that decarboxylation occurs to form 2-methyl-1,2,3,4-tetrahydroisoquinoline, which then reacts with benzaldehyde via the enamine pathway as before.

2. *Nucleophilic Attack at C-3*

a. *Yohimbines.* Julian and Magnani[120] reported that when the amide (**95**) was dehydrogenated with palladium black, the 1,2-dihydroisoquinoline (**96**) was formed which, with LAH, gave compound **97** (R = Me). However, Belleau[121] pointed out that the intermediate is not the expected **96**, but the oxindole (**98**). Reduction with LAH then occurs with rearrangement to the observed product

---

[119] S. Tachibawa, H. Matsuo, and S.-I. Yamada, *Chem. Pharm. Bull.* **16**, 414 (1968).
[120] P. L. Julian and A. Magnani, *J. Amer. Chem. Soc.* **71**, 3207 (1949).
[121] B. Belleau, *Chem. Ind.* 229 (1955).

[(**97**) R = Me]. There are now several examples of such rearrangements.¹²²

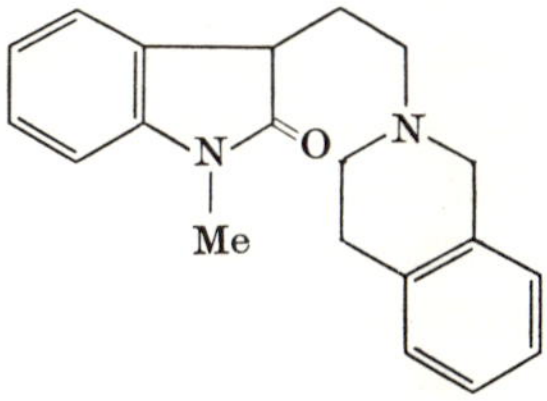

(95) 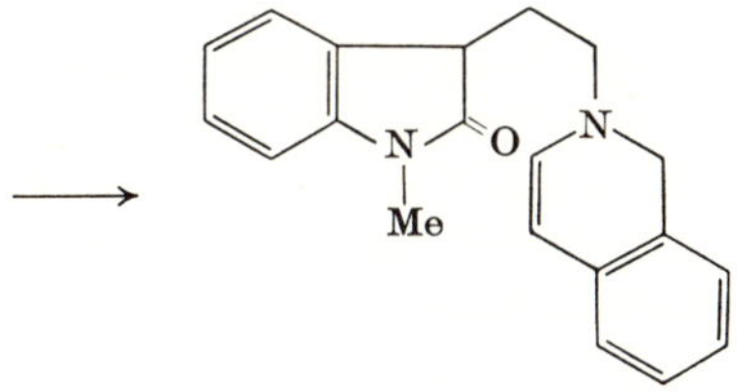 (96)

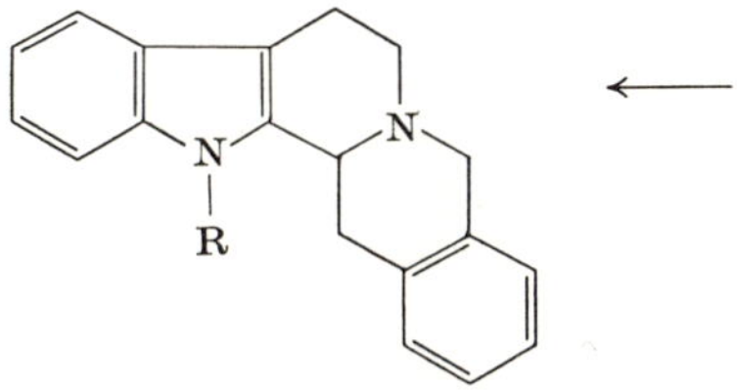

(97) 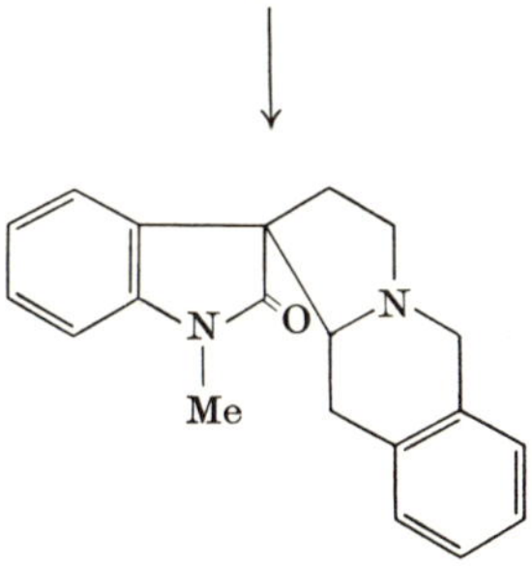 (98)

It was Belleau¹²¹ who first described the reduction of the isoquinolinium salt (**99**) to the 1,2-dihydroisoquinoline [(**100**) R = H] with LAH and its cyclization to **97** (R = H), via the imminium ion (**101**). The same reaction has also been described by others,¹²³,¹²⁴ the intermediate [(**100**) R = H] being claimed to be unstable by one

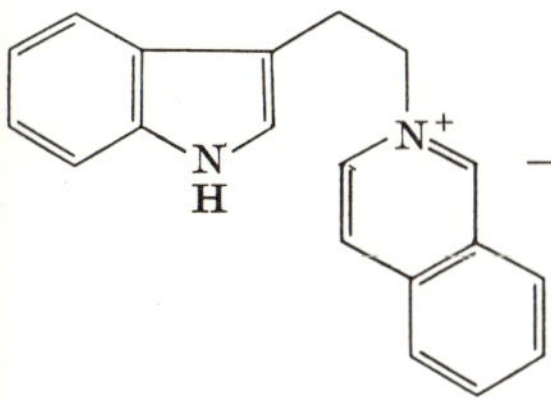

(99) 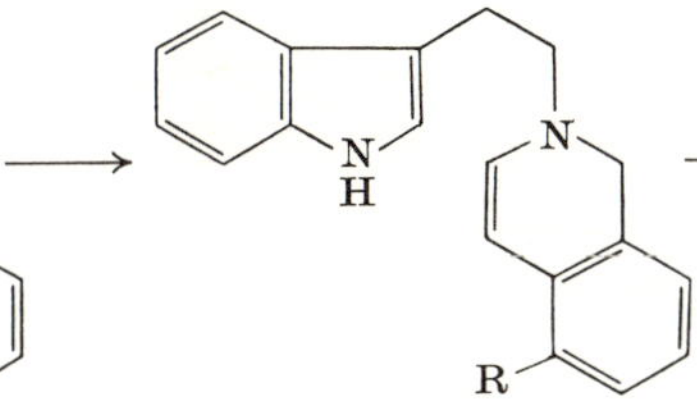 (100) 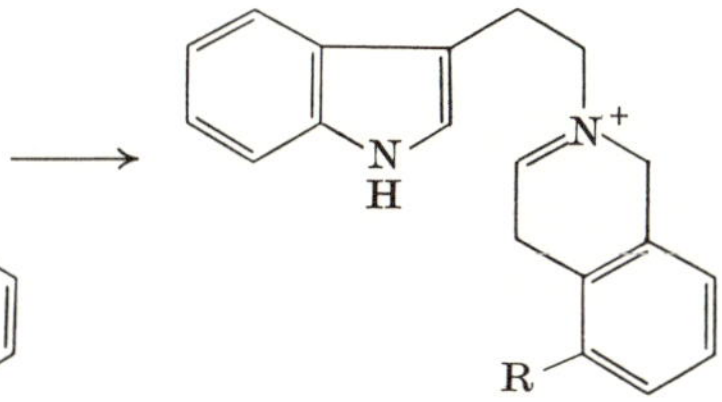 (101)

---

¹²² A. H. Jackson and B. Naidoo, *Tetrahedron* **25**, 4843 (1969) and references therein.
¹²³ K. T. Potts and R. Robinson, *J. Chem. Soc.* 2675 (1955).
¹²⁴ J. W. Huffman, *J. Amer. Chem. Soc.* **80**, 5193 (1958).

group,[124] but isolated crystalline by the other group.[123] A simplification[125] of the method of preparation of **100** (R = H) makes it a readily available compound. The overall synthetic sequence was quickly used[126] to synthesize the alkaloid alstonilinol [(**103**) R = CH_2OH] from the isoquinolinium salt (**102**). A recent modification[127] of this theme involves the reduction of **102** (but with H in place of OMe) with $NaBH_4$ in the presence of sodium cyanide. The 1,2-dihydroisoquinoline [(**100**) R = CO_2Et] is trapped as **104**. Treatment of this pseudocyanide with acid in a subsequent step regenerates the 1,2-dihydroisoquinoline, which is cyclized under the acid conditions to **103** (R = CO_2Et).

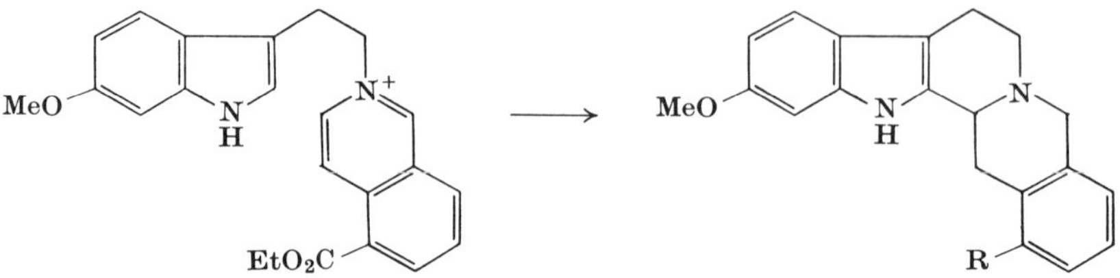

(**102**)

(**103**)

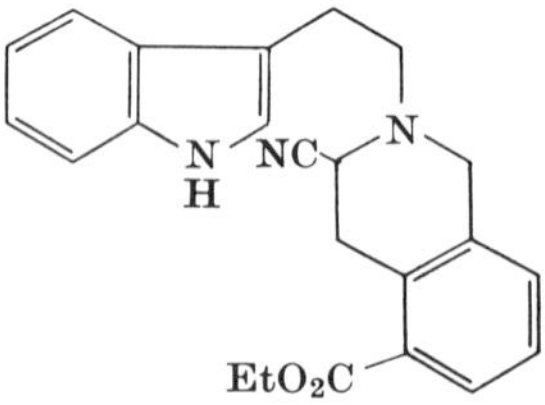

(**104**)

b. *Pavines*. When papaverine (**105**) is reduced with tin and hydrochloric acid, the major product is the expected 1,2,3,4-tetrahydropapaverine, but a small amount (ca. 5%) of another base $C_{20}H_{23}NO_4$, called[128] pavine, is also formed. Its structure was eventually shown,[84]

[125] D. R. Liljegren and K. T. Potts, *J. Org. Chem.* **27**, 377 (1962); K. T. Potts and D. R. Liljegren, *ibid.* **28**, 3202 (1963).
[126] R. C. Elderfield and B. A. Fischer, *J. Org. Chem.* **23**, 332 (1958).
[127] J. A. Beisher, *Tetrahedron* **26**, 1961 (1970).
[128] (a) H. Goldschmiedt, *Monatsh. Chem.* **7**, 485 (1886); **19**, 324 (1898); (b) F. L. Pyman, *J. Chem. Soc.* **95**, 1610 (1909); (c) F. L. Pyman and W. C. Reynolds, *J. Chem. Soc.* **97**, 1320 (1910).

by classical degradation methods, to be **106** (R = H). When papaverine methiodide is similarly reduced or better, is treated with LAH, followed by strong acid at elevated temperatures, *N*-methylpavine [(**106**) R = Me] is formed. It was suggested[84] that reduction of the aromatic isoquinoline gives the 1,2-dihydroisoquinoline [(**43**) R = H or Me] as intermediate, which is protonated at C-4 to yield **107** (R = H or Me) and then either undergoes further reduction to the tetrahydroisoquinoline or is cyclized by an intramolecular nucleophilic attack by the dimethoxyphenyl ring at C-3 (**105**) → (**43**) → (**107**) → (**106**). The parent compound [(**106**) R = Me], but with hydrogen in place of

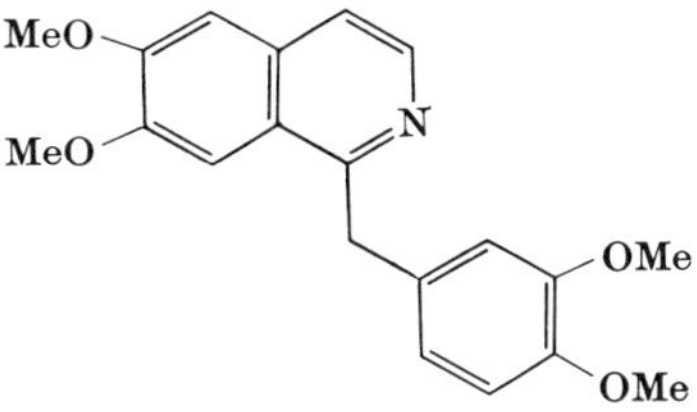

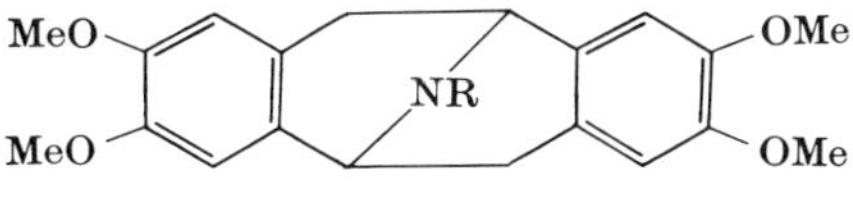

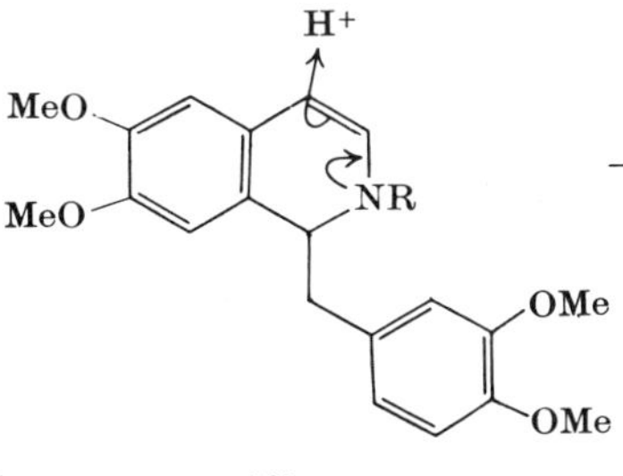

the four methoxyl groups, is also known,[129] and was prepared, in low yield, by heating 1-benzyl-2-methyl-1,2-dihydroisoquinoline with an excess of phosphoric acid.

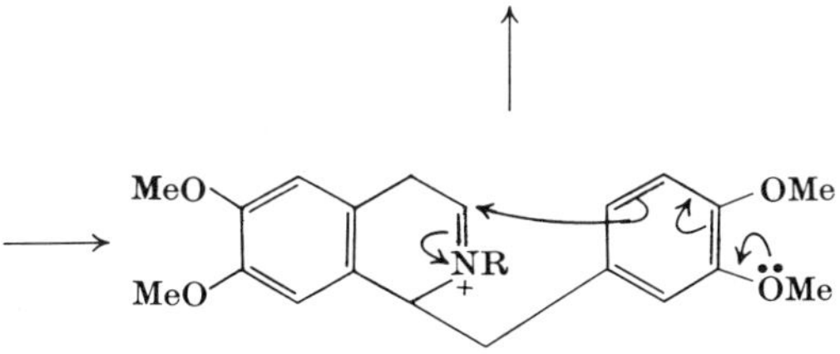

It is interesting to note that about nine alkaloids are now known[130] that possess the pavine type of nucleus and that two of these have been synthesized by the route outlined above.

[129] G. Wittig, H. Tenhaeff, W. Schoch, and G. Koenig, *Ann.* **572**, 1 (1951).
[130] T. Kametani, *in* "The Chemistry of the Isoquinoline Alkaloids," Chapter 4. Elsevier, Amsterdam, 1969.

c. *Berbines*. The logical extension [83] of the pavine cyclization reaction to the formation of the berbine skeleton from N-β-arylethylisoquinolinium salts (**109**) was reported independently by Battersby et al.[75] and Huffman and Miller.[131] This route to berbines, which has many advantages over other methods,[132] was applied by Battersby [75] to the synthesis of norcoralydine (**112**) from 6,7-dimethoxyisoquinoline (**108**) as indicated in (**108**) → (**112**). Reduction of (**109**)

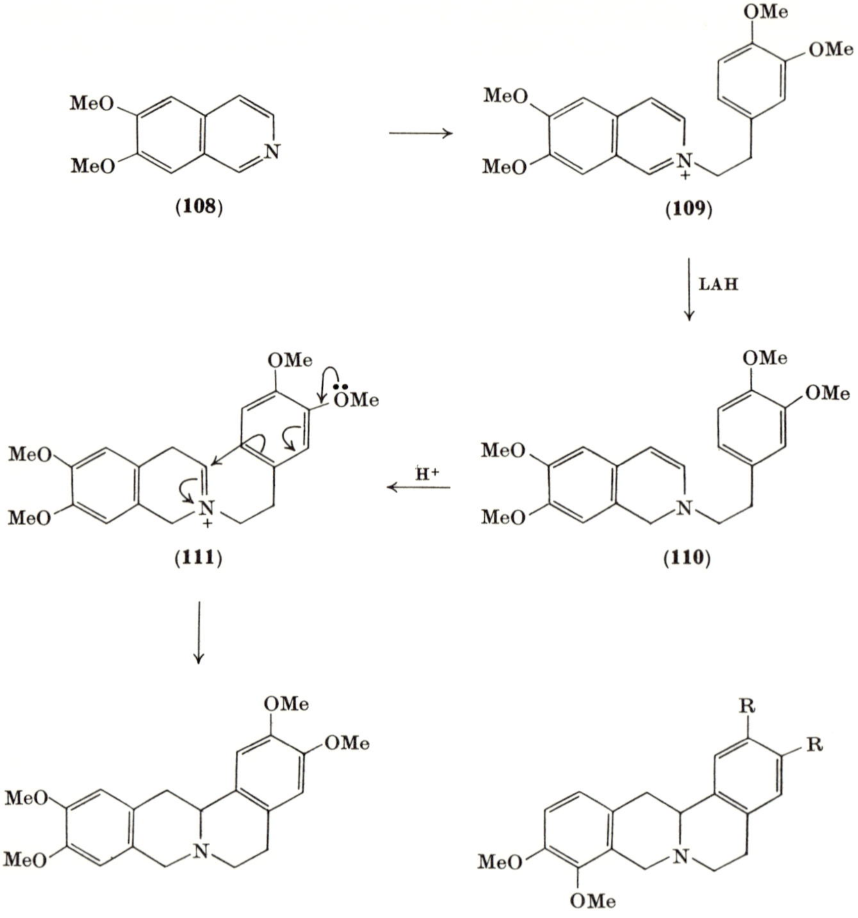

[131] J. W. Huffman and E. G. Miller, *J. Org. Chem.* **25**, 90 (1960).
[132] K. Peltz, *Chem. Listy* **57**, 1107 (1963).

with LAH gave the 1,2-dihydroisoquinoline (**110**), which with strong acid was protonated to (**111**) and cyclized. The yield of **112** from **108** was 40%. Now that 7,8-dimethoxyisoquinoline is easily available,[133] new syntheses of tetrahydropalmatine [(**113**) R = OMe] and tetrahydroberberine [(**113**) 2R = CH$_2$O$_2$] by the above route have been described.[133] A simplification of this route to berbines involves treating an N-β-arylisoquinolinium salt successively with concentrated aqueous alkali and concentrated hydrochloric acid.[26] The initially formed pseudobase (**114**) undergoes disproportionation (Section II,B) to yield the isocarbostyril (**115**) and the 1,2-dihydroisoquinoline, which is cyclized as for **109** above. A surprising discovery

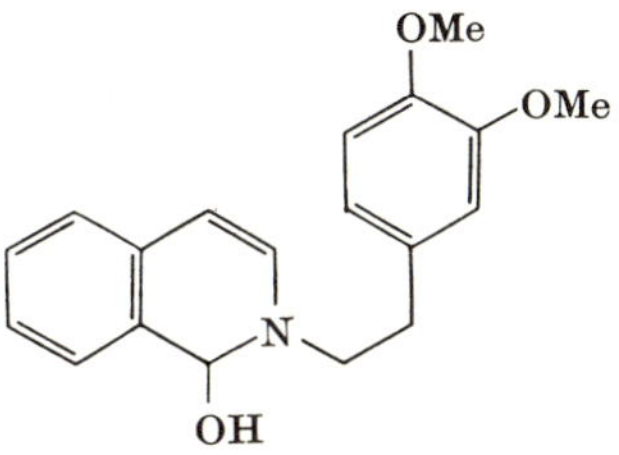

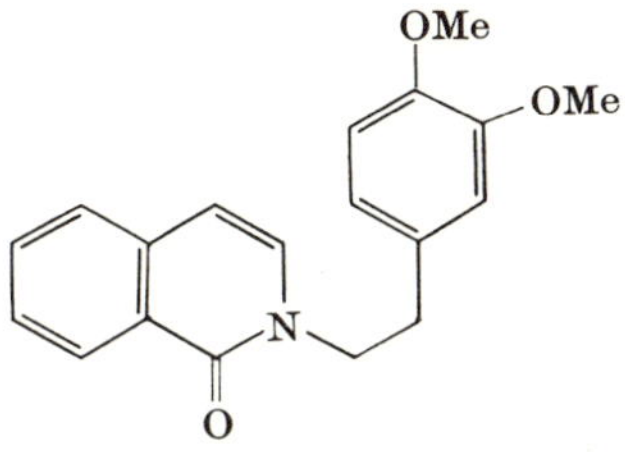

was that **115** is also cyclized by concentrated hydrochloric acid to yield the 8-oxoberbine (**116**). An attempt to cause[133] the double

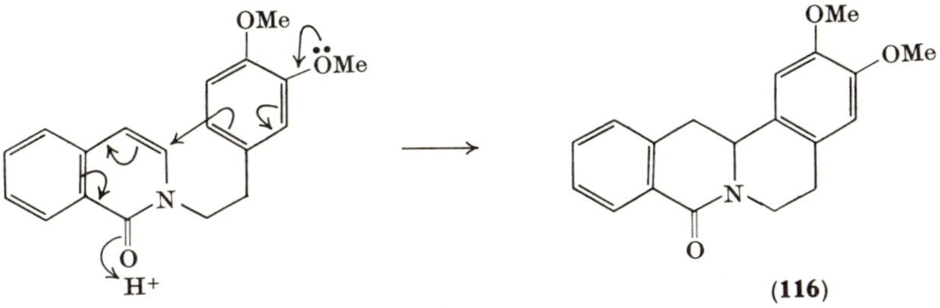

cyclization of an aminoacetal such as **117** to tetrahydropalmatine [(**113**) R = OMe] via the 1,2-dihydroisoquinoline, gave instead the bridged-ring compound (**118**).

[133] M. Sainsbury, D. W. Brown, S. F. Dyke, and G. Hardy, *Tetrahedron* **25**, 1881 (1969).

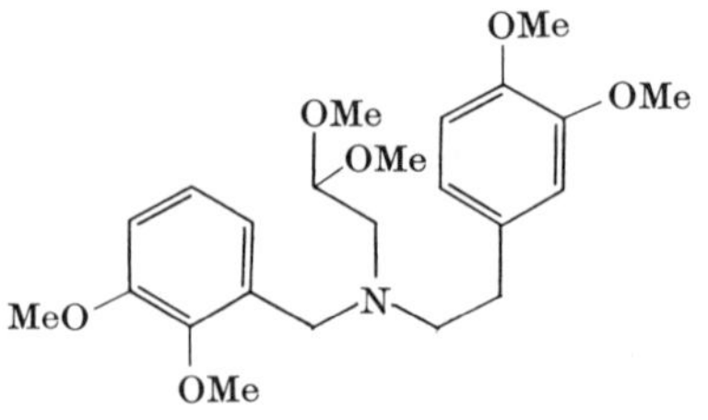

(117) 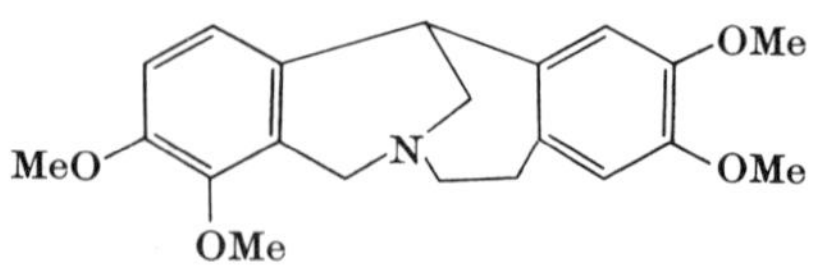 (118)

d. *Benzo[c]phenanthridines.* The benzo[c]phenanthridine nucleus is found in about a dozen alkaloids; the two main types are exemplified by sanguinarine chloride (119) and chelidonine (120). A possible

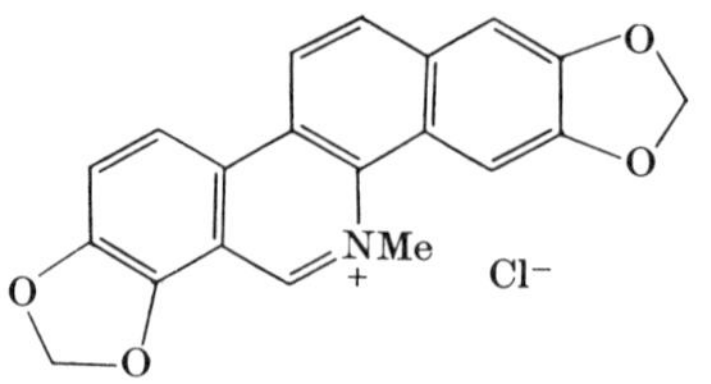

(119) 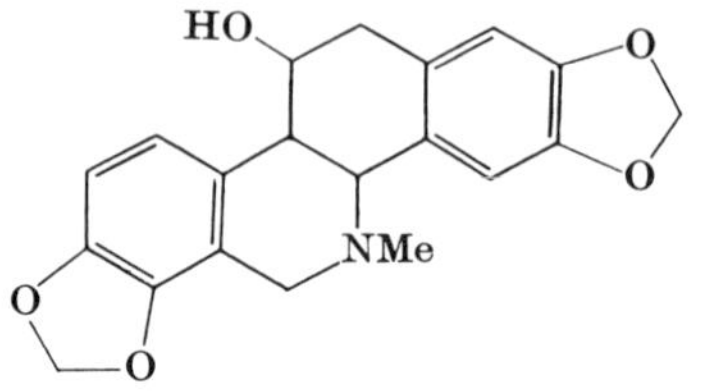 (120)

synthetic route would involve the acid-catalyzed cyclization of an appropriately substituted 1,2-dihydroisoquinoline, but compound 51 undergoes disproportionation and not cyclization when treated with

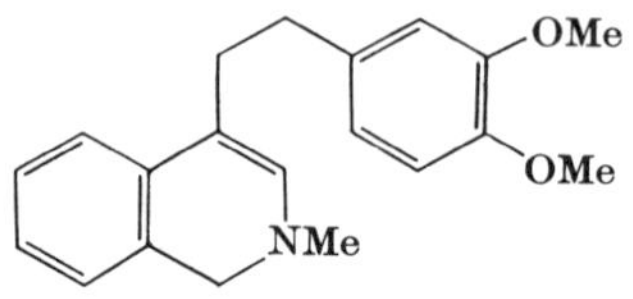

(51) 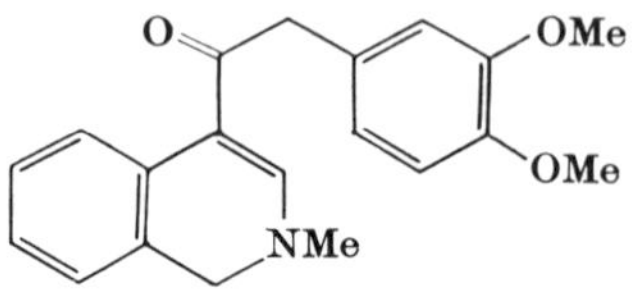 (121)

aqueous hydrochloric acid.[86] Some other routes to benzo[c]phenanthridines, also involving 1,2-dihydroisoquinolines, have been successfully developed.[114, 133a] Attempts to cyclize the 4-acyl-1,2-dihydroisoquinoline (121) failed.[8]

e. *Indeno[1,2-c]isoquinolines.* Gensler et al.[117] have reported that when the 4-benzylisoquinoline (123) was prepared by Bobbitt's method[113] using the benzylaminoacetal (122) and piperonal, a small

[133a] M. Sainsbury, S. F. Dyke, and B. J. Moon, *J. Chem. Soc. C* 1797 (1970).

Sec. V. D.2] 1,2-DIHYDROISOQUINOLINES 317

amount (ca. 5%) of the indeno[1,2-c]isoquinoline (**90**) was also formed. The formation of this latter product involves the initial generation of **124**, followed by intramolecular nucleophilic attack by the C-4 arylalkyl substituent at C-3.[87] By a careful choice of conditions,[87] the yield of **90** has been raised to 20–30%.

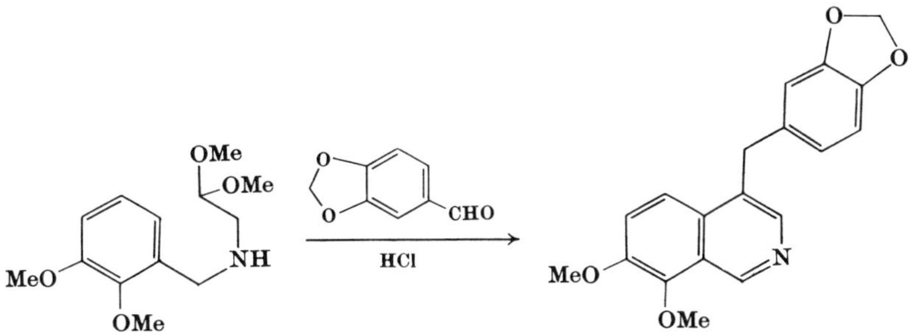

(**122**)   (**123**)

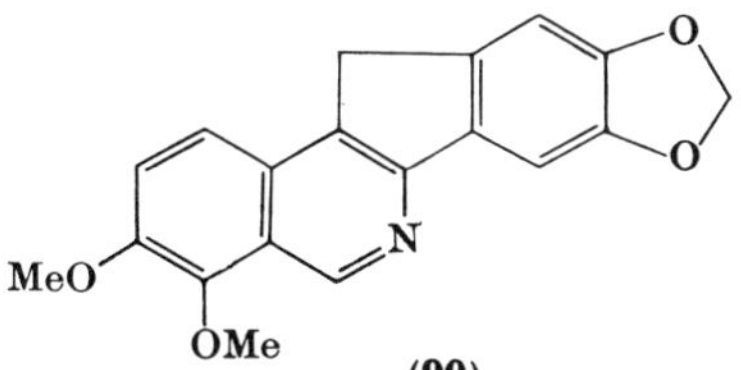

(**90**) 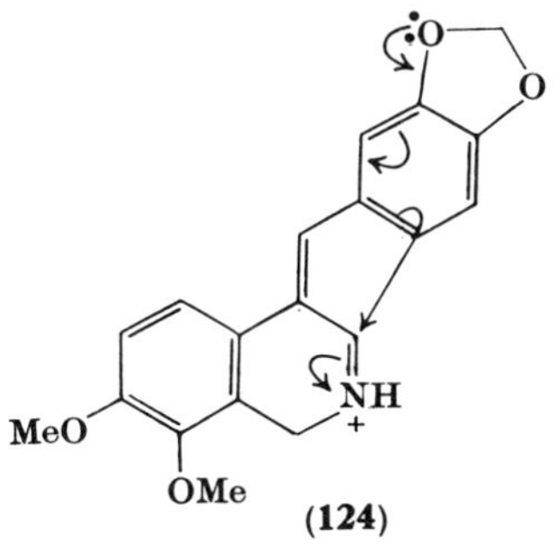   (**124**)

It has been claimed[117] that 2-methyl-4-benzyl-1,2-dihydroisoquinolines, e.g., **125**, are cyclized to the 5,6,12,13-tetrahydro-11*H*-indeno[1,2-c]isoquinoline (**126**) by mineral acid, but this reaction

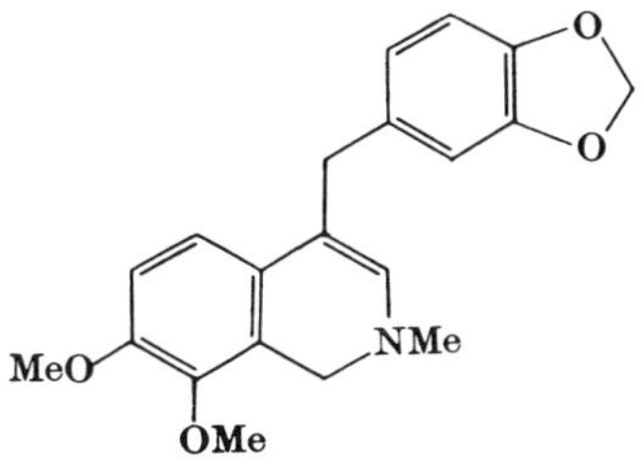

(**125**) 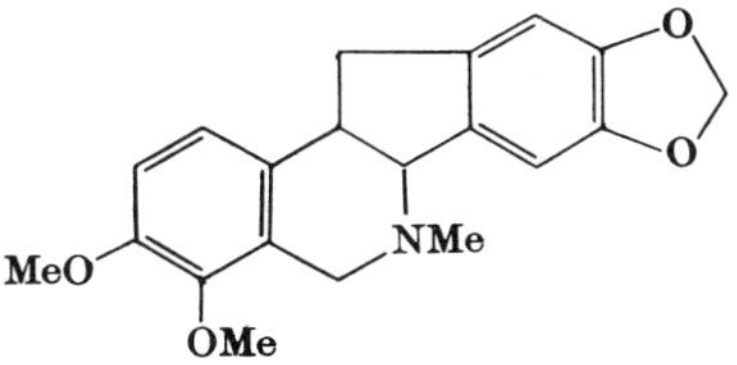   (**126**)

has not been substantiated.[87] Only disproportionation of **125** was observed, probably due to the fact that protonation of it occurs not at C-4, but at nitrogen. However, in an attempt to hydrogenate **127** in glacial acetic acid solution, a high yield of **128** was realized.[87]

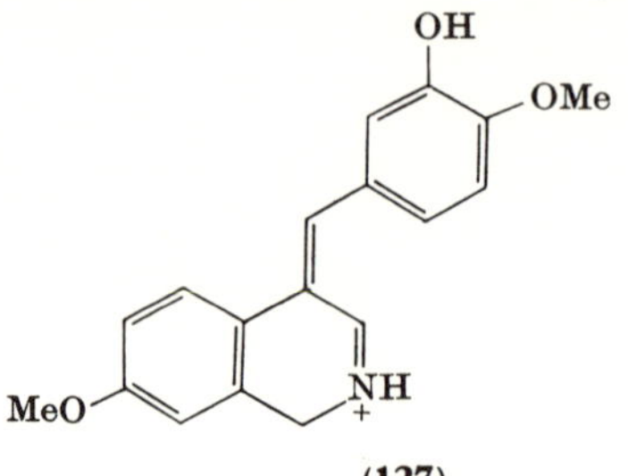

(127) 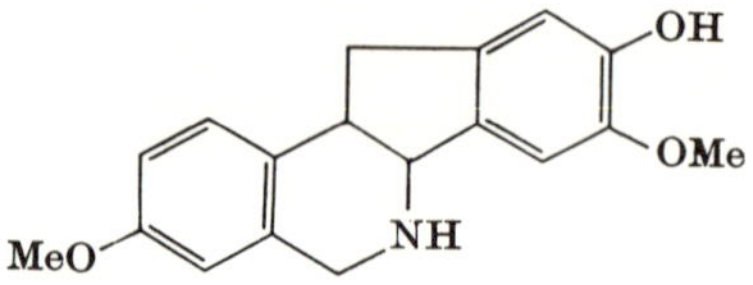   (128)

f. *Miscellaneous.* Bobbitt and Moore[134] reported that when benzylaminoacetals are reacted with methyl vinyl ketone, the products are the benzo[b]quinolizine derivatives (**131**). It was shown that the reaction proceeds through the 4-hydroxy-1,2,3,4-tetrahydroiso-

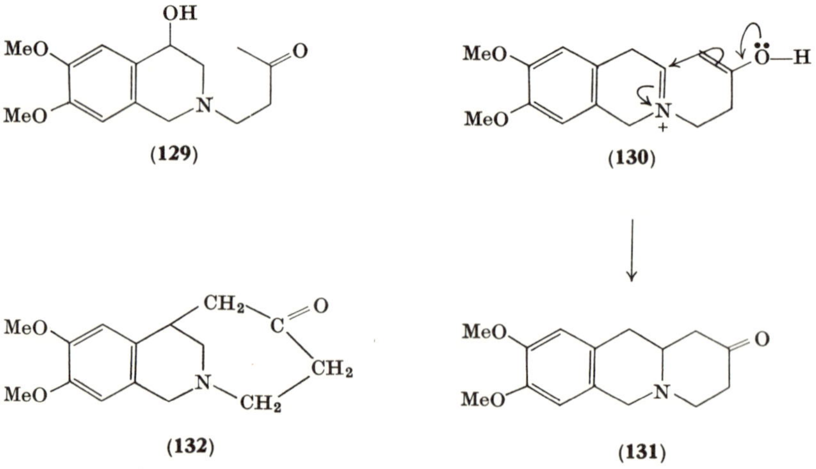

(129)   (130)

(132)   (131)

quinoline of type **129** as an intermediate. Presumably dehydration of **129** occurs and then, in the enol form of the iminium ion (**130**), cyclization occurs by nucleophilic addition to C-3. In view of the ease of displacement of the C-4 hydroxyl group by nucleophiles,[133,]

[134] J. M. Bobbitt and T. E. Moore, *J. Org. Chem.* **33**, 2958 (1968).

structure **132** is also possible for the products described by Bobbitt and Moore, but **131** has been shown to be correct by an alternative synthesis of a derivative.[137]

E. REARRANGEMENT

1. *Benzyl Migration from C-1 to C-3*

In 1963, it was reported[138] that when 2-methyl-1,2-dihydropapaverine [(**133**) R = OMe] is treated with dilute acid (2% HCl or acetic acid) at 100° for 30 minutes, a high yield of the 2-methyl-3-benzyl-3,4-dihydroisoquinolinium salt [(**134**) R = OMe] is formed.

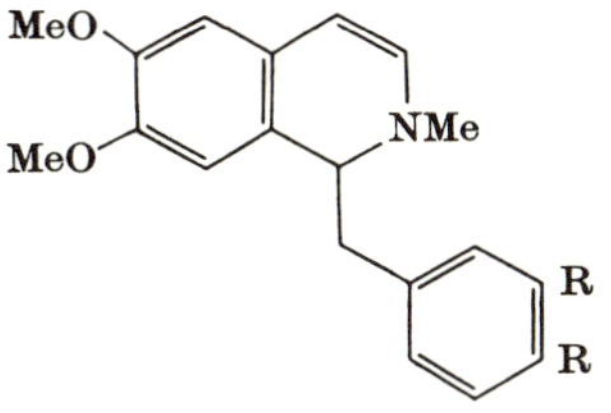

(**133**)

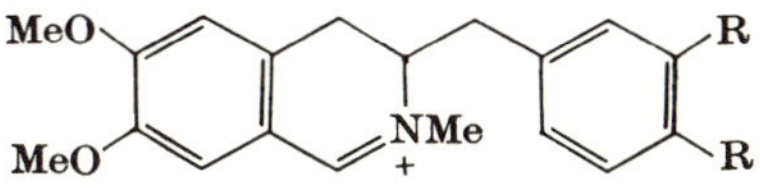

(**134**)

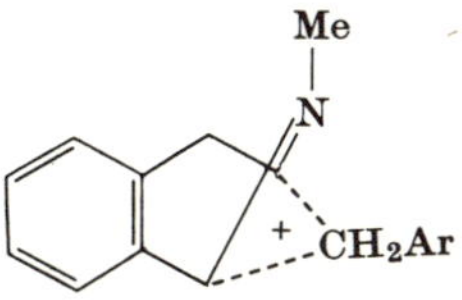

(**135**)

A derivative of **134** was prepared to prove its structure.[139] A similar rearrangement was observed with **133** (R + R = CH_2O_2), and it was suggested that the rearrangement involves an intermediate of the

[135] J. M. Bobbitt and S. Shibuya, *J. Org. Chem.* **35**, 1181 (1970).
[136] D. W. Brown, S. F. Dyke, G. Hardy, and M. Sainsbury, *Tetrahedron Lett.* 1515 (1969).
[137] S. F. Dyke and R. A. Kilminster, to be published.
[138] J. Knabe, J. Kubitz, and N. Ruppenthal, *Angew. Chem.* **75**, 981 (1963); *Angew. Chem. Int. Ed. Engl.* **2**, 689 (1963); J. Knabe and J. Kubitz, *Arch. Pharm. (Weinheim)* **297**, 129 (1964).
[139] J. Knabe and N. Ruppenthal, *Arch. Pharm. (Weinheim)* **297**, 141, 268 (1964).

type **135**.[13] However, it was subsequently reported[140] that when a mixture of **133** (R = OMe) and **136** was subjected to the rearrangement conditions, *four* products were detected, namely, **134** (R = OMe), **134** (R = OEt), **137** (R = OMe), and **137** (R = OEt), thus suggesting

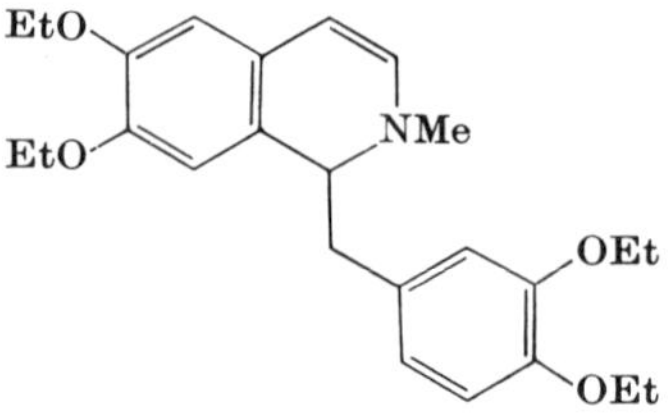

(**136**) 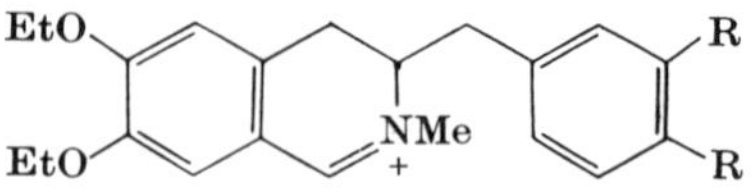   (**137**)

that the reaction is an intermolecular one. Although Dewar[141] has warned about the interpretation of cross-migration experiments, the thermal concerted process originally visualized is a disallowed process by application of the Woodward–Hoffmann rules.[142]

During the course of their study of 1-benzyl-2-methyl-1,2-dihydroisoquinoline and its cyclization to the pavine system (Section V, D, 2, b), Wittig *et al.*[129] described a quaternary salt $C_{17}H_{18}NI$, but did not identify it. This compound is now known[143] to be the rearrangement product 2-methyl-3-benzyl-3,4-dihydroisoquinolinium iodide.

In an effort to define the scope of the benzyl migration reaction, and to elucidate its mechanism, Knabe *et al.*[144] have treated a series of 1-substituted 1,2-dihydroisoquinolines with acids. The following reactions have been observed: (*a*) rearrangement, (*b*) elimination of the C-1 substituent, (*c*) disproportionation without rearrangement, (*d*) disproportionation after rearrangement, and (*e*) recovery of starting material (Scheme VIII). The results are summarized in

[140] J. Knabe and K. Detering, *Chem. Ber.* **99**, 2873 (1966).
[141] M. J. S. Dewar, *in* "Molecular Rearrangements" (P. de Mayo, ed.), Vol. I, p. 208. Interscience, New York, 1963.
[142] R. B. Woodward and R. Hoffmann, *in* "The Conservation of Orbital Symmetry." Academic Press, New York, 1970.
[143] M. Sainsbury, D. W. Brown, S. F. Dyke, R. G. Kinsman, and B. J. Moon, *Tetrahedron* **24**, 6695 (1968).
[144] (a) J. Knabe and N. Ruppenthal, *Naturwissenschaften* **51**, 482 (1964); (b) *Arch. Pharm. (Weinheim)* **299**, 159 (1966); (c) J. Knabe and K. Detering, *ibid.* **300**, 97 (1967); (d) J. Knabe, W. Krause, and K. Sierocks, *ibid.* **303**, 255 (1970).

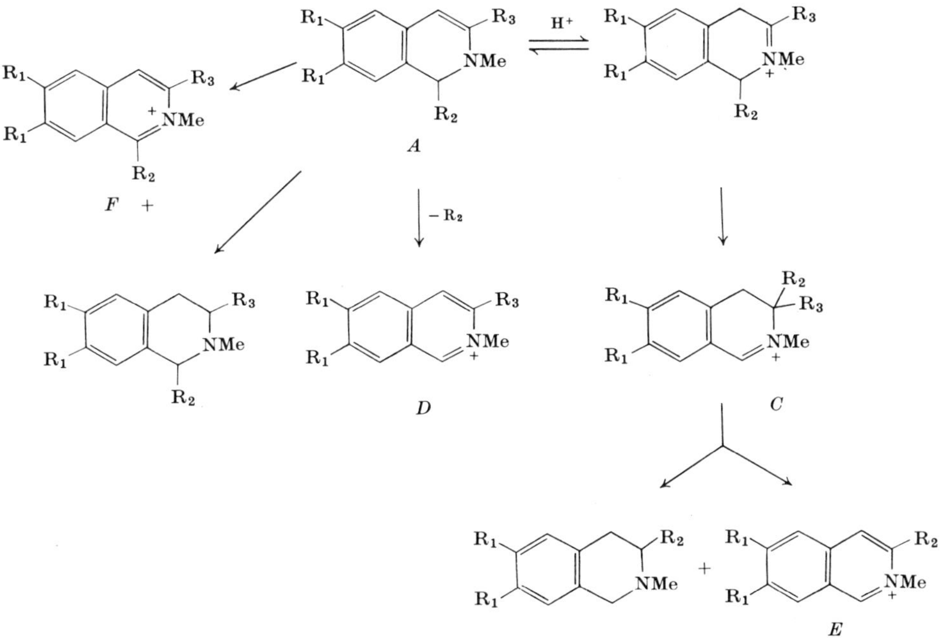

SCHEME VIII

Table VI. Knabe believes that the rearrangement proceeds by way of a carbanion intermediate, but when an optically active sample of **138** (R = Me, R′ = H) is rearranged, the product (**139**) obtained after reduction with $NaBH_4$ is inactive.[145]

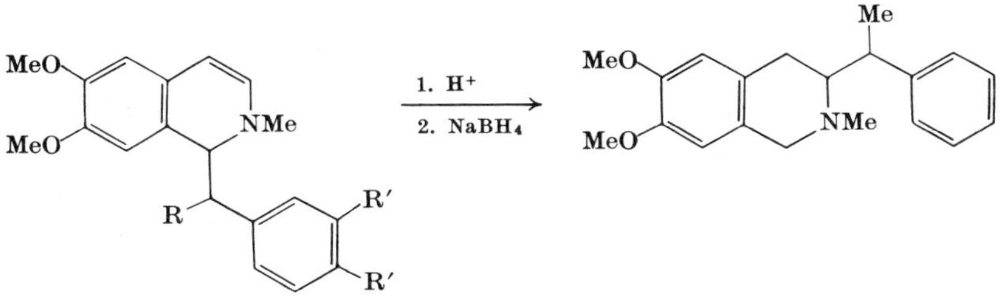

(**138**)   (**139**)

[145] J. Knabe and H. Rowilleit, *Arch. Pharm. (Weinheim)* **303**, 37 (1970).

TABLE VI

Reactions of 1-Substituted 1,2-Dihydroisoquinolines

Substituent			Yields (%)[a]				
R_1	R_2	R_3	A	C	D	E	F
MeO	Me	H	—	—	—	—	√
MeO	CH_2CH_2Ph	H	—	—	—	—	√
MeO	CH_2Ph	H	—	79	10	—	—
MeO	n-Bu	H	—	—	—	—	√
MeO	p-BrC$_6$H$_4$CH$_2$	H	12	46	20	√	—
MeO	3,4-(OMe)$_2$C$_6$H$_3$CH$_2$	H	—	√	—	—	—
MeO	PhCH$_2$	Et	85	—	15	—	—
MeO	p-MeOC$_6$H$_4$CH$_2$	Me	80	—	10	—	—
CH_2O_2	3,4-CH$_2$O$_2$C$_6$H$_3$CH$_2$	Me	95	—	5	—	—
H	p-NO$_2$C$_6$H$_4$CH$_2$	H	20	41	24	—	—
H	o-NO$_2$C$_6$H$_4$CH$_2$	H	15	41	38	—	—

[a] √ means that this product was observed, but yield is not quoted.

The 1-benzyl-1,2-dihydroisoquinoline [(138) R = OH, R′ = OMe] has been rearranged,[146] and the structure of the product (140) proved by synthesis.[147] The threshold for pavine formation versus migration

(138)   R = OH   dilute
        R′ = OMe ──────→
                    HAc

(140)

has been examined,[143] using 1-benzyl-2-methyl-1,2-dihydroisoquinoline, but no clear-cut distinction could be found.

The 1,2-dihydroisoquinoline (141) cyclizes very readily to 142; rearrangement to 143 could only be observed under special conditions of acid treatment.[148]

[146] W. Wiegrebe and E. Roesel, *Arch. Pharm.* (*Weinheim*) **302**, 310 (1969).
[147] W. Wiegrebe, E. Roesel, and H. Budzikiewicz, *Arch. Pharm.* (*Weinheim*) **302**, 572 (1969).
[148] H. Zinnes, F. R. Zuleski, and J. Shavel, *J. Org. Chem.* **33**, 3605 (1968).

(141) (142) (143)

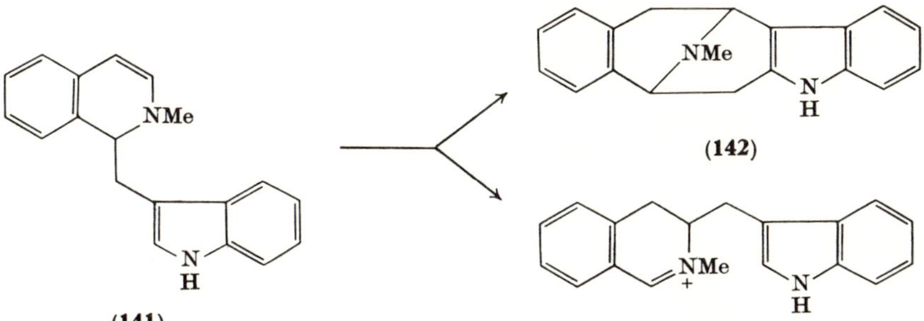

Some benzyl migrations have been observed in simple enamines,[3,149] e.g., (**144**) → (**145**) → (**146**); allyl migrations have also been observed,[3] e.g., (**147**) → (**148**).

(144) (145) (146)

(147) (148)

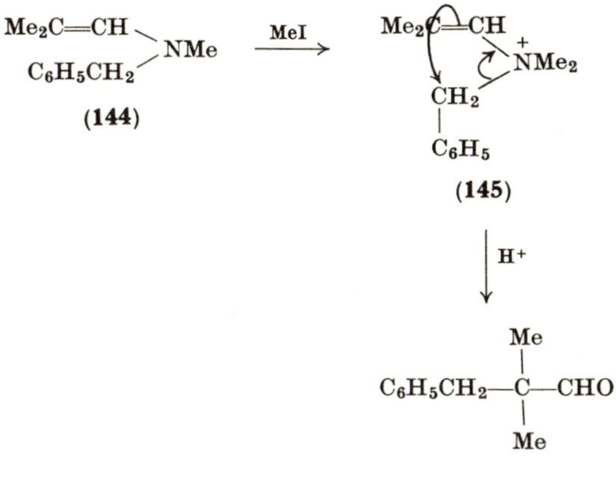

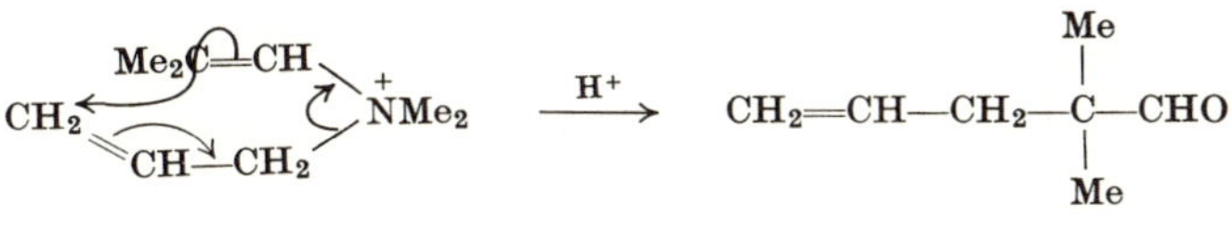

2. *Allyl Migrations from C-1 to C-3*

When 1-allyl-2-methyl-1,2-dihydroisoquinoline [(**149**) R = H] is treated with dilute mineral acids under the conditions of the benzyl

[149] R. K. Hull and N. W. Gilman, *Tetrahedron Lett.* 1421 (1967).

migrations, rearrangement to **150** (R = H) occurs.[143] A similar reaction has been reported[150] for **149** (R = OMe). The aminoacetal

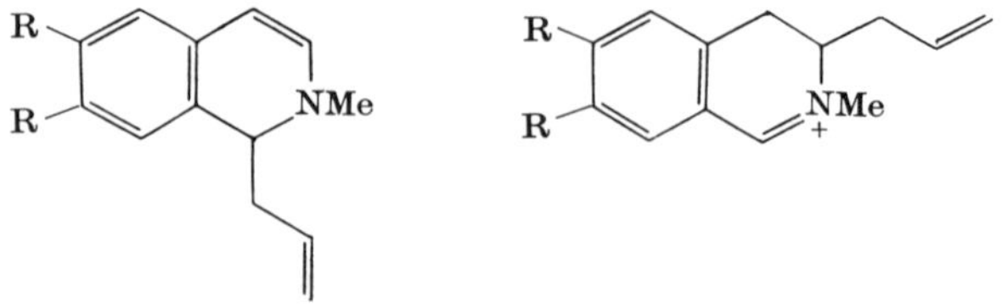

(149)   (150)

derivative (**151**), when treated with acid, undergoes cyclization to **152**, followed by dehydration to the 1,2-dihydroisoquinoline and rearrangement to the 3-allyl-3,4-dihydroisoquinoline in yields approach-

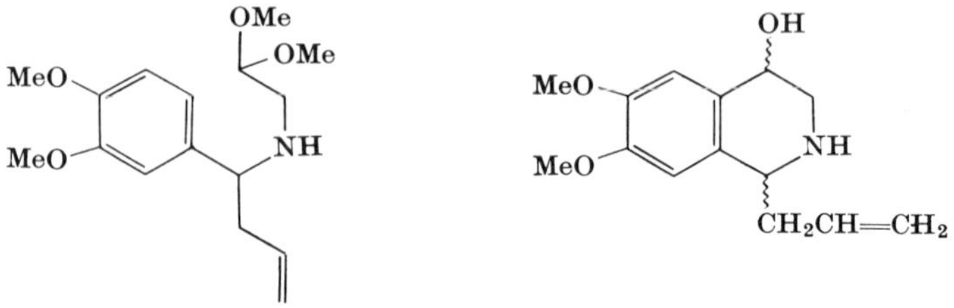

(151)   (152)

ing 90%.[151] The allyl rearrangement reaction has been shown[46] to be an intramolecular process and can be viewed as a suprafacial sigmatropic [3,3]-migration. 1-Allyl-2-methyl-1,2-dihydropapaverine (**153**) rearranges to give **154** exclusively.[152] It is interesting to note

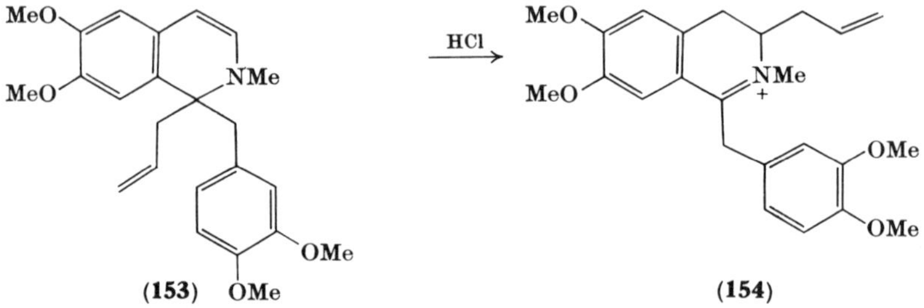

(153)   (154)

[150] J. Knabe and H.-D. Holtje, *Tetrahedron Lett.* 433 (1969).
[151] D. W. Brown, S. F. Dyke, R. G. Kinsman, and M. Sainsbury, *Tetrahedron Lett.* 1731 (1969).
[152] S. F. Dyke, R. G. Kinsman, J. Knabe, and H. D. Holtje, *Tetrahedron*, in press.

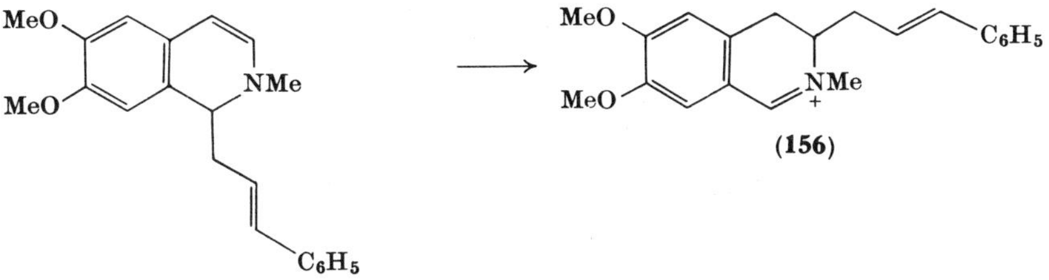

Sec. V. E.3] 1,2-DIHYDROISOQUINOLINES 325

(155)

(156)

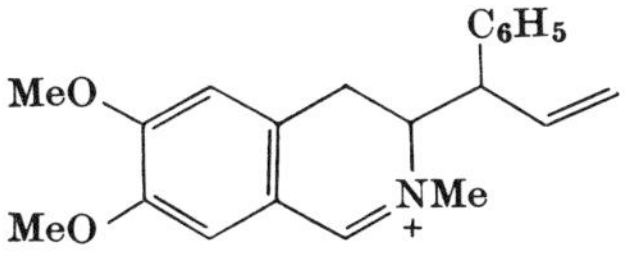

(157)

that the 1-cinnamyl-1,2-dihydroisoquinoline (155) rearranges[153] as a vinylogous benzyl group to give 156; in view of the reported[154] Claisen rearrangements of cinnamyl groups, some of the allyl-type migration product (157) might have been anticipated.

A C-1 to C-3 migration has been achieved[46] with the 1-propargyl-1,2-dihydroisoquinoline (158); the product was the expected allene (159).

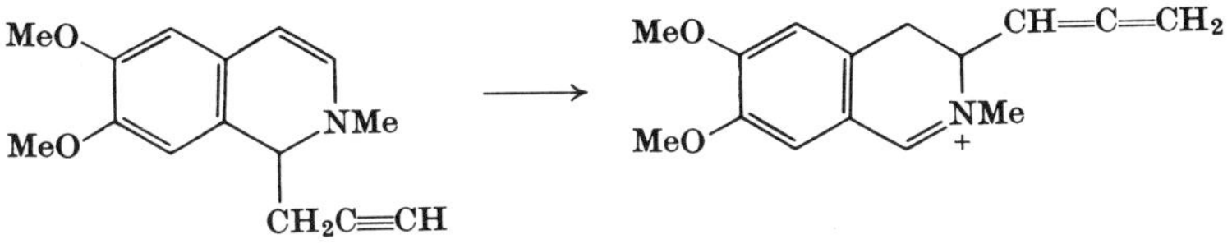

(158)

(159)

3. N-*Tosyloxyisoquinolines to 4-Tosylisoquinolines*

When isoquinoline N-oxide [(160) R = H] or 3-methylisoquinoline N-oxide [(160) R = Me] are treated with p-toluenesulfonyl chloride, the products are the 4-tosylisoquinolines [(162) R = H or Me].[155] The

[153] J. Knabe and H.-D. Holtje, *Arch. Pharm. (Weinheim)* **303**, 404 (1970).
[154] W. N. White and W. K. Fife, *J. Amer. Chem. Soc.* **83**, 3846 (1961).
[155] V. J. Traynelis, *in* "Mechanism of Molecular Migrations" (B. S. Thyagarajan, ed.), Vol. 2. Academic Press, New York, 1969.

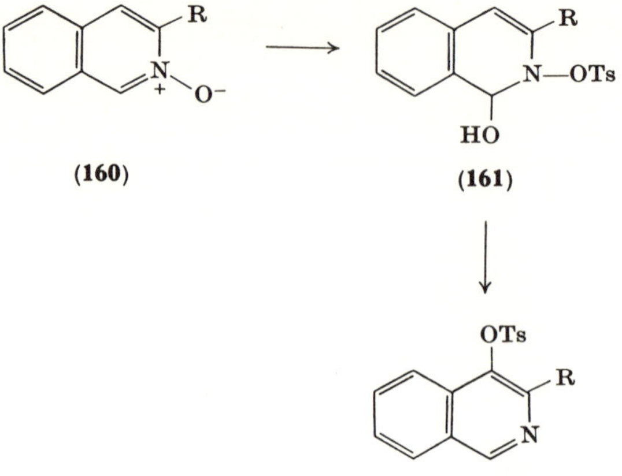

(160) → (161)

↓

(162)

mechanism of the reaction is believed[156] to involve the 1,2-dihydroisoquinoline (161), but little is known about the scope of this reaction.

F. 1,2-Dihydroisoquinolines in Biosynthesis

It has been shown by Battersby et al.[157] that (+)-reticuline (163) is the precursor of stylopine (164) and chelidonine (120), and by a detailed study of the fate of variously labeled samples of 163, it was concluded[157] that a 1,2-dihydroisoquinoline, such as 165, is involved as an intermediate.

Norlaudanosoline [(166) R = H] has for a long time been regarded, in schematic terms, as the biogenetic precursor of the isoquinoline alkaloids, and since it is now known[158] that noradrenaline [(167) R = OH] is incorporated into berberastine (168) in *Hydrastis canadensis* L., presumably via 4-hydroxynorlaudanosoline [(166) R = OH], it is possible[159] that a series of alkaloids exists based on 166 (R = OH) as parent. It is significant in this context that the alkaloids imenine

---

[156] K. Ogino, S. Kozuka, and S. Oae, *Tetrahedron Lett.* 3559 (1969).
[157] A. R. Battersby, R. J. Francis, E. A. Ruveda, and J. Staunton, *Chem. Commun.* 89 (1965).
[158] I. Monkovic and I. D. Spenser, *J. Amer. Chem. Soc.* 87, 1137 (1965); *Can. J. Chem.* 43, 2017 (1965).
[159] P. W. Jeffs, *in* "The Alkaloids" (R. H. Manske, ed.), Vol. 9, p. 101. Academic Press, New York, 1967.

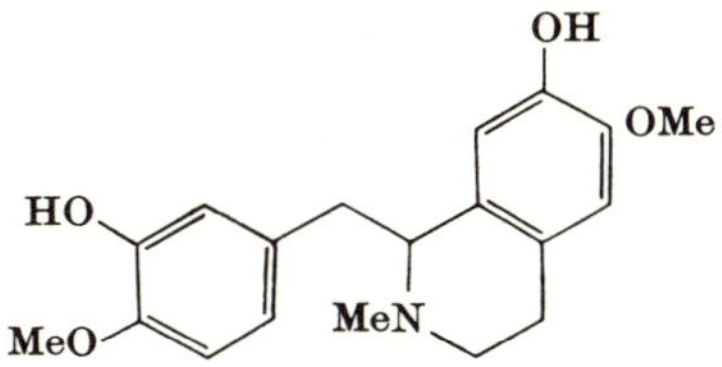

(169),[160] erythrinine (170),[161] and erythristemine (171)[162] have been described recently.

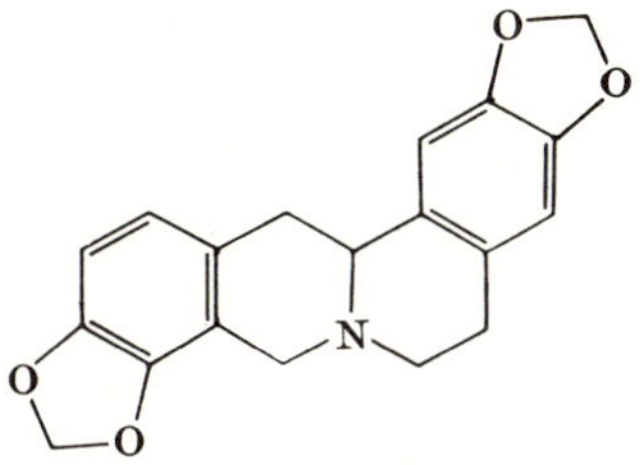

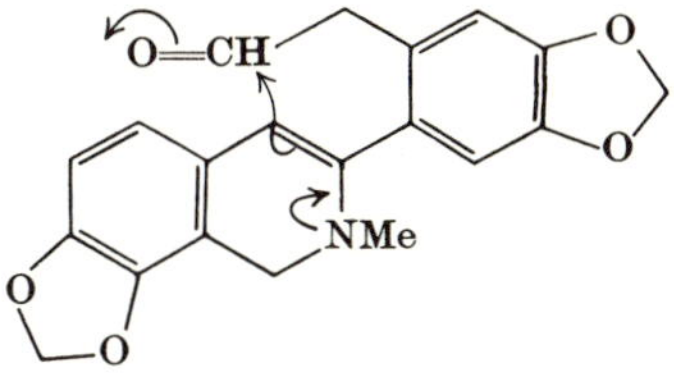

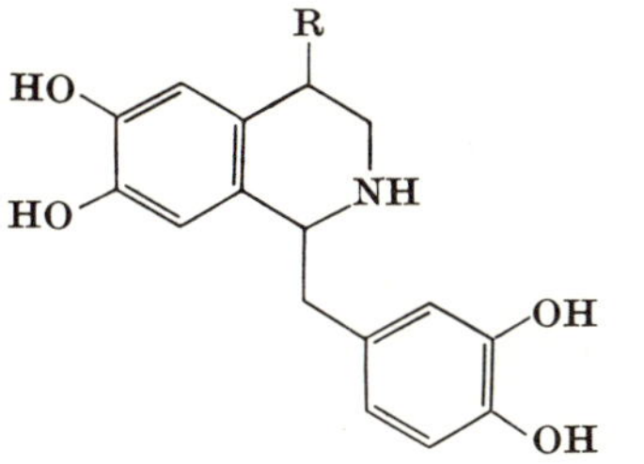

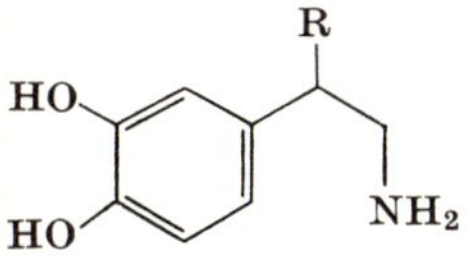

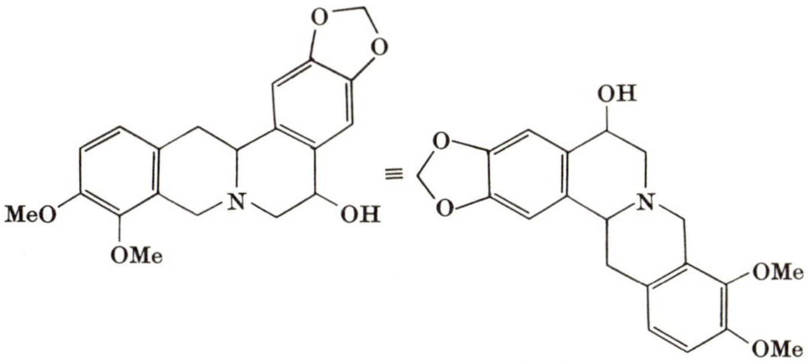

---

[160] M. D. Glick, R. E. Cook, M. P. Cava, M. Srinivasan, and J. Kunitomo, *Chem. Commun.* 1217 (1969).
[161] K. Ito, H. Furukawa, and H. Tanaka, *Chem. Commun.* 1076 (1970).
[162] D. H. R. Barton, P. N. Jenkins, R. Letcher, D. A. Widdowson, E. Hough, and D. Rogers, *Chem. Commun.* 391 (1970).

It is quite likely[136] that the isopavine alkaloids, e.g., amurensine (**173**), are formed *in vivo* from a 1-benzyl-4-hydroxytetrahydroisoquinoline, e.g., **172**. Since chelidonine is derived from stylopine,

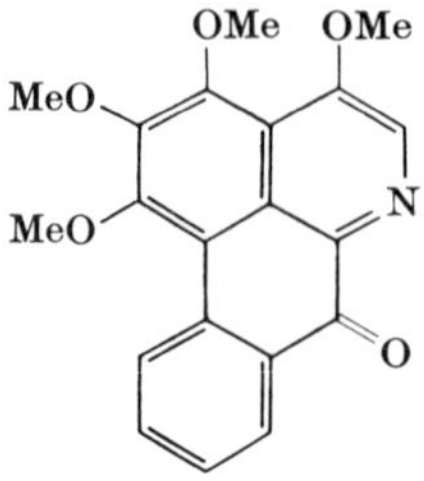

(**169**) 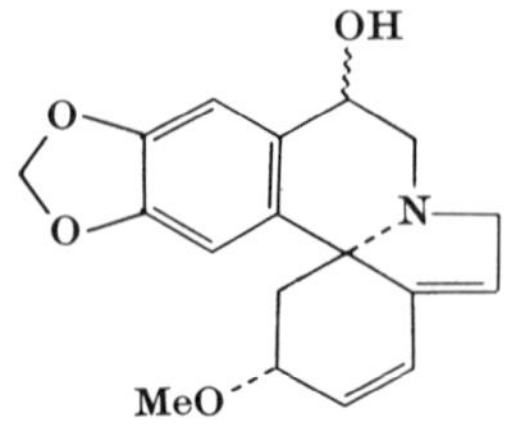   (**170**) 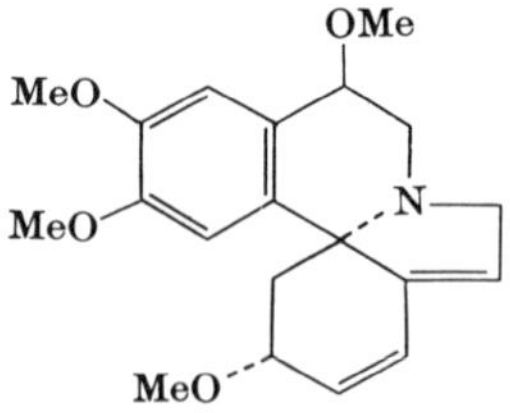   (**171**)

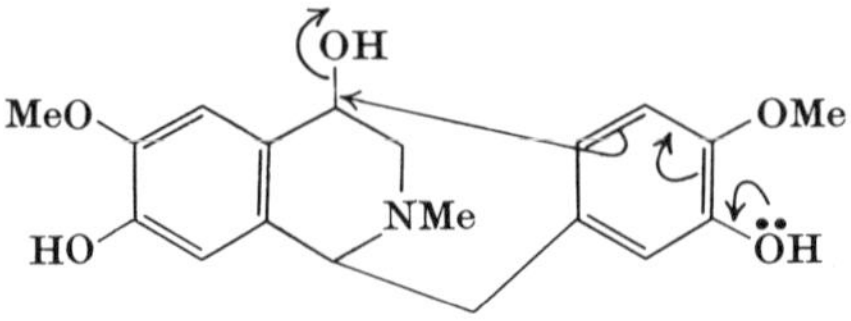

(**172**) 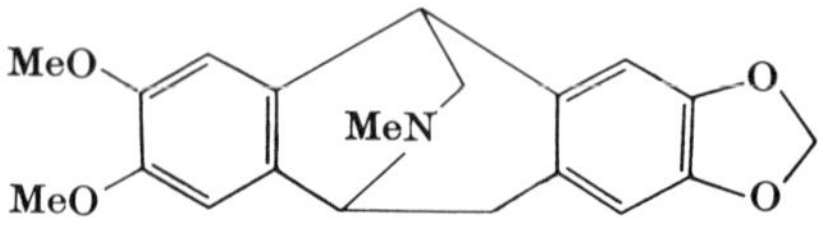   (**173**)

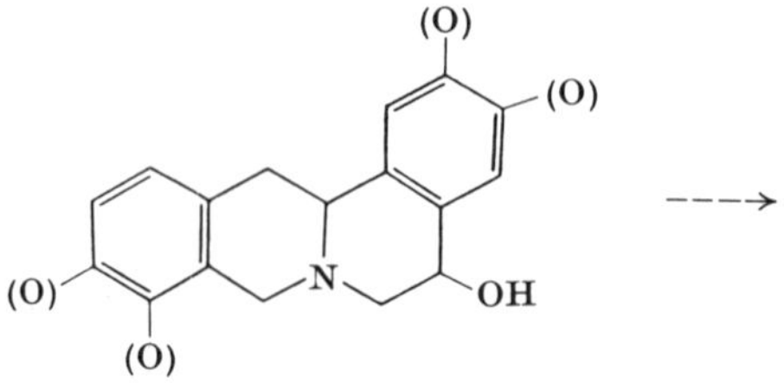

(**174**) 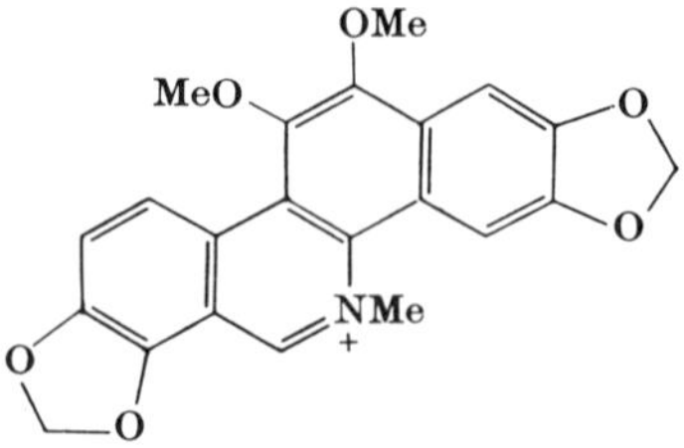   (**175**)

it is possible[136] that marcarpine (**175**), recently isolated[163] from *Eschscholtzia douglasii*, arises in the plant from a 5-hydroxyberbine (**174**).

It has been suggested[164] that (+)-reticuline (**163**) is the precursor of the pavine alkaloids, but the racemate was not incorporated into

[163] J. Slavik, L. Slavikova, and K. Haisova, *Collect. Czech. Chem. Commun.* **32**, 4420 (1967).
[164] F. R. Stermitz, S. Y. Lwo, and G. Kallos, *J. Amer. Chem. Soc.* **85**, 1551 (1963); F. R. Stermitz and K. D. McMurtry, *J. Org. Chem.* **34**, 555 (1969).

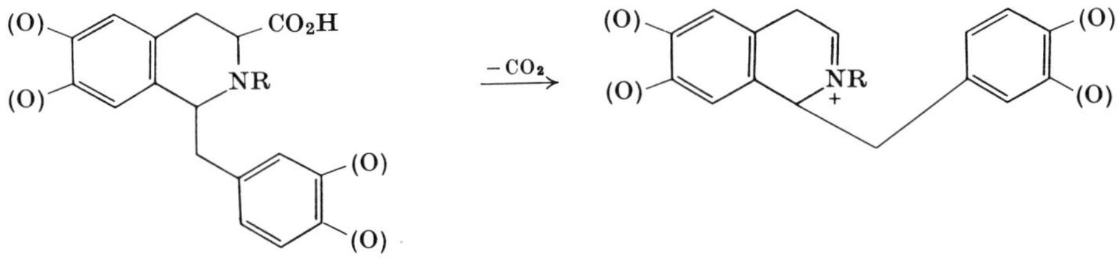

(176) → (177)

argemonine when it was fed to argemone plants.[16] The immediate precursor is almost certainly a 1,2-dihydroisoquinoline, and although it is feasible[165] that this can be generated from the α-amino acid (176), via the iminium ion (177), it is also possible[136] that dehydration of a 1-benzyl-4-hydroxytetrahydroisoquinoline, e.g., 172, is involved instead.

ACKNOWLEDGMENTS

The author thanks Dr. P. A. Bather and Mr. R. G. Kinsman for reading and checking the manuscript and for assisting with the proofreading.

[165] E. E. van Tamelen, V. B. Haarstad, and R. L. Orvis, *Tetrahedron* **24**, 687 (1968).

Benzo[c]thiophenes

B. IDDON

Department of Chemistry and Applied Chemistry, University of Salford, Salford, Lancashire, England

I. Introduction	331
II. Theoretical Aspects	333
III. Hydrobenzo[c]thiophenes	335
A. 1,3-Dihydrobenzo[c]thiophenes (2-Thiaindanes)	335
B. 4,5-Dihydrobenzo[c]thiophenes	341
C. 4,5,6,7-Tetrahydrobenzo[c]thiophenes	342
D. 1,3,4,7-Tetrahydrobenzo[c]thiophenes	347
E. Octahydrobenzo[c]thiophenes (2-Thiahydrindanes)	348
F. Miscellaneous Hydrobenzo[c]thiophenes	349
IV. Benzo[c]thiophene	350
V. Alkyl and Aryl Derivatives of Benzo[c]thiophene	355
A. Alkyl Derivatives	355
B. Aryl Derivatives	356
VI. Hydrobenzo[c]thiophene 2-Oxides and 2,2-Dioxides	360
A. 2-Oxides	361
B. 2,2-Dioxides	362
VII. 2-Thiophthalide, Phthalic Thioanhydride, and Related Compounds	368
A. 2-Thiophthalide	368
B. Phthalic Thioanhydride	375
Note Added in Proof	381

I. Introduction

Previously we[1] had reviewed the chemistry of benzo[b]thiophene (**1**); the present review covers the chemistry of the isomeric system, (**2**), through 1970. Although the currently accepted name for **2** in *Chemical Abstracts* (*CA*) is benzo[c]thiophene, isothianaphthene (or isothionaphthene), isobenzothiophene, and 2-thiaindene are also commonly used, the latter particularly in the Russian literature. In *CA* (Ring Index No. 854; RRI No. 1354) the ring system is currently

---

[1] B. Iddon and R. M. Scrowston, *Advan. Heterocycl. Chem.* **11**, 177 (1970).

numbered as shown in **2**, but the alternative system of numbering as shown in **3** is sometimes used, particularly in the earlier literature.

(1)   (2)

(3)

1,3-Dihydrobenzo[c]thiophene (Section III, A) is also known as 1,3-dihydroisothianaphthene, 2-thiaindane, o-xylylene sulfide (commonly used), thiophthalan, 3,4-benzothiacyclopentane, and 3,4-benzo-2,5-dihydrothiophene. In this case we have used the first-mentioned name throughout; other problems of nomenclature are discussed as they arise. Since 1,3-dihydrobenzo[c]thiophenes are important intermediates in this field, their chemistry is discussed in detail. With other "derivatives" such as 2-thiophthalide (Section VII, A) and phthalic thioanhydride (Section VII, B), which may be regarded as 1,3-dihydrobenzo[c]thiophen-1-one and 1,3-dihydrobenzo[c]thiophen-1,3-dione, respectively, our coverage of the literature may be incomplete, since we have only included references which arose directly or indirectly (i.e., as cross references) by searching benzo[c]thiophene or equivalent terms in *CA*.

Of the analogous parent heterocycles, benzo[c]pyrrole (isoindole),[2] benzo[c]furan, and benzo[c]thiophene, only the last has proved sufficiently stable for isolation (see Sections II and IV).* Although the first reference to the benzo[c]thiophene system in the literature was Leser's[3] synthesis of 1,3-dihydrobenzo[c]thiophene in 1884 (Section

---

* *Added in proof:* Benzo[c]furan has since been isolated [D. Wege, *Tetrahedron Lett.* 2337 (1971) and R. N. Warrener, *J. Amer. Chem. Soc.* **93**, 2346 (1971)]. It is an extremely unstable colorless crystalline solid, m.p. ca. 20°.

[2] J. D. White and M. E. Mann, *Advan. Heterocycl. Chem.* **10**, 113 (1969).
[3] G. Leser, *Chem. Ber.* **17**, 1824 (1884).

III, A) and the first example of the fully aromatic system, namely, 1,3-diphenylbenzo[c]thiophene, was synthesized in 1922 by Bistrzycki and Brenken[4] (Section V, B), the parent heterocycle (2) remained unknown until 1962 when it was synthesized[5,6] (Section IV) by a route suggested earlier by Hartough and Meisel.[7] Hartough and Meisel's other suggestion, namely, the possibility of preparing 2 by heating o-xylene with sulfur,[7] does not appear to have been investigated. Because benzo[c]thiophene is unstable, it is unlikely to withstand the conditions necessary for most electrophilic substitution reactions (Section IV). The only electrophilic substitution reactions reported in this field to date are the nitration reactions of 1,3-diphenyl-benzo[c]thiophene[8] (Section V, B) and 1,3-dihydrobenzo[c]thiophene 2,2-dioxide[9] (Section VI, B) and the Friedel–Crafts acetylation of 1-methyl-4,5,6,7-tetrahydrobenzo[c]thiophene[10] (Section III, C).

Benzo[c]thiophenes do not appear to be of any commercial importance at the present time, although some have been claimed to exhibit biological activity.

II. Theoretical Aspects

Several calculations have been made of the electronic structure of benzo[c]thiophene.[11–15] The earliest calculation[11] based on the

[4] A. Bistrzycki and B. Brenken, *Helv. Chim. Acta* **5**, 20 (1922).
[5] R. Mayer, H. Kleinert, S. Richter, and K. Gewald, *Angew. Chem. Int. Ed. Engl.* **1**, 115 (1962).
[6] R. Mayer, H. Kleinert, S. Richter, and K. Gewald, *J. Prakt. Chem.* **20**, 244 (1963).
[7] H. D. Hartough and S. L. Meisel, *in* "Compounds With Condensed Thiophene Rings" (A. Weissberger, ed.), p. 167. Wiley (Interscience), New York, 1954.
[8] C. Ecary, *C. R. Acad. Sci.* **224**, 1828 (1947).
[9] R. F. Watson, Ph.D. Thesis, Univ. of Tennessee (1963); *Diss. Abstr.* **25**, 795 (1964).
[10] P. Cagniant, A. Reisse, and D. Cagniant, *Bull. Soc. Chim. Fr.* 991 (1969).
[11] J. De Heer, *J. Amer. Chem. Soc.* **76**, 4802 (1954).
[12] R. Zahradník, C. Párkányi, V. Horák, and J. Koutecký, *Collection Czech. Chem. Commun.* **28**, 776 (1963).
[13] R. Zahradník, *Advan. Heterocycl. Chem.* **5**, 1 (1965).
[14] R. Zahradník, J. Fabian, A. Mehlhorn, and V. Kvasnička, *in* "Organosulfur Chemistry" (M. J. Janssen, ed.), p. 203. Wiley (Interscience), New York, 1967.
[15] N. Trinajstić and A. Hinchliffe, *Z. Phys. Chem. (Frankfurt)* [N.S.] **59**, 271 (1968).

Longuet–Higgins LCAO method, those published more recently and based on the simple MO-LCAO (HMO) procedure,[12, 13] and the recently published calculations[14, 15] based on the more sophisticated Pariser–Parr–Pople semiempirical method (LCI-SCF-MO-LCAO) predict a high free valence value for the equivalent 1- and 3-positions. However, whereas the later calculations[14, 15] predict that the 5- and 6-positions should undergo electrophilic substitution in preference to the 4- and 7-positions, the earlier calculations[11, 12] predict the opposite result. Using the HMO procedure, a model in which the two $3d$ orbitals of the sulfur atom are fully involved as $3pd^2$ π hybrid orbitals and a model in which they are not involved at all can both be made to give similar results.[12] The calculated delocalization energy for benzo[c]thiophene is close to, but less than, that of benzo[b]thiophene and is significantly different from that of thiophene.[12] Apart from the Diels–Alder cycloaddition reactions of benzo[c]thiophene (Section IV), which occur in the 1- and 3-positions as predicted, there is little experimental evidence to support the theoretical predictions. The instability of the parent heterocycle, undoubtedly due to the high reactivity of the 1- and 3-positions and reflected in the fact that the molecule possesses an o-benzoquinodimethane structure (2),[12, 13] suggests that such evidence will be difficult to obtain. Nevertheless, further calculations supported by experimental evidence on derivatives of benzo[c]thiophene are awaited with interest.

Although the HMO procedure has allowed satisfactory agreement to be reached between the wavenumbers of the first long wavelength bands in the electronic absorption spectrum of benzo[c]thiophene and the corresponding $N-V_1$ transition energies, the shortcomings of this procedure do not allow a full interpretation of the spectrum.[12] Greater success has been achieved using the Pariser–Parr–Pople semiempirical method.[14, 15]

There has been no systematic study of the spectra (mass, UV, IR, or NMR) of benzo[c]thiophene derivatives, although spectroscopic data have been recorded for individual compounds in the more recent literature. The spectroscopic data for benzo[c]thiophene itself have been recorded,[5, 6, 12, 15, 16] the IR spectrum of 1,3-diphenylbenzo[c]thiophene has been studied in some detail,[17] and the excitation and

[16] B. D. Tilak, H. S. Desai, and S. S. Gupte, *Tetrahedron Lett.* 1953 (1966).
[17] F. R. McDonald and G. L. Cook, *U.S. Bur. Mines Rep. Invest. No.* **6911** (1967); *Chem. Abstr.* **67**, 58846 (1967).

emission wavelength and phosphorescence intensity (relative to carbazole) have been recorded for 1,3-dihydrobenzo[c]thiophene.[18]*

(4) CHPh +

It is noteworthy that HMO calculations have been made of the energy characteristics of the carbonium ion (4).[19]

III. Hydrobenzo[c]thiophenes

Since hydrobenzo[c]thiophenes are often used as intermediates for the preparation of benzo[c]thiophene and its derivatives, it is convenient to review the chemistry of these compounds first. Hydrobenzo[c]thiophene 2-oxides and 2,2-dioxides are reviewed in Section VI and 1,3-dihydrobenzo[c]thiophen-1-one and 1,3-dihydrobenzo[c]-thiophen-1,3-dione are covered in Section VII.

A. 1,3-Dihydrobenzo[c]thiophenes (2-Thiaindanes)

The parent compound (5) may be prepared by heating o-xylylene dibromide (α,α'-dibromo-o-xylene) with sodium or potassium sulfide, usually in aqueous ethanol (see Table I for references), or by reducing phthalic thioanhydride (Section VII,B) with aluminum hydride or diborane;[20] the former procedure has been used to prepare a number of derivatives of 5 (Table I). Disodium o-xylylenedimercaptide (6) can be isolated from the reaction between o-xylylene dibromide and

---

* *Added in proof:* For a recent study of the spectroscopic properties of this compound see M. G. Voronkov, A. N. Pereferkovich, V. A. Pestunovich, A. Ginvalde, and I. V. Turovskii, *Khim. Geterotsikl. Soedin.* 885 (1970); *Chem. Abstr.* **74**, 42337 (1971).

[18] H. V. Drushel and A. L. Sommers, *Anal. Chem.* **38**, 10 (1966).
[19] V. Horák, C. Párkányi, J. Pecka, and R. Zahradník, *Collection Czech. Chem. Commun.* **32**, 2272 (1967).
[20] R. H. Schlessinger and I. S. Ponticello, *Chem. Commun.* 1013 (1969).

sodium sulfide and gives **5** on being heated in a sealed tube at 120°–130°.[21] When o-xylylene dibromide is treated with sodium methanethiolate, it gives the methylsulfonium bromide [(**7**) R = H];[9] the 5-nitro derivative [(**7**) R = NO$_2$] and the corresponding ethylsulfonium bromide may be similarly prepared. The reduction of 1,3-diarylbenzo[c]thiophenes to their 1,3-dihydro derivatives is discussed in Section V,B.

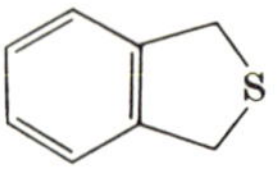

(5) 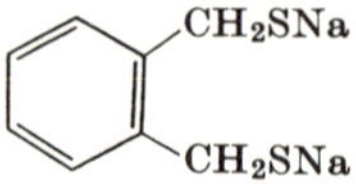   (6)

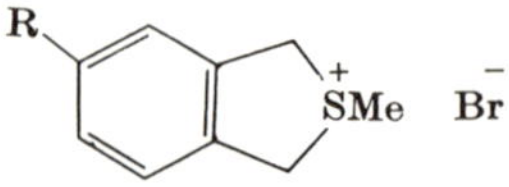

(7)

When 2-thiophthalide (Section VII,A) is heated with phosphorus pentachloride, 1,1,3,3-tetrachloro-1,3-dihydrobenzo[c]thiophene (1,1,3,3-tetrachlorothiophthalan) [(**8**) R = H; X = Cl] is obtained, together with o-chloromethylbenzoyl chloride.[22] Derivatives of **8** (R = F, Cl, Br, NO$_2$; X = Cl) (Table I) may be similarly prepared. When the 1,1,3,3-tetrachloro compounds [(**8**) X = Cl and R as before] are treated with anhydrous hydrofluoric acid, they give the corresponding tetrafluoro compounds (Table I).[22] 5-Cyano-1,1,3,3-tetrafluoro-1,3-dihydrobenzo[c]thiophene [(**8**) R = CN; X = F] may be prepared by treating the 5-bromo compound with cuprous cyanide, and converted into the corresponding acid [(**8**) R = CO$_2$H; X = F] and carbonamide by standard procedures.[22]

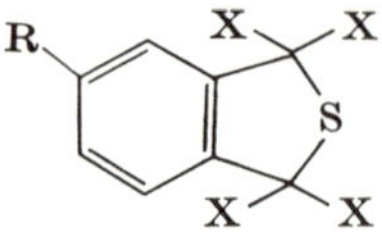

(8)

[21] W. Autenrieth and A. Brüning, *Chem. Ber.* **36**, 183 (1903).
[22] L. M. Yagupol'skii and R. V. Belinskaya, *J. Gen. Chem. USSR* **36**, 1421 (1966).

TABLE I

1,3-DIHYDROBENZO[c]THIOPHENES

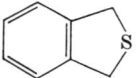

Substituents	M.p. and/or b.p. (°C)	Yield (%)	Ref.
None	22.5–23.5 (94.7/5 mm)	81, 87	28, 34a
	25 (106–107/9 mm, 241.5/760 mm)	67	29[a]
	(65–70/0.3 mm)	51	9
	26 (108/14 mm)	40	27
	(62–72/2 mm)	44	87
	—	100	20
Methiodide	175	—	27
	160–161	—	29
	154–155	—	31
Methobromide	188–190	39	9
Methohydroxide	—	—	31
$HgCl_2$ adduct	131.5–132.5	—	29, 31
1-Me	(115–116/16 mm)	70, 30	24, 25
Methiodide	137	—	24
$HgCl_2$ adduct	121	—	24
1,3-Me_2	43–44	69.5	88
1-Ph	60–62 (135–140/0.1 mm)	—	23
1,3-Ph_2, trans isomer	106.5–107.5	29	4 (cf. 68)
	107–108	64.5	69
1-Ph, 3-C_6H_4Me-p	103–104	—	74
1-Me, 3-Ph	(138–148/0.1 mm)	—	23
1,1-Me_2, 3-Ph	70–72	—	23
1,1-Me_2, 3-CO_2H	86–87	—	23
1,1-Me_2, 3-CO_2Et	(108–111/0.3 mm)	—	23
1,1-Me_2, 3-$CONH_2$	124–125	—	23
1,1-Me_2, 3-Ph, 3-OH	84–85	—	23
1-Ph, 1-$(CH_2)_3NMe_2$[b], HCl salt	200–202	—	23
4,5,6,7-F_4	33–34	—	30
4,5,6,7-Br_4	214–216	58	95
1,1,3,3-F_4	—	—	22
1,1,3,3,5-F_5	(68/25 mm)	55	22
1,1,3,3-F_4, 5-Cl	(95/30 mm)	33	22
1,1,3,3-F_4, 5-Br	(108/30 mm)	37	22
1,1,3,3-F_4, 5-CN	53–54	28.5	22
1,1,3,3-F_4, 5-CO_2H	212–213	62.5	22
1,1,3,3-F_4, 5-$CONH_2$	134–135	—	22

(*continued*)

TABLE I—continued

Substituents	M.p. and/or b.p. (°C)	Yield (%)	Ref.
1,1,3,3-Cl_4	112–113	50	22
1,1,3,3,5-Cl_5	48–49	77	22
1,1,3,3-Cl_4, 5-F	45–46	67	22
1,1,3,3-Cl_4, 5-Br	43–44	78	22
1,1,3,3-Cl_4, 5-NO_2	109–110	83	22
5-NO_2			
Methobromide	Decomposes	73	9
Ethobromide	122–123	—	9
4,7-$(OMe)_2$	92	35	c
5-CO_2Me	64.5–65.5	70	54

[a] See also S. F. Birch, *J. Inst. Petrol.* **39**, 185 (1953).
[b] Related compounds are described in the same patent.[23]
[c] L. Horner, P. V. Subramaniam, and K. Eiben, *Ann. Chem.* **714**, 91 (1968).

1,1-Dimethyl-3-phenyl-1,3-dihydrobenzo[c]thiophene [(**9**) R = H] may be prepared by treating the sulfide (**10**) with sodium hydride in dimethyl sulfoxide or by treating the alkene (**11**) with hydrogen sulfide.[23] 1,1 - Dimethyl- 1,3-dihydrobenzo[c]thiophene - 3 - carboxylic acid may be similarly prepared by the former procedure, and the amine [(**9**) R = $(CH_2)_3NMe_2$] by the latter, and also by direct alkylation of **9** (R = H). In the presence of molybdenum sesquisulfide, cyclization of **11** with hydrogen sulfide gives the alcohol [(**9**) R = OH], which gives **9** (R = H) on being heated with formic acid. Compound **9** [R = $(CH_2)_3NMe_2$] and related compounds are claimed[23] to be potentiating agents for adrenaline and noradrenaline and are said to exhibit sedative and anticholinergic activity; they are also reported to be useful in the treatment of endogenic depression.*

(**9**) (**10**)

Added in proof: See also, N. Lassen and T. Ammitzboell, *Acta Chem. Scand.* **25**, 2879 (1971).

[23] P. V. Petersen, N. Lassen, and T. Ammitzboell, S. African Patent 6,800,199 (1969); *Chem. Abstr.* **72**, 90271 (1970).

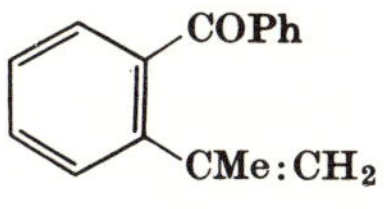

(11) — structure with COPh and CMe:CH₂

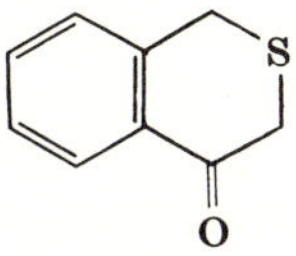

(12) — isothiochroman-4-one structure

1-Methyl-1,3-dihydrobenzo[c]thiophene arises by rearrangement during the Clemmensen reduction of isothiochroman-4-one (12),[24, 25] and the spiran (14) is one of three products that arise when the thiopyran (13) is treated with aqueous perchloric acid [Eq. (1)].[26] The formation of these three products has been rationalized by the suggestion that the interesting bridgehead sulfonium ion species (17)

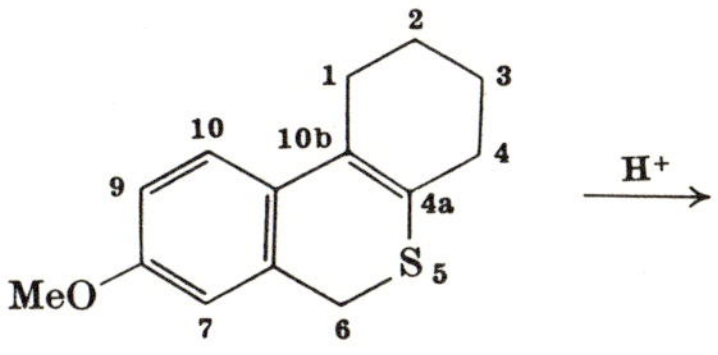

(13)

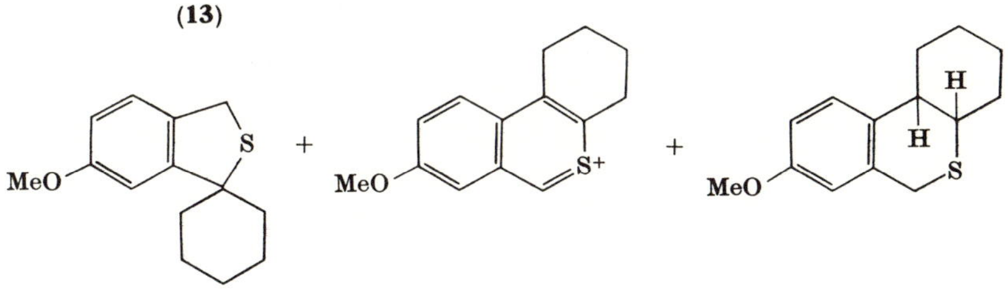

(14) (7.6%)   (15) (45%)   (16) (34.3%)   (1)

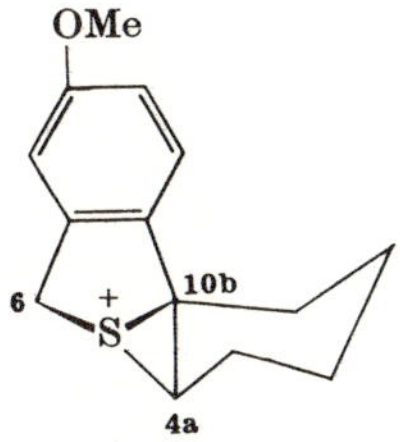

(17)

[24] J. von Braun and K. Weissbach, *Chem. Ber.* **62B**, 2416 (1929).
[25] N. J. Leonard and J. Figueras, *J. Amer. Chem. Soc.* **74**, 917 (1952).
[26] E. R. De Waard, W. J. Vloon, and H. O. Huisman, *Chem. Commun.* 841 (1970).

is formed by protonation of **13** at C-4a followed by interaction of the sulfur d orbitals with the empty orbital left at C-10b (see also Section V,A). Transfer of a hydride ion from C-6 of the starting material (**13**) to C-10b or C-4a of **17** gives **15**, **16**, and **14**, respectively.

1,3-Dihydrobenzo[c]thiophene (**5**) is an intermediate in the interconversion reactions of phthalic acid, or its anhydride, and o-xylene with hydrogen sulfide or a mixture of sulfur and water, respectively (Section VII,A).

1,3-Dihydrobenzo[c]thiophene is an unstable, low-melting solid with a thiophenelike odor. It is stable for several months in the absence of air, but it forms a deep red liquid in air from which an amorphous blue solid of unknown identity is finally precipitated.[9,27-29] Some derivatives of 1,3-dihydrobenzo[c]thiophene also appear to be unstable; for example, 4,5,6,7-tetrafluoro-1,3-dihydrobenzo[c]thiophene decomposes rapidly in air, and even under nitrogen.[30]

1,3-Dihydrobenzo[c]thiophene forms a sulfoxide, a sulfone (Section VI), adducts with mercuric chloride, platinum chloride, and bromine, and a methylsulfonium iodide which gives the hydroxide with moist silver oxide.[29,31] Unlike 2,3-dihydrobenzo[b]thiophene, it readily forms a water-soluble adduct with mercuric acetate from which the parent compound is regenerated with hydrochloric acid.[32] It may be dehydrogenated to benzo[c]thiophene (Section IV), undergoes reductive cleavage to o-xylene with calcium hexammine,[33] gives o-xylene and a trace of 1,2-dimethylcyclohexane with hydrogen in the presence of molybdenum disulfide,[34] and gives a magenta color

[27] J. von Braun, O. Bayer, and W. Kaiser, *Chem. Ber.* **58B**, 2166 (1925).
[28] J. A. Oliver and P. A. Ongley, *Chem. Ind.* (*London*) 1024 (1965).
[29] S. F. Birch, R. A. Dean, and E. V. Whitehead, *J. Inst. Petrol.* **40**, 76 (1954).
[30] P. L. Coe, B. T. Croll, and C. R. Patrick, *Tetrahedron* **23**, 505 (1967).
[31] E. Hjelt, *Chem. Ber.* **22**, 2904 (1889).
[32] S. F. Birch and D. T. McAllan, *J. Inst. Petrol.* **37**, 443 (1951).
[33] J. van Schooten, J. Knotnernus, H. Boer, and P. M. Duinker, *Rec. Trav. Chim.* **77**, 935 (1958).
[34] S. Landa and A. Mrnková, *Sb. Vysoke Skoly Chem.-Technol. Praze, Technol. Paliv* **11**, 5 (1966); *Chem. Abstr.* **67**, 13492 (1967).

Sec. III. B.] BENZO[c]THIOPHENES 341

with concentrated sulfuric acid.[29] With chlorine an aqueous suspension of 1,3-dihydrobenzo[c]thiophene (5) gives the sulphonyl chloride (17a).[34a] The latter compound yields 3H-2,3,4-benzothiadiazepine 2,2-dioxide (17b) on treatment with hydrazine (see also Section VI, B). The IR and UV spectra of 5 have been recorded.[29]

B. 4,5-DIHYDROBENZO[c]THIOPHENES

Successive reduction of 1,3-dimethyl-4,5,6,7-tetrahydrobenzo[c]-thiophen-4-one (Section III,C) with sodium borohydride and dehydration of the resulting alcohol with polyphosphoric acid gives the unstable 1,3-dimethyl derivative [(18) R = H] in excellent yield,[35] and successive treatment of the same ketone with methylmagnesium iodide and acid gives the 1,3,7-trimethyl derivative [(18) R = Me] (Table IV) and a second compound which is reported to be a "dimer."[36] When 1,3-dimethyl-4,5,6,7-tetrahydrobenzo[c]thiophen-4-one is treated with a mixture of phosphorus oxychloride and dimethylformamide (Vilsmeier–Haack formylation), it gives compound 19, which yields 20 on being treated successively with thioglycolic acid and concentrated aqueous potassium hydroxide.[37]

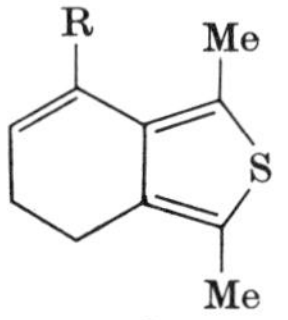

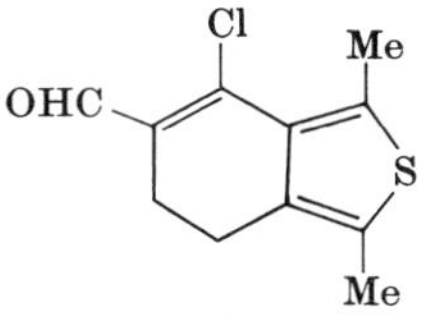

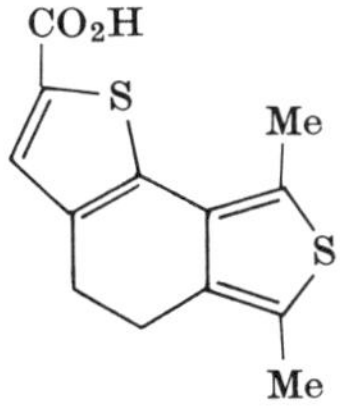

[34a] J. F. King, A. Hawson, B. L. Huston, L. J. Danks, and J. Komery, *Canad. J. Chem.* **49**, 943 (1971).
[35] P. Cagniant, G. Merle, and D. Cagniant, *Bull. Soc. Chim. Fr.* 322 (1970).
[36] N. P. Buu-Hoi, N. Hoan, and N. H. Khoi, *Rec. Trav. Chim.* **69**, 1053 (1950).
[37] A. Ricci, D. Balucani, C. Rossi, and A. Croisy, *Boll. Sci. Fac. Chim. Ind. Bologna* **27**, 279 (1969).

C. 4,5,6,7-Tetrahydrobenzo[c]thiophenes

The sulfide (21) condenses with cyclohexane-1,2-dione (22) in the presence of sodium ethoxide to give a mixture of the mono- (23a) and diethyl ester (23b) of 4,5,6,7-tetrahydrobenzo[c]thiophene-1,3-dicarboxylic acid (23c).[6] The free acid (23c) is obtained by hydrolysis of these esters and it undergoes decarboxylation on being heated to 360° to give 4,5,6,7-tetrahydrobenzo[c]thiophene.[6] This compound is also known as 3,4-tetramethylenethiophene and as 8-thiabicyclo-[4.3.0]nona-6,9-diene.

$S(CH_2CO_2Et)_2$ + (cyclohexane-1,2-dione) $\xrightarrow{NaOEt}$ (tetrahydrobenzo[c]thiophene-1,3-dicarboxylate)

(21)   (22)   (23a) $R^1 = H$, $R^2 = Et$
(23b) $R^1 = R^2 = Et$
(23c) $R^1 = R^2 = H$

A similar condensation reaction between ethyl thioglycolate and 2-ethoxymethylenecyclohexanone (24a) gives compound 25, the methyl ester of which gives methyl 4,5,6,7-tetrahydrobenzo[c]thiophene-1-carboxylate on being heated with p-toluenesulfonic acid.[16, 38] Hydrolysis of this ester with barium hydroxide and decarboxylation of the resulting acid also gives 4,5,6,7-tetrahydrobenzo[c]thiophene in excellent yield.[16, 38] Likewise, thioglycolic acid condenses with 2-hydroxymethylenecyclohexanone (24b) to give compound 26.[39] The methyl and ethyl esters of 26 cyclize on being treated with sodium methoxide to give the corresponding esters of 4,5,6,7-tetrahydrobenzo[c]thiophene-1-carboxylic acid.[39]

+ $HSCH_2CO_2Et$ $\xrightarrow{NaOEt}$

(24a) R = Et
(24b) R = H
(25)

[38] B. D. Tilak and S. S. Gupte, *Indian J. Chem.* **7**, 9 (1969).
[39] H. Fiesselmann and H. Habicht, German Patent 1,092,929 (1960); *Chem. Abstr.* **57**, 5894 (1962).

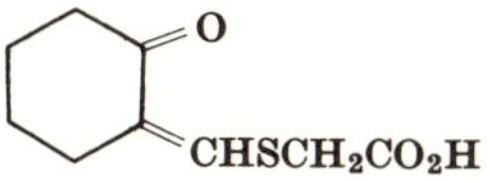

(26)

Friedel–Crafts acetylation of 1-methyl-4,5,6,7-tetrahydrobenzo[c]-thiophene is reported[10] to give 1-acetyl-3-methyl-4,5,6,7-tetrahydrobenzo[c]thiophene. Successive treatment of methyl 4,5,6,7-tetrahydrobenzo[c]thiophene-1-carboxylate with N-bromosuccinimide and sodium methoxide gives benzo[c]thiophene-1-carboxylic acid.[16, 38]

TABLE II

4,5,6,7-TETRAHYDROBENZO[c]THIOPHENES

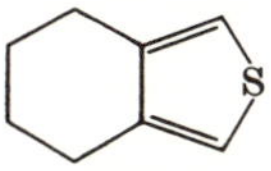

Substituents	M.p. and/or b.p. (°C)	Yield (%)	Ref.
None	(75/5 mm)	86	16, 38
	(97–99/16 mm)	30	6
1-Me	(105/18 mm)	—	10
1,3-Me$_2$	(107.5–118.5/18 mm, 245/760 mm)	70–85, low	35, 41
1,4,4-Me$_3$	21–22 (66–70/0.3 mm)	—	40
Chloromercuri derivative	197–199	—	40
Mercury dithienyl derivative	179–180	—	40
Acetoxymercuri derivative	163–165	—	40
1,3-Et$_2$	(131.5–132/9 mm)	82	42
1-Me, 3-COMe	57 (175–176/18 mm)	—	10
DNPh[a]	155	—	10
Oxime derivative	111	—	10
1,3-Me$_2$, 4-OH	108	100	35
Phenylurethane derivative	149	—	35
1-CO$_2$H	203–204	92, 88	16, 38
	205–206	—	39
1-CO$_2$Me	25 (126–127/0.1 mm)	—	39
	(118/2.5 mm)	82, 75	16, 38
1-CO$_2$Et	(118/0.05 mm)	—	39
1,3-(CO$_2$H)$_2$	320 (dec.)	80, 66	6
1-CO$_2$H, 3-CO$_2$Et	195	—	6
1,3-(CO$_2$Et)$_2$	138	14	6

[a] 2,4-Dinitrophenylhydrazone derivative.

This ester cannot be dehydrogenated with palladium–charcoal, selenium, or chloranil;[16] similar attempts to aromatize 4,5,6,7-tetrahydrobenzo[c]thiophene have failed.[6,16] Diethyl 4,5,6,7-tetrahydrobenzo[c]thiophene-1,3-dicarboxylate forms an adduct with maleic anhydride.[6]

Derivatives of 4,5,6,7-tetrahydrobenzo[c]thiophene are listed in Table II. 1,4,4-Trimethyl-4,5,6,7-tetrahydrobenzo[c]thiophene (**27**) has been detected in a kerosine extract derived by refining a Middle-East crude oil.[40]

(27)

Several derivatives of 4,5,6,7-tetrahydrobenzo[c]thiophen-4-one (Table III) have been prepared by the sequence of reactions exemplified by Eq. (2) or by a closely related sequence of reactions. When

$$\text{(1) Succinic anhydride/SnCl}_4 \quad \text{(2) N}_2\text{H}_4 \text{ reduction} \quad \text{(3) SOCl}_2 \quad \text{(4) SnCl}_4/\text{CS}_2 \tag{2}$$

the acid chloride (**28**) is treated with stannic chloride or aluminum chloride in carbon disulfide, cyclization occurs normally.[10] With aluminum chloride in benzene, however, loss of the *t*-butyl group occurs to give 1-methyl-4,5,6,7-tetrahydrobenzo[c]thiophen-4-one.[10]

4,5,6,7-Tetrahydrobenzo[c]thiophen-4-one and its derivatives may be reduced to the corresponding 4,5,6,7-tetrahydrobenzo[c]thiophene (Table II) by the Clemmensen[41] or Wolff–Kishner–Huang Minlon[10,35,42] procedure; catalytic reduction of the parent ketone gives the saturated alcohol (**29**).[43] Several reactions of 1,3-dimethyl-4,5,6,7-

[40] S. F. Birch, T. V. Cullum, R. A. Dean, and D. G. Redford, *Tetrahedron* **7**, 311 (1959).
[41] W. Steinkopf, I. Poulsson, and O. Herdey, *Ann. Chem.* **536**, 128 (1938).
[42] P. Cagniant and P. Cagniant, *Bull. Soc. Chim. Fr.* 713 (1953).
[43] V. Georgian, U.S. Patent 2,858,314 (1958); *Chem. Abstr.* **53**, 10258 (1959).

TABLE III
4,5,6,7-Tetrahydrobenzo[c]thiophen-4-ones

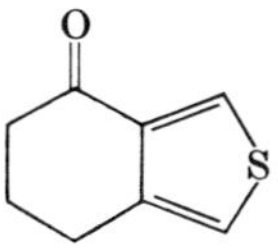

Substituents	M.p. and/or b.p. (°C)	Yield (%)	Ref.
None	—	—	43
1-Me	49–50 (157–158/17 mm)	—	10
DNPh[a]	275	—	10
Oxime derivative	113	—	10
1,3-Me$_2$	46 (152–153/13 mm)	Good	36
	39.5–41	33	37, 41
	40	80	35
DNPh[a]	280	—	35
Ketazine derivative	244	—	35
1,3,5-Me$_3$	(158–158.5/17 mm)	82	35
DNPh[a]	252	—	35
Oxime derivative	135	—	35
1,3,6-Me$_3$	57.5	70–80	35
DNPh[a]	278	—	35
Oxime derivative	153	—	35
Semicarbazone derivative	226	—	35
1-Me, 3-t-Bu	52 (175/16 mm)	Good	10
Oxime derivative	121	—	10
1,3-Et$_2$	(140/3.5 mm, 162/10.5 mm)	91	42
DNPh[a]	217	—	42
Oxime derivative	122	—	42
Ketazine derivative	137–137.5	—	42

[a] 2,4-Dinitrophenylhydrazone derivative.

tetrahydrobenzo[c]thiophen-4-one have already been described in Section III, B.

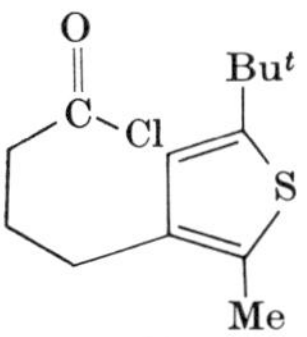

(28) 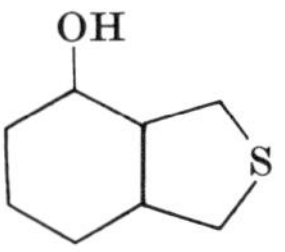   (29)

In a Pfitzinger–Borsche reaction with isatin and 5-bromoisatin, 1,3-dimethyl-4,5,6,7-tetrahydrobenzo[c]thiophen-4-one gives compounds **30a,b**, respectively, and it condenses with various phenylhydrazines to give the corresponding phenylhydrazones, which may be successively cyclized and aromatized to give compounds **31a–f**, **32**, and **33**.[36] 1-Methyl-4,5,6,7-tetrahydrobenzo[c]thiophen-4-one also undergoes a Pfitzinger–Borsche reaction with 5-methylisatin to give compound **30c**.[10] It is noteworthy that 3-*t*-butyl-1-methyl-4,5,6,7-tetrahydrobenzo[c]thiophen-4-one will not undergo this reaction, presumably due to steric hindrance by the *t*-butyl group.[10]

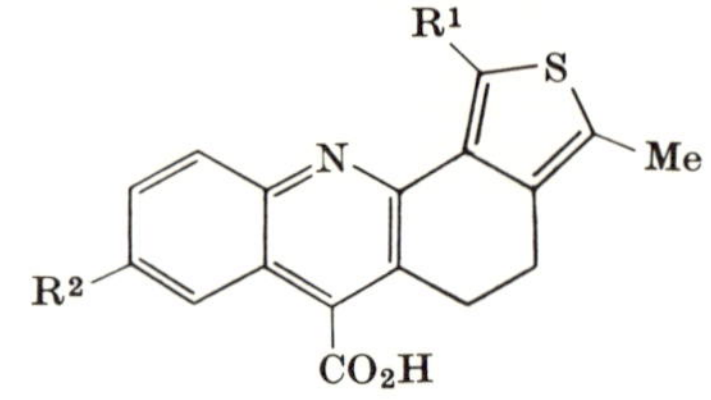

(**30a**) R^1 = Me, R^2 = H
(**30b**) R^1 = Me, R^2 = Br
(**30c**) R^1 = H, R^2 = Me

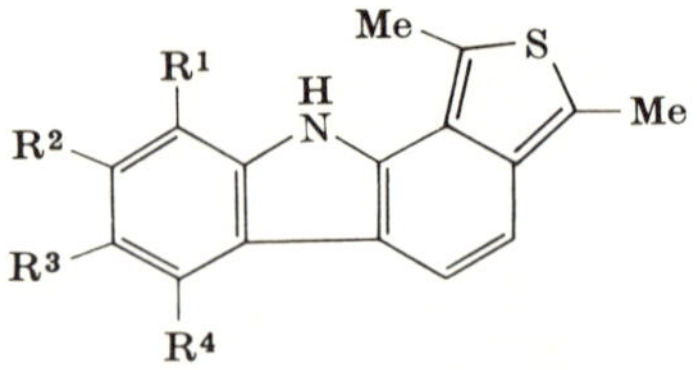

(**31a**) $R^1 = R^2 = R^3 = R^4$ = H
(**31b**) R^1 = Me, $R^2 = R^3 = R^4$ = H
(**31c**) R^2 = Me, $R^1 = R^3 = R^4$ = H
(**31d**) R^3 = Me, $R^1 = R^2 = R^4$ = H
(**31e**) $R^1 = R^4$ = Me, $R^2 = R^3$ = H
(**31f**) $R^1 = R^2$ = Me, $R^3 = R^4$ = H

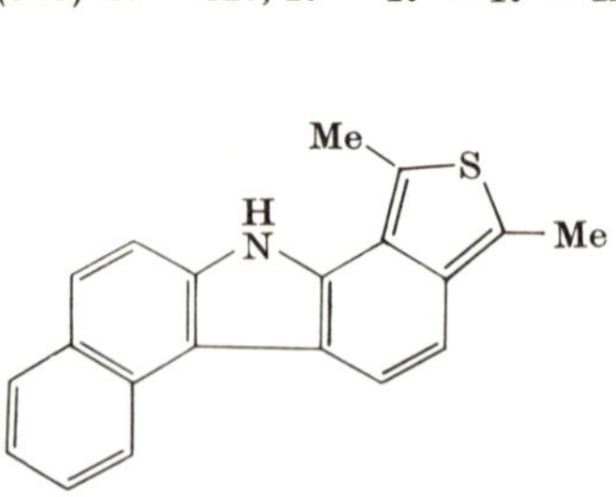

(**32**)

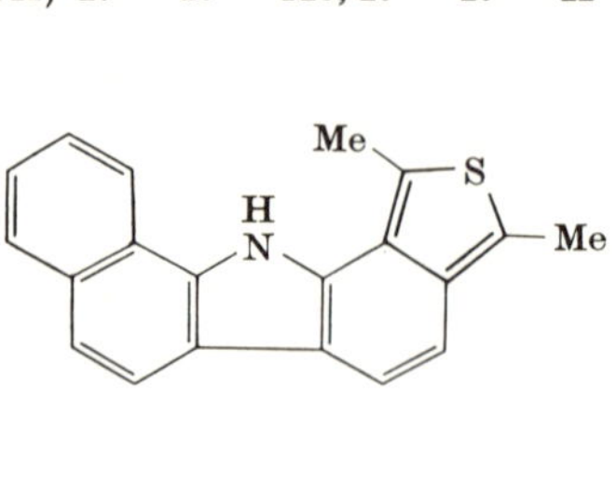

(**33**)

D. 1,3,4,7-TETRAHYDROBENZO[c]THIOPHENES

The parent compound **34** (Table IV) is referred to as 8-thia-*trans*-bicyclo[4.3.0]non-3-ene in the literature and may be prepared in 51%

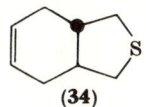

(34)

TABLE IV
OCTAHYDRO- AND OTHER HYDROBENZO[c]THIOPHENES

Compound	M.p. and/or b.p. (°C)	Yield (%)	Ref.
1,3-Me$_2$, 4,5-H$_2$ [(**18**) R=H][a]	(115.5–116/17 mm)	80–90	35
1,3,7-Me$_3$, 4,5-H$_2$ [(**18**) R=Me]	(135–140/15 mm)	Moderate	36
1,3-Me$_2$, 4,5-H$_2$, 6-CHO, 7-Cl (**19**)	Red oil	—	37
Semicarbazone derivative	236	—	37
1,3,4,7-H$_4$ (**34**)	83	51	44
Octahydrobenzo[c]thiophene (**35**)	—	53, 69	48
Cis isomer	(108.5/21 mm)	75	45[b]
	(46/1 mm, 95.5/16.5 mm)	By-product	46
HgCl$_2$ adduct	187–188	—	45, 46
Methiodide	142.5–144 (dec.)	—	45
Trans isomer	(104.5–105.5/20 mm)	91	45[b]
HgCl$_2$ adduct	212.5–213.5	—	45
Methiodide	164.5–166 (dec.)	—	45
(−)-(8*R*,9*R*)- trans isomer	(80–100/28 mm)	47	47
HgCl$_2$ adduct	207–209	—	47
Octahydrobenzo[c]thiophen-4-one (**37**)	—	—	43
4-Hydroxyoctahydrobenzo-[c]thiophene (**29**)	—	—	43
(**25**) Methyl ester	(110/0.002 mm)	46, 39	16, 38
(**39**)	52–52.5 (100/2.0 mm)	43	49
(**41**)	(73/15 mm)	50	49
(**42**)	85–85.5	85	49

[a] Unstable.
[b] See also S. F. Birch, *J. Inst. Petrol.* **39**, 185 (1953).

yield by treating the bistoluene-*p*-sulfonate of *trans*-4,5-di(hydroxymethyl)cyclohexene with sodium sulfide.[44]

E. OCTAHYDROBENZO[c]THIOPHENES (2-THIAHYDRINDANES)

Octahydro- (we have used this nomenclature throughout) or perhydrobenzo[c]thiophene (35) exists as the cis and trans isomers and is more commonly named 2-thiahydrindane or 8-thiabicyclo-[4.3.0]nonane. Each stereoisomer may be prepared by treating the corresponding isomer of 1,2-di(bromomethyl)cyclohexane with sodium sulfide (details of the products are given in Table IV).[45] Cyclization of the cis isomer may be effected partially with thiourea via the formation of *cis*-hexahydro-*o*-xylylenebis(isothiouronium bromide); with sodium disulfide it gives a mixture of *cis*-octahydrobenzo[c]thiophene and *cis*-2,3-dithiadecalin.[46] Optically pure (−)-(8*R*,9*R*)-*trans*-octahydrobenzo[c]thiophene has been prepared from (+)-*trans*-cyclohexane-1,2-dicarboxylic acid via reduction of the diacid to the diol, tosylation, and ring closure of the bistosylate with sodium sulfide.[47]

When 1,2,3,6-tetrahydrophthalic anhydride (4-cyclohexene-1,2-dicarboxylic acid anhydride) (36) is heated with a mixture of hydrogen sulfide and hydrogen in the presence of cobalt sulfide, it gives octahydrobenzo[c]thiophene (53%) and a trace of 4,5,6,7-tetrahydrobenzo[c]thiophene.[48] Similar treatment of hexahydrophthalic anhydride gives a mixture (69%) of *cis*- and *trans*-octahydrobenzo[c]-thiophene, which may be separated by VPC.[48]

(35)   (36)

[44] B. J. Auret, D. R. Boyd, and H. B. Henbest, *J. Chem. Soc. C* 2374 (1968).
[45] S. F. Birch, R. A. Dean, and E. V. Whitehead, *J. Org. Chem.* **19**, 1449 (1954).
[46] A. Lüttringhaus and A. Brechlin, *Chem. Ber.* **92**, 2271 (1959).
[47] P. Laur, H. Häuser, J. E. Gurst, and K. Mislow, *J. Org. Chem.* **32**, 498 (1967).
[48] R. W. Campbell, U.S. Patent 3,345,381(1967); *Chem. Abstr.* **68**, 29587 (1968).

We have referred to the preparation of compound **25** in Section III,C; Oppenauer oxidation of the saturated alcohol (**29**) (Section III,C) gives the corresponding ketone (**37**).[43] This ketone and its sulfoxide and sulfone condense with bromo-*p*-tolylhydrazines to give hydrazones, which undergo cyclization in hot acetic acid to give the carbazolenines (**38**) and their sulfoxides and sulfones, respectively. These are claimed to be useful intermediates in the synthesis of morphinelike analgesics.[43]

(37)   (38)

The structures of *cis*- and *trans*-octahydrobenzo[*c*]thiophene have been confirmed by Raney nickel desulfurization to the corresponding isomers of 1,2-dimethylcyclohexane.[45] Both isomers form water-soluble mercuric acetate adducts from which the parent compound may be regenerated with hydrochloric acid.[45] They also form adducts with mercuric chloride, give methylsulfonium iodides (Table IV), and may be oxidized to the corresponding sulfones with peracetic acid.[45] Their IR spectra and other physical properties have been recorded.[45] An interesting feature of the stereochemistry of *cis*- and *trans*-octahydrobenzo[*c*]thiophene is that, whereas the trans compound should give only one sulfoxide, the cis compound should give two stereoisomeric sulfoxides. However, the hygroscopic nature of the products prevented their separation.[45]

In a study[47] of the ORD and circular dichroism properties of (−)-(8*R*,9*R*)-*trans*-octahydrobenzo[*c*]thiophene and A-nor-2-thiacholestane it was shown that the sulfur chromophore is useful for making stereochemical correlations. The two compounds show optically active transitions at 244 and 205 nm; the signs of these transitions reflect the chirality of the neighboring centers.

F. Miscellaneous Hydrobenzo[*c*]thiophenes

4,7-Dihydro-1,3-dimethyl-4,7-ethanobenzo[*c*]thiophene (**39**) (Table IV) may be prepared by passing a mixture of hydrogen sulfide and hydrogen chloride into an ethanolic solution of the diketone (**40**).[49] Compounds **41** and **42** may be similarly prepared. The Diels–Alder adducts of 2,5-dihydrothiophene with 1,2,3,4-tetrachloro-5,5-dimethyl-, 1,2,3,4,5-pentabromo-5-methyl-, and 1,2,3,4-tetrachloro-5-(3-methoxypropyl)cyclopentadiene may be further halogenated to give compounds **43**, **44**, and **45**, respectively.[50, 51] These products are claimed to exhibit bactericidal, insecticidal, nematocidal, and herbicidal activity.

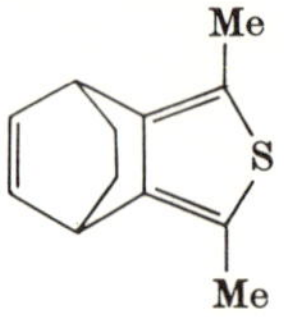

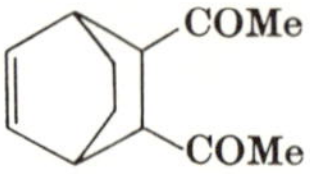

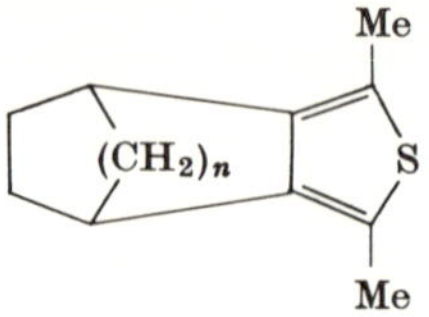

IV. Benzo[c]thiophene

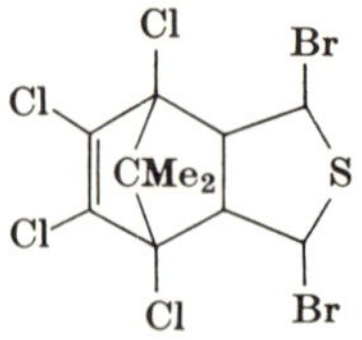

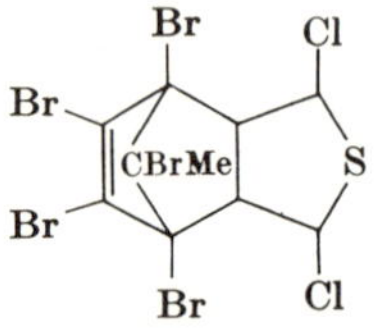

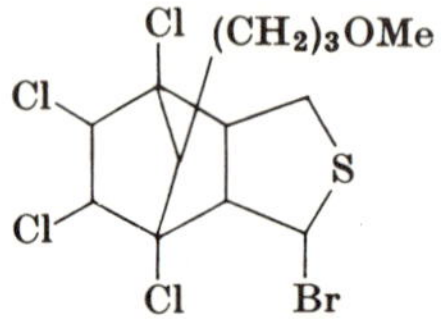

Benzo[c]thiophene may be prepared by low-pressure (20 mm) vapor-phase catalytic dehydrogenation of 1,3-dihydrobenzo[c]thiophene (Section III, A) at 330° under nitrogen,[5, 6] by decarboxylation of benzo[c]thiophene-1-carboxylic acid (Section III, C) with copper in quinoline[16, 38] or by dehydration of 1,3-dihydrobenzo[c]thiophene 2-oxide (Section VI, A) in acetic anhydride or over aluminum oxide at 20 mm Hg and 100°–125° in a sublimation tube.[52] A trace of water appears to be beneficial to the first reaction, and it has been suggested[53]

[49] H. Wynberg and A. J. H. Klunder, *Rec. Trav. Chim.* **88**, 328 (1969).
[50] V. Mark, U.S. Patent 3,256,299 (1966); *Chem. Abstr.* **65**, 5444 (1966).
[51] V. Mark, U.S. Patent 3,281,317 (1966); *Chem. Abstr.* **66**, 4390 (1967).
[52] M. P. Cava and N. M. Pollack, *J. Amer. Chem. Soc.* **88**, 4112 (1966).
[53] J. M. Holland and D. W. Jones, *J. Chem. Soc. C* 537 (1970).

that the dehydration of 1,3-dihydrobenzo[c]thiophene 2-oxide in acetic anhydride proceeds via a Pummerer rearrangement to give the 1-acetoxy compound (46) which then loses acetic acid. Unlike the 1,3-dihydro compound, the following compounds, 4,5,6,7-tetrahydrobenzo[c]thiophene (Section III,C),[6,16] ethyl 4,5,6,7-tetrahydrobenzo[c]thiophene-1-carboxylate (Section III,C),[16,38] and octahydrobenzo[c]thiophene (Section III, E),[6] are not dehydrogenated by the usual reagents.

(46)

Cava and Pollack's elegant synthesis of benzo[c]thiophene[52] has been extended to the synthesis of, for example, naphtho[1,2-c]thiophene (47),[52] methyl benzo[c]thiophene-5-carboxylate,[54] and 1,3-dimethyl- (48)[55] and 1,3,4,6-tetraphenylthieno[3,4-c]thiophene (49).[56] The last compound is of particular interest because it is a remarkably stable (cf. compound 48 which only exists transiently) "nonclassical" thiophene containing ten π electrons for which the only uncharged resonance structure (49) contains a tetravalent sulfur atom. When the sulfoxide (50) is pyrolyzed over aluminum oxide, it gives the parent cyclic sulfide and 51 by disproportionation.[57] However, when 50 is heated in acetic anhydride in the presence of N-phenylmaleimide,

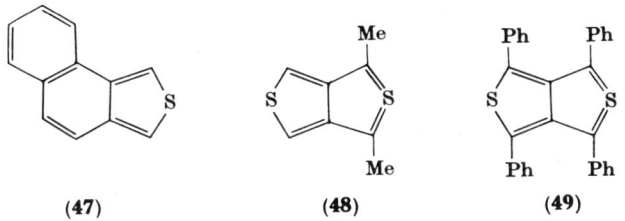

(47) (48) (49)

[54] H. Wynberg, T. Feijen, and D. J. Zwanenburg, Rec. Trav. Chim. 87, 1006 (1968).
[55] M. P. Cava and N. M. Pollack, J. Amer. Chem. Soc. 89, 3639 (1967).
[56] M. P. Cava and G. E. M. Husbands, J. Amer. Chem. Soc. 91, 3952 (1969).
[57] M. P. Cava, N. M. Pollack, M. J. Mitchell, and O. A. Mamer, personal communication.

naphtho[2,3-c]thiophene (**52**) is generated as a transient intermediate and trapped as a mixture of adducts.[57]*

(50)   (51)   (52)

(53)   (3)

The low-yield (14%) Wittig-type synthesis of the unusual compound **53** [Eq. (3)] is noteworthy, since this compound was the first known heterocyclic analog of biphenylene.[58]

When N,N-dimethylisoindolinium bromide is treated with phenyllithium, it gives N-methylisoindole via the ylid (**54**).[2,59,60] An attempt to prepare benzo[c]thiophene via the analogous ylid (**55**) failed. Thus, when 1,3-dihydrobenzo[c]thiophene methylsulfonium iodide was treated with phenyllithium, it gave a mixture of methyl phenyl sulfide, spiro[5.5]-1-methylthio-2,3-benzo-6-methylthiomethyleneundeca-7,9-diene (**56**), and 3,4-bis(methylthio)-1,2;5,6-dibenzo-1,5-cyclooctadiene (**57**).[59,60] The formation of methyl phenyl sulfide may be explained by the formation and ring cleavage of compound **58**, and compounds **56** and **57** arise by Diels–Alder dimerization of the o-quinodimethane (**59**) formed by ring cleavage of the ylid (**55**).

(54)   (55)

---

* *Added in proof:* See also D. W. H. MacDowell, A. T. Jeffries, and M. B. Meyers, *J. Org. Chem.* **36**, 1416 (1971); A. T. Jeffries, Ph.D. Thesis, University of West Virginia (1970); *Diss. Abstr.* **32B**, 169 (1971).

[58] P. J. Garratt and K. P. C. Vollhardt, *Chem. Commun.* 109 (1970).
[59] J. Bornstein and J. H. Supple, *Chem. Ind. (London)* 1333 (1960).
[60] J. Bornstein, J. E. Shields, and J. H. Supple, *J. Org. Chem.* **32**, 1499 (1967).

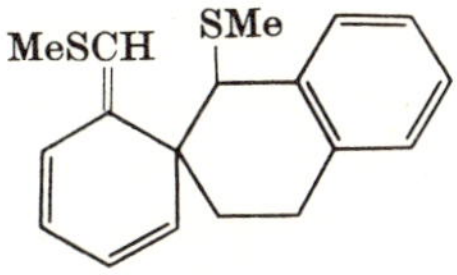

(56) 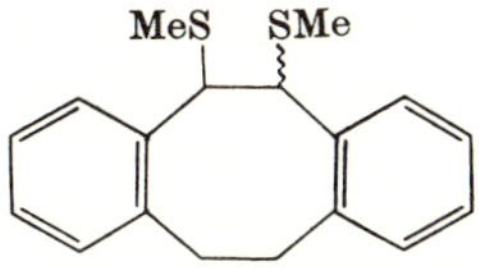 (57) 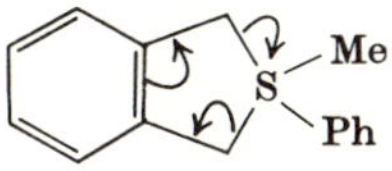 (58) 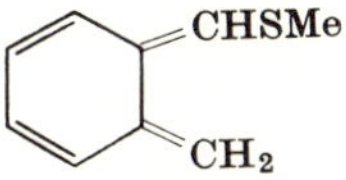 (59)

In the mass spectrum of thieno[3,4-*d*]thiepin (**60**), its sulfoxide, and sulfone (**61**), a major fragment ion arises at m/e 134, which may be attributed to the formation of benzo[*c*]thiophene.[61] The formation of benzo[*c*]thiophene from these compounds is confirmed by the fact that they give the adduct **62** on being heated with *N*-phenylmaleimide. The parent system (**60**) also gives dimethyl naphthalene-2,3-dicarboxylate on being heated with dimethyl acetylenedicarboxylate. However, unlike benzo[*c*]thiophene, which gives a 1:1 mixture of the exo and endo isomers of the adduct **62** on being heated with *N*-phenylmaleimide at various temperatures, the thiepin (**60**) and its oxidized derivatives give mixtures of the isomers in which the exo isomer always predominates. The amount of exo isomer formed increases with temperature. At 150°, **60** gives only the exo adduct, although mixtures are obtained at lower temperatures. The reason for this unusual behavior is not clear, although suggestions have been made.[61]

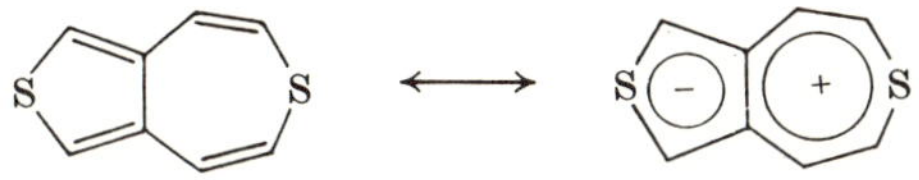

(60a) ⟷ (60b)

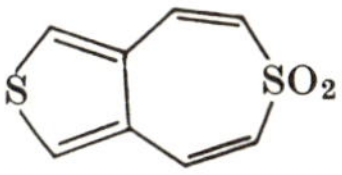

(61) 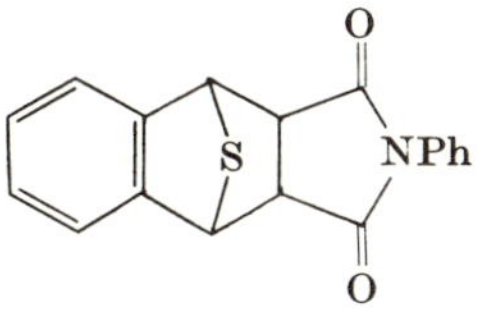 (62)

[61] R. H. Schlessinger and G. S. Ponticello, *J. Amer. Chem. Soc.* **89**, 7138 (1967); *Tetrahedron Lett.* 3017 (1968).

Benzo[c]thiophene is a colorless low-melting solid (Table V) with a strong naphthalenelike odor. It is stable only as a solid under nitrogen for a few days at $-30°$ (see Section II). The benzo[c]thiophene ring system can be stabilized, however, by suitable substitution, particularly at the highly reactive 1- and 3-positions. Thus, for

TABLE V

BENZO[c]THIOPHENES

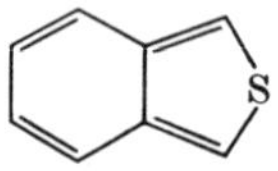

Substituents	M.p. and/or b.p. (°C)	Yield (%)	Ref.
None	50–51	65, 80	5, 6, 16, 38
	53–55	94	52
1,3,4-Me$_3$ ^{a}	(148–153/15 mm)	Low	36
1,3,4,7-Me$_4$	150–152	42	63
Picrate derivative	135–136	—	63
5-Me, 1,3-Ph$_2$	116–118	61	79
5,6-Me$_2$, 1,3-Ph$_2$	182.5–183	78	79
1-Me, 5-OMe	91–94	19	64
1-Me, 5,6-(OMe)$_2$	110–113	—	64
1-C$_6$H$_4$Me-p (?)	217	8	66
1-C$_6$H$_4$Cl-p (?)	241–242	12	66
1,3-Ph$_2$	117–118	24	4, 75
	—	—	20
	118–119	50, 80, 77	68, 70, 73
	119–120	60	69
1-Ph, 3-C$_6$H$_4$Me-p	99.5–100.5	33	74, 75
1-Ph, 3-C$_6$H$_4$Cl-o	89	12	74, 75
1-Ph, 3-C$_6$H$_4$Cl-m	102–103.5	11	74, 75
1-Ph, 3-C$_6$H$_4$Cl-p	99.5–100.5	33	74, 75
1-Ph, 3-C$_6$H$_4$Br-p	135–136	37	75
1-Ph, 3-C$_6$H$_4$NO$_2$-m	142	6	74, 75
	—	—	8
1,3,4,7-Ph$_4$	283	—	77
1,3,5,6-Ph$_4$	227–228	60–70	78
1,3-Ph$_2$, 5,6-Me$_2$	182.5–183	90	79
1-CO$_2$H	198	78, 89	16, 38
1-CO$_2$Me	(110/0.01 mm)	89	16, 38
5-CO$_2$Me	69.5–70.5	52	54

^{a} Product was not conclusively identified.

example, methyl benzo[c]thiophene-5-carboxylate (Table V) is stable in air for 2 days at room temperature and at −20° for at least a month.[54] 1,3-Disubstituted derivatives (e.g., the 1,3-diphenyl derivative) are usually quite stable.

Benzo[c]thiophene readily forms adducts with maleic anhydride (m.p. 153°–154°,[5, 6] 148°–152°[52]), dimethyl acetylenedicarboxylate (see above),[61] and N-phenylmaleimide (to give **62**).[52, 61] In the last case the exo and endo isomers of the adduct may be separated by fractional crystallization.[52] Benzo[c]thiophene reacts with acetyl nitrate by 1,3-addition in preference to substitution.[6] It exhibits halochromy with concentrated sulfuric acid, trichloroacetic acid, and stannic chloride.[62]

V. Alkyl and Aryl Derivatives of Benzo[c]thiophene

A. Alkyl Derivatives

1,3,4,7-Tetramethylbenzo[c]thiophene (Table V) is obtained in 42% yield when 2,5-dimethylthiophene is condensed with acetonyl acetone in anhydrous hydrofluoric acid [Eq. (4)].[63] Its structure has

$$\text{(4)}$$

been confirmed by Raney nickel desulfurization, which gives 2,3-diethyl-1,4-dimethylbenzene, and by the formation of an adduct with maleic anhydride.[63] An attempt to aromatize 1,3,7-trimethyl-4,5-dihydrobenzo[c]thiophene [(**18**) R = Me] (Section III,B) to 1,3,4-trimethylbenzo[c]thiophene gave a low yield of an oil which may have been the desired product.[36]

When the enamines [(**63**) (R = H or OMe] are heated at 160°/2 × 10⁻⁵ mm Hg for 40 minutes, they give the corresponding benzo[c]thiophene (**64**) possibly via formation of the thiiranium ions (**65**), followed by

---

[62] G. Tsatsaronis, *Chim. Chronica, Spec. Ed.* 57 (1957); *Chem. Abstr.* **55**, 8414 (1961).
[63] O. Dann, M. Kokorudz, and R. Gropper, *Chem. Ber.* **87**, 140 (1954).

hydride transfer to give **66** (R = H or OMe) (see also Section III, A).[64] The thiiranium ions (**65**) may be considered to arise by protonation of the enamines and stabilization of the resulting iminium ions by participation of the sulfur. The benzo[c]thiophenes [(**64**) R = H or OMe] give adducts with N-phenylmaleimide.[64]

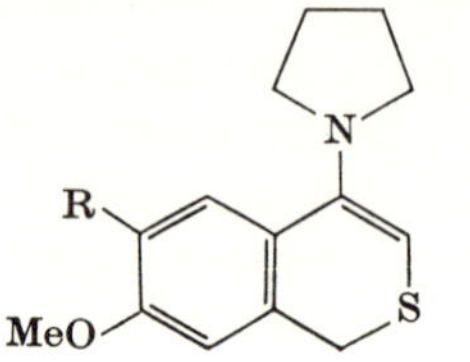

(63) 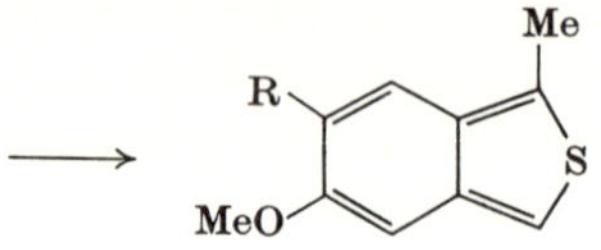   (64)

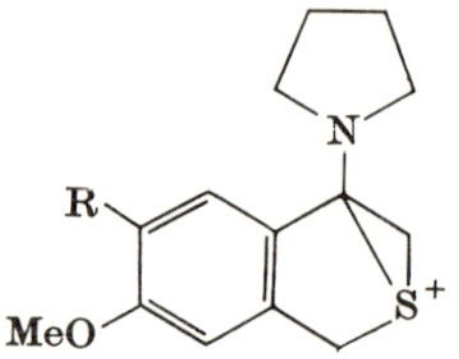

(65) 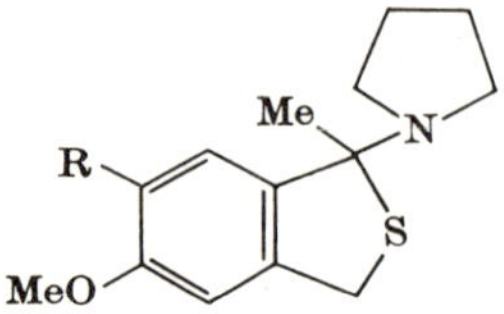   (66)

It is noteworthy that 1,3-dimethylnaphtho[1,2-c]thiophene (1,3-dimethyl-4,5-benzisothianaphthene in the literature) (**67**) has been prepared; its properties have been compared with those of the benzanthracene derivative (**68**).[65]

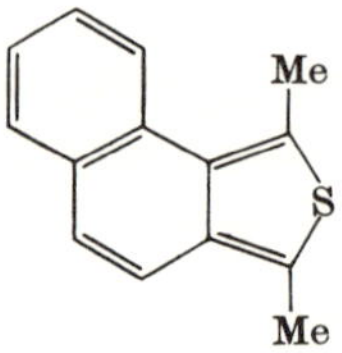

(67) 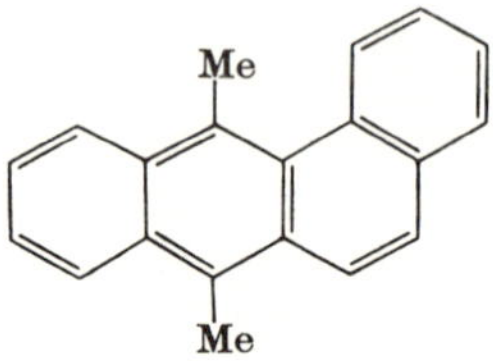   (68)

B. Aryl Derivatives

It has been claimed that 1-phenylbenzo[c]thiophene (2-phenyl-3,4-benzothiophene in the literature) (**69a**) may be prepared by treating

[64] F. H. M. Deckers, W. N. Speckamp, and H. O. Huisman, *Chem. Commun.* 1521 (1970).
[65] O. Dann and H. Distler, *Chem. Ber.* **87**, 365 (1954).

o-benzoylbenzoic acid with phosphorus pentasulfide in boiling xylene, and that other 1-arylbenzo[*c*]thiophenes (Table V) may be similarly prepared.[66] More recently, however, Bordwell and Cutshall[67] have shown that the compound which was originally claimed to be **69a** is, in fact, the dimer, 3,3'-diphenyl-1,1'-bibenzo[*c*]thienyl (**69c**; Ar = 3-phenyl-1-benzo[*c*]thienyl). This observation casts some doubt on the other work described in the original paper (see Section VII).[66]

1,3-Diphenylbenzo[*c*]thiophene (**69b**) is formed when 1,3-diphenylbenzo[*c*]furan is heated with phosphorus pentasulfide.[68-70] 1,3-Diphenylnaphtho[2,3-*c*]thiophene (**70**) may be prepared similarly.[71,72] When 1,2-dibenzoylbenzene is treated successively with phosphorus pentachloride and potassium ethyl xanthate, 1,3-diphenylbenzo[*c*]-thiophene is formed via intermediates **71** and **72**.[73] It may also be

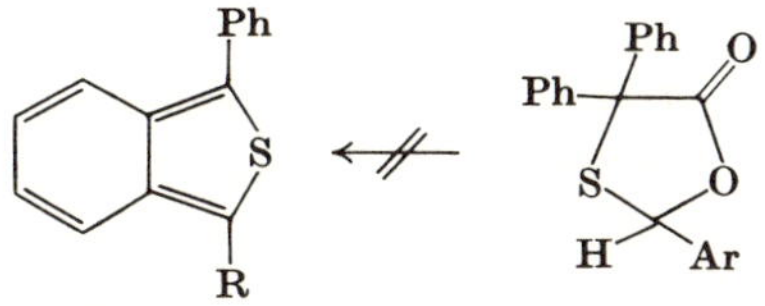

(**69a**) R = H
(**69b**) R = Ph
(**69c**) R = Ar

(**73**)

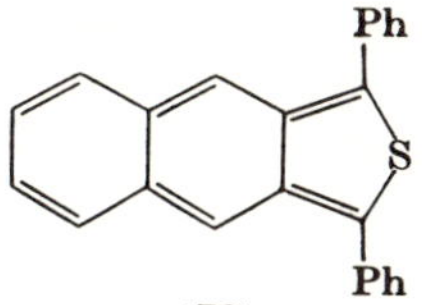

(**70**)

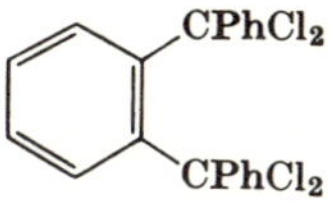

(**71**) 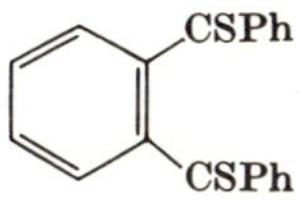   (**72**)

---

[66] J. O'Brochta and A. Lowy, *J. Amer. Chem. Soc.* **61**, 2765 (1939).
[67] F. G. Bordwell and T. W. Cutshall, *J. Org. Chem.* **29**, 2019 (1964).
[68] C. Dufraisse and D. Daniel, *Bull. Soc. Chim. Fr.* **4**, 2063 (1937).
[69] M. P. Cava, M. J. Mitchell, and A. A. Deana, *J. Org. Chem.* **25**, 1481 (1960).
[70] G. Wittig, E. Knauss, and K. Niethammer, *Ann. Chem.* **630**, 10 (1960).
[71] M. P. Cava and J. P. VanMeter, *J. Org. Chem.* **34**, 538 (1969).
[72] M. P. Cava and J. P. VanMeter, *J. Amer. Chem. Soc.* **84**, 2008 (1962).
[73] A. Schönberg and E. Frese, *Chem. Ber.* **101**, 701 (1968).

prepared from phthalic thioanhydride (Section VII,B).[20] The 2-aryl-4,4-diphenyl-1,3-oxathiolan-5-ones [(73), for substituents see Table V] give the corresponding 1-aryl-3-phenylbenzo[c]thiophene (69b,c) on being treated with concentrated sulfuric acid.[4, 68, 74, 75] 9-Aryl-9,10-*endo*-sulfido-9,10-dihydroanthracenes are not intermediates in these reactions (cf. Hartough and Meisel[7]). Attempts to prepare compounds 69b,c by pyrolysis of the oxathiolanones (73) failed.[76]

1,3,4,7-Tetraphenylbenzo[c]thiophene may be prepared by treating its oxygen analog or the dihydro compound (74) with sulfur.[77] The adduct (75) formed between 1,2-dibenzoylethylene and 2,3-diphenylbuta-1,3-diene reacts with sulfur at high temperatures to give 1,3,5,6-tetraphenylbenzo[c]thiophene.[78] Related to the latter reaction is the recent report by Mann and White[79] of a general synthesis of isoindoles, benzo[c]furans, and benzo[c]thiophenes, which involves treating the Diels–Alder adduct (76) formed between dibenzoylacetylene and 2,3-dimethylbuta-1,3-diene and related adducts with an amine, acetic anhydride, or phosphorus pentasulfide, respectively. Thus, for example, with phosphorus pentasulfide in boiling toluene, 76 gives 5,6-dimethyl-1,3-diphenylbenzo[c]thiophene (90%).

(74)   (75)   (76)

1,3-Diphenylbenzo[c]thiophene (69b) (like rubrene) is reported[80] to catalyze the photochemically induced precipitation of an insoluble form of sulfur from a solution of sulfur in carbon disulfide. It is reduced by zinc amalgam in acetic acid to *trans*-1,3-diphenyl-1,3-

[74] A. Bistrzycki and A. Traub, *Helv. Chim. Acta* **7**, 935 (1924).
[75] C. T. Pedersen, *Acta Chem. Scand.* **20**, 2314 (1966).
[76] C. T. Pedersen, *Acta Chem. Scand.* **22**, 247 (1968).
[77] E. D. Bergmann, S. Blumberg, P. Bracha, and S. Epstein, *Tetrahedron* **20**, 195 (1964).
[78] C. F. H. Allen and J. W. Gates, *J. Amer. Chem. Soc.* **65**, 1283 (1943).
[79] J. D. White, M. E. Mann, H. D. Kirshenbaum, and A. Mitra, *J. Org. Chem.* **36**, 1048 (1971); M. E. Mann and J. D. White, *Chem. Commun.* 420 (1969).
[80] C. Dufraisse, C. Pinazzi, and J. Baget, *C. R. Acad. Sci.* **219**, 278 (1944).

dihydrobenzo[c]thiophene (cf. Ref. 4);[69,81] with zinc and fuming hydrochloric acid[4] or zinc amalgam in acetic acid[81] it may also be reduced to 1,2-dibenzylbenzene, and it gives 9-phenylanthracene and anthracene with zinc dust at 400° or 500°, respectively (cf. Bistrzycki and Brenken[4]).[68] Likewise, with zinc and acid, 1-phenyl-3-p-tolyl-benzo[c]thiophene is reduced to its 1,3-dihydro derivative.[74] Nitration of 1,3-diphenylbenzo[c]thiophene is reported[8] to give 1-m-nitrophenyl-3-phenylbenzo[c]thiophene (cf. Bistrzycki and Brenken[4] and Pedersen[75]), which may be oxidized to 2-m-nitrobenzoylbenzophenone with chromic anhydride. Similar results are obtained by starting with 1,3-diphenylbenzo[c]furan.[8] With peracetic acid and aqueous potassium permanganate, respectively, 1,3-diphenyl- and 1-phenyl-3-p-tolyl-benzo[c]thiophene are also oxidized to benzophenones.[4,74]

1,3-Diarylbenzo[c]thiophenes form adducts (**77**) with maleic anhydride[70,75] and 1,3-diphenylbenzo[c]thiophene is also reported[70] to form the adduct (**78**) with benzyne, albeit in low yield. These reactions may be considered either as Diels–Alder cycloadditions to the 1,4-dienic system of the thiophene ring or as 1,3-dipolar cycloaddition reactions involving the polar mesomeric structure (**79**) of the diarylbenzo[c]thiophene.[75] In adduct-forming reactions benzo[c]furans are more reactive than the corresponding benzo[c]thiophenes; for example, when equimolar amounts of 1,3-diphenylbenzo[c]thiophene (**69b**) and its oxygen analog are allowed to compete for maleic anhydride, adduct formation occurs exclusively with the furan (see also Cava and VanMeter[72]).[71]

(**77**)   (**78**)   (**79**)

1,3,5,6-Tetraphenylbenzo[c]thiophene also forms an adduct with maleic anhydride, but it does not give a sulfone or a methylsulfonium iodide.[78] With nitric acid or chromic anhydride in acetic acid, ring cleavage occurs to give 1,2-dibenzoyl-4,5-diphenylbenzene.[78] Electrochemiluminescence absorption and emission data, polarographic

[81] I. Gillet, *Bull. Soc. Chim. Fr.* 1141 (1950).

half-wave oxidation and reduction potentials, and estimated anion- and cation-radical stabilities have been recorded for 1,3,4,7-tetraphenylbenzo[c]thiophene and compared with similar data for isoindoles and benzo[c]furans.[82] This tetraphenyl compound is sensitive to light, and irradiation in the presence of oxygen converts it into 1,2-dibenzoyl-3,6-diphenylbenzene.[77] Both of the tetraphenyl compounds mentioned here fluoresce intensely green,[77,78] and compound **69c** (Ar = m-$O_2NC_6H_4$) exhibits a very strong triboluminescence.[74]

VI. Hydrobenzo[c]thiophene 2-Oxides and 2,2-Dioxides

1,3-Diphenylbenzo[c]thiophene 2,2-dioxide is the only oxidized derivative of the parent aromatic system so far reported in the literature; although highly reactive, it appears to be appreciably stabilized by sulfonyl conjugation (**80a**) ↔ (**80b**). It may be prepared by treating 1-bromo-1,3-diphenyl-1,3-dihydrobenzo[c]thiophene 2,2-dioxide (see later) with diazabicyclononene in benzene (40% yield), or by the action of copper powder on 1,3-dibromo-1,3-diphenyl-1,3-dihydrobenzo[c]thiophene 2,2-dioxide in refluxing benzene.[83] Freshly prepared purple solutions of **80** in benzene may be purified by chromatography on basic alumina. The color of these solutions is stable for several hours at room temperature, and frozen benzene solutions of **80** are stable for about a week at −78°. As expected, **80** forms a 1:1 adduct with N-phenylmaleimide and reacts with hydrogen and bromine (at −67°) by addition to give 1,3-diphenyl- (88% yield) and 1,3-dibromo-1,3-diphenyl-1,3-dihydrobenzo[c]thiophene 2,2-dioxide (74%), respectively.[83]

(**80a**)   ↔   (**80b**)

[82] A. Zweig, G. Metzler, A. Maurer, and B. G. Roberts, *J. Amer. Chem. Soc.* **89**, 4091 (1967).
[83] M. P. Cava and J. McGrady, *Chem. Commun.* 1648 (1968).

A. 2-OXIDES

Several 1,3-dihydrobenzo[c]thiophenes have been oxidized to the corresponding sulfoxides (Table VI) with peracetic acid,[43,44] t-butyl hydroperoxide,[9] sodium periodate in aqueous ethanol,[52,55,56] or

TABLE VI
HYDROBENZO[c]THIOPHENE 2-OXIDES[a]

Compound	M.p. and/or b.p. (°C)	Yield (%)	Ref.
1,3-Dihydrobenzo[c]thiophene 2-oxide	82–84 (150–160/0.3 mm)	9	9, 52
	90	—	29
Methyl 1,3-dihydrobenzo[c]-thiophene-5-carboxylate 2-oxide	115.5–116.5	63	54
1,3,4,7-Tetrahydrobenzo[c]-thiophene (**34**) 2-oxide	97–98	95	44

[a] The stereoisomeric sulfoxides of **35** are hygroscopic[45] and the sulfoxide of compound **37** has been claimed in a patent.[43]

iodobenzene dichloride in aqueous pyridine.[54] t-Butyl hydroperoxide has also been used to oxidize *cis*- and *trans*-octahydrobenzo[c]thiophene to the corresponding sulfoxide.[45] Further oxidation of 1,3-dihydrobenzo[c]thiophene 2-oxide with peracetic acid gives the 2,2-dioxide.[52] Some stereoselectivity has been observed during the oxidation of the racemic compound **34** (Section III, D) to its sulfoxide and of the racemic sulfoxide to the corresponding sulfone (Table VII) using a growing culture of *Aspergillus niger*.[44]

Twelve sulfoxides, including that of 1,3-dihydrobenzo[c]thiophene, have been separated by TLC on alumina.[84]

We have discussed the preparation of benzo[c]thiophene and its derivatives from 1,3-dihydrobenzo[c]thiophene 2-oxides in Section IV.

In each case the ^{1}H NMR spectra of 1,3-dihydrobenzo[c]thiophene and its 2,2-dioxide exhibit a singlet for the four equivalent methylene protons. The spectrum of the unsymmetrical sulfoxide [shown in **81**

[84] E. N. Karaulova, T. S. Bobruiskaya, and G. D. Gal'pern, *Zh. Anal. Khim.* **21**, 893 (1966); *Chem. Abstr.* **65**, 16046 (1966).

as a side elevation], however, exhibits two doublets for the two nonequivalent groups of methylene protons.[9, 85] At 200° the signals do not coalesce to a singlet as might be expected if thermal inversion of the sulfoxide group begins to occur with increasing temperature. Thus, the energy barrier to inversion in **81** must be considerably greater than that for related sulfoxide systems. Inversion of the sulfoxide group in **81** occurs in the presence of acid and a mechanism has been proposed.[9] An interesting feature of the temperature dependence of the methylene signals of **81** is observed when a solution of **81** in carbon tetrachloride or chloroform is cooled to below $-5°$.[9, 85] The signals coalesce to a singlet at $-37°$. This effect has been rationalized[9, 85] by assuming that dimeric association (**82**) occurs at low temperatures. Coalescence of the signals is not observed in dimethyl sulfoxide, which prevents association. Amann and Kresze[86] have shown, however, that the similar NMR behavior of related sulfoxides can be attributed to solvent effects on chemical shifts.

(81)   (82)

B. 2,2-DIOXIDES

1,3-Dihydrobenzo[c]thiophene may be oxidized to its sulfone (Table VII) with peracetic acid,[28, 69, 87, 88] aqueous hydrogen peroxide,[9] or potassium permanganate,[21] and when benzocyclobutene and 1,2-diphenylbenzocyclobutene are heated with sulfur dioxide, they give 1,3-dihydrobenzo[c]thiophene 2,2-dioxide[89] and its cis-1,3-diphenyl derivative (**83**),[69, 90] respectively [Eq. (5)]. 1,3-Dihydrobenzo[c]thiophene 2,2-dioxide also arises when the thiadiazepine (**17b**) is heated just above its melting point (175°–180°).[34a]

When sulfur dioxide is passed into a solution of diphenyldiazo-

[85] R. F. Watson and J. F. Eastham, *J. Amer. Chem. Soc.* **87**, 664 (1965).
[86] W. Amann and G. Kresze, *Tetrahedron Lett.* 4909 (1968).
[87] M. P. Cava and A. A. Deana, *J. Amer. Chem. Soc.* **81**, 4266 (1959).
[88] M. P. Cava and M. J. Mitchell, *Rec. Chim., Acad. Rep. Populaire Roumaine* **7**, 737 (1962).
[89] F. R. Jensen, W. E. Coleman, and A. J. Berlin, *Tetrahedron Lett.* 15 (1962).
[90] F. R. Jensen and W. E. Coleman, *J. Amer. Chem. Soc.* **80**, 6149 (1958).

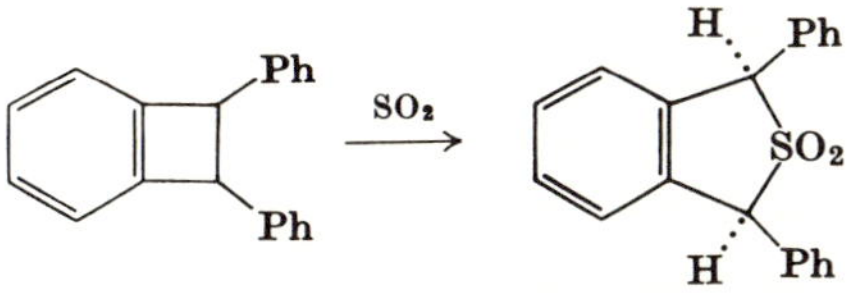

methane in light petroleum, a vigorous reaction occurs to give tetraphenylethylene sulfone (tetraphenylthiirane 1,1-dioxide) (**84**).[91-93] Decomposition of compound **84** gives products markedly

TABLE VII

1,3-DIHYDROBENZO[c]THIOPHENE 2,2-DIOXIDES [a]

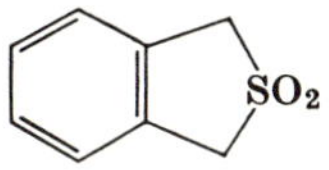

Substituents	M.p. and/or b.p. (°C)	Yield (%)	Ref.
None	148.5–149	90	94
	150–152	94, 89, 73, 69	28, 87,
		?, Good	9, 89, 29, 21
1,3-Me$_2$	112–113	87.5	88
1,3-Ph$_2$			
Cis isomer	232.5–234 (dec.)	100	69, 90
Trans isomer	200–201 (dec.)	90	69
1,1,3-Ph$_3$	174–174.5	—	91–93
4,5,6,7-Br$_4$	260–261	61	95
5-NO$_2$	167–169	55	9
5-NH$_2$	194–196	64	9
5-NHAc	200–203	33	9
4,7-(OMe)$_2$	182 [b]	90	c
1,3-Ph$_2$, 1-Br	198–200	—	83
1,3-Ph$_2$, 1,3-Br$_2$	225–227	—	83

[a] The sulfones of compounds **34**, m.p. 174°–175°,[44] **35**, m.p. cis isomer, 39.5°–41° and trans isomer, 105°–105.5°,[45] (**37**),[43] and the sulfone (**85**), m.p. 159°–160°,[92] have also been prepared.
[b] Private communication from the authors.
[c] L. Horner, P. V. Subramaniam, and K. Eiben, *Ann. Chem.* **714**, 91 (1968).

[91] H. Staudinger and F. Pfenninger, *Chem. Ber.* **49**, 1941 (1916).
[92] H. J. Backer and H. Kloosterziel, *Proc. Koninkl. Nederland Akad. Wetenschap.* **53**, 1507 (1950).
[93] H. Kloosterziel and H. J. Backer, *Rec. Trav. Chim.* **71**, 1235 (1952).

dependent on the solvent used. Thus, in water at 100°, it gives sulfur dioxide and tetraphenylethylene, in hot carbon disulfide or benzene it gives the sulfone (**85**), and in methanol or acetic acid it gives the isomeric sulfone (**86**).[91-93] The sulfone (**85**) rearranges to **86** in methanol, acetic acid, or in the presence of amines.

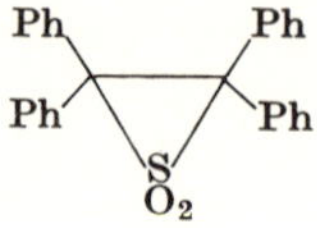

(**84**) 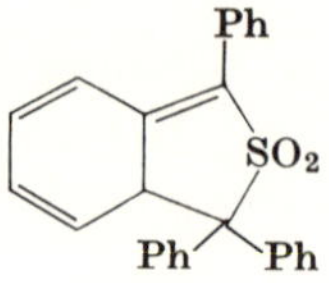 (**85**) 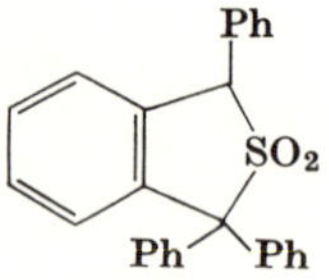 (**86**)

Nitration of 1,3-dihydrobenzo[c]thiophene 2,2-dioxide with fuming nitric acid is reported[9] to give a good yield of the 5-nitro derivative (**87**), which may be reduced to the corresponding amine. This amine reacts normally with acetic anhydride, but attempted diazonium reactions have not led to any useful products.[9]

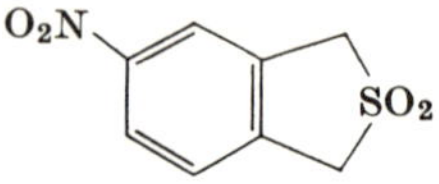

(**87**)

Light-catalyzed bromination of 1,3-diphenyl-1,3-dihydrobenzo[c]-thiophene 2,2-dioxide gives either the 1-monobromo or the 1,3-dibromo derivative depending on the amount of bromine used.[83]

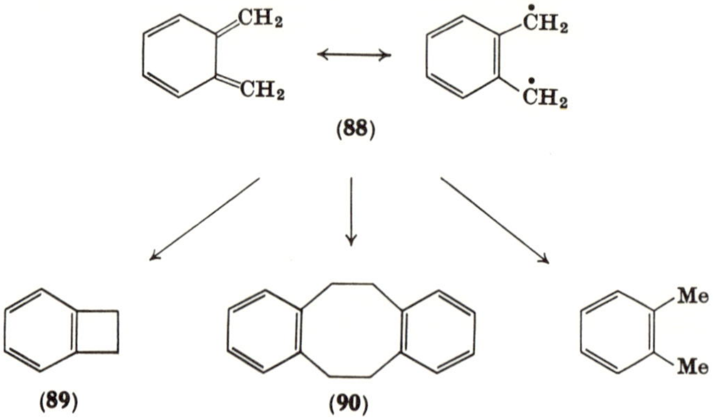

(**88**)

(**89**) (**90**)

SCHEME I

1,3-Dihydrobenzo[c]thiophene 2-oxide loses water on being heated to give benzo[c]thiophene (Section IV). In contrast, 1,3-dihydrobenzo[c]thiophene 2,2-dioxide loses sulfur dioxide to give o-quinodimethane (88) as the initial product (Scheme I).[87, 94] The fate of 88 depends on the reaction conditions. Thus, pyrolysis of the sulfone in the molten state at 280° gives a mixture (16% yield) of benzocyclobutene (89) with less o-xylene (formed by hydrogen abstraction from the starting material by 88), and 1,2;5,6-dibenzocyclooctadiene (90) (formed by dimerization of 88).[87] When the sulfone is heated in boiling diethyl phthalate (300°), the formation of o-xylene is suppressed and 90 is the major product (48% yield).[87] Alternatively, low-pressure gas-phase pyrolysis of the sulfone at 770° under nitrogen gives benzocyclobutene (89) as the major product (74 and 67% yields).[28, 87] Evidence that 88 is an intermediate in these reactions is provided by the fact that pyrolysis of the sulfone in the molten state at 260° in the presence of N-phenylmaleimide gives the adduct 91.[87]

(91)

In the same way that benzocyclobutene (89) may be prepared by desulfonylation of 1,3-dihydrobenzo[c]thiophene 2,2-dioxide, the following compounds may also be prepared: 3,4,5,6-tetrabromobenzocyclobutene;[95] naphtho[b]cyclobutene (92a) (>60% yield);[96]

(92a) $R^1 = R^2 = H$
(92b) $R^1 = R^2 = Ph$

(93)

---

[94] Y. Odaira, K. Yamaji, and S. Tsutsumi, *Bull. Soc. Chem. Japan* **37**, 1410 (1964).
[95] R. B. Fechter, Ph.D. Thesis, The Ohio State University (1966); *Diss. Abstr.* **27B**, 3034 (1967).
[96] M. P. Cava and R. L. Shirley, *J. Amer. Chem. Soc.* **82**, 654 (1960).

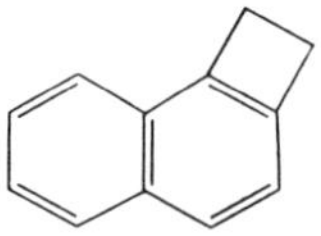

(94) 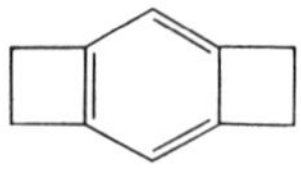   (95a)

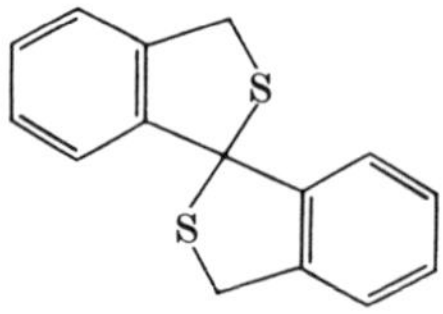

(95b) 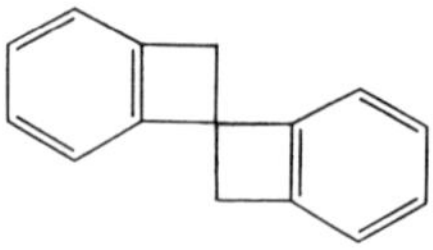   (95c)

3,8-diphenylnaphtho[b]cyclobutene (92b) (77%);[71] 3,10-diphenylanthra[b]cyclobutene (93) (43%);[71] naphtho[a]cyclobutene (94) (51%);[97] and benzo[1,2;4,5]dicyclobutene (95a) (40%).[98] Cava and Kuczkowski[99] have recently synthesized the spirosulfide (95b) and pyrolyzed its bisulfone in the gas phase to give, among other products, 1,1'-spirobibenzocyclobutene (95c). The desulfonylation reactions of sulfones have been reviewed by Kice.[100]

When 1,3-diphenyl-1,3-dihydrobenzo[c]thiophene 2,2-dioxide is heated at 250° in diethyl phthalate, 1,2-diphenylbenzocyclobutene is not formed. Instead, rearrangement [Eq. (6)] of the initially formed o-quinodimethane derivative (96) occurs to give 9-phenyl-9,10-

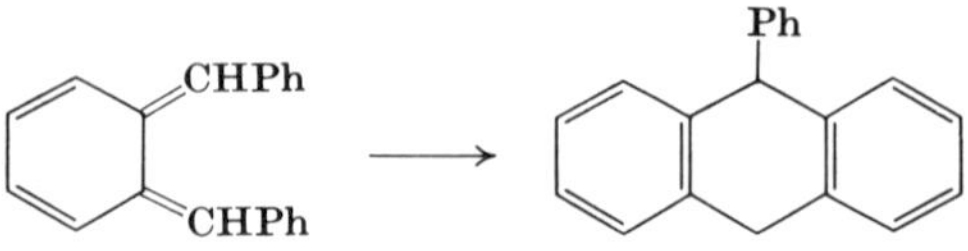

(96) ⟶ [structure]   (6)

dihydroanthracene (94% yield).[69, 90] The behavior of this sulfone is similar to that of 86 which, on being heated at 150°, gives a mixture of 9,10-diphenylanthracene (65%), 9,10-diphenyl-9,10-dihydroanthracene (6%), and a trace of 9,9-diphenyl-9,10-dihydroanthracene (Scheme II).[91–93] It is noteworthy that, on being heated, 1,2-diphenylbenzocyclobutene undergoes a ring-cleavage reaction to give the

[97] M. P. Cava, R. L. Shirley, and B. W. Erickson, J. Org. Chem. 27, 755 (1962).
[98] M. P. Cava, A. A. Deana, and K. Muth, J. Amer. Chem. Soc. 82, 2524 (1960).
[99] M. P. Cava and J. A. Kuczkowski, J. Amer. Chem. Soc. 92, 5800 (1970).
[100] J. Kice, in "The Chemistry of Organic Sulfur Compounds" (N. Kharasch and C. Y. Meyers, eds.), Vol. 2, p. 115. Pergamon Press, Oxford, 1966.

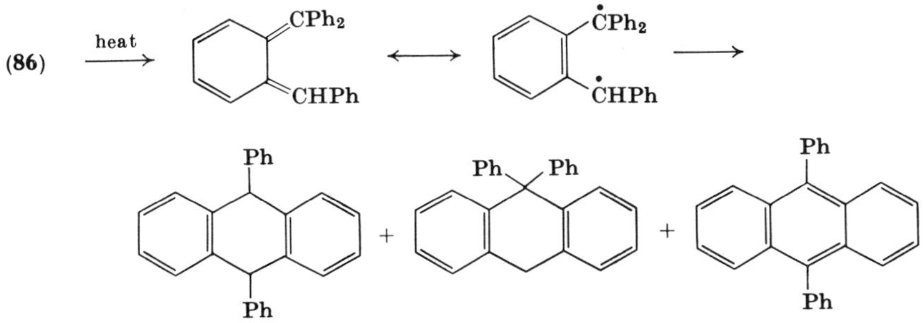

BENZO[c]THIOPHENES

(86) [scheme showing heat-induced rearrangement through o-quinodimethane intermediates to give three anthracene products: 9,10-diphenyl-9,10-dihydroanthracene, 9,9-diphenyl-9,10-dihydroanthracene, and 9,10-diphenylanthracene]

SCHEME II

o-quinodimethane derivative (**96**), which reacts further to give 9-phenyl-9,10-dihydroanthracene [Eq. (6)].[90] Such ring-cleavage reactions of benzocyclobutenes are general.[89, 94]

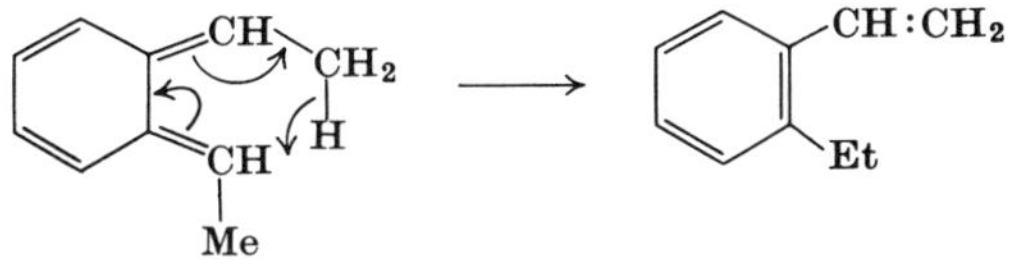

[structure (97) showing o-quinodimethane with Me and H, 1,5-hydride shift arrow] → [o-ethylstyrene structure with CH:CH$_2$ and Et] (7)

(**97**)

1,3-Dimethyl-1,3-dihydrobenzo[c]thiophene 2,2-dioxide also fails to give a benzocyclobutene on being heated. In this case the intermediate o-quinodimethane derivative (**97**) undergoes a rapid 1,5-hydride shift to give o-ethylstyrene [Eq. (7)].[88, 89]

Although it has been reported[101] that 1,3-dihydrobenzo[c]thiophene 2,2-dioxide is unaffected by light of wavelength 220–400 nm, Odaira et al.[94] have shown that its thermal desulfonylation in the molten state at low pressures is accelerated by UV irradiation. The relative amount of **89** and **90** formed is unchanged by irradiation, and although the rate of desulfonylation is independent of pressure, the fate of the o-quinodimethane (o-xylylene diradical) (**88**) formed is strongly dependent on the pressure. Lower pressures favor the formation of **89** and higher pressures the formation of **90**.

We have already mentioned that 1,3-diphenyl-1,3-dihydrobenzo-[c]thiophene 2,2-dioxide gives 9-phenyl-9,10-dihydroanthracene as the major product [Eq. (6)] on being heated. Unlike 1,3-dihydro-

[101] M. P. Cava, R. H. Schlessinger, and J. P. VanMeter, *J. Amer. Chem. Soc.* **86**, 3173 (1964).

benzo[c]thiophene 2,2-dioxide, this sulfone may also be photochemically desulfonylated by light of wavelength 220–280 nm to give *trans*-1,2-diphenylbenzocyclobutene.[101] Prolonged irradiation, however, gives the cyclobutene contaminated with 9-phenyl-9,10-dihydroanthracene.

Thus, the ease of desulfonylation of a 1,3-dihydrobenzo[c]thiophene 2,2-dioxide and therefore the products obtained is dependent on the stability of the intermediate *o*-quinodimethane. The photochemical decomposition of 1,3-dihydrobenzo[c]thiophene 2,2-dioxide probably involves an excitation energy of > 74 kcal/mole and is not possible under the conditions already mentioned, whereas its 1,3-diphenyl derivative may be photochemically desulfonylated under these conditions because the intermediate *o*-quinodimethane derivative (**96**) is stabilized relative to *o*-quinodimethane itself (see also Cava and Shirley[96]).[101]

VII. 2-Thiophthalide, Phthalic Thioanhydride, and Related Compounds

A. 2-Thiophthalide

Although this compound (**98**) may be regarded as 1,3-dihydrobenzo[c]thiophen-1-one, we have used the more commonly accepted thiophthalide nomenclature. Strictly speaking this compound should be known as 2-thiophthalide to distinguish it from the isomeric system, 1-thiophthalide (**99b**). At one time (cf. Graebe[102] and Day and Gabriel[103]) it was difficult to decide whether 2-thiophthalide had structure **98** or **99b**. However, X-ray crystallographic analysis has confirmed structure **98** and suggests that the C–S bond adjacent to the carbonyl group possesses 35% double-bond character.[104] That the dipole moments in the series phthalide → 2-thiophthalide → 2-selenophthalide decrease significantly from left to right is evidence for resonance interaction between the carbonyl group of **98** and the ring sulfur atom.[105] In the series **99a** → **99b** → **99c** the dipole moments increase slightly from left to right as expected.[105]

[102] C. Graebe, *Ann. Chem.* **247**, 289 (1888).
[103] A. W. Day and S. Gabriel, *Chem. Ber.* **23**, 2478 (1890).
[104] E. Shefter, *J. Pharm. Sci.* **57**, 175 (1968).
[105] I. Wallmark, M. H. Krackov, S.-H. Chu, and H. G. Mautner, *J. Amer. Chem. Soc.* **92**, 4447 (1970).

(**98**)

(**99a**) X = O
(**99b**) X = S
(**99c**) X = Se

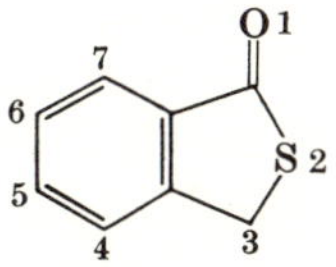

1-Thiophthalide (**99b**), which may be prepared by treating phthalide with phosphorus pentasulfide in boiling xylene, rearranges to 2-thiophthalide on being heated with aniline in a sealed tube; prolonged heating or heating at higher temperatures gives *N*-phenylphthalimidine.[106, 107] The rearrangement of **99b** to **98** has been studied in some detail and the important intermediate stages have been identified.[107, 108] In a related reaction phthalide gives 2-thiophthalide on being heated with ammonium sulfide; 3-phenyl- and 3,3-diphenyl-2-thiophthalide may be similarly prepared.[108]

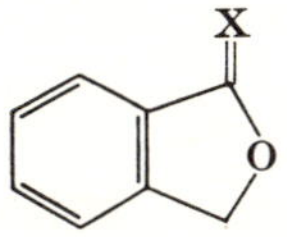

Compound **100** (R = H) may be prepared by treating the thiocyanate (**101a**) with sulfuric acid,[103, 109, 110] by treating *o*-chloromethylbenzonitrile with alcoholic potassium hydrogen sulfide (see later),[103, 110] by cleavage of *o*-benzylthiomethylbenzonitrile (**101b**)

(**100a**) (**100b**) (**100c**)

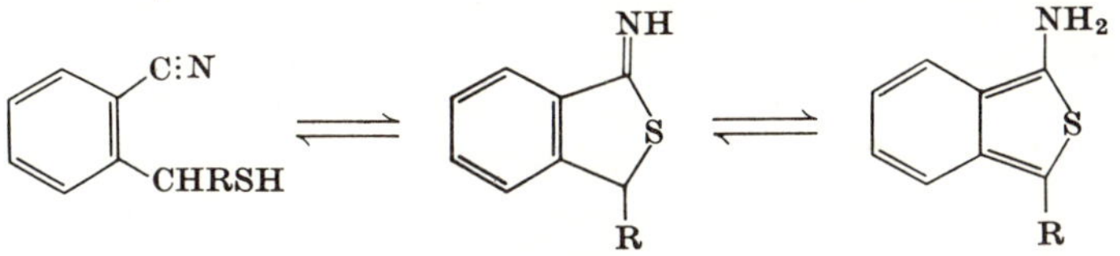

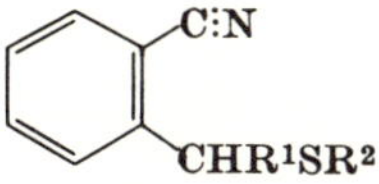

(**101a**) $R^1 = H, R^2 = CN$
(**101b**) $R^1 = H, R^2 = CH_2Ph$
(**101c**) $R^1 = H, R^2 = \overset{+}{C}(NH_2)_2\ \bar{B}r$
(**101d**) $R^1 = Ph, R^2 = \overset{+}{C}(NH_2)_2\ \bar{B}r$

106 V. Prey and P. Kondler, *Monatsh* **89**, 505 (1958).
107 V. Prey, B. Kerres, and H. Berbalk, *Monatsh* **91**, 319 (1960).
108 V. Prey, B. Kerres, and H. Berbalk, *Monatsh* **91**, 774 (1960).
109 M. Renson and R. Collienne, *Bull. Soc. Chim. Belges* **73**, 491 (1964).
110 S. Gabriel and E. Leupold, *Chem. Ber.* **31**, 2646 (1898).

with a Lewis acid (e.g., $AlBr_3$ or HCl),[111, 112] or by alkaline hydrolysis of the isothiouronium bromide (101c).[111] The phenyl derivative [(100) R = Ph] may be prepared by alkaline hydrolysis of the isothiouronium bromide (101d).[111] The products are conveniently isolated as the hydrochloride salts of the imines [(100b) R = H or Ph].[111] Hydrolysis of 100 (R = H) gives 2-thiophthalide.[109] Earlier reports[103, 109, 110] suggested that compound 100 (R = H) may exist as a tautomeric mixture, 100a ⇌ 100b. Stacy and his co-workers,[111, 113, 114] however, have found that the open-chain tautomers of compounds 100 (R = H or Ph) cannot be detected spectroscopically by NMR or IR analysis (see also Ref. 109), nor did they find any evidence for the existence of the alternative tautomers [(100c) R = H or Ph]. They had hoped that the phenyl ring of 100 (R = Ph) would stabilize the aminobenzo[c]thiophene structure [(100c) R = Ph].[113]*

(102)   (103)

An attempted cyclization of o-(benzylmercaptomethyl)benzoic acid (102) with polyphosphoric acid gave an excellent yield of 2-thiophthalide instead of the expected product (103).[115]

2-Thiophthalide is also formed when o-xylene is oxidized with a mixture of sulfur and water at high temperatures; prolonged oxidation under these conditions gives phthalic acid.[116, 117] The formation of 2-thiophthalide is favored if the reaction mixture is cooled under the

* See also G. W. Stacy, A. J. Papa, F. W. Villaescusa, and S. C. Ray, *J. Org. Chem.* **29**, 607 (1964).

[111] G. W. Stacy and L. M. Phipps, personal communication.
[112] D. L. Eck and G. W. Stacy, *J. Heterocycl. Chem.* **6**, 147 (1969), and references cited therein.
[113] G. W. Stacy and A. Y. Orita, *Abstr. 160th Amer. Chem. Soc. Meeting, Chicago,* Sept., 1970, Paper **Orgn. 85**.
[114] G. W. Stacy, D. L. Eck, and T. E. Wollner, *J. Org. Chem.* **35**, 3495 (1970).
[115] M. Protiva, M. Rajšner, E. Alderová, V. Seidlová, and Z. J. Vejdělek, *Collection Czech. Chem. Commun.* **29**, 2161 (1964).
[116] W. G. Toland, *J. Org. Chem.* **26**, 2929 (1961).
[117] W. G. Toland, U.S. Patent 3,031,429 (1962); *Chem. Abstr.* **57**, 11169 (1962).

generated hydrogen sulfide pressure prior to work-up. Sulfur dioxide and sulfuric acid may also be used.[116, 117] 2-Thiophthalide is also formed when phthalic acid, or its anhydride, is reduced with hydrogen

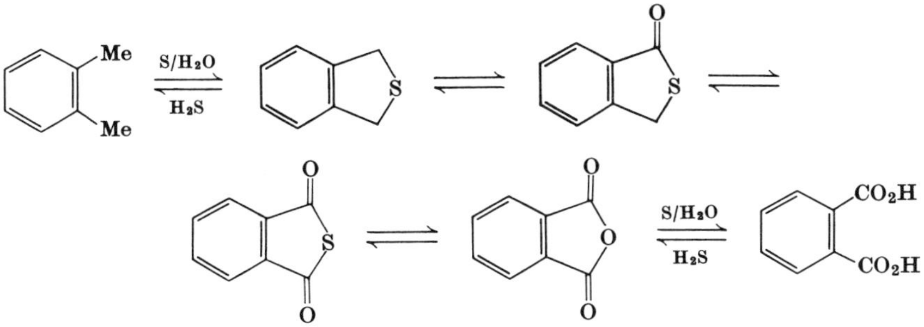

SCHEME III

sulfide or, preferably, with a mixture of hydrogen sulfide and hydrogen.[118, 119] The reversibility of these reactions (Scheme III) has been established.[116–119] The yield of any one of the intermediates may be enhanced by a proper choice of conditions or by recycling the undesired intermediates. Pyromellitic acid dihydrate is similarly reduced by hydrogen sulfide to 2-thiophthalide-5,6-dicarboxylic acid (**105**).[119]

An attempt to scale up the synthesis of 2-thiophthalide by the reduction of phthalic anhydride with hydrogen sulfide gave 3,3′-bithiophthalide (**106a**), which arises by condensation of 2-thiophthalide (**98**) with phthalic thioanhydride (**104**) (Scheme III).[118]

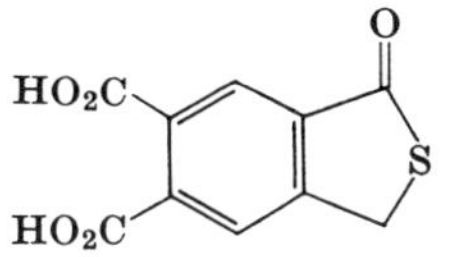

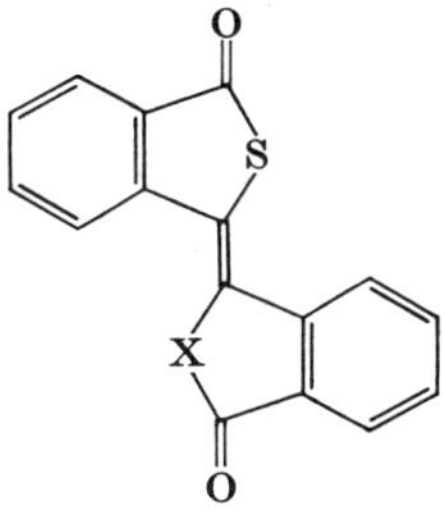

(106a) X = S
(106b) X = O

Oxidation of **106a** with sulfur and water gives phthalic acid.[118]

[118] W. G. Toland and R. W. Campbell, *J. Org. Chem.* **28**, 3124 (1963).
[119] W. G. Toland, U.S. Patent 3,061,612 (1962); *Chem. Abstr.* **58**, 4477 (1963).

1-(**107**) and 4-Cyclohexene-1,2-dicarboxylic acid anhydride (**108**) are also reduced to 2-thiophthalide in high yield by hydrogen sulfide.[118] When a mixture of hydrogen sulfide and hydrogen is used to reduce **108**, a mixture of 2-thiophthalide (23%) and the tetrahydro

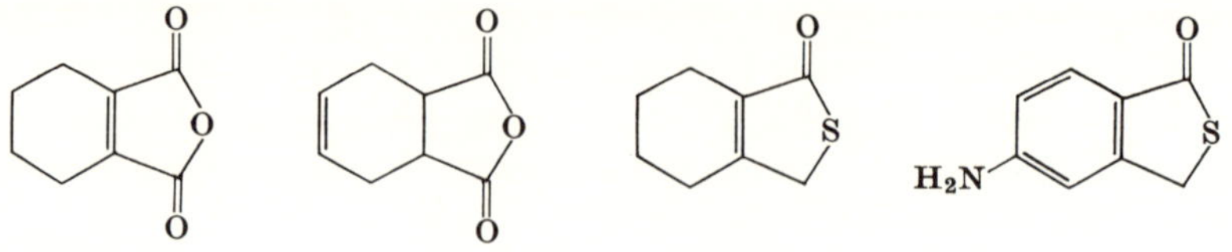

(**107**)   (**108**)   (**109**)   (**110**)

TABLE VIII

2-THIOPHTHALIDES [a]

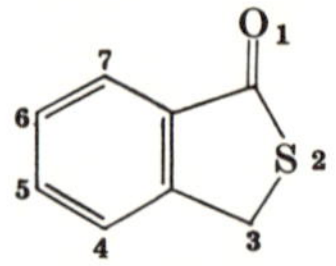

Substituents	M.p. and/or b.p. (°C)	Yield (%)	Ref.
None	58–60 (115–120/0.4 mm)	75, ?	115, 103
	60	30, ?	109, 102
	58–59 (144/7 mm, 150/8 mm)	62, 94	116–119
	59–60	Good, 95	106–108
3-Ph	107	28	108
3,3-Ph$_2$ (**112a**)	162 (?)	—	66, 121
	108–109	52	108
5-F	87–88	43	22
5-Cl	143–144	61	22
5-Br	173–174	60	22
5-NO$_2$	196–197	60	22
5-NH$_2$ (**110**)	168–169	80	22
5-OH	154–155	72	22
5-OMe	103–104	53	22
5-CN	180–181	46	22
5,6-(CO$_2$H)$_2$	—	11	119

[a] Compound **109** has also been prepared (25–47% yield), m.p. 42°–43°.[118, 120]

[120] R. W. Campbell, U.S. Patent 3,335,153 (1967); *Chem. Abstr.* **68**, 21828 (1968).

[121] W. Kleibacker, Ph.D. Thesis, Univ. of Pittsburgh (1938) (cf. O'Brochta and Lowy[66]).

compound (109) is formed. A similar mixture is obtained when cyclohexane-1,2-dicarboxylic acid anhydride is treated with hydrogen sulfide alone or admixed with hydrogen.[118, 120]

5-Amino-2-thiophthalide (110) may be prepared by treating 5-aminophthalide with an alcoholic solution of sodium hydrogen sulfide and hydrogen sulfide in an autoclave.[22] A number of derivatives of 2-thiophthalide (Table VIII) may be prepared from this amine by diazonium reactions. 2-Thiophthalide reacts with nitrating mixture to give 6-nitro*phthalide*.[22]

It is claimed[66, 121] that the reaction of o-benzoylbenzoic acid with phosphorus pentasulfide in boiling benzene gives the sulfide (111), which reacts further with benzene in the presence of aluminum chloride to give 3,3-diphenyl-2-thiophthalide (thiodiphenyl phthalide in the literature) (112a). However, the melting point of the latter compound differs significantly from that reported elsewhere[108] (see Table VIII), and since other work described by these authors has been shown to be wrong (Section V,B), we believe that this work should be reinvestigated. The authors[66] also claim that 112a gives the oxygen analog (112b) with peracetic acid and that both compounds 112a and 112b give compound 113a on being heated with phosphorus pentasulfide; 113a is reconverted to 112b with hydrogen peroxide. Compound 113b is formed when either 100 (R = H)[109] or o-chloromethylbenzonitrile[110] is treated with ethanolic hydrogen sulfide.

(111)

(112a) X = S
(112b) X = O

(113a) R¹ = R² = Ph
(113b) R¹ = R² = H

3-Benzylidene-2-thiophthalide (114b) (Table IX) may be prepared by decarboxylation of the acid (114c) (Section VII,B).[122] It adds bromine across its side-chain double bond (see also Section VII,B) and the product loses hydrogen bromide on crystallization to give 114d.[122] The latter compound is also formed when 114b is brominated

---

[122] V. P. Oshkaya and D. Kh. Mutsenietse, *J. Org. Chem. USSR* **5**, 1812 (1969).

at high temperatures in acetic acid.[122] 3-Benzylidene-2-thiophthalide (**114b**) is hydrolyzed by concentrated alkali to the acid (**115**); in the presence of dimethyl sulfate the *S*-methyl derivative of **115** is

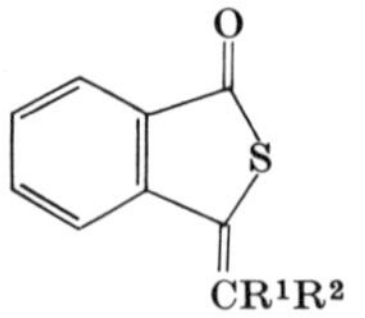

(**114a**) R^1 = Et, R^2 = H
(**114b**) R^1 = Ph, R^2 = H
(**114c**) R^1 = Ph, R^2 = CO_2H
(**114d**) R^1 = Ph, R^2 = Br

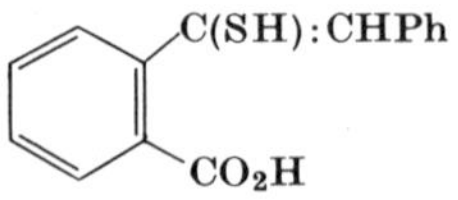

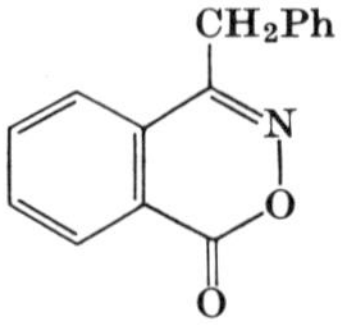

formed.[123] With hydroxylamine in the presence of concentrated alkali **114b** gives 4-benzyl-1*H*-2,3-benzoxazin-1-one (**116**) through formation of **115** as an intermediate.[123]

2-Thiophthalide gives the expected product (**117**) with vinyllithium.[124] Hydrolysis of **117** gives 4,5,6,7-tetrahydro-2*H*-benzo[*c*]-thiepin-5-one (**118**) via formation of the vinyl ketone (**119**).[124]

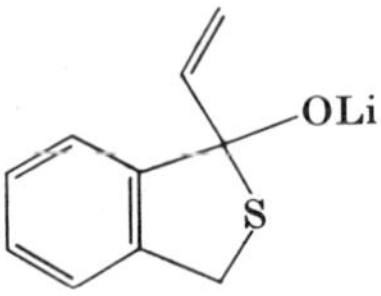

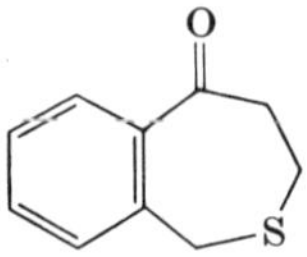

Various uses of 2-thiophthalide have been claimed,[117, 119] including its use to prepare *o*-mercaptomethylbenzoic acid, which arises when

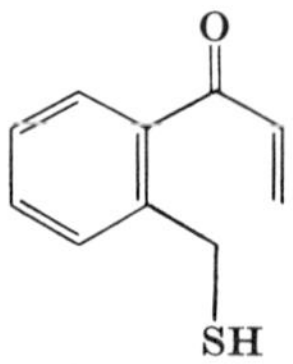

[123] D. Kh. Mutsenietse and V. P. Oshkaya, *J. Org. Chem. USSR* **6**, 592 (1970).
[124] W. C. Lumma, G. A. Dutra, and C. A. Voeker, *J. Org. Chem.* **35**, 3442 (1970).

2-thiophthalide is hydrolyzed with alkali. 3,3-Dichloro-4,5,6,7-tetrahydro-2-thiophthalide is claimed to exhibit insecticidal properties.[120] It may be prepared by treating **109** with chlorine.

B. Phthalic Thioanhydride

More commonly, but incorrectly, named thiophthalic anhydride and sometimes referred to as phthaloyl monosulfide or *o*-phthalyl sulfide, this compound (**104**) may be prepared by heating phthalic anhydride with sodium sulfide[125] or by heating phthaloyl chloride with potassium hydrogen sulfide,[126] hydrogen sulfide,[127] or ethanolic sodium disulfide.[128] The reaction between phthaloyl chloride and hydrogen sulfide also gives phthalic anhydride and di-*o*-phthaloyl disulfide.[127] Early attempts to prepare **104** have been described by Chakravarti.[126] 4-Nitrophthalic thioanhydride may be prepared by hydrolysis of **8** (R = NO$_2$; X = Cl) (Section III, A).[22]

At one time phthalic thioanhydride was thought to have the structure **120a** and derivatives (e.g., **120b**) were thought to have similar structures. X-Ray crystallographic data for 4,5-dimethyl-

(**120a**) R^1 = R^2 = H
(**120b**) R^1 = R^2 = Me

phthalic thioanhydride, however, have shown conclusively that it must be regarded as a derivative of **104** and not as **120b**.[129, 130]

Phthalic thioanhydride forms charge-transfer complexes with *N*,*N*-diethylaniline and various aromatic hydrocarbons (e.g., a 1:1 complex with naphthalene).[131] When oxygen and sulfur are members

---

[125] A. Reissert and H. Holle, *Chem. Ber.* **44**, 3027 (1911).
[126] G. C. Chakravarti, *J. Indian Chem. Soc.* **5**, 405 (1928).
[127] L. Szperl, *Roczniki Chem.* **10**, 652 (1930); *Chem. Abstr.* **25**, 938 (1931).
[128] Yu. O. Gabel and L. F. Shpeïer, *J. Gen. Chem. USSR* **17**, 2277 (1947); *Chem. Abstr.* **42**, 4976 (1948).
[129] W. T. Eeles, *Acta Cryst.* **7**, 649 (1954).
[130] W. T. Eeles, *Acta Cryst.* **9**, 365 (1956).
[131] G. P. Naletova, L. V. Osintseva, and S. V. Tronov, *J. Gen. Chem. USSR* **37**, 1694 (1967).

TABLE IX
Phthalic Thioanhydride and Related Compounds

Compound	M.p. and/or b.p. (°C)	Yield (%)	Ref.
Phthalic thio-anhydride	114	68.5	125, 126
	109–110	12.5	128
	113–114	—	127
	111–112	—	141, 142
4-Nitrophthalic thioanhydride	105–106	100	22
(122)	29 (81–83/0.1 mm)	42	136, 137
(125)	ca. 110–120	—	138
(121)	34–35	35	135
(124b)			
Endo isomer	134–135	100	137
Exo isomer	59–61	—	137
(124a)	185–190	—	138
Endo isomer	81–82	100	137
(114a)	56–57 (155–160/6 mm)	31	140
(114b)	105–106	31.5	122, 141
(114c)	232–234	71.5	122
Methyl ester	165–166	88	122
Ethyl ester	158–159	90.5	122
(114d)	143–144	60	122
(113b)	67	77	109
(113a)	—	—	66, 121
(100) (R = H)	60, 54–57	94, ?, 80–95	103, 109, 113, 111
(100) (R = Ph)	98–99	80	111, 113

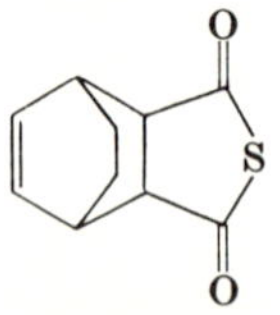

of a ring in analogous systems, the systems invariably exhibit isomorphism. However, although phthalic thioanhydride forms mixed crystals with its selenium analog, it does not form mixed crystals with phthalic anhydride, despite the fact that the two compounds distill at the same temperature.[132]

[132] G. Cilento, *J. Phys. Coll. Chem.* **54**, 564 (1950).

The polarographic reduction of phthalic thioanhydride has been studied.[133, 134]

When 3,4,5,6-tetrahydrophthalic anhydride is treated with sodium sulfide at room temperature, it gives the sulfur analog (**121**).[135] The isomeric compound (**122**) may be prepared similarly; it undergoes a retrodiene cleavage reaction to give maleic thioanhydride (**123a**) on being heated.[136, 137] Dichloromaleic thioanhydride (**123b**) is obtained when tetrachlorothiophene is oxidized with nitric acid.[138] Both **123a** and **123b** form adducts which may be regarded as derivatives of benzo[*c*]thiophene. Thus, **123b** gives the adducts **124a** and **125**, respectively, with cyclopentadiene and buta-1,3-diene.[138] Compound

(**121**)   (**122**)   (**123a**) R = H   (**124a**) R = Cl
                         (**123b**) R = Cl   (**124b**) R = H

(**125**)

123a forms adducts with cyclopentadiene (**124b**) and cyclohexa-1,3-diene.[137]

Phthalic thioanhydride is reduced by aluminum hydride or diborane to 1,3-dihydrobenzo[*c*]thiophene (100%) (Section III,A),[20] by one equivalent of lithium aluminum hydride to *o*-phthalaldehyde

---

[133] L. Tirouflet and R. Dabard, *C. R. Acad. Sci.* **242**, 2839 (1956).
[134] R. Dabard and L. Tirouflet, in "Advances in Polarography," (I. S. Longmuir, ed.), Vol. 1, p. 288. Pergamon Press, New York, 1960.
[135] R. Mayer, G. Daebritz, G. Ullrich, H. Werner, G. Ulbricht, and K. Gewald, *J. Prakt. Chem.* **21**, 80 (1963).
[136] H.-D. Scharf and M. Verbeek, *Angew. Chem. Int. Ed. Engl.* **6**, 874 (1967).
[137] M. Verbeek, H.-D. Scharf, and F. Korte, *Chem. Ber.* **102**, 2471 (1969).
[138] O. Scherer and F. Kluge, *Chem. Ber.* **99**, 1973 (1966).

(50–70%),[20] with an excess of lithium aluminum hydride to the hydroxythiol (126),[20] and with sodium amalgam to phthalide.[125] With sodium borohydride in benzene–methanol it also gives phthalide (80%), and its successive treatment with three equivalents of phenylmagnesium bromide and acid gives a mixture of o-dibenzoylbenzene, 1,3-diphenylbenzo[c]furan, and 1,3-diphenylbenzo[c]thiophene.[20] Successive treatment of phthalic thioanhydride with the same quantity of phenylmagnesium bromide followed by oxidative hydrolysis of the product gives exclusively o-dibenzoylbenzene (80%) (phenyllithium gives the same product in a lower yield), and with one equivalent of the same reagent followed by oxidative hydrolysis of the product it gives o-benzoylbenzoic acid (50%).[20] With methanolic hydrogen peroxide, phthalic thioanhydride is reported to give methyl phthalate.[139]

CH_2OH
CH_2SH
(126)

In the Gabriel modification of the Perkin reaction with a mixture of propionic anhydride and sodium propionate at 160°, phthalic thioanhydride gives the 3-ethylidene derivative (114a), which turns orange on exposure to light.[140] Phthalic thioanhydride condenses with phenylacetic acid in acetic anhydride in the presence of triethylamine to give α-phenyl-α-thiophthalidylideneacetic acid (114c).[122] If it is fused with phenylacetic acid in the presence of sodium acetate, it gives 3-benzylidene-2-thiophthalide (114b).[122]

Phthalic thioanhydride (104) reacts some thousand times faster with hot triethyl phosphite (TEP) to give *trans*-3,3'-bithiophthalide (106a) than does phthalic anhydride to give 3,3'-biphthalide.[141, 142] In contrast, N-methyl- and N-phenylphthalimide do not react at all with TEP.[141] The reactivity sequence $S > O \gg NR$ is that expected for nucleophilic attack by TEP on the carbonyl oxygen atom, and Bird and Wong[141] have suggested that the carbene (127) is initially formed and then trapped to give the phosphorane (128), which

[139] R. Kitamura, *J. Pharm. Soc. Japan* **58**, 29 (1938).
[140] D. T. Mowry, E. L. Ringwald, and M. Renoll, *J. Amer. Chem. Soc.* **71**, 120 (1949).
[141] C. W. Bird and D. Y. Wong, *Chem. Commun.* 932 (1969).
[142] J. H. Markgraf, C. I. Heller, and N. L. Avery, *J. Org. Chem.* **35**, 1588 (1970).

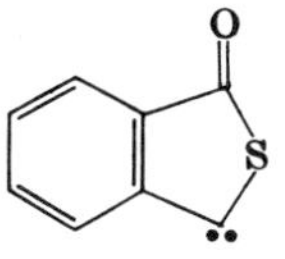

(127) 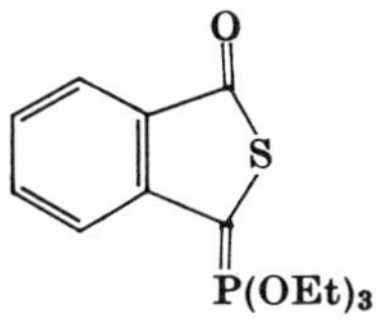   **(128)**

proceeds to **106a** via a Wittig reaction. That brief treatment of **104** with TEP in the presence of phthalic anhydride gives **106b** and that a similar reaction in the presence of benzaldehyde gives 3-benzylidene-2-thiophthalide (**114b**) seems to support this mechanism.[141] However, an alternative mechanism (Scheme IV) is possible, which involves the formation of the intermediate (**129**) [this is also the expected precursor of the carbene (**127**)].[142] The trans structure of **106a** has been confirmed by spectroscopic and X-ray analysis,[142] and its mass

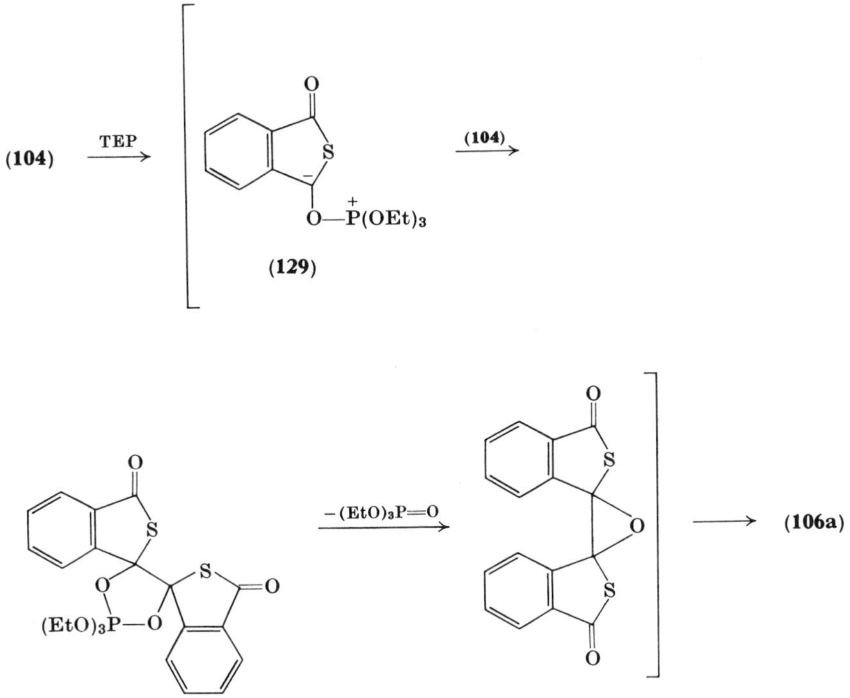

SCHEME IV

spectral breakdown has been studied recently.[143] It is noteworthy that, whereas early reports[110,125] depicted 3,3'-bithiophthalide (**106a**) as the cis isomer and later reports[118,141] depicted it as the trans isomer, no comment on its stereochemistry was made until the work of Markgraf et al.[142] was published. In the previous reports[110,118,125] the compound is described as green-yellow needles, m.p. 335°, which sublime to orange-yellow needles (m.p. not given). The compound prepared by Markgraf et al.[142] was orange-yellow, m.p. 351°, and, therefore, it appears that the sublimate of the earlier workers was the trans isomer (**106a**).

Phthalic thioanhydride (**104**) does not react with triphenylphosphine, but it reacts with tris(dimethylamino)phosphine to give a

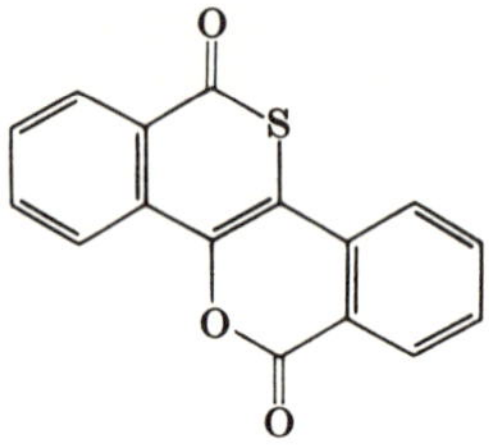

(**130**)

mixture of **130** and **106b**.[142] A mechanism different from the one already given for the formation of **106a** from **104** with TEP, but involving the phosphorane **128**, has been given to account for this result.

A number of other reactions of phthalic thioanhydride have been described by Reissert and Holle.[125]

ACKNOWLEDGMENTS

The author wishes to thank Dr. Carl Th. Pedersen (University of Copenhagen) and Dr. R. M. Scrowston (University of Hull) for critical discussions during the preparation of the manuscript, and Professor L. Horner, Dr. Gardner W. Stacy and Dr. L. M. Phipps, and Professor Michael P. Cava for making available unpublished data.

---

[143] C. W. Koch and J. H. Markgraf, *J. Heterocycl. Chem.* **8**, 225 (1971).

NOTE ADDED IN PROOF: The following references appeared too late for inclusion in the text: M. P. Cava, N. M. Pollack, O. A. Mamer, and M. J. Mitchell, *J. Org. Chem.* **36**, 3932 (1971); J. Gourier and P. Canonne, *Bull. Soc. Chim. Fr.* 3299 (1971); J. D. Andose, A. Rauk, R. Tang, and K. Mislow, *Int. J. Sulfur Chem.*, Part A **1**, 66 (1971); J. McGrady, Ph.D. Thesis, Wayne State University (1969); *Diss. Abstr.* **32B**, 2607 (1971).

Author Index

Numbers in parentheses are reference numbers and indicate that an authors's work is referred to, although his name is not cited in the text.

A

Abe, Y., 145, 196, 198
Abraitys, V. Y., 192
Acheson, R. M., 137, 138(202)
Adachi, T., 100
Adam, W., 106
Adams, A., 2, 3(3), 16(3), 17(3), 21(3), 23(3), 24(3), 29(3), 36(3), 38(3), 51, 55(29), 77(29)
Adams, R., 227, 244(29), 245(29), 246
Adamson, J., 124
Adembri, G., 108
Adomeit, M., 238
Agosti, A., 49, 50(18), 53(18), 60(18), 61(18), 78(18), 79(18), 80(18)
Agui, H., 288
Ainsworth, D. P., 254
Akabori, S., 241
Akahoshi, M., 219
Akerkar, A. S., 14, 18(69), 21(69), 29(69), 30(69), 31(69, 139)
Akimove, L. N., 186
Akkerman, A. A., 131, 209
Albeit, A., 107, 119, 151
Albreckt, E., 305
Alderová, E., 370, 372(115)
Algerino, A., 61, 77(69)
Allen, C. F. H., 58, 63(46), 88(46), 90(46), 354(78), 358, 359(78), 360(78)
Allen, C. W., 285, 286(39), 287(39)
Allison, C. G., 120, 163
Alt, G. H., 302, 303
Amann, W., 362
Ambler, A. P., 108
Ambrosoli, G., 62, 63(74)
Ammitzboell, T., 337(23), 338
Amonetti, L., 56, 58, 62(52, 53, 54, 55), 63(38), 86(38), 88(56), 89(56), 90(56), 92(52, 53, 54, 55, 56), 93(52, 53, 54, 55, 56)
Anderson, A. A., 118
Anderson, C. G., 238
Anderson, R. C., 15, 18(76, 80), 37(80)
Andreetti, G. D., 56, 85(39)
Andrievskaya, O. I., 264
Angier, R. B., 12
Anishina, L. N., 269
Anthony, W. C., 304
Armand, J., 122
Armbrust, H., 66, 75(102), 95(102), 96(102)
Arnold, D. R., 192
Asai, A., 41
Asai, M., 140, 181, 193
Ashby, J., 68, 94(107a)
Asheshov, I. N., 101
Ashley, J. N., 63, 78(94), 89(93, 94)
Ashley, K., 104
Asker, W., 60
Assaad, N. E. N., 267
Atherton, N. M., 111
Auerbach, J., 199
Auret, B. J., 347(44), 348, 361(44)
Austel, V., 121
Autenrieth, W., 336, 362(21), 363(21)
Avery, N. L., 376(142), 378, 379(142), 380(142)
Awamura, M., 184
Ayres, F. D., 104

B

Baba, H., 109
Backer, H. J., 363, 364(92, 93), 366(92, 93)

Badger, G. M., 170
Badstübner, W., 45, 50(3), 77(3), 81(3)
Baget, J., 358
Baker, A. D., 106
Baker, B. R., 62
Baker, C., 106
Balakrishnan, P. V., 104
Balucani, D., 341, 345(37), 347(37)
Bambas, L. L., 44, 51, 52, 54(1), 58, 76(1), 77(1), 78(1), 80, 84(1), 88(1), 92(1)
Bamberger, E., 70
Bapat, J. B., 100, 203(9)
Barbieri, W., 167
Barlin, G. B., 125
Barlow, R. B., 259
Barnes, R. K., 185
Barraclough, D. J., 257
Barry, R. D., 282
Bartaccini, G., 61
Bartok, W, 288
Barton, D. H. R., 281, 327, 329(16)
Barton, J. W., 142, 152
Baruffini, A., 56, 58, 59, 62(51, 57, 62, 63), 63(37), 76(63), 85(37), 86(37), 89(57), 90(57), 91(51, 62, 63, 80, 86), 92(50, 51, 89), 93(50, 51, 85, 89)
Bass, P., 65, 69(101), 70(101), 72(101), 75(101), 95(101), 96(101)
Bather, P. A., 304
Batka, H., 187
Battersby, A. R., 294, 296, 312(84), 313 (84), 314(83), 326
Bauer, L., 148, 197
Baumgarten, E., 108
Baxter, G. E., 117, 128, 155, 156(283), 157(283)
Baxter, I., 273
Baxter, R. A., 119, 154(134), 160, 166
Bayer, O., 337(27), 340
Bayerlein, F., 251
Beak, P., 192
Beck, L. H., 140
Becke, F., 47, 77(11), 79(11), 80(11), 82(11)
Beech, G., 104
Begun, G. M., 108
Behringer, H., 36

Behun, J. D., 130, 131(178), 132, 161
Beisher, J. A., 312
Beissner, G., 122
Belinskaya, R. V., 336, 337(22), 338(22), 372(22), 373(22), 375(22), 376(22)
Bell, S. L., 107
Belleau, B., 310
Bellobono, I. R., 109
Benkelman, T. E., 193
Benschop, H. P., 30, 31(137), 33(137), 41(137)
Bennett, B. A., 23, 25(119), 29(119), 31 (119), 37(119)
Berbalk, H., 369, 372(107, 108), 373 (108)
Berends, W., 128, 139, 141, 168, 182, 183, 191
Berg, H., 122
Bergmann, E., 285
Bergmann, E. D., 354(77), 358, 360(77)
Beringer, M., 9, 18(37), 20(37), 23(37)
Berkowitz, J., 124, 193(155), 196
Berlin, A. J., 362, 363(89), 367(89)
Bernardi, L., 140, 153, 154, 158, 159(273, 296), 161, 162(273, 307), 163(296), 167, 172(271), 193, 194(392), 196 (392, 398)
Berry, R. O., 12, 15(60)
Bersch, H. W., 300
Bertaccini, G., 50, 53(24), 61(24), 77(24), 78(24), 79(24), 80(24), 82(24), 88(24)
Bertazzoli, C., 102
Berteccini, F., 49, 50(19), 53(19), 60(19), 61(19), 78(19)
Bertini, F., 125
Bertini, V., 129
Bethune, V. G., 148
Bichaut, P., 281, 289(17), 305(17), 289, 292(13), 293
Bicking, J. B., 149, 170(254, 255, 256, 257)
Binks, R., 296, 312(84), 313(84), 314(83)
Birch, S. F., 337(29), 340, 343(40), 344, 347(45), 348, 349(45), 361(29, 45), 363(29)
Bird, C. W., 376(141), 378, 379(141), 380 (141)
Birhofer, L., 304, 305

AUTHOR INDEX

Bistrzycki, A., 333, 337(4, 74), 354(4, 74), 358(4), 359(74), 360(74)
Black, A. L., 124
Black, D. R., 103, 111(30), 113(30)
Black, D. St.C., 100, 203(9)
Black, L. L., 9, 29(42)
Black, P. J., 105
Blackburn, S., 241
Blackwell, J. T., 270
Blaha, K., 279
Blake, K. W., 119, 124(135), 154(135), 188, 193, 198(393), 201(393)
Blomstrom, D. C., 2, 24(5)
Blosick, G. J., 242, 243(69)
Blumberg, S., 354(77), 358, 360(77)
Blum-Bergmann, O., 285
Blumenthal, H., 279
Bobbitt, J. M., 285, 286(37, 38, 39), 287(39), 299(38), 305, 307, 318, 319
Bobruiskaya, T. S., 361
Bode, G., 289
Bode, K. D., 66
Boekelheide, V., 139
Boer, H., 340
Böshagen, H., 48, 50, 51(27), 52(26, 27), 53, 56(32), 60(16), 63(30), 77(14, 31), 79(31), 80(31), 81(14, 16), 82(16), 83(14), 84(26), 85(32), 86(32), 87(30)
Boggs, J. E., 111
Bonanomi, M., 153, 162(270), 175(270)
Bonzom, A., 15, 129
Bordwell, F. G., 357
Borgna, P., 56, 59, 62(62), 63(37), 85(37), 86(37), 91(62), 93(89)
Bornstein, J., 352
Borodkin, V. F., 59, 88(60)
Borowitz, I. J., 295
Bothner-By, A. A., 110
Boudet, R., 46, 77(5)
Boulton, A. J., 68
Bourgoin-Legay, D., 46, 77(5)
Bouscasse, L., 16
Boveri, S., 114, 141, 166(115)
Boyd, D. R., 347(44), 348, 361(44)
Bracha, P., 354(77), 358, 360(77)
Brachwitz, H., 157, 167(292)
Bradley, W., 264, 284, 290, 292(1), 293

Bradsher, C. K., 181
Bramwell, A. F., 103, 110, 125, 130(31)
Braunholtz, J. T., 56, 63(35)
Brechlin, A., 347(46), 348
Bredereck, H., 143
Brenken, B., 333, 337(4), 354(4), 358(4), 359
Brewer, R. G., 109
Bridges, J. W., 38
Bridgland, B. E., 104
Brintzinger, H., 191
Broeke, J. T., 157
Brooks, J. R., 288, 290(53), 292(4), 293
Brothuhn, G., 46
Brown, D. J., 119
Brown, D. W., 281, 282(18), 283(18), 285, 286, 288, 289, 290(18, 41), 291(18), 294(87, 105), 298, 299(27, 57), 300(69), 301(66), 302(66), 303(87), 304, 305(27), 306(18, 41), 307, 308(18), 309(18, 27, 65, 87), 315(26), 317(87), 318(87, 133), 319, 320, 322(143), 324(46, 143), 325(46), 328(136), 329(136)
Brown, R. D., 105
Brown, R. F. C., 100, 144, 203(9)
Brown, W. V., 125
Bruckhausen, F., 299
Brüning, A., 336, 362(21), 363(21)
Brundle, C. R., 106
Bruno, A., 5, 32(19)
Bucci, P., 17
Buchardt, O., 203
Buck, J. S., 289
Buckley, R. K., 69, 72(111), 94(111), 96(111), 98(111)
Budzikiewicz, H., 111, 322
Buogo, A., 102
Bup, O., 50, 53(22), 60(22), 82(22)
Buraczewska, M., 151
Burdeska, K., 186, 187(381)
Burrell, J. W. K., 103, 130(31)
Burrows, E. P., 304, 305(110)
Bursey, M. M., 111
Bustard, T. M., 106
Buttery, R. G., 103, 111(30), 113(30)
Buttimore, D., 8, 14, 17(70), 18(70), 20(70), 21, 22(112), 23(70), 24(70, 106,

112), 25(70), 28(70, 112, 113), 29 (70, 113), 30(70), 31(70), 38(112)
Buu-Hoi, N. P., 341, 345(36), 346(36), 347(36), 354(36), 355(36)

C

Caccialanza, P., 62, 89(82)
Cagniant, D., 333, 341, 343(10, 35), 344 (10, 35), 345(10, 35), 346(10), 347 (35)
Cagniant, P., 333, 341, 343(10, 35, 42), 344(10, 35), 345(10, 35, 42), 346(10), 347(35)
Cain, C. K., 104, 171(36)
Cairns, T. L., 2, 24(5)
Califano, S., 15, 108
Camerino, B., 165, 167(312), 198
Cameron, D. W., 271, 272, 273, 277(123), 278(124)
Campbell, R. D., 301
Campbell, R. W., 347(48), 348, 381, 372 (118), 373(118, 120), 375(120), 380 (118)
Capps, D. B., 61, 77(71)
Caputo, J. A., 133
Capuzzi, R., 150
Carbon, J. A., 177
Carmichael, W. M., 104
Carrara, A., 118, 155(133)
Carrington, A., 111
Carrol, F. I., 270
Caton, M. P. L., 14, 18(72), 19(72), 21 (72), 23(72), 24, 28(113), 29(113), 30, 31(121, 140), 32(121), 34(123), 36, 38(121, 123, 140), 40(123)
Catteau, J. P., 37
Cava, M. P., 327, 337(69, 87, 88), 350, 351 (52), 352(57), 354(52, 69), 355(52), 359(69, 71), 360, 361(52, 55, 56), 362(69), 363(69, 83, 87, 88), 364(83), 365(87), 366(69, 71), 367(88, 101)
Celadnik, M., 99, 102, 105(3)
Cenker, M., 117, 128, 141
Cervinka, O., 279, 303
Chakrabartty, S. K., 133, 135, 136

Chakravarti, G. C., 375, 376(126)
Chambers, R. D., 107, 120, 155, 157(279), 158(279), 163(279), 172(279)
Chan, A. W. K., 27, 28(129), 35(133)
Chan, T. L., 159
Chandler, J. A., 105
Chaplen, P., 33, 37(144)
Cheburkov, Y. A., 120
Cheeseman, G. W. H., 123, 153(147), 159 (147), 165(147), 166(147), 172, 174 (336, 338a), 176, 177(147, 336), 179 (147), 180(147), 215
Chen, S.-J., 182
Cheney, L. C., 5
Chieli, T., 102
Chigira, Y., 116, 205(124), 206(124)
Chipman, W. B., 303
Chivers, G. E., 194
Christiani, A. F., 285
Christmann, A., 186
Chu, S.-H., 368
Cilento, G., 376
Ciuti, M. P., 62, 63(74)
Clark, J., 120
Clark-Lewis, J. W., 259, 261(99, 100), 262
Clementi, E., 106
Clipperton, R. D. J., 294(105), 299, 304
Cobb, R. L., 291
Coe, P. L., 337(30), 340
Cogliano, J. A., 152
Cohen, E., 150
Coleman, W. W., 362, 363(89, 90), 366 (90), 367(89, 90)
Collienne, R., 369, 370(109), 372(109), 373(109), 376(109)
Collier, H. O. J., 65, 69(101), 70(101), 72(101), 75(101), 95(101), 96(101)
Collins, K. H., 155
Collins, R. F., 63, 78(94), 89(93, 94)
Condorelli, P., 13, 32(65)
Cook, A. G., 279
Cook, A. H., 102
Cook, G. L., 334
Cook, R. E., 327
Cooke, G. W., 288
Corbin, U., 5
Corr, D. H., 124
Corradi, A., 56, 85(39)

Coulson, C. A., 105
Cox, R. H., 110
Cragoe, E. J., 147, 149, 150, 152, 157, 159 (265), 162, 170(254, 255, 256, 257, 258, 259, 260)
Cramer, K.-E., 256, 258(82)
Crawford, R. J., 47
Crenshaw, R. R., 6
Croci, M., 58, 92(50), 93(50)
Croisy, A., 341, 345(37), 347(37)
Croll, B. T., 337(30), 340
Cross, A. D., 108
Crossley, M. L., 165
Crow, W. D., 9, 26(38), 27, 28(129), 34 (132, 134), 35(133), 127, 144
Cruanes, J., 282, 292(14), 293, 305
Cruickshank, P. A., 59
Cullen, J., 199
Cullum, T. V., 343(40), 344
Cummins, B. L., 65, 69(101), 70(101), 72 (101), 75(101), 95(101), 96(101)
Curphey, T. J., 123
Cutshall, T. W., 357

D

Dabard, R., 377
Daebritz, G., 376(135), 377
Dähne, S., 278
Dal Monte Casoni, D., 61, 77(69)
Damiens, R., 143, 147(228), 151(228)
Danfee, B., 209
Daniel, D., 337(68), 354(68), 357, 358 (68), 359(68)
Danishefsky, S., 136
Danks, L. J., 337(34a), 341, 362(34a)
Dann, O., 354(63), 355, 356
D'Aria, L., 61, 77(69)
Das, B. K., 108
Davis, M., 11, 50, 63, 64, 65(25), 66, 67 (106), 68(105, 106), 69(106), 70 (106), 71(110), 72(111), 78(94), 89 (93, 94), 94(105, 106, 110, 111), 95 (110), 96(106, 110, 111, 114), 97 (114), 98(111)
Davis, P., 143
Day, A. R., 114, 171(117), 172(117)

Day, A. W., 368, 369(103), 370(103), 372 (103), 376(103)
Dean, R. A., 337(29), 340, 343(40), 344, 347(45), 348, 349(45), 361(29, 45), 363(29)
Deana, A. A., 337(69, 87), 354(69), 357, 359(69), 362(69), 363(69, 87), 365 (87), 366(69)
de Bie, D. A., 25
Deckers, F. H. M., 354(64), 356
Dede, L., 209
De Graw, J. I., 303
De Heer, J., 333, 334(11)
Delaby, R., 143, 147(228), 151(228)
Del Bene, J., 106
deNet, R. W., 130
Denisov, V. Ya., 267, 269
Denton, D. A., 215
Desai, H. S., 334, 342(16), 343(16), 344 (16), 347(16), 350(16), 351(16), 354 (16)
Detering, K., 320
Detsina, A. M., 272
de Vries, G., 209
De Waard, E. R., 339
Dewar, M. J. S., 320
Deys, W. B., 239
Dickore, K., 180
Di Lauro, C., 108
DiNenno, F., 291, 302(73)
Distler, H., 356
Djerassi, C., 111
Döpp, D., 238
Doerr, F., 109
Domiano, P., 56, 85(39)
Dorofeenko, G. N., 304
Dos Santos-Veiga, J., 111
Dou, H. J., 20, 37
Drushel, H. V., 335
Duffin, G. F., 123
Dufraisse, C., 337(68), 354(68), 357, 358 (68), 359(68)
Duinker, P. M., 340
Dumanovic, D., 111
Dunn, G., 100
Dutcher, J. D., 101
Dutra, G. A., 374
Dutta, C. P., 287

AUTHOR INDEX

Dutta, R. L., 104
Dyke, S. F., 280, 281, 282(18), 283(18), 285, 286, 288, 289, 290(18, 41, 53), 291(291,18), 292(4, 5), 293, 294(8, 87, 105), 296, 298, 299(27, 57), 300(69), 301(8), 302(8), 303(87), 304, 305(27), 306(18, 41), 307, 308(18), 309(18, 27, 65, 87, 114), 315(26), 316(8, 86, 114), 317(87), 318(87, 133), 319, 320(13), 322(143), 324(46, 143), 325(46), 328(136), 329(136)

E

Eastham, J. F., 362
Ebermann, R., 285, 286(38), 299(38)
Ecary, C., 333, 354(8), 359(8)
Eck, D. L., 370
Edgar, J. A., 259
Edwards, D. A., 104
Eeles, W. T., 375
Efros, L. S., 195
Egerton, G. S., 267
Eiben, K., 363
Eishold, K., 46
Elderfield, R. C., 174, 312
Elina, A. S., 194, 196, 197(406), 200
Elliot, I. W., 295
Elliott, R. D., 104, 209
Ellis, A. C., 296
Ellis, A. F., 129
El-Sayed, M. A., 106, 109
El Shanta, M. S., 48, 50(13)
Elslager, E. F., 61, 77(71)
El'tsov, A. V., 257
Elwood, T. A., 111
Engelhardt, V. A., 146
Engels, H. D., 304, 305
English, J. P., 165
Epstein, P. F., 55, 85(33)
Epstein, R., 280
Epstein, S., 354(77), 358, 360(77)
Erdey, L., 104
Erickson, B. W., 366
Erickson, A. E., 153
Erlenmeyer, H., 9, 18(37), 20(37), 23(37), 191
Ernst, W., 69

Esche, J., 167
Essery, J. M., 6
Evans, J. R., 306

F

Fabian, J., 56, 74, 333, 334(14)
Fabris, D., 118, 155(133)
Fahrenholtz, K., 139
Fanghänel, E., 31, 57, 85(45)
Faust, J., 4, 6, 7(27), 24(27), 33(27), 49, 50(21), 77(21), 83(20, 21)
Favini, G., 109
Fechter, R. B., 337(95), 363(95), 365
Fedrick, J. L., 124
Feijen, T., 338(54), 351, 354(54), 355(54), 361(54)
Feld, R. S., 121
Felder, E., 114, 140
Fellman, J. H., 121
Feltkamp, H., 53, 56(32), 85(32), 86(32)
Fielden, R., 233, 235(39), 237(40)
Fiesselmann, H., 342, 343(39)
Fife, W. K., 325
Figueras, J., 337(25), 339
Finar, I. L., 229
Finckh, K., 258
Finley, J. H., 17, 18(91)
Fischer, B. A., 312
Fischer, R., 58, 62(48), 88(83), 89(83), 90(83), 92(48, 83), 93(83)
Fishman, M., 59
Flament, I., 113
Fleischer, K., 300
Flock, F. H., 7, 8(31), 16(32), 17(32), 18(32), 21(32), 24(32), 31(32)
Flow, H., 101
Fokin, E. P., 264, 265, 266, 267, 269, 272
Foks, H., 151, 194, 197
Forman, S. E., 131
Fort, G., 118, 155(131, 132)
Foster, J. P., 238
Fowler, F. W., 182
Fowles, G. W. A., 104
Foxton, M. W., 137, 138(202)
Francis, R. J., 326
Franz, J. E., 9, 29(42)
Freedman, L., 209

Freeman, W. A., 38
Freifelder, M., 130
Frenzel, H., 177
Frese, E., 354(73), 357
Freund, M., 289, 300
Fries, K., 46
Fries, M., 270
Fukumoto, K., 280, 288(12), 289(12), 290(12), 292(10), 293
Fujii, S., 195, 206
Fujimura, H., 41
Fujisawa, K., 18, 24(99), 25(99), 31(99), 37(99)
Fujiwara, E. J., 161
Fukuda, Y., 162, 180(308)
Furukawa, H., 327

G

Gabel, Yu. O., 375, 376(127)
Gabriel, S., 63, 68(96, 97), 70(96, 97), 71(97), 94(96, 97), 147, 185, 368, 369(103), 370(103, 110), 372(103), 373(110), 376(103), 380(110)
Gadamer, J., 283
Gaetani, E., 49, 50(18), 53(18), 60(18), 61(18), 78(18), 79(18), 80(18)
Gagan, J. M. F., 137
Gagliardi, G. N., 181
Gagliostro, D. E., 143
Gainer, H., 128, 141(171), 155, 156(281), 157(281)
Galli, R., 125
Gal'pern, G. D., 361
Gal'pern, M. G., 147
Ganorkar, M. C., 104
Gapski, G. R., 121
Gardini, G. P., 125
Garner, R., 215, 221, 222(20), 223(20), 245
Garratt, P. J., 352
Garratt, S., 294, 314(75)
Garside, P., 289, 305(58)
Gates, J. W., 354(78), 358, 359(78), 360(78)
Gault, R., 302

Gaze, R., 285
Geiger, W., 50, 51(27), 52(26, 27), 53, 56(32), 63(30), 77(31), 79(31), 80(31), 84(26), 85(32), 86(32), 87(30)
Geisel, W., 45, 50(3), 77(2, 3), 81(2, 3)
Gelas, J., 125
Gensler, W. J., 283, 285(25), 287, 292(42), 293, 295, 309, 316, 317(117)
Georgian, V., 344, 345(43), 347(43), 349(43), 361(43)
Gewald, K., 4, 11, 35(47), 75, 333, 334(5, 6), 342(6), 343(6), 344(6), 350(5, 6), 351(6), 354(5, 6), 355(5, 6), 37(135), 377
Ghione, M., 102
Ghosh, S., 104
Gialdi, F., 56, 58, 59, 62(51, 57, 62, 63), 63(37), 76(63), 85(37), 86(37), 89(57, 82), 90(57), 91(51, 62, 63, 80, 86), 92(50, 51), 93(50, 51, 85)
Gibson, P. A., 161
Gil, V. M. S., 110
Giles, R. G. F., 271, 272, 277(123), 278(124)
Gillet, I., 359
Gillham, R. A., 61, 77(72), 79(72)
Gilman, N. W., 323
Ginvalde, A., 335
Glick, M. D., 327
Glover, E. E., 124, 183
Gnad, J., 6, 29(25)
Godwin, R. A., 153, 158(272), 172(272), 174
Goerdeler, J., 3, 4, 5, 6(15), 8(18), 17(18), 21(18), 23(6), 24(118), 26(12, 18, 118), 27(18, 118), 28(18), 29(25), 32(12), 38(12), 45, 55(4), 65(4), 74, 76(122), 77(4), 85(4), 94(4)
Goggins, A. E., 111
Goldman, I. M., 67
Goldschmiedt, H., 312
Golombok, E., 143
Goodman, L., 109
Goodwin, H. A., 105
Gordon, J. M., 198
Gosney, I., 9, 27, 28(129), 34(132, 134), 35(133)
Goto, T., 123, 169, 205(327)

Gougoutas, J. Z., 14, 21(68), 30(68), 31 (68), 36(68)
Govindachari, T. R., 288
Grabitz, E. B., 114, 140, 141
Graebe, C., 368, 372(102)
Grabowski, E. J. J., 156, 164(285)
Gränacher, C., 186
Grahe, G., 110
Graner, G., 289, 290, 299(54)
Grant, M. S., 20, 23(107), 28(107), 29(107), 37(107)
Grantham, R. K., 239, 240(54), 256, 257(84), 260, 261(85, 101), 277(81)
Grasz, I., 209
Graziani, O., 51, 77(28)
Grewe, R., 289, 295(55), 305
Griffiths, J. H., 15
Grimmett, M. R., 184
Grimson, A., 106
Gross, A., 186
Gropper, R., 354(63), 355
Grudzinski, S., 145
Guadagni, D. G., 103, 111(30), 113(30)
Gubler, H. U., 62
Gulbenk, A. H., 157
Gulland, J. M., 288
Gumprecht, W. H., 123, 193
Gupte, S. S., 334, 342(16), 343(16, 38), 344(16), 347(16, 38), 350(16, 38), 351(16, 38), 354(16, 38)
Gurst, J. E., 347(47), 348, 349(47)
Gutowski, G. E., 26
Guyader, M. L., 231
Gylys, J. A., 102

H

Haarstad, V. B., 329
Haase, W. H., 301
Haberland, G., 285
Habicht, H., 342, 343(39)
Häuser, H., 347(47), 348, 349(47)
Hagimiwa, J., 288, 292(15), 293
Haisova, K., 328
Halczenko, W., 150
Hale, J. M., 111
Hall, S. F., 141, 144(219), 146(219)
Hanifin, J. W., 150

Hanna, C., 63
Hansel, W., 7, 16(32), 17(32), 18(32), 21(32), 24(32), 31(32)
Harcourt, D. N., 288, 290(53), 292(4), 293
Hardy, G., 282, 288, 315, 318(133), 319, 328(136), 329(136)
Harris, J. G., 103, 111(30), 113(30)
Harrison, D., 199
Harrison, D. R., 137
Hartke, K., 11, 21(53, 54), 27(53), 38(53, 54)
Hartmann, H. J., 6, 33(26), 74, 76(127)
Hartough, H. D., 333, 358
Hasek, W. R., 146
Haskins, R. H., 101
Hasselquist, H., 175
Hatchard, W. R., 10, 11(45), 22(51), 24(44, 51), 26(44), 31(44, 45, 51), 32(51), 34(51), 36(51)
Hauck, F. D., 297
Hausen, H. V., 133
Hauser, C. R., 141
Hautala, R. R., 238
Hawson, A., 332(34a), 341, 362(34a)
Hayasaka, T., 288
Heffernan, M. L., 105
Heine, H. W., 242, 243(69)
Heinze, H., 7, 16(32), 17(32), 18(32), 21(32), 24(32), 31(32)
Heitmueller, R., 74
Heller, C. I., 376(142), 378, 379(142), 380(142)
Helmers, R., 301
Henbest, H. B., 347(44), 348, 361(44)
Herdey, O., 343(41), 344, 345(41)
Hershenson, F. M., 148
Hershenson, H. M., 108
Hertz, H. S., 160, 161(300)
Hesse, R. H., 281, 329(16)
Hester, J. B., 219
Hetman, N. E., 193, 195(391), 196(391)
Heydkamp, W., 251
Hickmott, P. W., 301
Hicks, A. E., 109
Hiiragi, M., 288
Hill, J. H., 100, 203(10)
Hinchliffe, A., 333, 334(15)
Hinkel, L. E., 143

AUTHOR INDEX

Hinton, A. J., 62
Hinton, I. G., 292(2), 293
Hirsch, A. L., 197
Hirsch, S. S., 144
Hirschberg, A., 136, 146(196), 155, 156 (282), 158, 161, 176
Hishmat, O. H., 60
Hjelt, E., 337(31), 340
Hoan, N., 341, 345(36), 346(36), 347(36), 354(36), 355(36)
Hoffmann, R., 320
Hofmann, H., 7, 8(31), 16(32), 17(32), 18(32), 21(32), 24(32), 31(32)
Holland, A., 21, 22(112), 24(112), 28(112), 38(112)
Holland, J. M., 350
Holle, H., 375, 376(125), 378(125), 380
Holtje, H.-D., 292(6), 293, 324, 325
Holtzer, L., 180
Holz, W. J., 149, 170(260)
Honma, Y., 203
Hopkins, B. J., 301
Horák, V., 333, 334(12), 335
Horn, H., 4, 38(10)
Horne, D., 156, 157(286), 163, 164(286, 311), 180(311)
Horner, L., 186, 363
Horning, D. G., 304
Horsfall, I. G., 62, 84(90)
Hoshi, H., 12, 24(57), 25(57), 26(57)
Hossack, D., 38
Hough, E., 327
Hoyer, G. A., 31, 32(143)
Hubmann, M., 280, 281(10)
Huckel, W., 289, 290, 299(54)
Hübenett, F., 7, 8(31), 16(32), 17(32), 18(32), 21(32), 24(32), 31(32)
Hügel, H., 68, 69(109)
Huffman, J. W., 311, 312(124), 314
Huisgen, R., 178, 230, 251
Huisman, H. O., 339, 354(64), 356
Hull, R. K., 323
Hultquist, M. E., 175
Humphrey, S. A., 107, 110(67c), 112
Hurd, C. D., 148
Hurni, H., 62, 88(83), 89(83), 90(83), 92(83), 93(83)
Husbands, G. E. M., 351, 361(56)

Huston, B. L., 337(34a), 341, 362(34a)
Huyser, H. W., 238

I

Iddon, B., 331
Iio, A., 37
Ikekawa, N., 203
Ikena, T., 241
Illuminati, G., 124
Impicciatore, M., 50, 53(24), 61(24), 77(24), 78(24), 79(24), 80(24), 82(24), 88(24)
Ing, H. R., 259
Inglis, J., 38
Innes, K. K., 105, 108, 109
Inoue, S., 169
Iqbal, A. F. M., 182
Irick, G., 238
Ishiguro, T., 128, 184
Ishii, T., 280
Ishimaru, H., 288
Iskander, Y., 64
Isobe, M., 123
Isogai, Y., 104
Ito, K., 327
Itok, M., 198
Ivanoff, N., 125
Iyer, V. S., 14, 18(69), 21(69), 29(69), 30(69), 31(69)

J

Jackman, L. M., 280, 290, 292(3), 293
Jackson, A. H., 311
Jackson, D. R., 117, 128
Jaffe, H. H., 106
Jaffe, W., 280
Jeffrey, S., 284, 290, 292(1), 293
Jeffries, A. T., 6, 352
Jeffs, P. W., 326
Jenkins, P. N., 327
Jennings, A. L., 111
Jensen, F. R., 362, 363(89, 90), 366(90), 367(89, 90)
Jentzsch, J., 6, 33(26), 74, 75(125), 76(125, 127)
Jezo, I., 184

Joffe, D. V., 195
Johnson, D. W., 121
Johnson, M. D., 291
Johnson, W. S., 281
Johnston, H., 154, 156, 157(286), 164 (286), 194
Joiner, R. R., 165
Jones, D. G., 30, 33(138)
Jones, D. H., 14, 17(70), 18(70, 72), 19 (72), 20(70), 21(71, 72), 22, 23(70, 72), 24(70), 25(70, 119), 28(70, 71), 29(70, 119), 30(70, 71), 31(70, 71, 119, 140), 37(119), 38(140)
Jones, D. W., 350
Jones, G., 30, 33(138), 183
Jones, J. H., 147, 149, 150, 152, 159(265), 162, 170(254, 259, 260)
Jones, L. B., 238
Jones, R. A., 108
Jones, R. G., 113, 143(112)
Jones, T. E., 103
Julian, P. L., 310
Jung, J. A., 301
Junghans, W., 270

K

Kabacinski, F. F., 160, 161(300)
Kaczynski, T., 145
Kaiser, W., 337(27), 340
Kakemi, K., 145
Kakoi, H., 169, 205(327)
Kallos, G., 328
Kalontarov, I. Y., 266
Kamal, M. R., 131, 132(181), 133(179, 181), 136
Kamath, S. D., 104
Kamei, H., 110
Kamel, M., 60
Kametani, T., 280, 288(12), 289(12), 290 (12), 292(10, 12), 293, 294(64), 313
Kamiya, H., 100
Kan, P. T., 161
Kanda, Y., 108
Kandler, J., 45, 55(4), 65(4), 77(4), 85(4), 94(4)
Kanematsu, K., 160
Kano, H., 233, 240

Kanzaki, K., 141
Kappe, T., 186, 187(381)
Karasawa, K., 100
Karaulova, E. N., 361
Karmas, G., 143, 153, 173(269), 176
Karpetsky, T. P., 170
Karrer, P., 186, 280, 281(10), 290(7), 292 (7, 8), 293, 294(7)
Karten, M. J., 209
Katagi, T., 280, 288(12), 289(12), 290 (12), 292(10), 293
Kato, K., 141
Katritzky, A. R., 108, 195
Katz, L., 59, 61, 76(58), 93(58)
Kaufman, A., 143
Kauffmann, Th., 122
Kaufmann, H., 227, 243
Kawaguchi, H., 18, 24(99), 25(99), 31 (99), 37(99)
Kawasaki, N., 160, 180(303)
Kenkyuso, R., 203
Kennedy, J. G., 303
Kerr, G. H., 250
Kerres, B., 369, 372(107, 108)
Keuser, U., 4, 23, 26(12), 32(12), 74, 76(122)
Keyworth, D. A., 107
Khanna, K. L., 285, 286(37)
Khoi, N. H., 341, 345(36), 346(36), 347 (36), 354(36), 355(36)
Kice, J., 366
Kiely, J. M., 285, 286(37, 38), 299(38), 305, 307
Kigasawa, K., 288
Kilger, J., 36
Kilminster, R. A., 319
Kim, S. M., 304, 305
Kimura, T., 139, 141
King, F. E., 259
King, J. F., 337(34a), 341, 362(34a)
King, K. F., 148
Kinsman, R. G., 288, 290, 292(5), 293, 320, 322(143), 324(46, 143), 325(46)
Kintzinger, J. P., 15
Kirby, G. W., 281, 329(16)
Kirshenbaum, H. D., 354(79), 358
Kiryukhina, Z. V., 186
Kiseleva, N. N., 266

Kitamura, R., 378
Kitao, T., 251
Kitaoka, Y., 251
Kleibacker, W., 371, 373, 376(121)
Klein, B., 124, 134, 193(155), 195(391), 196(391), 198, 199
Kleinert, H., 333, 334(5, 6), 342(6), 343(6), 344(6), 350(5, 6), 351(6), 354(5, 6), 355(5, 6)
Klesper, H., 62
Kless, M., 31, 32(143)
Kliegman, J. M., 185
Klingsburg, E., 33
Kloosterziel, H., 363, 364(92, 93), 366(92, 93)
Klug, R., 304
Kluge, F., 376(138), 377
Klunder, A. J. H., 349(49), 350
Klutchko, S., 155, 172(284), 199
Kluyver, A. J., 102
Klyuev, V. N., 57, 59, 84(41, 42), 88(60)
Knabe, J., 292(6), 293, 294(144, 145), 319, 320, 321, 324, 325
Knauss, E., 354(70), 357, 359(70)
Knopf, R. J., 152
Knotnernus, J., 340
Kobatake, H., 195
Koch, C. W., 380
Kochetkov, N. K., 16
Koelsch, C. F., 123
Koenig, G., 313, 320(129)
Kofman, H., 209
Kohno, K., 241
Kojima, M., 15, 37(84)
Kokorudz, M. 155, 156(281), 157(281), 354(63), 355
Kolsaker, P., 278
Komery, J., 337(34a), 341, 362(34a)
Kondler, P., 369, 372(106)
Kondo, Y., 300
Konno, T., 112
Koppelmann, E., 122
Koros, E., 104
Korshak, V. V., 144
Korte, F., 376(137), 377
Kosasayama, A., 112
Kosasayama, E., 198
Koshisue, K., 139

Kosuge, T., 100
Kotelko, A., 145
Koudijs, A., 159
Koutecký, J., 333, 334(12)
Kozuka, S., 326
Krackov, M. H., 368
Krause, W., 294(144), 320
Kresze, G., 362
Krohnke, F., 309
Krone, U., 4
Kruger, W., 289, 295(55), 305(55)
Kubala, E., 102
Kubitz, J., 294, 319
Kuczkowski, J. A., 366
Kudrna, J. C., 238
Kulpe, S., 271
Kunitomo, J., 327
Kuniyoshi, I., 162, 180(308)
Kushida, H., 195
Kuznetsov, S. G., 221, 222(22), 224
Kvasnička, V., 333, 334(14)
Kwiatkowski, S., 105
Kwong, S. F., 149, 170(254, 255)

L

Lablanche-Combier, A., 37
Lahmani, F., 125
Lamchen, M., 190, 203
Landa, S., 340
Landesberg, J. M., 12, 15(60), 33, 35(145)
Landquist, J. K., 99, 191
Langdon, W. K., 117, 128, 155, 156(281), 157(281)
Langheinrich, K., 62
Langkammerer, C. M., 229
Larèze, F., 104
Larini, G., 140, 154, 159(273), 162(273), 193, 194(392), 196(392)
Laseter, A. G., 209
Lassen, N., 337(23), 338
Lauer, W. M., 229
Launay, J. P., 122
Laur, P., 347(47), 348, 349(47)
Lauterbur, P. C., 110
Layton, A., 18, 30(101), 31(101)
Lazarus, S., 155, 172(284), 199
Leaver, D., 12

Le Count, D. J., 294, 314(75)
Le Coustumer, G., 13, 14(64), 33(64)
Lederer, K., 289
Ledig, K. W., 104
Lee, F. T., 15, 24(86), 31, 34(86), 35
Lee, W. W., 62
Leete, E., 264
Legler, J., 122
Legrand, L., 74, 76(123, 124)
Lehn, J. M., 15
Lemal, D. M., 121
Lenard, K., 166
Lenk, C. T., 161
Leonard, N. J., 9, 26(38), 285, 297, 337 (25), 339
Leone, A., 118, 154, 155(133), 158, 159 (273, 296), 162(273), 163(296), 193, 194(392), 196(392)
Leser, G., 332
Letcher, R., 327
Letsinger, R. L., 238
Leubner, G. W., 285
Leupold, E., 71
Lever, A. B. P., 104, 123
Levine, A. W., 121
Levine, R., 130, 131(178), 132(181), 133 (179, 181), 135, 136
Levins, P. L., 281, 292(11), 293, 294(20)
Levis, W. W., Jr., 113, 128, 155, 156 (283), 157(283)
Lewis, I. M., 259
Lewis, J., 104, 123
Lewis, J. W., 295, 304
Li, B. W., 15, 24(86), 34(86)
Light, R. J., 141
Liljegren, D. R., 312
Limpricht, J., 229
Ling-Tse Yoh, L., 104
Lions, F., 105
Litmanowitsch, M., 114
Loadman, M. J. R., 124
Lohr, D. H., 181
Lohrmann, R., 177
Lont, P. J., 159
Looyenga, H., 105
Lorenz, W., 60, 61(65), 62, 63, 78(65), 79(65)
Loridan, G., 20

Lovell, J. B., 181
Lowrie, G. B., 242, 243(69)
Lowy, A., 354(66), 357, 371(66), 373(66), 376(66)
Lozac'h, N., 74, 76(123, 124)
Lucas, R. A., 103, 130(31)
Lüttringhaus, A., 26, 347(46), 348
Lugton, W. G. D., 281, 282(18), 283(18), 290(18), 291(18), 293, 299(27), 305 (27), 306(18), 308(18), 309(18, 27)
Luini, F., 155
Lukens, R. J., 62, 84(90)
Lukes, G. E., 55, 85(33)
Luk'yanets, E. A., 147
Lumma, W. C., 374
Lunt, E., 18, 30(101), 31(101)
Lutz, W. B., 155, 161, 172(284), 173, 199
Luzak, I., 184
Lwo, S. Y., 328
Lynch, J., 266, 267(110), 268(110, 111), 269(110), 270(111), 271(111)

M

Maas, W., 281
McAllan, D. T., 340
MacBride, J. A. H., 120, 155, 157(279), 158(279), 163(279), 172(279)
McCarthy, M., 281, 292(11), 293, 294(20)
McDonald, F. R., 334
MacDonald, J. C., 101, 102, 205, 208
McDowell, C. A., 105, 111
MacDowell, D. W. H., 352
McEwen, W. E., 291
McFadzean, J. A., 38
McGrady, J., 360, 363(83), 364(83)
McGregor, D. N., 5
Machón, Z., 23, 24(117), 38, 41(117)
McIntyre, P. S., 107
McKinnon, D. M., 12, 13, 15(60), 33(63)
McLean, R. A., 291, 293
McLeod, D., Jr., 192
McMurtry, K. D., 328
Maeda, M., 15, 37(84)
Maeder, H., 178
Magat, M., 125
Magee, R. J., 181

Mager, H. I. X., 128, 139, 141, 168, 182, 183, 191
Magnani, A., 310
Maguire, J., 171
Maier, H. G., 103
Maisey, R. F., 264
Majee, B., 108
Malen, C., 209
Mallett, S. E., 171
Mamer, O. A., 351, 352(57)
Mamola, K., 108
Manley, T. R., 15
Mann, F. G., 292(2), 293
Mann, M. E., 332, 352(2), 354(79), 358
Mannschreck, A., 15, 18(77)
Manowska, W., 151
Manske, R. H. F., 291
Mantovani, P., 49, 50(18), 53(18), 60(18), 61(18), 78(18), 79(18), 80(18)
Marburg, S., 293, 309, 316(117), 317(117)
Marchand, A., 285, 286(39), 287(39)
Marcot, B., 282, 292(14), 293, 305(22)
Marion, J. P., 113
Mark, V., 350
Markgraf, J. H., 376(142), 378, 379(142), 380
Marshall, A. R., 280, 294(8), 301(8), 302(8), 316(8)
Marshall, J. A., 281
Marshall, P. R., 104
Martani, A., 46, 51, 77(8, 9, 28), 78(8, 9), 79(8, 9)
Martin, G. C. J., 24, 31(121), 32(121), 38(121)
Martin, J., 236
Martinez, A. P., 62
Masaki, M., 116, 205(124), 206(124, 125)
Mason, J. W., 149, 170(254, 259)
Mason, S. F., 108, 144
Mathews, B. W., 261
Mathios, A., 110
Matsumura, M., 128, 184
Matsuo, H., 310
Matsuo, J., 165, 184
Maurer, A., 360
Mautner, H. G., 368
Mayer, R., 6, 33(26), 56, 74, 75(125), 76(125, 127), 333, 334(5, 6), 342(6), 343(6), 344(6), 350(5, 6), 351(6), 354(5, 6), 355(5, 6), 376(135), 377
Medenwald, H., 50, 51, 52(26, 27), 63(30), 84(26), 87(30)
Mehlhorn, A., 333, 334(14)
Meisel, S. L., 333, 358
Meisenhelder, J. E., 61, 77(71)
Meller, A., 187
Meltzer, R. I., 133, 155, 172(284), 173, 199
Menozzi, M. G., 62, 63(74)
Menzel, A. E. O., 101
Merkle, M., 55
Merle, G., 341, 343(35), 344(35), 345(35), 347(35)
Merritt, J. A., 105
Meth-Cohn, O., 213, 216(7), 229, 230(32), 231(32), 233, 235(39), 236, 237(32, 40), 239, 240(54), 241, 244(32, 41), 247(41), 248(41), 249(41), 250, 251, 253(32), 254, 256, 257(84), 261(85), 266, 267(110), 268(110, 111), 269(110), 270(111), 271(111), 277(75, 81), 278
Metzger, J., 16, 20, 37
Metzler, G., 360
Metzner, R., 177
Meyer, R., 49, 83(20)
Meyer, R. F., 65, 69(101), 70(101), 72(101), 75(101), 95(101), 96(101)
Meyers, A. I., 302
Meyers, M. B., 352
Micetich, R. G., 17, 18, 19(103), 20(95), 21(95), 23(94), 24(94), 25(94), 27(94), 28(95), 29(95), 30(102), 31(102), 37(94, 95), 38(95), 101, 205
Michalkiewicz, D. M., 103, 130(31)
Michels, J. F., 131, 209
Miesel, J. L., 192
Mikolajewska, H., 145
Millar, I. T., 230
Millard, B. J., 15, 16(81)
Miller, E. G., 314
Miller, J., 159
Mingiardi, M. R., 62, 63(74)
Minisci, F., 125
Minkin, V. I., 304
Mirza, R., 281
Mislow, K., 347(47), 348, 349(47)

Mitchell, M. J., 337(69, 88), 351, 352(57), 354(69), 357, 359(69), 362(69), 363(69, 88), 366(69), 367(88)
Mitra, A., 354(79), 358
Mittag, T., 190, 203
Mittler, W., 5, 8(18), 17(18), 21(18), 26(18), 27(18), 28(18)
Mochalov, S. S., 236
Mollier, Y., 13, 14(64), 33(64)
Monkovic, I., 326
Montgomery, J. A., 104, 209
Moody, K., 259, 261(99, 100), 262(99)
Moomaw, W. R., 109
Moon, B. J., 286, 294(106), 298, 304, 307, 316(86), 320, 322(143), 324(143)
Moore, T. E., 318
Moreau, R. C., 38
Moritz, K. L., 72, 75(119), 98(119)
Morley, J. S., 59, 62(59), 88(59), 89(59), 90(59)
Morotomi, Y., 108
Mortimer, C. T., 104
Mossini, F., 50, 53(24), 56, 58, 61(24), 62(52, 53, 54, 55), 63(38), 77(24), 78(24), 79(24), 80(24), 82(24), 86(38), 88(24), 92(52, 53, 54, 55), 93(52, 53, 54, 55)
Mowry, D. T., 376(140), 378
Mrnková, A., 340
Muccia, P. M. R., 102
Muchlmann, F. L., 114, 171(117), 172(117)
Muchowski, J. M., 304
Muggleton, D. F., 16, 23(89)
Mullock, E. B., 215
Munk, M. E., 188
Murai, H., 128
Murakoshi, I., 288, 292(15), 293
Murata, M., 160
Murray, K. E., 103
Murrell, J. N., 110
Musatova, T. S., 194, 196, 197(406), 200
Musatti, A., 56, 85(39)
Mustafa, A. 60
Musgrave, W. K. R., 107, 120, 155, 163
Muth, K., 366
Mutsenietse, D. Kh., 373, 374(122), 376(122), 378(122)
Myers, P. L., 304

N

Nagagawa, S., 209
Nagarajan, K., 30, 31(139)
Nagoya, T., 139
Naidoo, B., 311
Nair, M. D., 227, 244(29), 245(29), 246
Naqui, M. A., 256, 261(85)
Naito, T., 7, 12, 15, 17, 18(29), 24(57, 99), 25(57, 99), 26(29, 57), 28(29), 30(29), 31(99), 32(29), 37(99), 209
Najarian, H., 61, 77(71)
Nakagawa, S., 3, 7, 12, 17(9), 18(9, 29), 20(9), 24(9, 57, 99), 25(9, 57, 99), 26(29, 57), 28(29), 30(29), 31(9, 99), 32(29), 37(99)
Nakamura, S., 101
Naletova, G. P., 375
Neadle, D. J., 241
Neath, G., 120
Negoro, H., 108
Nelson, P. J., 170
Nelson, S. M., 104
Neubert, M., 136
Neumann, W. P., 280
Neumeyer, J. L., 281, 292(11), 293, 294(20)
Neumüller, O.-A., 236, 237(43)
Newbold, G. T., 100, 166
Newman, H., 12
Niethammer, K., 354(70), 357, 359(70)
Nigam, I. C., 117
Novacek, L., 99, 102, 105(3), 208
Nuhn, P., 76
Nyholm, R. S., 104, 123

O

Oae, S., 251, 326
Obe, Y., 288, 292(15), 293
Oberlin, M., 283
O'Brochta, J., 354(66), 357, 371(66), 373(66), 376(66)
Ochiai, E., 195
Ochwat, P., 270
Odaira, Y., 363(94), 365, 367
O'Donnell, E., 198, 199

O'Donnell, M. E., 193, 195(391), 196 (391)
Offen, G., 181
Ogasawara, K., 290, 292(12), 293, 294(64)
Ogata, M., 109
Ogino, K., 326
Ohashi, M., 37
Ohta, A., 206
Ohta, K., 151
Ohta, M., 116, 205(124), 206(124, 125)
Okada, S., 112, 145, 196, 198
Okada, Y., 241
Okamoto, T., 104, 170, 198
Okuda, N., 162, 180(308)
Okumuria, J., 7, 12, 18(29), 24(57), 25 (57), 26(29, 57), 28(29), 30(29), 32 (29)
Oliva, M. L., 51, 77(28)
Oliver, J. A., 337(28), 340, 362(28), 363 (28), 365(28)
Olofson, R. A., 12, 15(60), 33, 35(145)
Omata, S., 203
Ongley, P. A., 337(28), 340, 362(28), 363 (28), 365(28)
Orita, A. Y., 370, 376(113)
Orvis, R. L., 329
Osdene, T. S., 142
Oshkaya, V. P., 373, 374(122), 376(122), 378(122)
Osintseva, L. V., 375
Owen, N. L., 15
Ozasayama, E., 145, 196

P

Pacifici, J. G., 238
Packham, D. I., 280, 290, 292(3), 293
Pagani, G., 59
Pages, R. A., 139, 144(207), 156(207)
Pain, D. L., 18, 20, 23(107), 24, 28(107), 29(107), 31(121), 32(121), 37(107), 38(121)
Paju, R., 193
Palamidessi, G., 115, 153, 155, 158, 159 (296), 161, 162(270), 307, 163(296), 165, 167(312), 168, 172(271), 175 (120, 270), 193, 194(392), 196(392, 398), 198

Palat, K., 99, 102, 105(3)
Palfreyman, M. N., 286, 290, 294(87), 298, 301(66), 302(66), 303(87), 307, 309 (65, 87), 317(87), 318(87)
Palmer, M. H., 105, 107
Panizzi, L., 115, 175(120)
Papa, A. J., 370
Pappalardo, G., 13, 32(65)
Párkányi, C., 333, 334(12), 335
Parkin, J. E., 109
Parkinson, J., 38
Parnell, E. W., 18
Partyka, R. A., 6
Pascand, X., 209
Pato, G., 209
Paton, D., 171
Patrick, C. R., 337(30), 340
Patterson, M. A., 75
Paudler, W. W., 107, 110(67c), 112, 121
Paul, H., 278
Paulik, F., 104
Pauling, P., 104
Paulus, K. F. G., 111
Pavlov, A. I., 144
Pay, M. H., 271, 277(123), 278(124)
Payne, L. S., 125
Pearce, A. A., 295
Pecka, J., 335
Pedersen, C. T., 358, 359
Peer, H. G., 103
Peltier, D., 231
Peltz, K., 314
Pereferkovich, A. N., 335
Perera, C., 245, 248(69a)
Perkampus, H. H., 108
Perkin, W. H., 291, 294
Peshkar, L., 11, 21(53, 54), 27(53), 38 (53, 54)
Pestunovich, V. A., 335
Peterkofsky, A., 161
Petersen, P. V., 337(23), 338
Petrov, A. S., 220, 221, 222(17, 22), 224 (21)
Pfenninger, F., 363, 364(91), 366(91)
Pfleiderer, W., 152, 177
Phan-Tan-Luu, R., 16
Phipps, L. M., 370, 376(111)
Piacenti, F., 15, 17

Pianka, M., 10, 31(46)
Piloty, O., 258
Pinazzi, C., 358
Pinnow, J., 212, 213(1, 2, 3, 4, 5, 6), 226(2), 229(2)
Pino, P., 17
Pinson, J., 115, 122
Piston, G., 212, 213(1)
Pitre, D., 114, 140, 141, 166(115)
Platenburg, D. H. J. M., 30, 31(137), 33(137), 41(137)
Plazzi, V., 56, 59, 63(38), 86(38), 88(56), 89(56), 90(56), 92(56), 93(56)
Plyler, E. K., 108
Pobiner, H., 288
Pogorelova, T. G., 144
Pohland, H. W., 3, 4, 5, 23(6)
Poisel, H., 190
Poite, J. C., 15, 20
Pollack, N. M., 350, 351(52), 352(57), 354(52), 355(52), 361(52, 55)
Pollak, P. I., 147, 156, 164(285)
Pollet, A., 37
Pollitt, R. J., 241
Polonovski, M., 249
Ponci, R., 49, 50(19), 53(19), 56, 58, 59, 60(19), 61(19), 62(51, 52, 53, 54, 57, 62, 63), 63(37), 76(63), 78(19), 85(37), 86(37), 89(57, 82), 90(57), 91(51, 62, 63, 80, 86), 92(50, 51, 52, 54, 89), 93(50, 51, 52, 53, 54, 85, 89)
Ponticello, G. S., 353, 355(61)
Ponticello, I. S., 335, 337(20), 354(20), 358(20), 377(20), 378(20)
Porter, A. E. A., 178, 188
Posner, T., 63, 68(96), 70(96), 94(96)
Postovskii, I. Ya., 272
Potts, K. T., 170, 171, 311, 312
Poulsson, I., 343(41), 344, 345(41)
Pratt, Y. T., 99
Prey, V., 369, 372(106, 107, 108), 373(108)
Price, D., 233, 235(39)
Price, R., 252, 253(78)
Prijs, B., 9, 18(37), 20(37), 23(37), 191
Protiva, M., 370, 372(115)
Prudchenko, E. P., 272
Purdon, R. A., 75

Purrello, G., 5, 12, 32(19)
Pyman, F. L., 312

Q

Quelet, R., 288

R

Raap, R., 8, 17, 18, 20(95), 21(95), 23(94), 24(94), 25(94), 27(94), 28(95), 29(95), 30(102), 31(102), 37(94, 95), 38(95)
Raecke, B., 140
Rafiq, M., 176
Rajappa, S., 14, 18(69), 21(69), 29(69), 30(69), 31(69, 139)
Rajšner, M., 370, 372(115)
Rake, G., 101
Ramage, G. R., 99
Rambaud, R., 125
Ranft, J., 278
Rao, A. R., 38
Ratajczyk, J. D., 177
Rausch, R., 304
Ray, S. C., 370
Readhead, M. J., 304
Reading, H. W., 38
Redford, D. G., 343(40), 344
Reeve, A. H., 302
Reid, D. H., 33
Reisse, A., 333, 343(10), 344(10), 345(10), 346(10)
Reisser, F., 177
Reissert, A., 49, 50(17), 88(17), 375, 376(125), 378(125), 380
Renoll, M., 376(140), 378
Renson, M., 369, 370(109), 372(109), 373(109), 376(109)
Renwick, J. D., 21, 28(113), 29(113)
Reynolds, W. C., 312
Riad, Y., 64
Ricci, A., 46, 51, 77(8, 9, 28), 78(8, 9), 79(8, 9), 341, 345(37)
Richards, G. O., 143
Richter, S., 333, 334(5, 6), 342(6), 343(6), 344(6), 350(5, 6), 351(6), 354(5, 6), 355(5, 6)

Riezebos, G., 103, 110, 125, 130(31)
Ringwald, E. L., 376(140), 378
Rist, H., 230
Ritchie, A. C., 289, 305(58)
Rizzi, G. P., 134
Robak, E. A., 13, 33(63)
Robb, C. M., 149, 170(254, 256)
Robba, M., 38, 143, 147(228), 151 (228)
Roberts, B. G., 360
Robertson, G. B., 104
Robertson, W. A. H., 12, 15(60)
Robinson, G. M., 300
Robinson, R., 279, 287, 289(43), 295, 300, 311, 312(123)
Rodriguez, G. 106
Roegler, M., 23, 24(118), 26(118), 27 (118)
Roesel, E., 322
Rogers, D., 327
Rogers, V., 23
Roggero, J., 15
Roland, J. R., 2, 24(5)
Roloff, H., 294
Rose, F. L., 171, 209
Rose, J. D., 104, 209
Rosenmund, P., 301
Rosenthal, A., 129
Rosmus, P., 56
Ross, L. O., 62
Ross, S. T., 66, 75(103), 95(103), 96(103), 98(103)
Rossi, C., 341, 345(37), 347(37)
Roth, J. J., 238
Rowilleit, H., 294(145), 312
Roy, D. N., 285, 286(39), 287(39)
Rudy, H., 256, 258(82)
Ruetman, S. H., 154
Ruff, F., 104
Rugheimer, L., 305
Rumpf, P., 281, 282, 289(17), 292(13, 14), 293, 305(17, 56)
Ruppenthal, N., 294(144), 319, 320
Russell, D. W., 241
Russkikh, V. V., 264, 265, 266
Rutner, H., 146, 158, 160, 165(299), 173 (299)
Ruveda, E. A., 326

S

Saemann, C., 226
Sagura, J. H., 58, 63(46), 88(46), 90 (46)
Saikachi, H., 165, 184
Sainsbury, M., 280, 281, 282(18), 283(18), 285, 286, 288, 290(18, 41), 291(18), 294(8, 87, 105), 298, 299(27), 301 (8, 66), 302(8, 66), 303(87), 304, 305 (27), 306(18, 41), 307, 308(18), 309 (18, 27, 65, 87, 114), 315, 316(8, 86, 114), 317(87), 318(87, 133), 319, 320 (13), 322(143), 324(46, 143), 325 (46), 328(136), 329(136)
Sakai, F., 12, 24(57), 25(57), 26(57)
Sakamoto, M., 192
Sammes, P. G., 119, 124(135), 154(135), 178, 188, 193, 198(393), 201(393)
Sandmeyer, T., 219
Sanger, F., 241
Sasaki, T., 160
Sasse, K., 180
Sasso, W. D., 281
Sato, M., 104
Sawlewicz, J., 151, 194, 197
Sbrana, G., 15, 108
Schaaf, K. H., 153
Scharf, H.-D., 376(136, 137), 377
Scheinbaum, M. L., 195
Schenk, G. O., 236, 237(43)
Scherer, O., 376(138), 377
Schleigh, W. R., 285
Schlessinger, R. H., 335, 337(20), 353, 354(20), 355(61), 358(20), 367, 377 (20), 378(20)
Schmid, H., 280, 290(7), 292(7), 293, 294(7)
Schmidt, U., 190
Schmötzor, G., 143
Schneider, R., 26
Schneller, S. W., 171
Schoch, W., 313, 320(129)
Schoeler, A., 283
Schönberg, A., 236, 237(43), 354(73), 357
Schoenewaldt, E. F., 143
Schonfelder, M., 122

Schopf, C., 292(9), 293, 304
Schrader, G., 63
Schroeder, W., 59, 61, 76(58), 93(58)
Schut, W. J., 168
Schutt, H., 140
Scrowston, R. M., 48, 50(13), 331
Searby, R., 215
Seefelder, M., 66, 75(102), 95(102), 96 (102)
Seibl, J., 231
Seidlová, V., 370, 372(115)
Seifeit, R. M., 103, 111(30), 113(30)
Shabarov, Y. S., 236
Shalaby, A. F. A., 60
Shamasundar, K. T., 293, 294, 309, 316 (117), 317(117)
Shamma, M., 283
Shannon, J. S., 259
Shapiro, S. L., 209
Sharefkin, D. M., 147, 166
Sharp, W., 172
Shavel, J., 322
Shefter, E., 368
Sheridan, J., 15
Shepard, K. L., 149, 150, 170(259)
Shepherd, R. G., 124
Shibuya, S., 319
Shields, J. E., 352
Shigeta, Y., 145, 196, 198
Shih Chiuen Chia, A., 107
Shimanskaya, M. V., 118
Shindo, H., 108, 194(71)
Shinoda, H., 162, 180(308)
Shinoda, J., 295
Shipton, J., 103
Shirley, R. L., 365, 366, 368
Shpeïer, L. F., 375, 376(128)
Sidgwick, N. V., 230
Sierocks, K., 294(144), 320
Sih, J. C., 285
Simmons, H. E., 24
Simmons, J. D., 108
Singerman, G. M., 136
Sirett, N. E., 63, 78(94), 89(94)
Skinner, W. A., 303
Slack, R., 2, 3(3), 8, 14, 16(3), 17(3, 70), 18(70, 72), 19(72), 20(70), 21(3, 71, 72), 22(112), 23(3, 70, 72, 107), 24(3, 70, 106, 112), 25(70, 119), 28(70, 71, 107, 112, 113), 29(3, 70), 107, 113, 119), 30(70, 71), 31(70, 71, 119, 140), 33, 34(123), 36(3), 37(107, 119, 144), 38(3, 112, 123, 140), 40(123), 51, 55 (29), 77(29)
Slater, C. A., 102
Slavik, J., 328
Slavikova, L., 328
Slotta, W., 285
Slusarchyk, W. A., 283
Smalley, R. K. 215, 245, 248, 251, 277 (75)
Smirnova, V. A., 144
Smith, C., 120
Smith, C. L., 110
Smith, D. P., 136, 146(196)
Smith, H. E., 109
Smith, R. E., 3, 21(8), 22(8), 23(8), 24(8), 31(8), 34(8), 38(8)
Smith, W. C., 146
Snegireva, F. P., 57, 84(41, 42)
Snowling, G., 11
Söderbäck, E., 10, 11, 31(43, 49, 50), 35 (49, 50)
Soell, D., 177
Sohn, A., 147
Sokolov, S. D., 16
Solly, R. K., 144
Somin, I. N., 220, 221, 222(17, 22), 224 (21)
Sommers, A. L., 335
Speciale, A. J., 302
Speckamp, W. N., 354(64), 356
Spence, G. G. 203
Spenser, I. D., 326
Spiegel, L., 227, 243
Spoerri, P. E., 134, 139, 141, 143, 144 (207, 219), 146(219), 153, 155, 156 (207, 282), 158, 160, 161(300), 165 (299)
Spring, F. S., 100, 119, 143, 154(134), 160, 166, 172
Springall, H. D., 230
Sprung, M. M., 229
Squires, S., 30, 31(140), 38(140)
Srinivasan, M., 327
Srivastava, K. C., 120

Srivastava, K. S. L., 69, 72(111), 94(111), 96(111), 98(111)
Staab, H. A., 15, 18(77)
Stacey, G. J., 209
Stacy, G. W., 370, 376(111, 113)
Stamhuis, E. J., 281
Staudinger, H., 363, 364(91), 366(91)
Staunton, J., 326
Steinfatt, F., 62
Steinfeld, A. S., 287
Steinkopf, W., 343(41), 344, 345(41)
Stelzner, R., 63, 68(97), 70(97), 71(97), 94(97)
Stepanov, I. P., 236
Stermitz, F. R., 328
Stevens, C. L., 188
Stiddard, M. H. B., 104
Stidham, H. D., 105
Stocks, I. D. H., 17, 24(96), 27(96), 28(96)
Stoll, M., 113
Stollé, R., 45, 50(3), 55, 64(2), 77(2, 3), 81(2, 3)
Strelitz, F., 101
Strem, M. E., 133
Stroh, J., 116
Strumillo, J., 145
Stubbs, J. K., 137, 138(202)
Subramaniam, P. V., 363
Sugahara, H., 288
Sugiura, S., 169
Sugujama, M., 116, 206(125)
Sulzbach, R. A., 182
Summers, L. A., 124
Supple, J. H., 352
Suschitzky, H., 68, 94(107a), 194, 213, 216(7, 8), 217, 218, 221, 222(20), 223(20), 227, 231, 232, 235(39), 236, 237(40), 244(41), 245, 247(41), 248(30, 41), 249(41, 70), 251, 254, 277(75)
Sutherland, R., 38
Sutton, M. E., 227, 231, 232, 247, 248(30), 249(70)
Swigor, J. E., 5
Symon, J. D., 33
Syrova, G. P., 194
Szmuskovicz, J., 279, 323(3)
Szperl, L., 375, 376(127)

T

Tabner B. J., 111
Tacchi Venturi, M., 167
Tachibawa, S., 310
Taft, E., 154
Taito, T., 3, 17(9), 18(9), 20(9), 24(9), 25(9), 31(9)
Takahashi, H., 108
Takahashi, K., 3, 7, 17(9), 18(9, 29), 20(9), 24(9, 99), 25(9, 99), 26(29), 28(29), 30(29), 31(9, 99), 32(29), 37(99)
Takahashi, S., 233, 240
Takasaki, R., 140
Takemoto, T., 300
Tanaka, H., 327
Tarailo, S. D., 117
Taube, C., 72, 75(119), 98(119)
Taylor, E. C., 114, 142(116), 152, 166, 203
Taylor, K. G., 188
Taylor, M. K., 102
Taylor, P. J., 209
Temple, C., 104, 209
Tenhaeff, H., 313, 320(129)
Terao, M., 100
Ternai, B., 68, 69(109)
Thierfelder, K., 292(9), 293
Thomas, G. A., 62
Thomas, O, 143
Thomason, H. J., 63
Thompson, P. E., 61, 77(71)
Thompson, M. J., 259, 261(99), 262(99)
Thompson, T. W., 209
Thorpe, J. G., 107
Thrift, R. I., 294, 314(75)
Thuillier, G., 281, 282, 289(17), 292(13, 14), 294, 305(17, 22, 56)
Tilak, B. D., 334, 342(16), 343(16, 38), 344(16), 347(16, 38), 350(16, 38), 351(16, 38), 354(16, 38)
Tiley, E. P., 291, 301(66), 302(66)
Tirouflet, L., 377
Törzs, E. S. G., 172, 174(336), 177(336)
Toland, W. G., 370, 371(116, 117), 372(116, 117, 118, 119), 373(118), 374(117, 119), 380(118)
Tomimatsu, Y., 192

AUTHOR INDEX

Tong, E., 62
Tonkyn, W. R., 299
Tori, K., 109
Torigoe, Y., 104
Torikoshi, Y., 170
Tornetta, B., 13, 32(65)
Tota, Y. A., 174
Traber, W., 280, 281(10)
Tramièr, B., 129
Traub, A., 337(74), 354(74), 358, 359(74), 360(74)
Traynelis, V. J., 325
Trefilova, L. F., 272
Trenholm, G., 183
Trumble, R. F., 104, 107
Trinajstić, N., 333, 334(15)
Tristram, E. W., 156, 164(285)
Tronov, S. V., 375
Tsatsaronis, G., 355
Tseitlin, G. M., 144
Tsutsumi, S., 363(94), 365, 367(94)
Tsyrulnikova, L. G., 200
Tuck, B., 215
Tull, R. J., 147, 156, 157, 164(285)
Turner, D. W., 106
Turner, J. N., 62
Turovskii, I. V., 335
Tutin, F., 115
Twigg, M. V., 48, 50(13)

U

Uchimaru, F., 112, 145, 196, 198
Ueda, T., 133
Ueno, T., 203
Uffindell, N. D., 267
Ulbricht, G., 376(135), 377
Ullmann, F., 270
Ullrich, G., 376(135), 377
Uyeo, S., 41
Uzzell, P. S., 296, 314(83)

V

Vahlberg, B., 46
Valenti, P. C., 109

van der Heijden, A., 103
van der Plas, H. C., 25, 159
van der Walt, J. P., 102
VanderWerf, C. A., 121
Vangermain, E., 289, 295(55), 305(55)
van Hooidonk, C., 30, 31(137), 33(137), 41(137)
van Leeuwen, G. C., 131, 209
VanMeter, J. P., 357, 359(71), 366(71), 367
van Oosten, A. M., 30, 31(137), 33(137), 41(137).
van Romburgh, P., 238, 239
van Schooten, J., 340
van Tamelen, E. E., 329
van Triet, A. J., 102
Vejdělek, Z. J., 370, 372(115)
Vankatasetty, H. V., 104
Verbeek, M., 376(136, 137), 377
Vernin, G., 20, 37
Vest, R. D., 2, 24(5)
Vilar, A., 289, 305(56)
Villadary, M., 115
Villaescusa, F. W., 370
Vincent, E. J., 16
Vinot, N., 115, 288
Vitali, T., 49, 50(18, 19), 53(18, 19, 24), 58, 59, 60(18, 19), 51(18, 19, 24), 62(52, 53, 54, 55), 77(24), 78(18, 19, 24), 79(18, 24), 80(18, 24), 82(24), 88(24, 56), 89(56), 90(56), 92(52, 53, 54, 55, 56), 93(52, 53, 54, 55, 56)
Vivaldi, R., 15
Vladimirtsev, I. F., 272
Vloon, W. J., 339
Voeker, C. A., 374
Vollhardt, K. P. C., 352
Voellmin, J., 231
Vogl, O., 114, 142(116)
Volke, J., 111
Volkova, V., 111
Volpp, G. P., 15, 17, 18(91), 19(105), 24(86), 29(105), 31, 34(86), 35
von Auwers, K., 69
von Braun, J., 337(24, 27), 339, 340
von Euler, H., 175
von Gizycki, U., 163
Voronkov, M. G., 335

W

Wagner, G., 76, 177, 180(351)
Waite, J. A., 17, 24(96), 27(96), 28(96)
Wallmark, I., 368
Walter, W., 66
Walton, R. A., 104
Warburton, E., 257
Ward, B. J., 16, 23(89)
Ward, P., 125
Ward, R. L., 111
Wardley, A., 15
Warren, T. F., 21, 22(112), 24(112), 28 (112), 38(112)
Warrener, R. N., 332
Watson, R. F., 333, 336(9), 337(9), 338 (9), 340(9), 361(9), 362(9), 363(9), 364(9)
Wege, D., 332
Wegler, R., 180
Weinhardt, K. W., 281, 292(11), 293, 294 (20)
Weisgraber, K. H., 287
Weiss, U., 101
Weissbach, K., 337(24), 339
Wells, R. D., 110, 125
Wendt, G. R., 104
Wentrup, C., 127
Werbel, L. M., 61, 77(71)
Werner, H., 376(135), 377
West, R. G., 105
Whaley, W. M., 288
Wheatley, P. J., 105
White, A. W., 66, 67(106), 68(105, 106), 69(106), 70(106), 71(110), 94(105, 106, 110), 95(110), 96(106, 110)
White, J. A., 15, 18(80), 37(80)
White, J. D., 332, 352(2), 354(79), 358
White, E. C., 100, 203(10)
White, E. H., 170
Whitehead, E. V., 337(29), 340, 347(45), 348, 349(45), 361(29, 45), 363(29)
Whitfield, F. B., 103
White, W. N., 325
Widdowson, D. A., 327
Wiedenmann, R., 36
Wiegrabe, W., 281, 322
Wiemann, J., 115

Wien, R., 38
Wiggins, D. W., 293, 294(87), 298, 303 (87), 309(87), 317(87), 318(87)
Wiggins, L. F., 111
Wilcox, R. D., 156, 157(286), 164(286)
Wildman, G. T., 143
Wilen, S. H., 121, 185
Wille, F., 8
Williams, D. A., 15
Williams, D. H., 111
Williams, G. J., 295
Williams, R. T., 38
Williams, V. E., 15
Willson, M. R., 23
Wilson, C. O., 103
Wilson, G. E., 26
Wilson, J. D., 106
Wilson, R. M., 291, 302(73)
Winch, R. W., 11
Winter, D. P., 286, 305, 307
Wintersteiner, O., 101
Wippel, H. G., 75, 97(130)
Wise, W. S., 111
Wittig, G., 279, 313, 320, 354(70), 357, 359(70)
Wittig, P., 74
Wojahn, H., 167
Wollner, T. E., 370
Woltersdorf, O. W., 149, 170(254, 255, 259)
Wong, D. Y., 376(141), 378, 379(141), 380(141)
Woo, C., 47
Wood, H. C. S., 119
Woodward, R. B., 320
Wooldridge, K. R. H., 2, 14, 17(70), 18 (70, 72), 19(72), 20(70), 21(71, 72), 22, 23(70, 72), 24(70, 96), 25(70, 119), 26(114), 27(96), 28(70, 71, 96), 29 (70, 119), 30(70, 71), 31(70, 71, 119, 140), 33, 37(119, 144), 38(140)
Woodward, R. B., 14, 18(67), 21(67), 29 (67), 30(67), 31(67), 36(67)
Worth, D. F., 61, 77(71)
Wubbels, G. W., 238
Wünsch, K.-H., 68
Wynberg, H., 281, 338(54), 349(49), 350, 351, 354(54), 355(54), 361(54)

Y

Yagi, H., 288
Yagupol'skii, L. M., 336, 337(22), 338 (22), 372(22), 373(22), 375(22), 376 (22)
Yamada, S., 133, 139, 141
Yamada, S.-I., 310
Yamaji, K., 363(94), 365, 367(94)
Yamanaka, T., 290, 292(12), 293, 294(64)
Yamanishi, Y., 160, 180(303)
Yandle, J. R., 111
Yencha, A. J., 106
Yonezawa, T., 37
Yoshida, S., 102
Yoshihira, K., 102
Yoshisue, K., 139
Young, P. C., 287, 289(43)
Yurel, S. P., 118

Z

Zahradník, R., 333, 334(12, 13, 14), 335
Zajaczkowska, B., 145
Zaugg, H. E., 130
Zavaglio, V., 102
Zbiral, E., 116
Zeigler, E., 186, 187(381)
Zeiler, A. G., 121
Zeiser, K., 285
Zell, R., 191
Ziegler, K., 285
Zinnes, H., 322
Ziolkowsky, B., 109
Zoltewicz, J. A., 110
Zuleski, F. R., 322
Zurawski, B., 105
Zwanenburg, D. J., 338(54), 354(54), 355 (54), 361(54)
Zweig, A., 360

Cumulative Index of Titles

A

Acetylenecarboxylic acids, reactions with heterocyclic compounds, **1**, 125
t-amino effect, **14**, 211
Aminochromes, **5**, 205
Anthranils, **8**, 277
Aromatic quinolizines, **5**, 291
Aza analogs, of pyrimidine and purine bases, **1**, 189
Azines, reactivity with nucleophiles, **4**, 145
Azines, theoretical studies of, physicochemical properties and reactivity of, **5**, 69
Azinoazines, reactivity with nucleophiles, **4**, 145
1-Azirines, synthesis and reactions of, **13**, 45

B

Benzisothiazoles, **14**, 43
Benzisoxazoles, **8**, 277
Benzoazines, reactivity with nucleophiles, **4**, 145
Benzofuroxans, **10**, 1
Benzo[b]thiophene chemistry, recent advances in, **11**, 178
Benzo[c]thiophenes, **14**, 331

C

Carbenes, reaction with heterocyclic compounds, **3**, 57
Carbolines, **3**, 79
Chemistry
  of benzo[b]thiophenes, **11**, 178
  of diazepines, **8**, 21
  of furans, **7**, 377
  of lactim ethers, **12**, 185
  of mononuclear isothiazoles, **14**, 1
  of phenanthridines, **13**, 315
  of phenothiazines, **9**, 321
  of 1,3,4-thiadiazoles, **9**, 165
  of thiophenes, **1**, 1
Claisen rearrangements, in nitrogen heterocyclic systems, **8**, 143
Complex metal hydrides, reduction of nitrogen heterocycles with, **6**, 45
Covalent hydration, in heteroaromatic compounds, **4**, 1, 43
Cyclic enamines and imines, **6**, 147
Cyclic hydroxamic acids, **10**, 199
Cyclic peroxides, **8**, 165

D

Development of the chemistry of furans, 1952–1963, **7**, 377
2,4-Dialkoxypyrimidines, Hilbert–Johnson reaction of, **8**, 115
Diazepines, chemistry of, **8**, 21
Diazomethane, reactions with heterocyclic compounds, **2**, 245
1,2-Dihydroisoquinolines, **14**, 279
Diquinolylmethane, and its analogs, **7**, 153
1,2- and 1,3-Dithiolium ions, **7**, 39

E

Electronic aspects of purine tautomerism, **13**, 77
Electronic structure of heterocyclic sulfur compounds, **5**, 1
Electrophilic substitutions of five-membered rings, **13**, 235

F

Ferrocenes, heterocyclic, **13**, 1
Five-membered rings, electrophilic substitutions of, **13**, 235
Furan chemistry, development of the chemistry of (1952–1963) **7**, 377

H

Halogenation of heterocyclic compounds, **7**, 1
Hammett equation, applications to heterocyclic compounds, **3**, 209
Hetarynes, **4**, 121
Heteroaromatic compounds, free-radical substitutions of, **2**, 131
  nitrogen, covalent hydration in, **4**, 1, 43
  prototropic tautomerism of, **1**, 311, 339; **2**, 1, 27
Heteroaromatic substitution, nucleophilic, **3**, 285
Heterocycles
  photochemistry of, **11**, 1
  by ring closure of ortho-substituted t-anilines, **14**, 211
Heterocyclic chemistry, literature of, **7**, 225
Heterocyclic compounds
  application of Hammett equation to, **3**, 209
  halogenation of **7**, 1
  mass spectrometry of, **7**, 301
  quaternization of, **3**, 1
  reactions of, with carbenes, **3**, 57
  reaction of acetylenecarboxylic acids with, **1**, 125
  reactions of diazomethane with, **2**, 245
  sulfur, electronic structure of, **5**, 1
N-Heterocyclic compounds, electrolysis of, **12**, 213
Heterocyclic diazo compounds, **8**, 1
Heterocyclic ferrocenes, **13**, 1
Heterocyclic pseudo bases, **1**, 167
Heterocyclic syntheses, from nitrilium salts under acidic conditions, **6**, 95
Hilbert–Johnson reaction of 2,4-dialkoxypyrimidines, **8**, 115

I

Imidazole chemistry, advances in, **12**, 103
Indole Grignard reagents, **10**, 43

Indoles, acid-catalyzed polymerization **2**, 287
Indoxazenes, **8**, 277
Isoindoles, **10**, 113
Isothiazoles, **4**, 107
Isoxazole chemistry, recent developments in, **2**, 365

L

Lactim ethers, chemistry of, **12**, 185
Literature of heterocyclic chemistry, **7**, 225

M

Mass spectrometry of heterocyclic compounds, **7**, 301
Metal catalysts, action on pyridines, **2**, 179
Monoazaindoles, **9**, 27
Mononuclear thiazoles, recent advances in chemistry of, **14**, 1
Monocyclic sulfur-containing pyrones, **8**, 219

N

Naphthyridines, **11**, 124
Nitrilium salts, heterocyclic syntheses involving, **6**, 95
Nitrogen heterocycles, reduction of, with complex metal hydrides, **6**, 45
Nitrogen heterocyclic systems, Claisen rearrangements in, **8**, 143
Nucleophiles, reactivity of azine derivatives with, **4**, 145
Nucleophilic heteroaromatic substitution, **3**, 285

O

1,3,4-Oxadiazole chemistry, recent advances in, **7**, 183
1,3-Oxazine derivatives, **2**, 311
Oxazolone chemistry, recent advances in, **4**, 75

P

Pentazoles, **3**, 373
Peroxides, cyclic, **8**, 165
Phenanthridine chemistry, recent developments in, **13**, 315
Phenothiazines, chemistry of, **9**, 321
Phenoxazines, **8**, 83
Photochemistry of heterocycles, **11**, 1
Physicochemical aspects of purines, **6**, 1
Physicochemical properties
    of azines, **5**, 69
    of pyrroles, **11**, 383
3-Piperideines, **12**, 43
Prototropic tautomerism of heteroaromatic compounds, **1**, 311, 339; **2**, 1, 27
Pseudo bases, heterocyclic, **1**, 167
Purine bases, aza analogs of, **1**, 189
Purines
    physicochemical aspects of, **6**, 1
    tautomerism, electronic aspects of, **13**, 77
Pyrazine chemistry, recent advances in, **14**, 99
Pyrazole chemistry, progress in, **6**, 347
Pyridazines, **9**, 211
Pyridine, effect of substituents in, **6**, 229
Pyridines, action of metal catalysts on, **2**, 179
Pyridopyrimidines, **10**, 149
Pyrimidine bases, aza analogs of, **1**, 189
Pyrones, monocyclic sulfur-containing, **8**, 219
Pyrroles
    acid-catalyzed polymerization, of, **2**, 287
    physicochemical properties of, **11**, 383
Pyrrolizidine chemistry, **5**, 315
Pyrrolopyridines, **9**, 27
Pyrylium salts, preparations, **10**, 241

Q

Quaternization of heterocyclic compounds, **3**, 1
Quinazolines, **1**, 253
Quinolizines, aromatic, **5**, 291
Quinoxaline chemistry, recent advances in, **2**, 203
Quinuclidine chemistry, **11**, 473

R

Reduction of nitrogen heterocycles with complex metal hydrides, **6**, 45
Reissert compounds, **9**, 1
Ring closure of ortho-substituted t-anilines, for heterocycles, **14**, 211

S

Selenazole chemistry, present state of, **2**, 343
Selenophene chemistry, advances in, **12**, 1
Synthesis and reactions of 1-azirines, **13**, 45

T

Tautomerism, prototropic, of heteroaromatic compounds, **1**, 311, 339; **2**, 1, 27
1,2,3,6-Tetrahydropyridines, **12**, 43
Theoretical studies of physicochemical properties and reactivity of azines **5**, 69
1,2,4-Thiadiazoles, **5**, 119
1,2,5-Thiadiazoles, **9**, 107
1,3,4-Thiadiazoles, chemistry of, **9**, 165
1,2,3,4-Thiatriazoles, **3**, 263
Thiophenes, chemistry of, recent advances in, **1**, 1
Three-membered rings, with two hetero atoms, **2**, 83
1,6,6aS^{IV}-Trithiapentalenes, **13**, 161

QD
400
A18
v.14
1972

JUL 26 1972